W9-CPF-874

WEATHER ATLAS of the UNITED STATES

Originally titled: CLIMATIC ATLAS OF THE UNITED STATES

U.S. DEPARTMENT OF COMMERCE
C. R. Smith, Secretary

ENVIRONMENTAL SCIENCE SERVICES ADMINISTRATION
Robert M. White, Administrator

ENVIRONMENTAL DATA SERVICE
Woodrow C. Jacobs, Director

JUNE 1968

Reprinted 1975 by
GALE RESEARCH COMPANY
Book Tower, Detroit, Michigan 48226

Library of Congress
Cataloging in Publication Data

United States. Environmental Data Service.
Weather atlas of the United States.

"The climatic maps in this atlas were prepared primarily by John L. Baldwin."
Reprint of the 1968 ed. published by the Service and sold by the Superintendent of Documents, U. S. Government Printing Office.
1. United States--Climate--Maps. I. Baldwin, John L. II. Title.
G1201.C8U55 1975 912'.1'5516973 74-11931
ISBN 0-8103-1048-1

CONTENTS

PREFACE

The purpose of this atlas is to depict the climate of the United States in terms of the distribution and variation of constituent climatic elements. Climate has a profound, often controlling, effect upon the life, mood, health, and activity of all of us.

Climate may be considered the collective state of the earth's atmosphere at a specific place for a long period of time (usually several decades). The short-term variations of the state of the atmosphere are called "weather." Weather is the product of the interaction of numerous natural elements; the long term statistical valuations of these various elements collectively define the climate. For many planning, engineering, and scheduling purposes it is more important to know the climate of a certain city, State, resort area, etc., than to know what the weather happens to be there today.

The Climatic Maps of the United States present in uniform format a series of analyses showing the national distribution of mean, normal, and/or extreme values of temperature, precipitation, wind, barometric pressure, relative humidity, dewpoint, sunshine, sky cover, heating degree days, solar radiation, and evaporation. The map projection has been standardized to allow accurate comparison and correlation of the various climatic elements and their patterns.

The individual analyses were originally prepared to meet the demand for climatic information from commercial, industrial, agricultural, research, and educational institutions, as well as from the general public. Each sheet, or set of sheets, was made available as soon as printed. Now the entire set - a total of 40 large sheets containing 271 climatic maps and 15 tables - has been collected and bound into this comprehensive atlas. (Individual sheets and sets may still be purchased separately).

It should be remembered that these analyses are not forecasts of temperature, precipitation, etc., but rather reflect collective atmospheric conditions that occurred over periods of years; often observed conditions for any given day, week, month - or even year - will differ sharply from those indicated in the analyses.

The climatic maps in this atlas were prepared primarily by John L. Baldwin, Chief of the Domestic Climatology Branch of the Environmental Data Service, ESSA, an agency of the U.S. Department of Commerce, with some map contributions from the Hydrologic Services Division and the former Solar Radiation Section of the Weather Bureau. Appreciation is due Dr. Helmut E. Landsberg, former Director of the Environmental Data Service, and to the National Academy of Science Advisory Committee on Climatology for advice and guidance.

TEMPERATURE

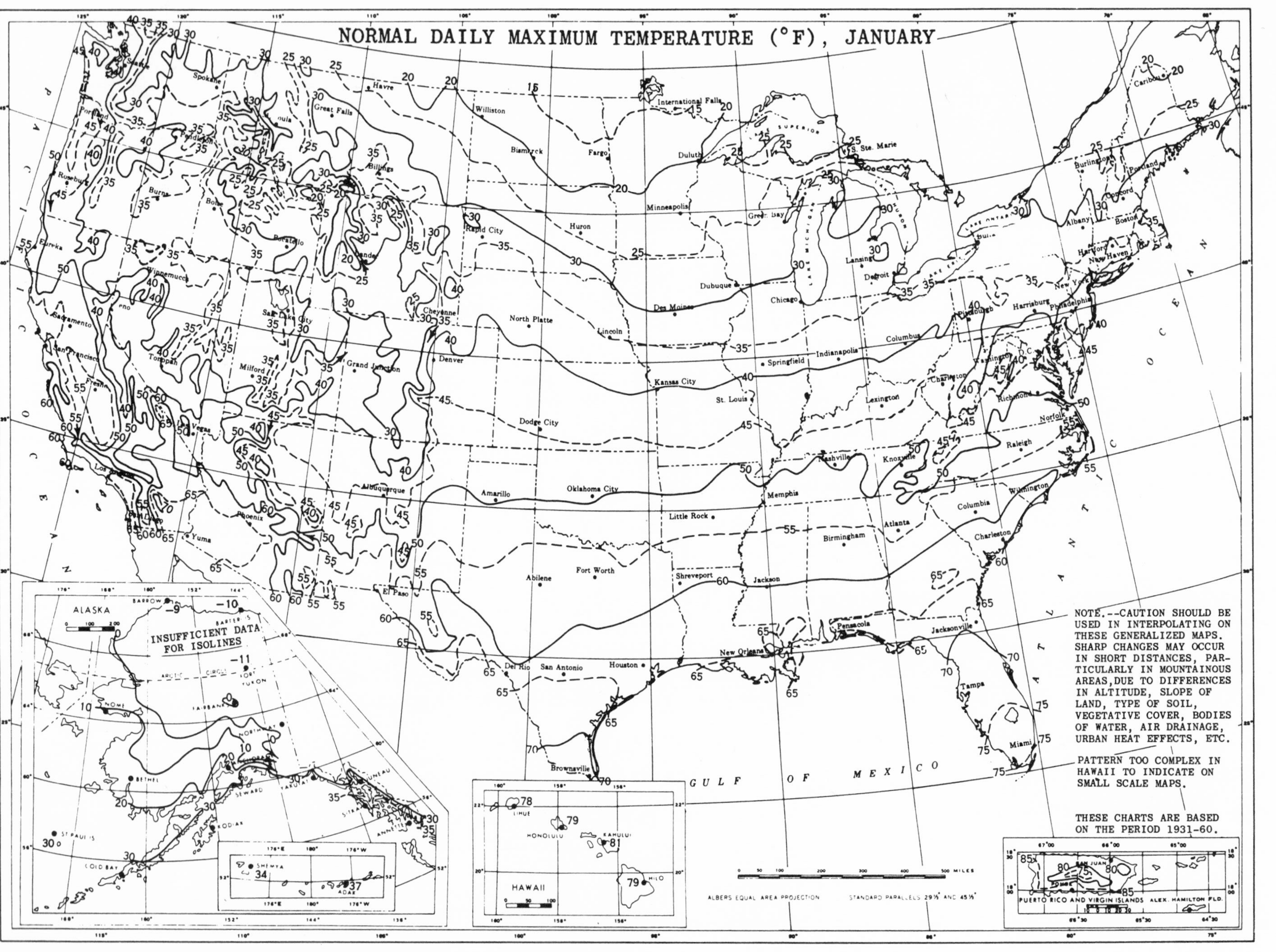
NORMAL DAILY MAXIMUM TEMPERATURE (°F), JANUARY
NOTE.--CAUTION SHOULD BE USED IN INTERPOLATING ON THESE GENERALIZED MAPS. SHARP CHANGES MAY OCCUR IN SHORT DISTANCES, PARTICULARLY IN MOUNTAINOUS AREAS, DUE TO DIFFERENCES IN ALTITUDE, SLOPE OF LAND, TYPE OF SOIL, VEGETATIVE COVER, BODIES OF WATER, AIR DRAINAGE, URBAN HEAT EFFECTS, ETC.
PATTERN TOO COMPLEX IN HAWAII TO INDICATE ON SMALL SCALE MAPS.
THESE CHARTS ARE BASED ON THE PERIOD 1931-60.
PACIFIC OCEAN
ATLANTIC OCEAN
GULF OF MEXICO
Spokane
Havre
Williston
International Falls
Great Falls
Bismarck
Fargo
Duluth
S. Ste. Marie
Caribou
Burlington
Portland
Concord
Albany
Boston
Hartford
New Haven
New York
Philadelphia
Harrisburg
Pittsburgh
Washington D.C.
Richmond
Norfolk
Raleigh
Wilmington
Charleston
Columbia
Atlanta
Jacksonville
Tampa
Miami
Pensacola
Birmingham
Knoxville
Nashville
Memphis
Little Rock
Jackson
New Orleans
Shreveport
Houston
San Antonio
Del Rio
Brownsville
Abilene
Fort Worth
Oklahoma City
Amarillo
Dodge City
Kansas City
St. Louis
Springfield
Indianapolis
Columbus
Lexington
Charleston
Chicago
Dubuque
Des Moines
Lincoln
North Platte
Minneapolis
Huron
Green Bay
Lansing
Detroit
Rapid City
Billings
Cheyenne
Denver
Grand Junction
Albuquerque
El Paso
Phoenix
Yuma
Las Vegas
Tonopah
Milford
Salt Lake City
Pocatello
Boise
Burns
Pendleton
Winnemucca
Reno
Portland
Roseburg
Eureka
Sacramento
San Francisco
Fresno
Los Angeles
San Diego
Seattle
LAKE SUPERIOR
LAKE MICHIGAN
LAKE HURON
LAKE ONTARIO
LAKE ERIE
ALASKA
INSUFFICIENT DATA FOR ISOLINES
BARROW
BARTER IS.
ARCTIC CIRCLE
FORT YUKON
FAIRBANKS
NOME
BETHEL
NORTHWAY
SEWARD
YAKUTAT
JUNEAU
SITKA
ANNETTE
KODIAK
ST. PAUL IS.
COLD BAY
SHEMYA
ADAK
HAWAII
LIHUE
HONOLULU
KAHULUI
HILO
PUERTO RICO AND VIRGIN ISLANDS
SAN JUAN
DORADO
ALEX. HAMILTON FLD.
ALBERS EQUAL AREA PROJECTION
STANDARD PARALLELS 29½° AND 45½°
MILES

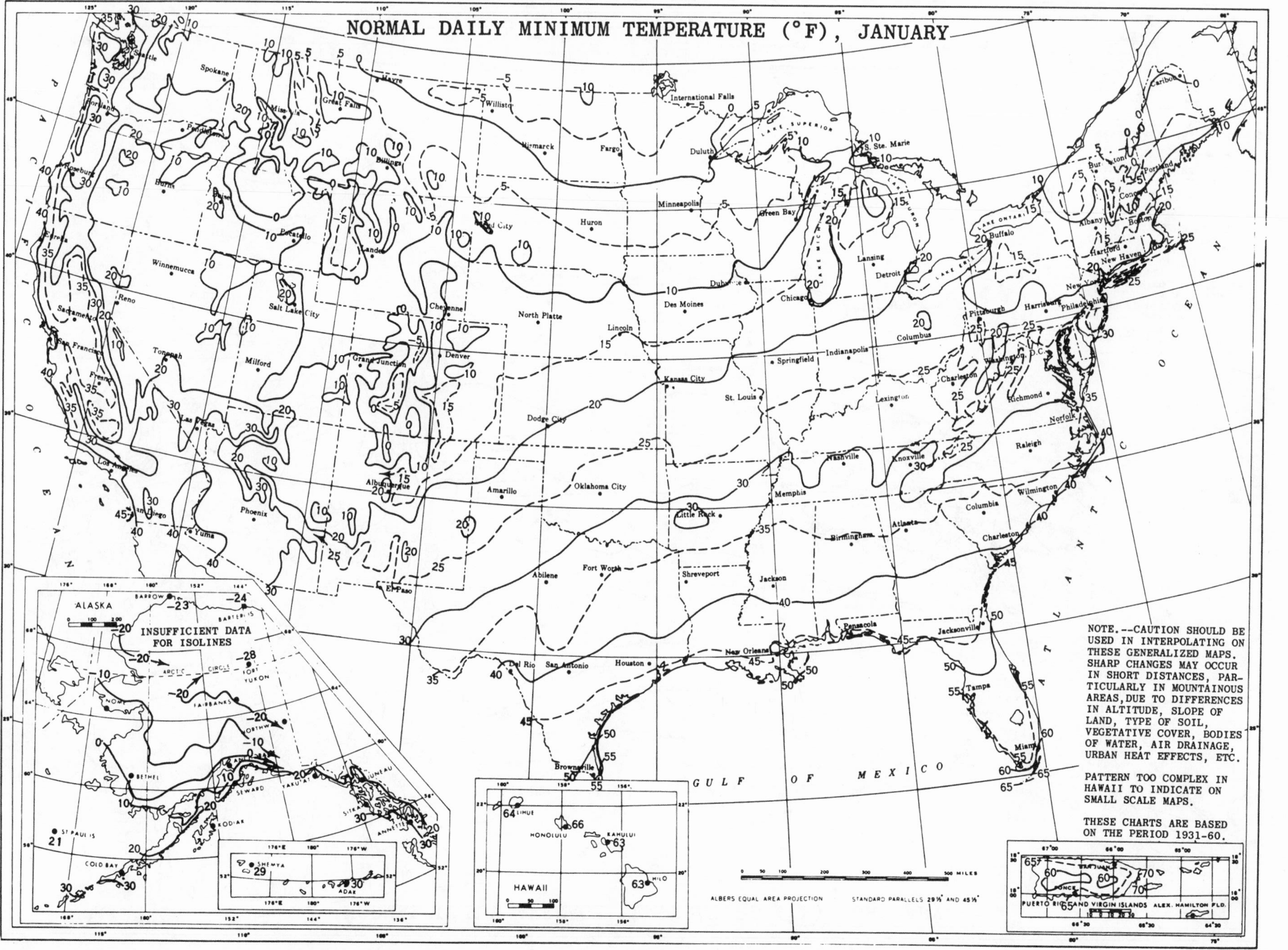
NORMAL DAILY MINIMUM TEMPERATURE (°F), JANUARY
NOTE.--CAUTION SHOULD BE USED IN INTERPOLATING ON THESE GENERALIZED MAPS. SHARP CHANGES MAY OCCUR IN SHORT DISTANCES, PARTICULARLY IN MOUNTAINOUS AREAS, DUE TO DIFFERENCES IN ALTITUDE, SLOPE OF LAND, TYPE OF SOIL, VEGETATIVE COVER, BODIES OF WATER, AIR DRAINAGE, URBAN HEAT EFFECTS, ETC.
PATTERN TOO COMPLEX IN HAWAII TO INDICATE ON SMALL SCALE MAPS.
THESE CHARTS ARE BASED ON THE PERIOD 1931-60.
PACIFIC OCEAN
ATLANTIC OCEAN
GULF OF MEXICO
ALASKA
INSUFFICIENT DATA FOR ISOLINES
HAWAII
PUERTO RICO AND VIRGIN ISLANDS
ALBERS EQUAL AREA PROJECTION
STANDARD PARALLELS 29½° AND 45½°
MILES

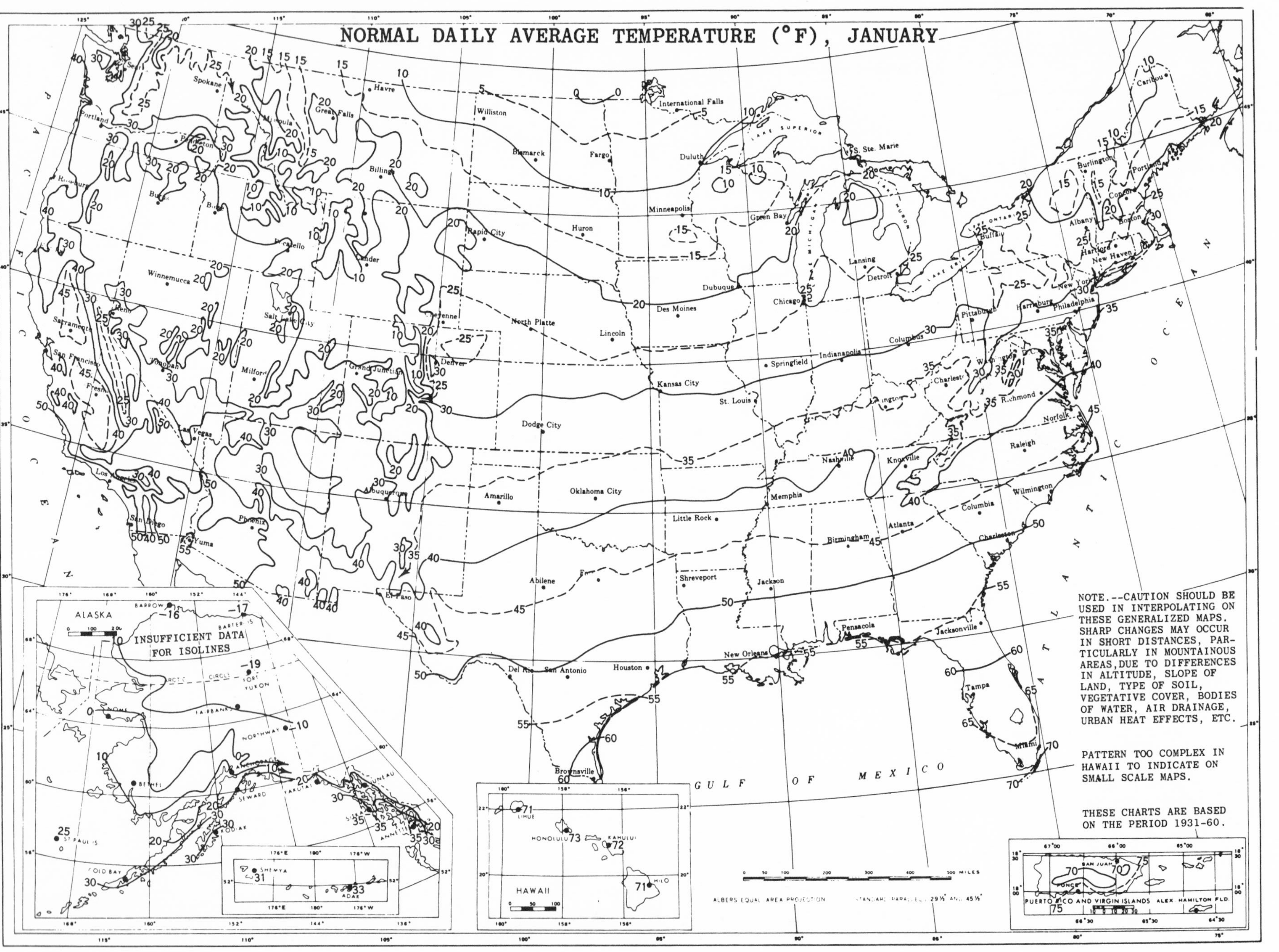
NORMAL DAILY AVERAGE TEMPERATURE (°F), JANUARY
NOTE.--CAUTION SHOULD BE USED IN INTERPOLATING ON THESE GENERALIZED MAPS. SHARP CHANGES MAY OCCUR IN SHORT DISTANCES, PARTICULARLY IN MOUNTAINOUS AREAS, DUE TO DIFFERENCES IN ALTITUDE, SLOPE OF LAND, TYPE OF SOIL, VEGETATIVE COVER, BODIES OF WATER, AIR DRAINAGE, URBAN HEAT EFFECTS, ETC.
PATTERN TOO COMPLEX IN HAWAII TO INDICATE ON SMALL SCALE MAPS.
THESE CHARTS ARE BASED ON THE PERIOD 1931-60.
ALASKA
INSUFFICIENT DATA FOR ISOLINES
HAWAII
PUERTO RICO AND VIRGIN ISLANDS
ALBERS EQUAL AREA PROJECTION
PACIFIC OCEAN
ATLANTIC OCEAN
GULF OF MEXICO

NORMAL DAILY RANGE OF TEMPERATURE (°F), JANUARY

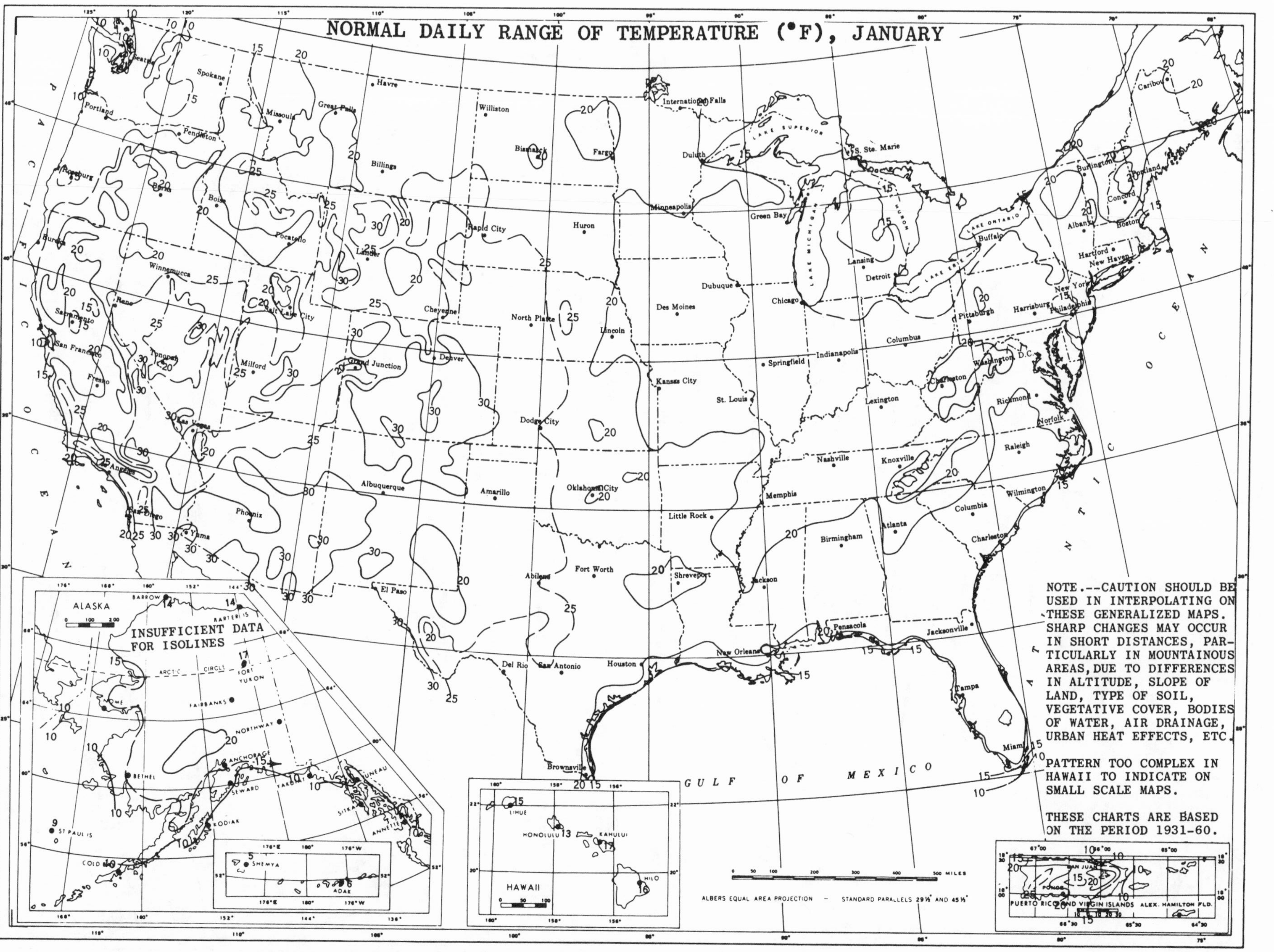

NORMAL DAILY MAXIMUM, AVERAGE, MINIMUM, AND EXTREME TEMPERATURES (°F), JANUARY

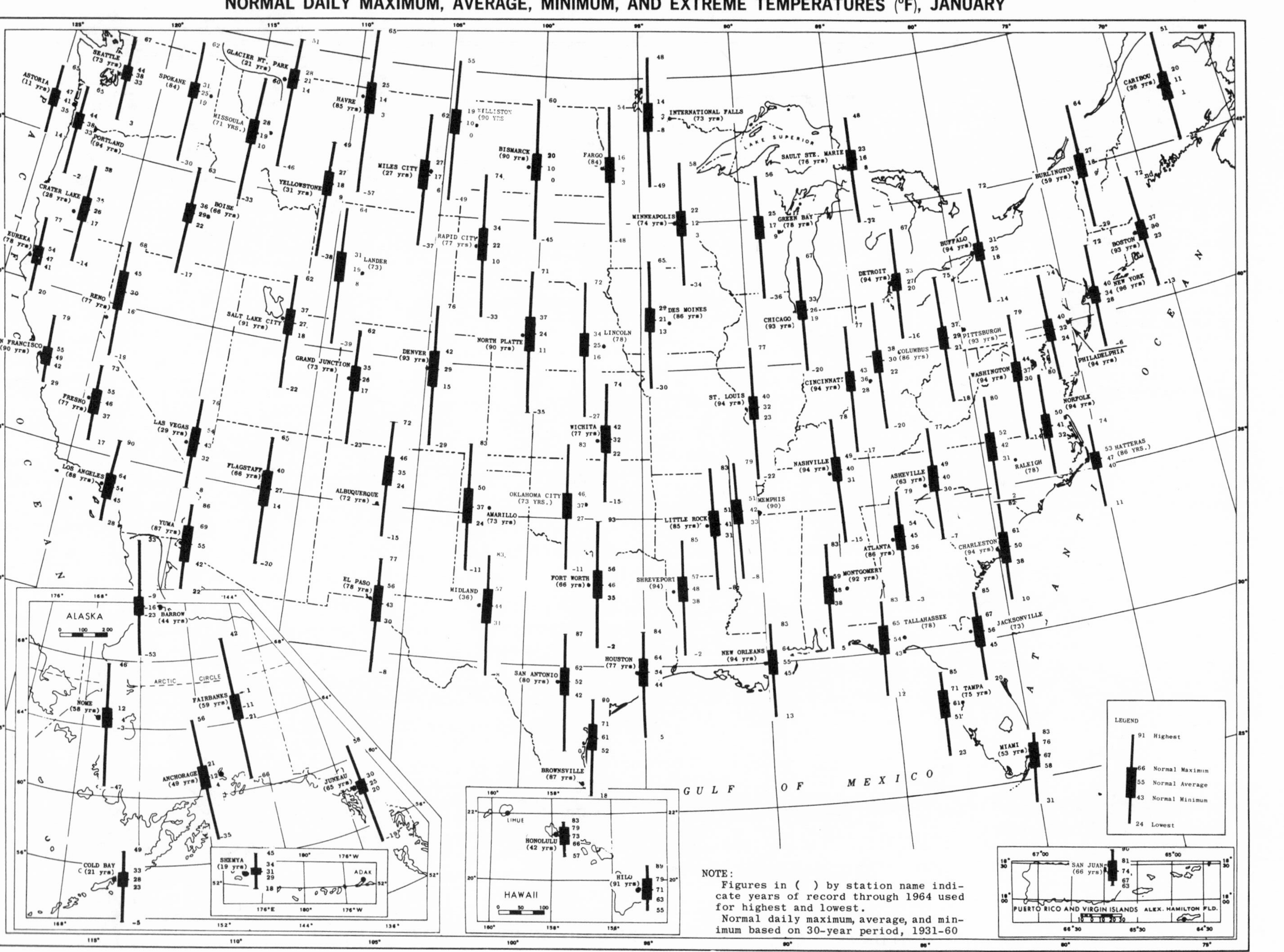

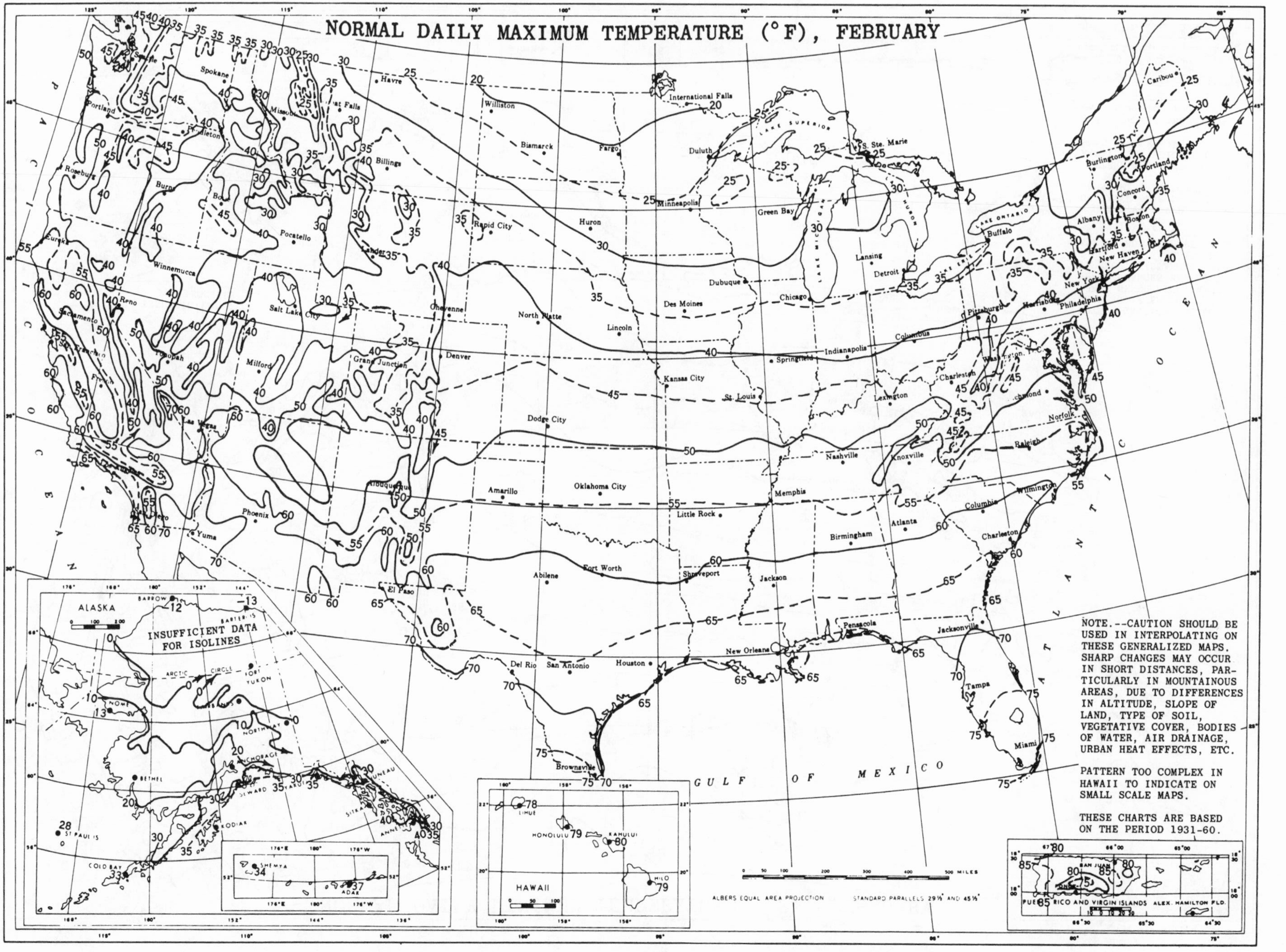

NORMAL DAILY MAXIMUM TEMPERATURE (°F), FEBRUARY
NOTE.--CAUTION SHOULD BE USED IN INTERPOLATING ON THESE GENERALIZED MAPS. SHARP CHANGES MAY OCCUR IN SHORT DISTANCES, PARTICULARLY IN MOUNTAINOUS AREAS, DUE TO DIFFERENCES IN ALTITUDE, SLOPE OF LAND, TYPE OF SOIL, VEGETATIVE COVER, BODIES OF WATER, AIR DRAINAGE, URBAN HEAT EFFECTS, ETC.
PATTERN TOO COMPLEX IN HAWAII TO INDICATE ON SMALL SCALE MAPS.
THESE CHARTS ARE BASED ON THE PERIOD 1931-60.
ALASKA
INSUFFICIENT DATA FOR ISOLINES
HAWAII
PUERTO RICO AND VIRGIN ISLANDS
ALBERS EQUAL AREA PROJECTION
STANDARD PARALLELS 29½° AND 45½°
GULF OF MEXICO
PACIFIC OCEAN
ATLANTIC OCEAN

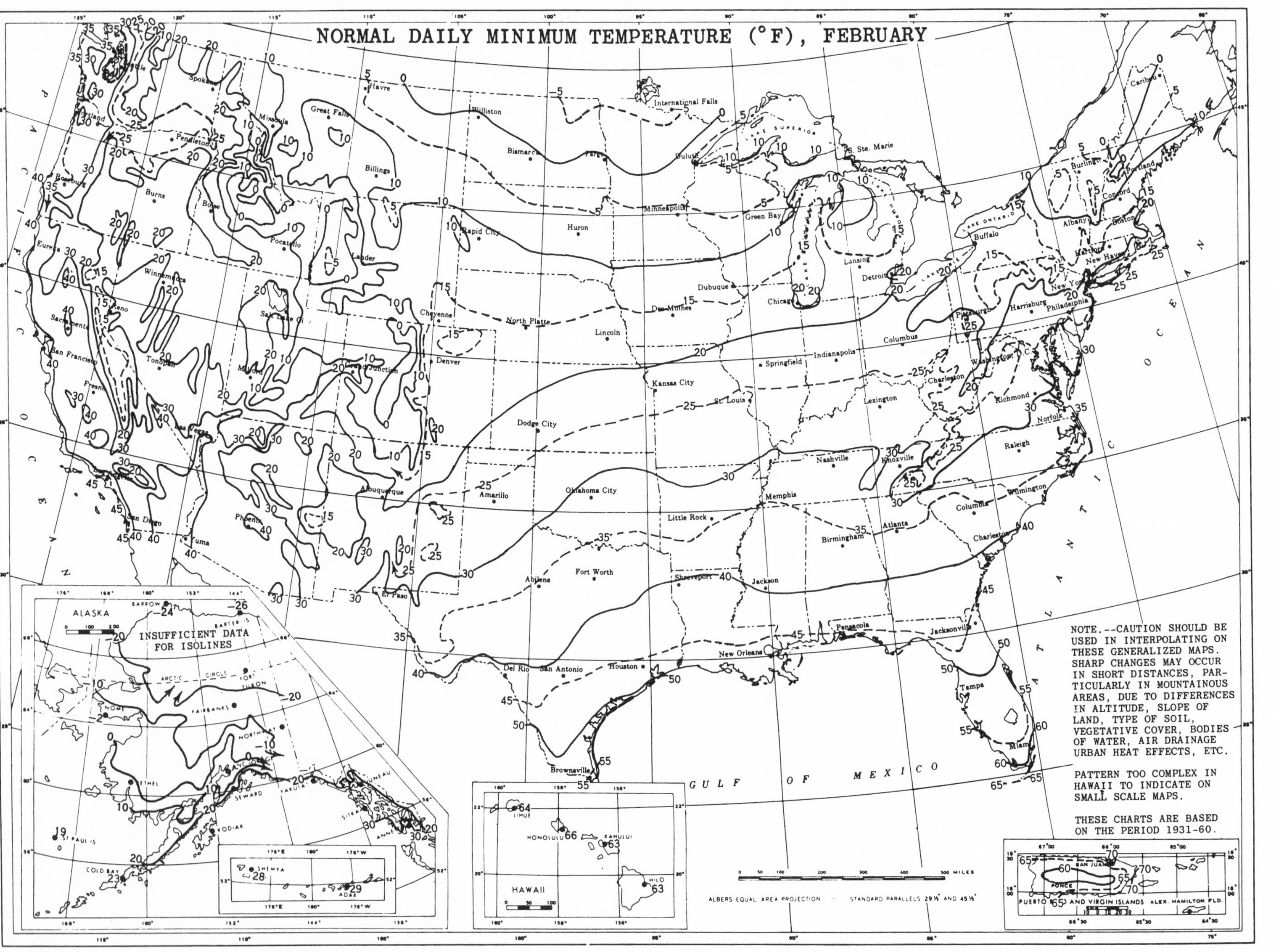
NORMAL DAILY MINIMUM TEMPERATURE (°F), FEBRUARY
NOTE.--CAUTION SHOULD BE USED IN INTERPOLATING ON THESE GENERALIZED MAPS. SHARP CHANGES MAY OCCUR IN SHORT DISTANCES, PARTICULARLY IN MOUNTAINOUS AREAS, DUE TO DIFFERENCES IN ALTITUDE, SLOPE OF LAND, TYPE OF SOIL, VEGETATIVE COVER, BODIES OF WATER, AIR DRAINAGE URBAN HEAT EFFECTS, ETC.
PATTERN TOO COMPLEX IN HAWAII TO INDICATE ON SMALL SCALE MAPS.
THESE CHARTS ARE BASED ON THE PERIOD 1931-60.
ALASKA
INSUFFICIENT DATA FOR ISOLINES
HAWAII
PUERTO RICO AND VIRGIN ISLANDS
PACIFIC OCEAN
ATLANTIC OCEAN
GULF OF MEXICO
ALBERS EQUAL AREA PROJECTION - STANDARD PARALLELS 29½° AND 45½°
0 50 100 200 300 400 500 MILES

NORMAL DAILY AVERAGE TEMPERATURE (°F), FEBRUARY

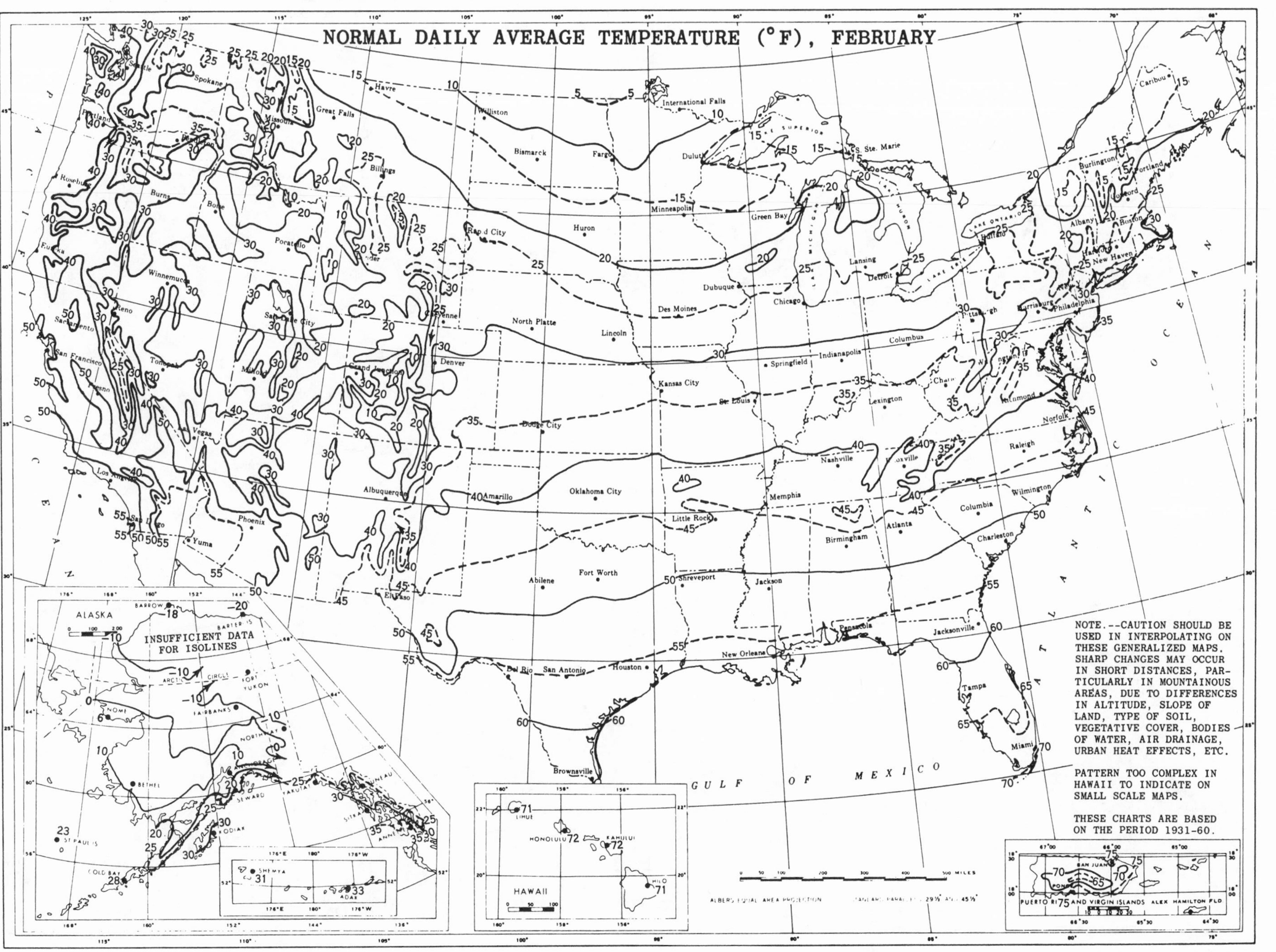

NORMAL DAILY RANGE OF TEMPERATURE (°F), FEBRUARY

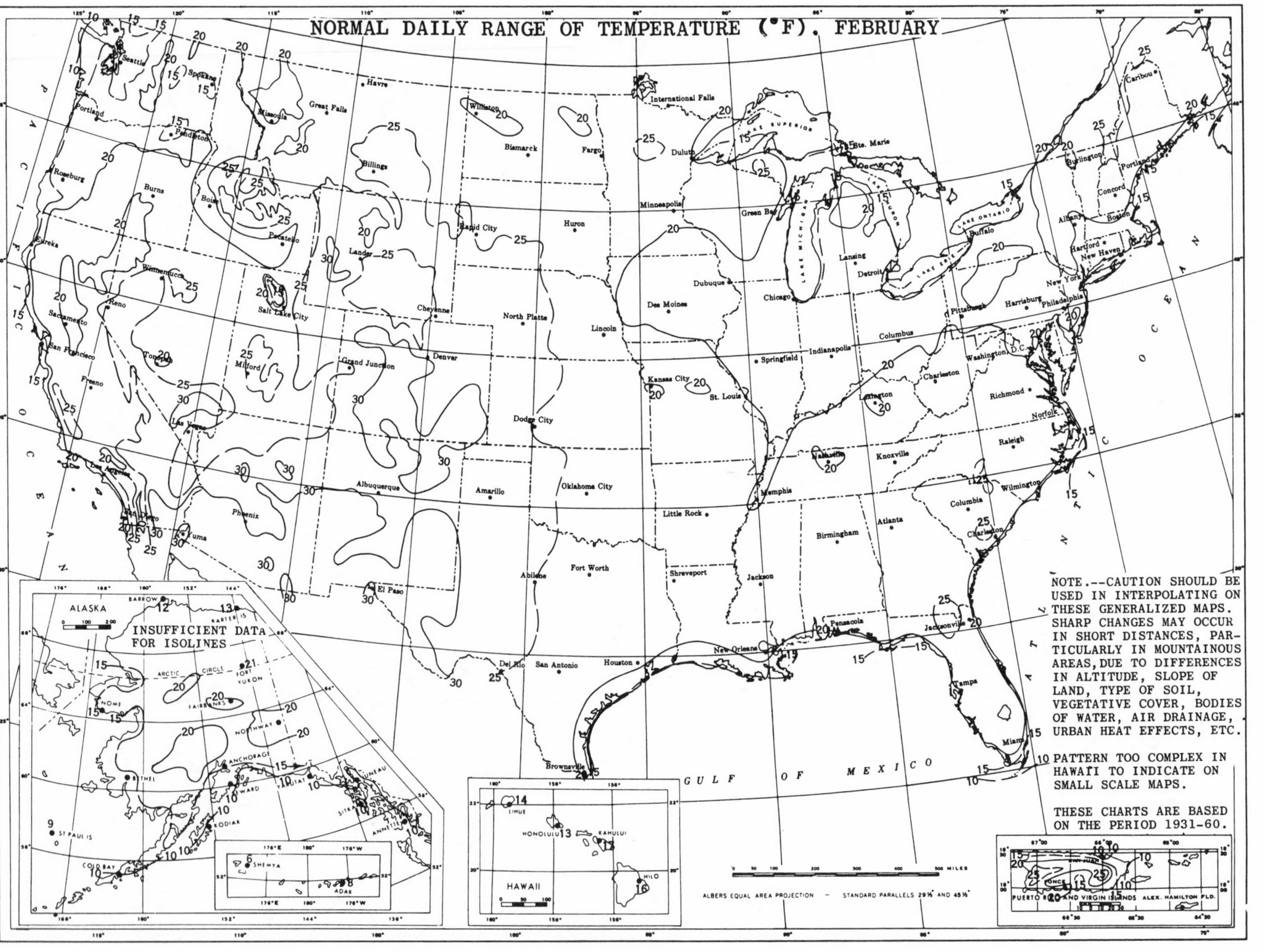
PACIFIC OCEAN
ATLANTIC OCEAN
GULF OF MEXICO
NOTE.--CAUTION SHOULD BE USED IN INTERPOLATING ON THESE GENERALIZED MAPS. SHARP CHANGES MAY OCCUR IN SHORT DISTANCES, PARTICULARLY IN MOUNTAINOUS AREAS, DUE TO DIFFERENCES IN ALTITUDE, SLOPE OF LAND, TYPE OF SOIL, VEGETATIVE COVER, BODIES OF WATER, AIR DRAINAGE, URBAN HEAT EFFECTS, ETC.
PATTERN TOO COMPLEX IN HAWAII TO INDICATE ON SMALL SCALE MAPS.
THESE CHARTS ARE BASED ON THE PERIOD 1931-60.
ALASKA
INSUFFICIENT DATA FOR ISOLINES
HAWAII
PUERTO RICO AND VIRGIN ISLANDS
ALBERS EQUAL AREA PROJECTION — STANDARD PARALLELS 29½° AND 45½°

NORMAL DAILY MAXIMUM, AVERAGE, MINIMUM, AND EXTREME TEMPERATURES (°F), FEBRUARY

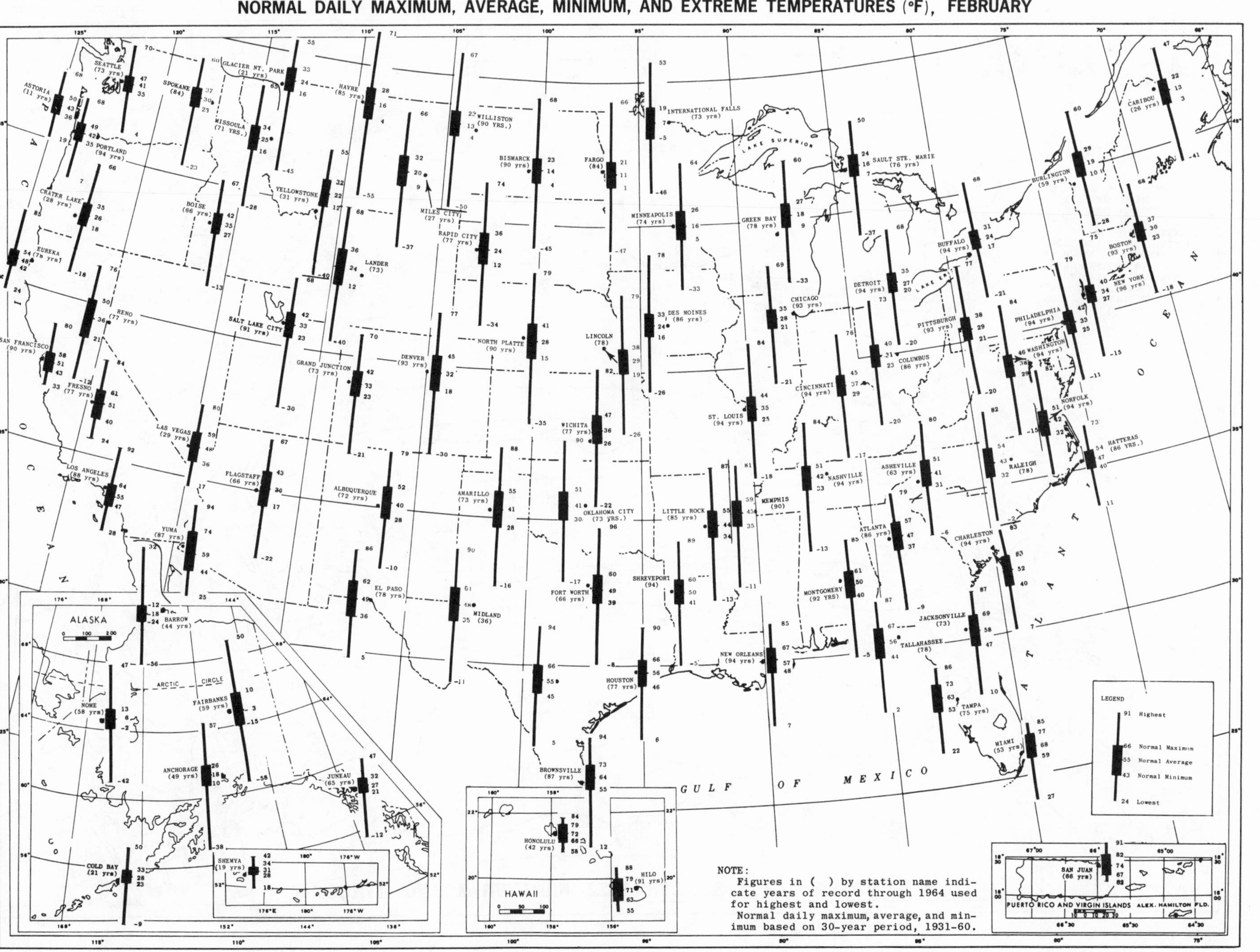

NORMAL DAILY MAXIMUM TEMPERATURE (°F), MARCH

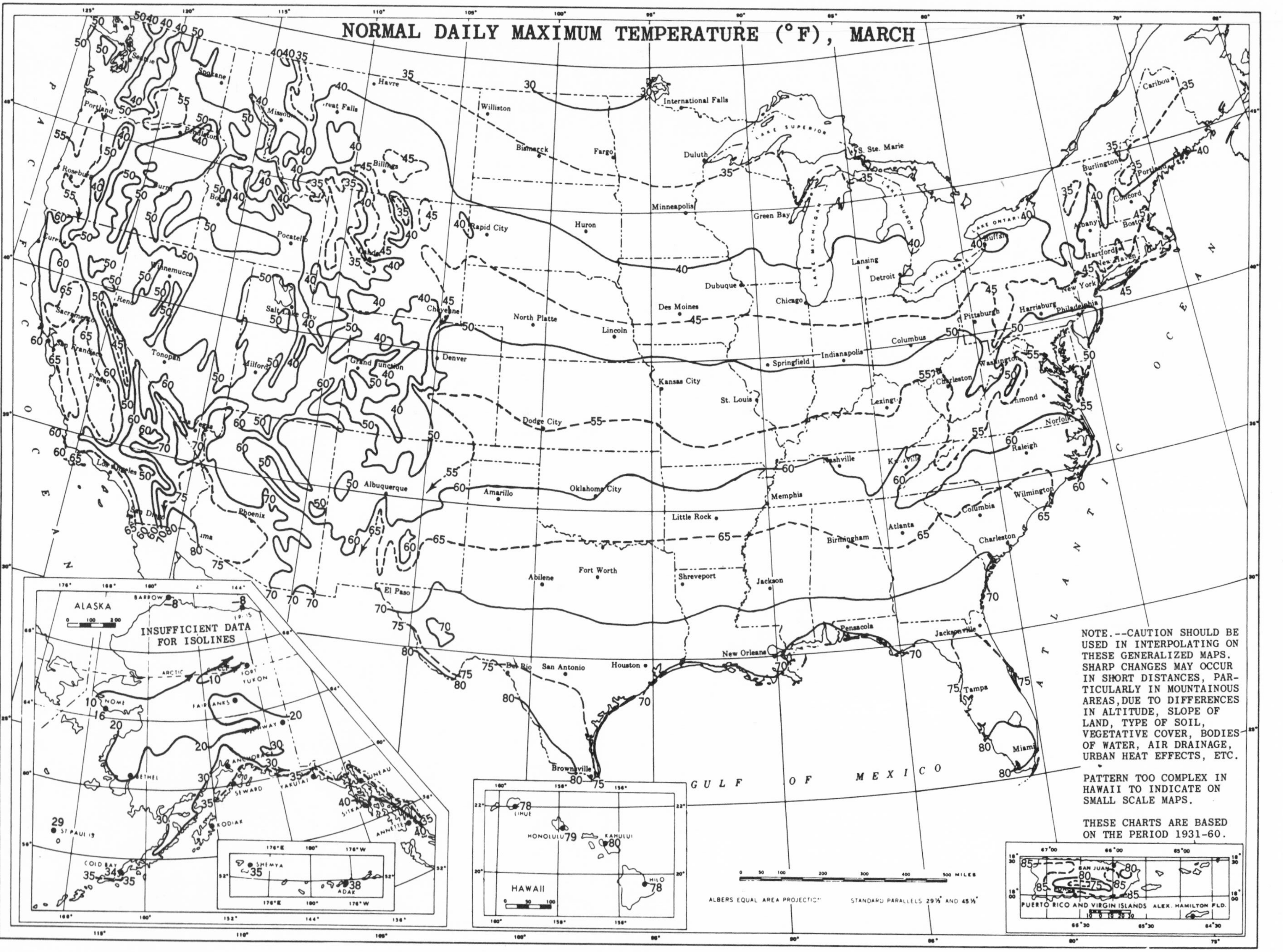

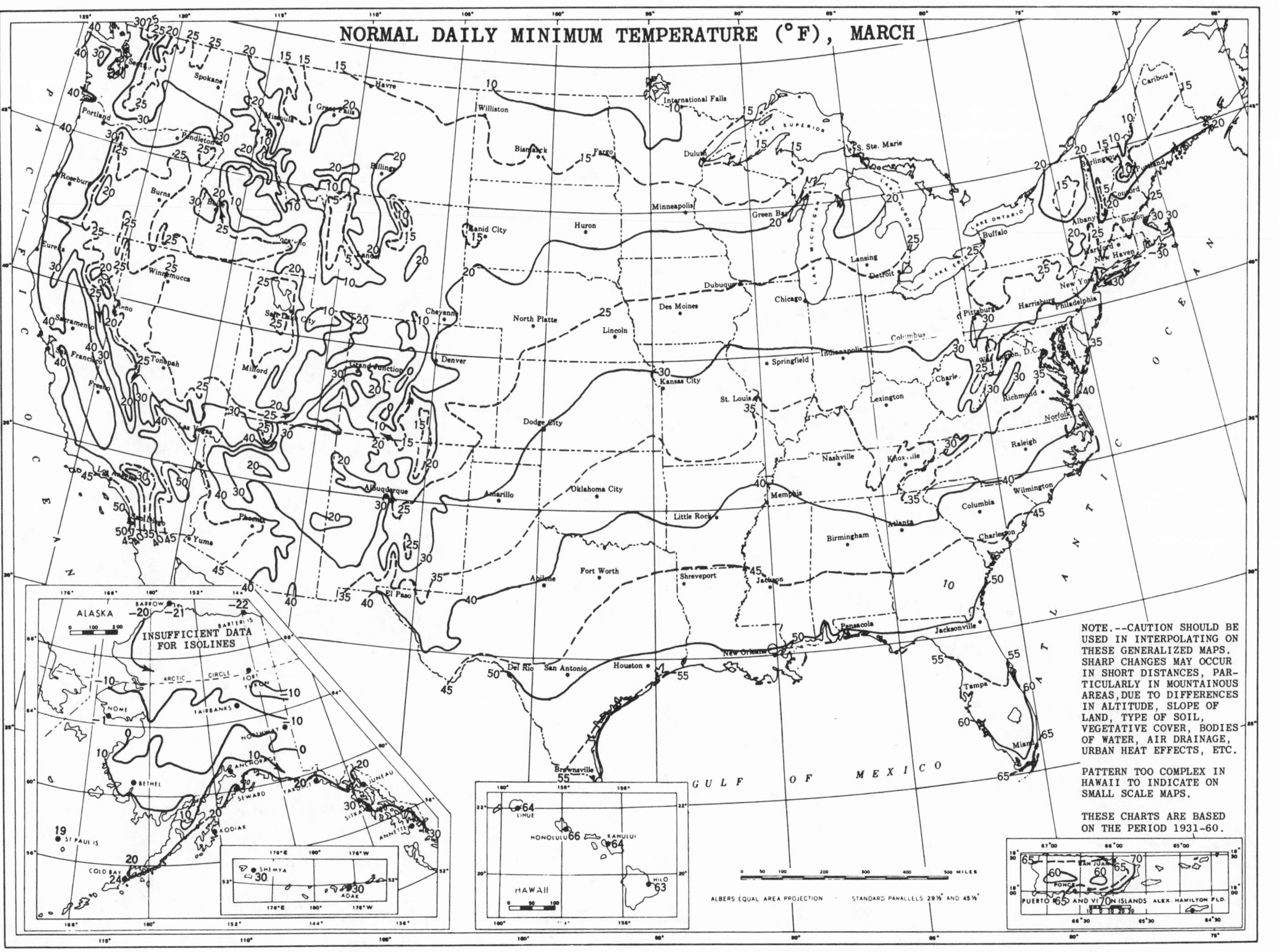
NORMAL DAILY MINIMUM TEMPERATURE (°F), MARCH
NOTE.--CAUTION SHOULD BE USED IN INTERPOLATING ON THESE GENERALIZED MAPS. SHARP CHANGES MAY OCCUR IN SHORT DISTANCES, PARTICULARLY IN MOUNTAINOUS AREAS, DUE TO DIFFERENCES IN ALTITUDE, SLOPE OF LAND, TYPE OF SOIL, VEGETATIVE COVER, BODIES OF WATER, AIR DRAINAGE, URBAN HEAT EFFECTS, ETC.
PATTERN TOO COMPLEX IN HAWAII TO INDICATE ON SMALL SCALE MAPS.
THESE CHARTS ARE BASED ON THE PERIOD 1931-60.
ALASKA
INSUFFICIENT DATA FOR ISOLINES
HAWAII
PUERTO RICO AND VIRGIN ISLANDS
ALBERS EQUAL AREA PROJECTION · STANDARD PARALLELS 29½° AND 45½°
PACIFIC OCEAN
ATLANTIC OCEAN
GULF OF MEXICO

NORMAL DAILY AVERAGE TEMPERATURE (°F), MARCH

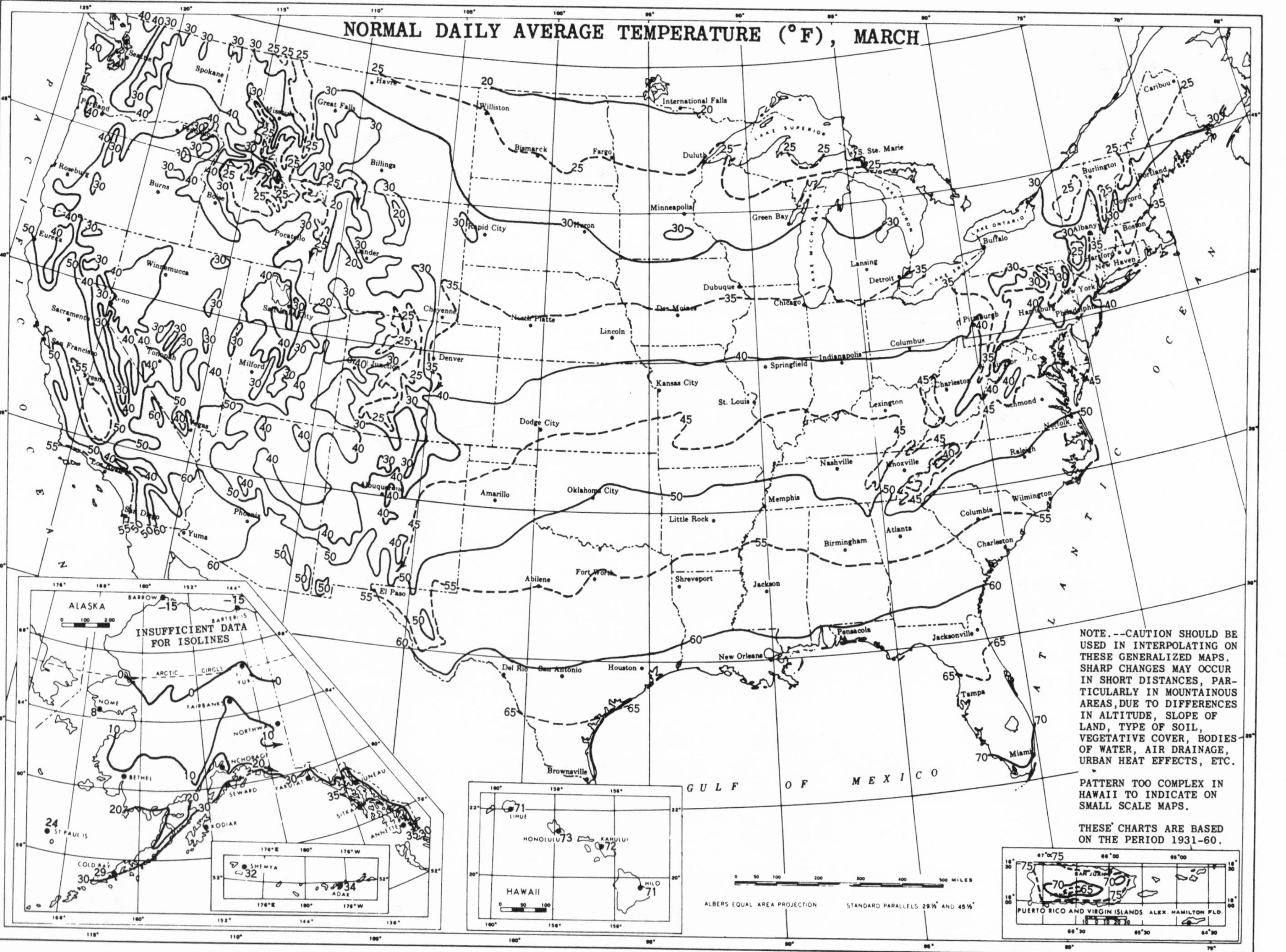

PACIFIC OCEAN
ATLANTIC OCEAN
GULF OF MEXICO
NOTE.--CAUTION SHOULD BE USED IN INTERPOLATING ON THESE GENERALIZED MAPS. SHARP CHANGES MAY OCCUR IN SHORT DISTANCES, PARTICULARLY IN MOUNTAINOUS AREAS, DUE TO DIFFERENCES IN ALTITUDE, SLOPE OF LAND, TYPE OF SOIL, VEGETATIVE COVER, BODIES OF WATER, AIR DRAINAGE, URBAN HEAT EFFECTS, ETC.
PATTERN TOO COMPLEX IN HAWAII TO INDICATE ON SMALL SCALE MAPS.
THESE CHARTS ARE BASED ON THE PERIOD 1931-60.
ALASKA
INSUFFICIENT DATA FOR ISOLINES
HAWAII
PUERTO RICO AND VIRGIN ISLANDS
ALBERS EQUAL AREA PROJECTION
STANDARD PARALLELS 29½° AND 45½°

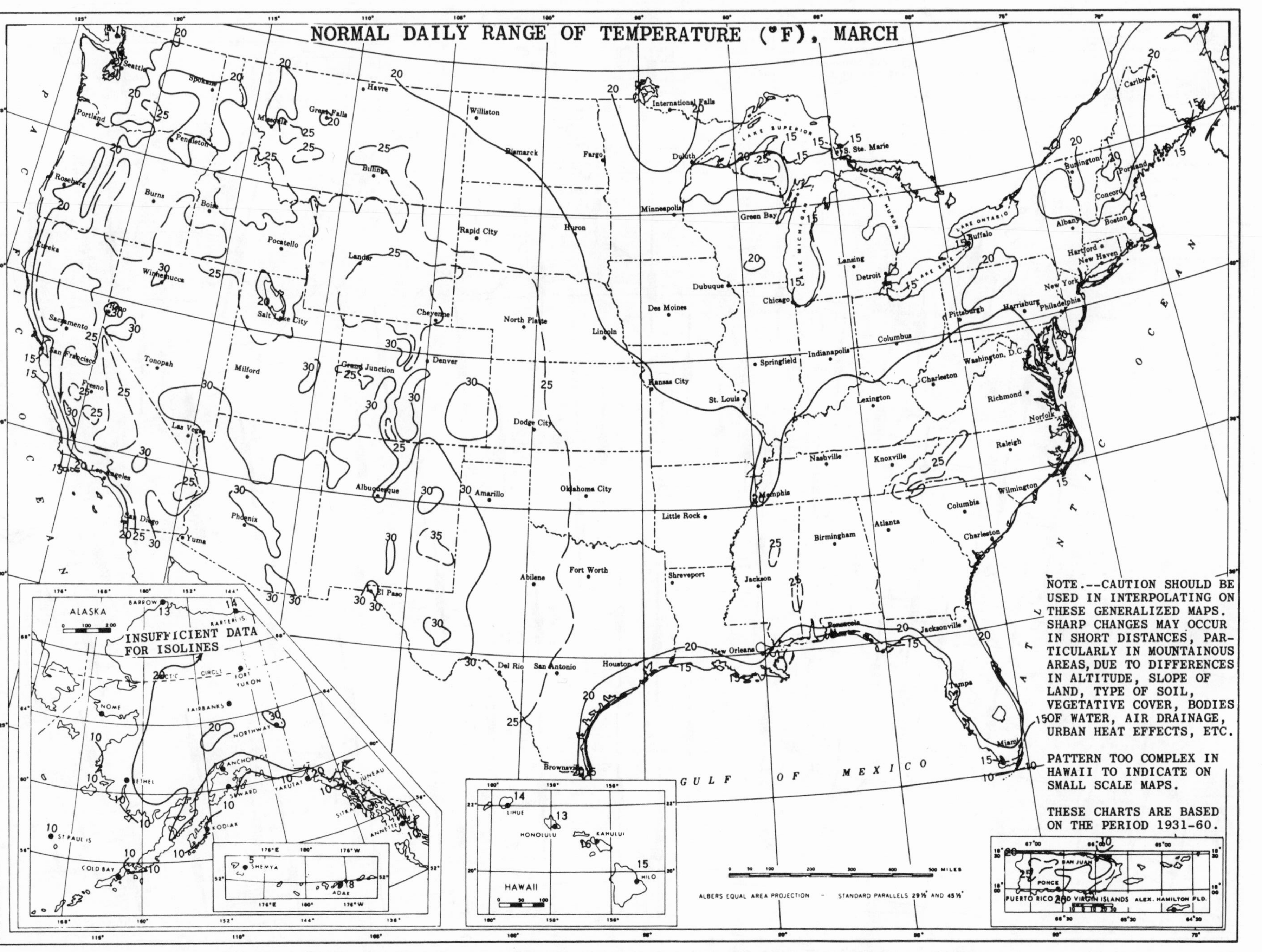
NORMAL DAILY RANGE OF TEMPERATURE (°F), MARCH
NOTE.--CAUTION SHOULD BE USED IN INTERPOLATING ON THESE GENERALIZED MAPS. SHARP CHANGES MAY OCCUR IN SHORT DISTANCES, PARTICULARLY IN MOUNTAINOUS AREAS, DUE TO DIFFERENCES IN ALTITUDE, SLOPE OF LAND, TYPE OF SOIL, VEGETATIVE COVER, BODIES OF WATER, AIR DRAINAGE, URBAN HEAT EFFECTS, ETC.
PATTERN TOO COMPLEX IN HAWAII TO INDICATE ON SMALL SCALE MAPS.
THESE CHARTS ARE BASED ON THE PERIOD 1931-60.
PACIFIC OCEAN
ATLANTIC OCEAN
GULF OF MEXICO
ALASKA
INSUFFICIENT DATA FOR ISOLINES
HAWAII
PUERTO RICO AND VIRGIN ISLANDS
ALBERS EQUAL AREA PROJECTION - STANDARD PARALLELS 29½° AND 45½°
0 50 100 200 300 400 500 MILES

NORMAL DAILY MAXIMUM, AVERAGE, MINIMUM, AND EXTREME TEMPERATURES (°F), MARCH

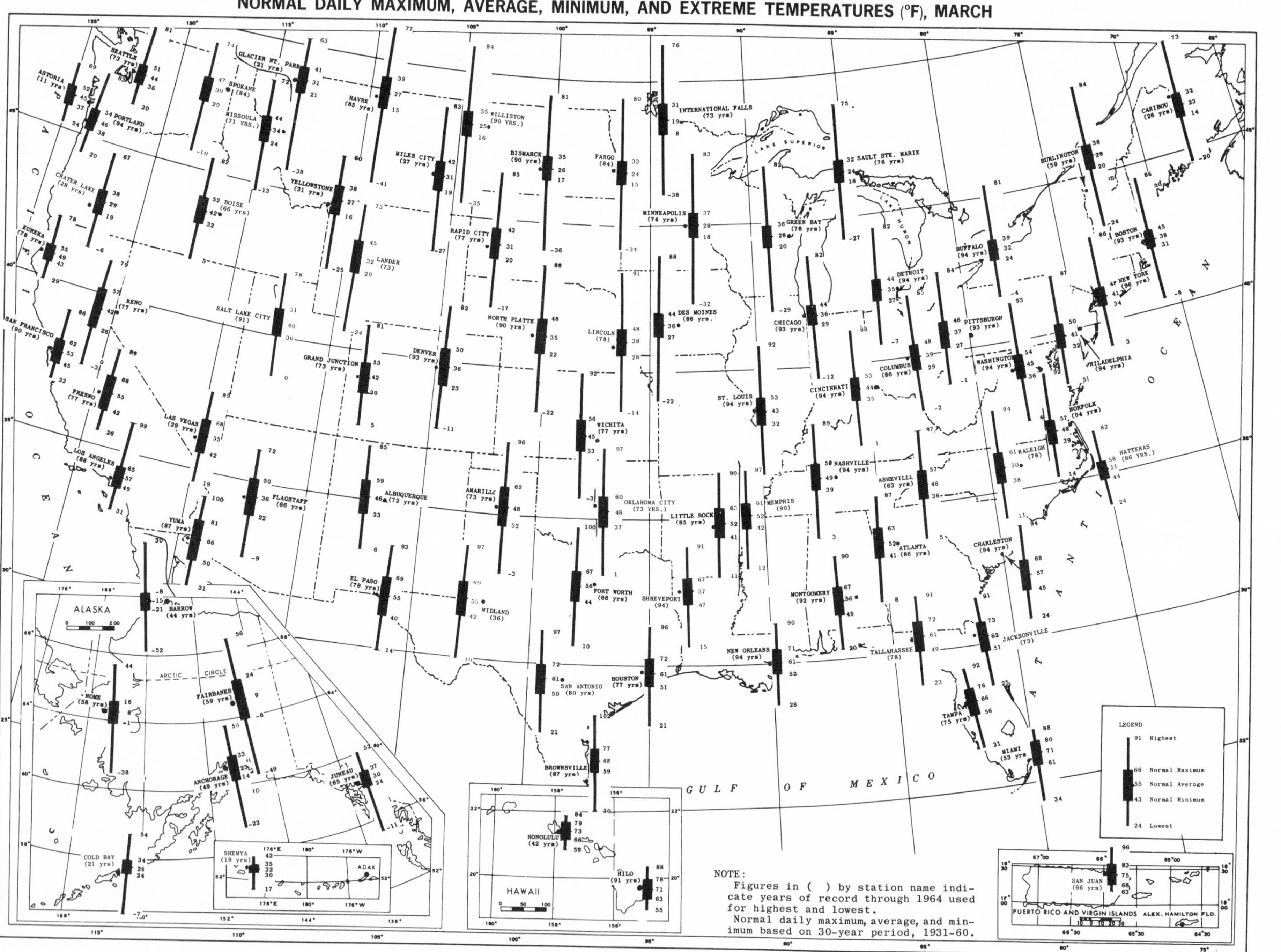

NORMAL DAILY MAXIMUM TEMPERATURE (° F), APRIL

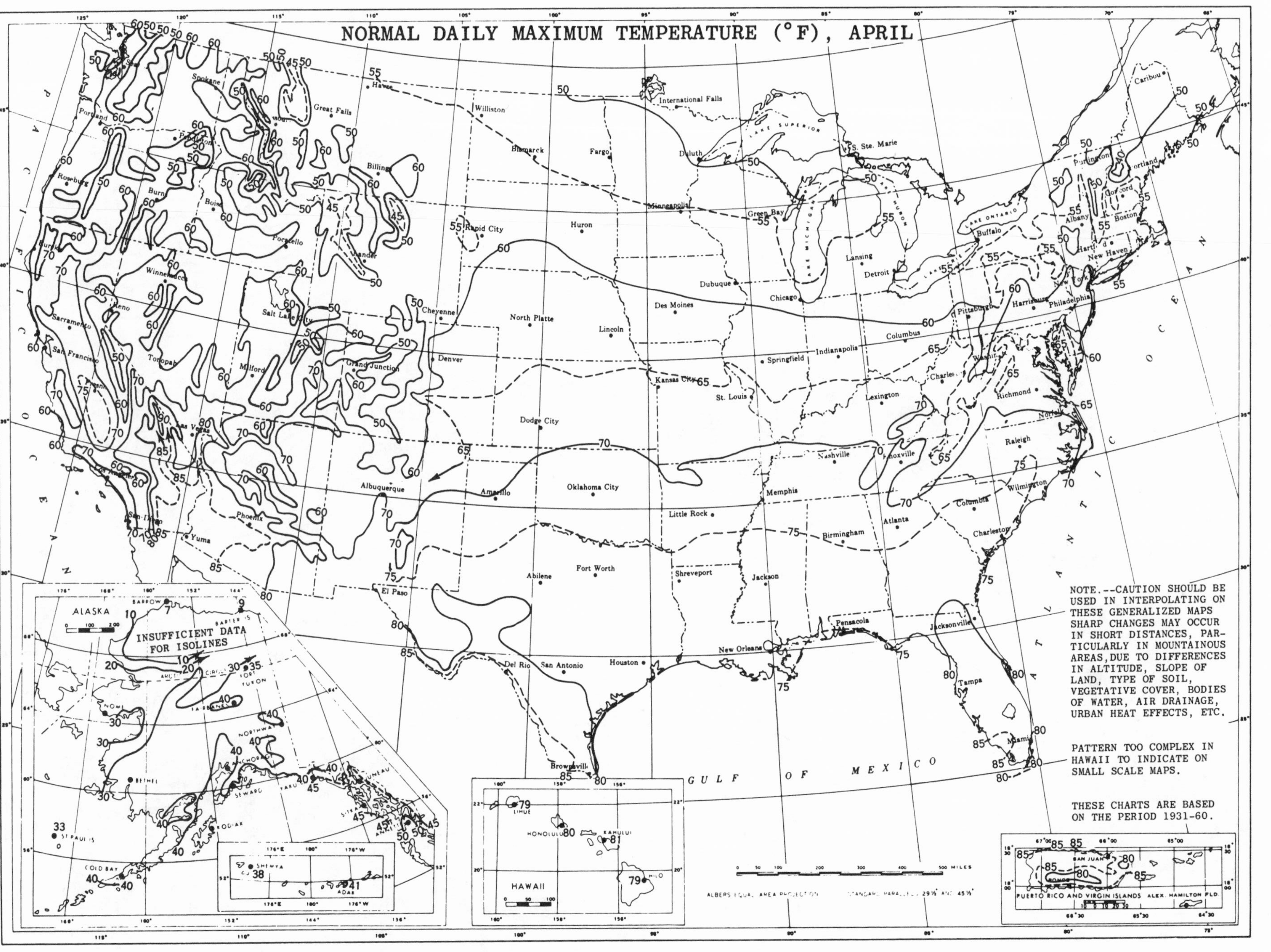

NOTE.--CAUTION SHOULD BE USED IN INTERPOLATING ON THESE GENERALIZED MAPS SHARP CHANGES MAY OCCUR IN SHORT DISTANCES, PARTICULARLY IN MOUNTAINOUS AREAS, DUE TO DIFFERENCES IN ALTITUDE, SLOPE OF LAND, TYPE OF SOIL, VEGETATIVE COVER, BODIES OF WATER, AIR DRAINAGE, URBAN HEAT EFFECTS, ETC.
PATTERN TOO COMPLEX IN HAWAII TO INDICATE ON SMALL SCALE MAPS.
THESE CHARTS ARE BASED ON THE PERIOD 1931-60.
ALASKA
INSUFFICIENT DATA FOR ISOLINES
HAWAII
PUERTO RICO AND VIRGIN ISLANDS
GULF OF MEXICO
PACIFIC OCEAN
ATLANTIC OCEAN

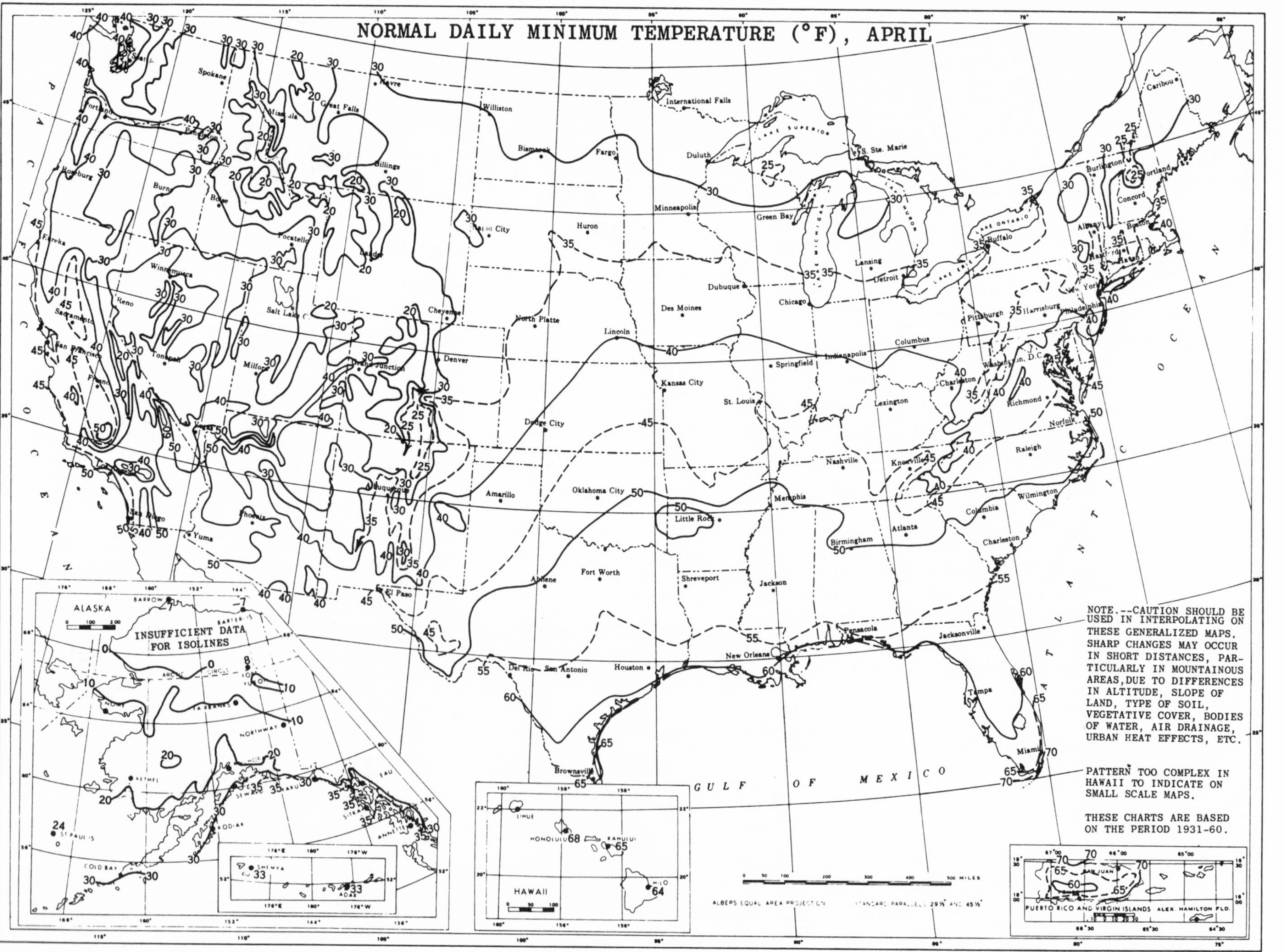
NORMAL DAILY MINIMUM TEMPERATURE (°F), APRIL
NOTE.--CAUTION SHOULD BE USED IN INTERPOLATING ON THESE GENERALIZED MAPS. SHARP CHANGES MAY OCCUR IN SHORT DISTANCES, PARTICULARLY IN MOUNTAINOUS AREAS, DUE TO DIFFERENCES IN ALTITUDE, SLOPE OF LAND, TYPE OF SOIL, VEGETATIVE COVER, BODIES OF WATER, AIR DRAINAGE, URBAN HEAT EFFECTS, ETC.
PATTERN TOO COMPLEX IN HAWAII TO INDICATE ON SMALL SCALE MAPS.
THESE CHARTS ARE BASED ON THE PERIOD 1931-60.
PACIFIC OCEAN
ATLANTIC OCEAN
GULF OF MEXICO
ALASKA
INSUFFICIENT DATA FOR ISOLINES
HAWAII
PUERTO RICO AND VIRGIN ISLANDS
ALBERS EQUAL AREA PROJECTION
Spokane
Great Falls
Havre
Williston
Bismarck
Fargo
International Falls
Duluth
Minneapolis
Green Bay
Huron
Billings
Boise
Burns
Roseburg
Eureka
Winnemucca
Reno
Sacramento
San Francisco
Fresno
Tonopah
Pocatello
Lander
Salt Lake C.
Milford
Grand Junction
Cheyenne
Denver
North Platte
Dodge City
Amarillo
Lincoln
Des Moines
Dubuque
Kansas City
St. Louis
Springfield
Chicago
Indianapolis
Lansing
Detroit
Columbus
Lexington
Nashville
Knoxville
Memphis
Little Rock
Oklahoma City
Fort Worth
Abilene
San Antonio
Del Rio
El Paso
Albuquerque
Phoenix
Yuma
San Diego
Houston
Brownsville
Shreveport
Jackson
New Orleans
Pensacola
Birmingham
Atlanta
Columbia
Charleston
Wilmington
Raleigh
Richmond
Norfolk
Washington, D.C.
Philadelphia
Harrisburg
Pittsburgh
New York
Buffalo
Albany
Hartford
Boston
Concord
Portland
Burlington
Caribou
Sault Ste. Marie
Jacksonville
Tampa
Miami
Lake Superior
Lake Michigan
Lake Huron
Lake Erie
Lake Ontario
Barrow
Barter Is.
Nome
Fairbanks
Northway
Bethel
Kodiak
St. Paul Is.
Cold Bay
Shemya
Adak
Lihue
Honolulu
Kahului
Hilo
San Juan
Alex. Hamilton Fld.

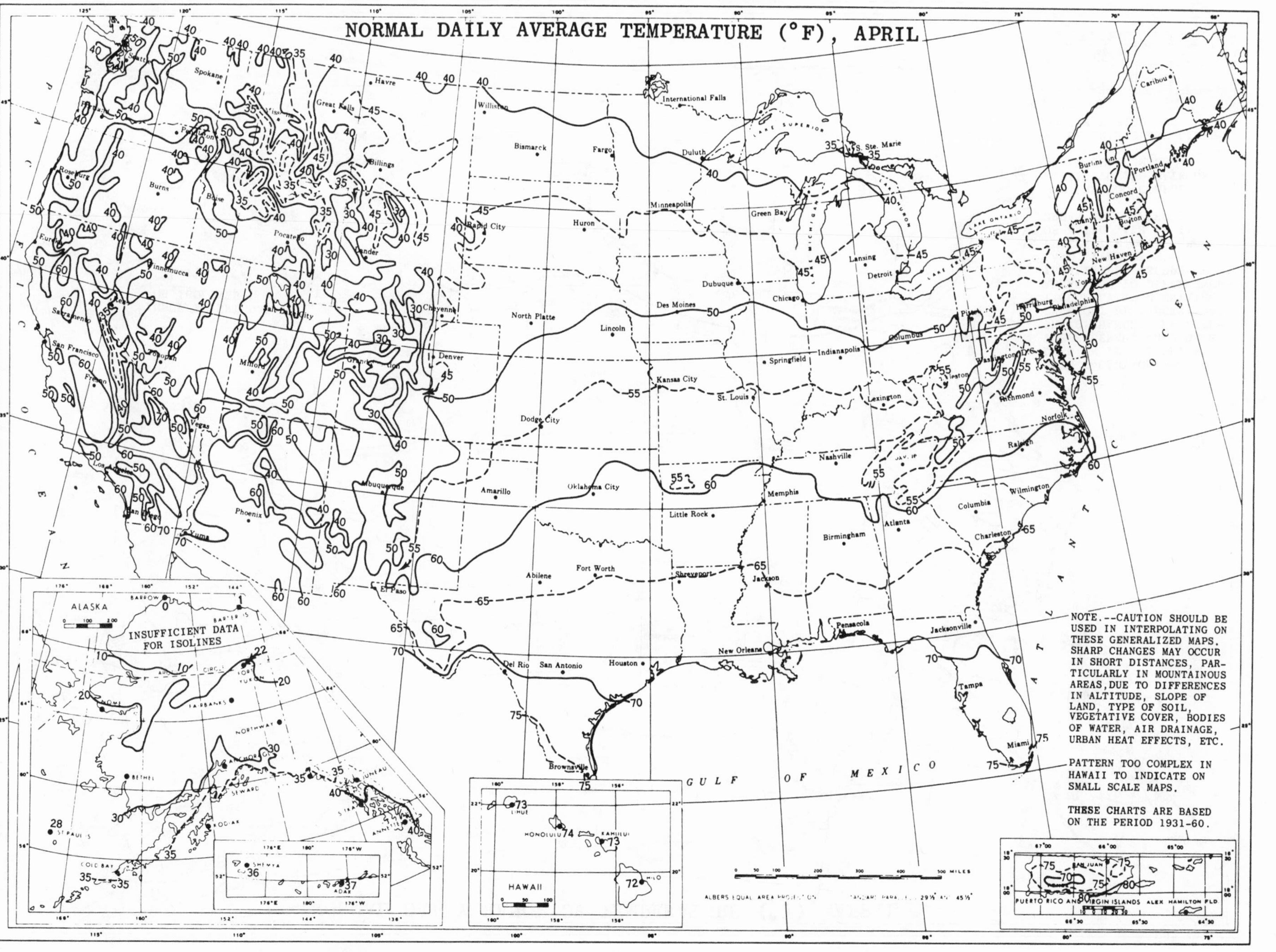
NORMAL DAILY AVERAGE TEMPERATURE (°F), APRIL
NOTE.--CAUTION SHOULD BE USED IN INTERPOLATING ON THESE GENERALIZED MAPS. SHARP CHANGES MAY OCCUR IN SHORT DISTANCES, PARTICULARLY IN MOUNTAINOUS AREAS, DUE TO DIFFERENCES IN ALTITUDE, SLOPE OF LAND, TYPE OF SOIL, VEGETATIVE COVER, BODIES OF WATER, AIR DRAINAGE, URBAN HEAT EFFECTS, ETC.
PATTERN TOO COMPLEX IN HAWAII TO INDICATE ON SMALL SCALE MAPS.
THESE CHARTS ARE BASED ON THE PERIOD 1931-60.
PACIFIC OCEAN
ATLANTIC OCEAN
GULF OF MEXICO
ALASKA
INSUFFICIENT DATA FOR ISOLINES
HAWAII
PUERTO RICO AND VIRGIN ISLANDS
ALBERS EQUAL AREA PROJECTION
Spokane
Havre
Great Falls
Billings
Williston
Bismarck
Fargo
International Falls
Duluth
S. Ste. Marie
Caribou
Portland
Burlington
Concord
Boston
New Haven
Roseburg
Burns
Boise
Pocatello
Lander
Rapid City
Huron
Minneapolis
Green Bay
Lansing
Detroit
Eureka
Winnemucca
Sacramento
Salt Lake City
Cheyenne
North Platte
Des Moines
Dubuque
Chicago
Harrisburg
Philadelphia
San Francisco
Fresno
Tonopah
Milford
Denver
Lincoln
Kansas City
Springfield
Indianapolis
Columbus
Pittsburgh
Washington
Richmond
Las Vegas
St. Louis
Lexington
Dodge City
Norfolk
Raleigh
Los Angeles
San Diego
Yuma
Phoenix
Albuquerque
Amarillo
Oklahoma City
Nashville
Memphis
Little Rock
Atlanta
Columbia
Wilmington
Birmingham
Charleston
El Paso
Abilene
Fort Worth
Shreveport
Jackson
Pensacola
Jacksonville
Del Rio
San Antonio
Houston
New Orleans
Tampa
Miami
Brownsville
Barrow
Fairbanks
Nome
Bethel
Kodiak
Cold Bay
Juneau
Northway
Anchorage
Seward
Annette
St. Paul
Shemya
Adak
Lihue
Honolulu
Kahului
Hilo
San Juan

NORMAL DAILY RANGE OF TEMPERATURE (°F), APRIL

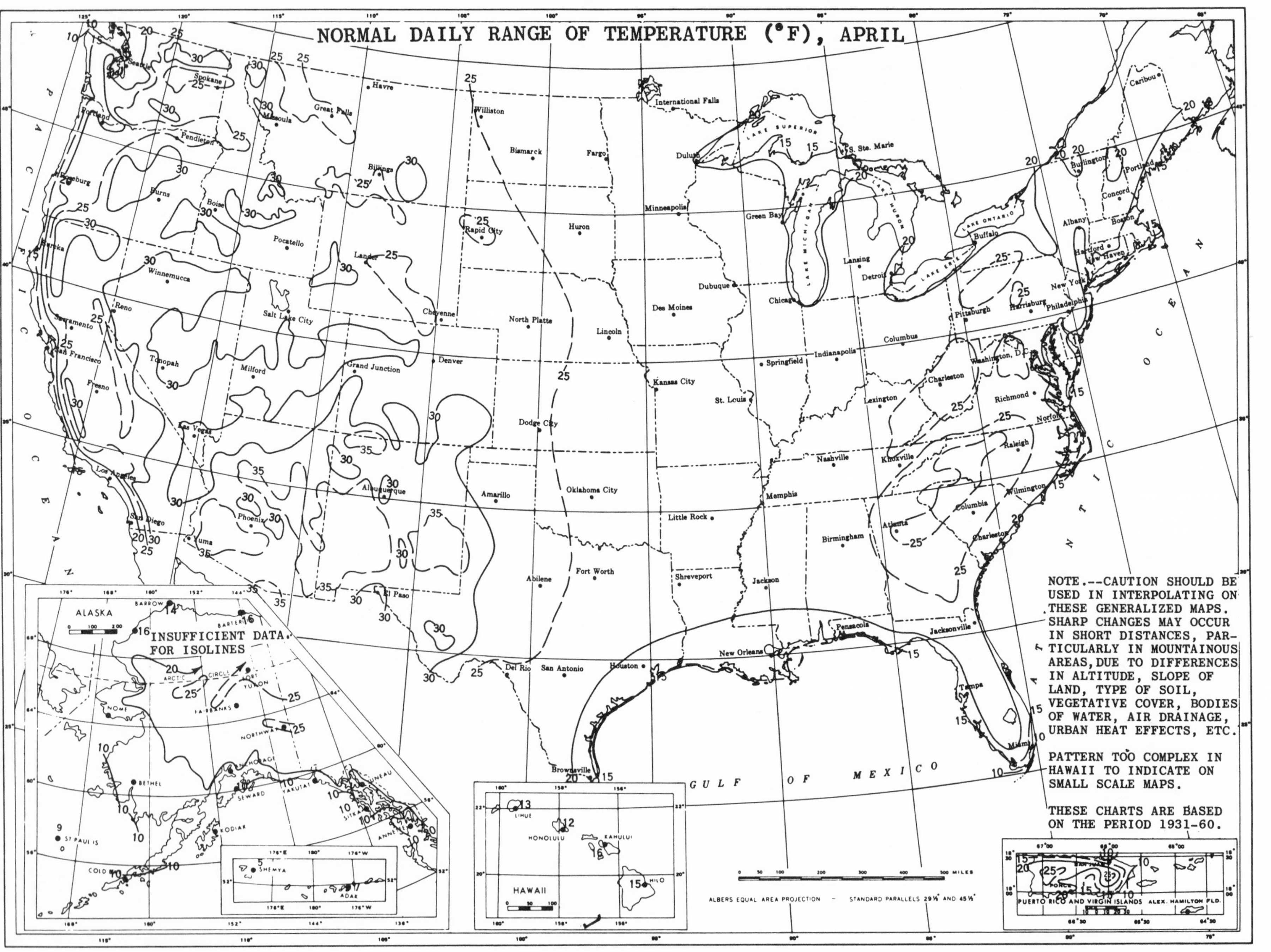

NOTE.--CAUTION SHOULD BE USED IN INTERPOLATING ON THESE GENERALIZED MAPS. SHARP CHANGES MAY OCCUR IN SHORT DISTANCES, PARTICULARLY IN MOUNTAINOUS AREAS, DUE TO DIFFERENCES IN ALTITUDE, SLOPE OF LAND, TYPE OF SOIL, VEGETATIVE COVER, BODIES OF WATER, AIR DRAINAGE, URBAN HEAT EFFECTS, ETC.
PATTERN TOO COMPLEX IN HAWAII TO INDICATE ON SMALL SCALE MAPS.
THESE CHARTS ARE BASED ON THE PERIOD 1931-60.
ALASKA
INSUFFICIENT DATA FOR ISOLINES
HAWAII
PUERTO RICO AND VIRGIN ISLANDS
ALBERS EQUAL AREA PROJECTION - STANDARD PARALLELS 29½° AND 45½°
PACIFIC OCEAN
ATLANTIC OCEAN
GULF OF MEXICO

NORMAL DAILY MAXIMUM, AVERAGE, MINIMUM, AND EXTREME TEMPERATURES (°F), APRIL

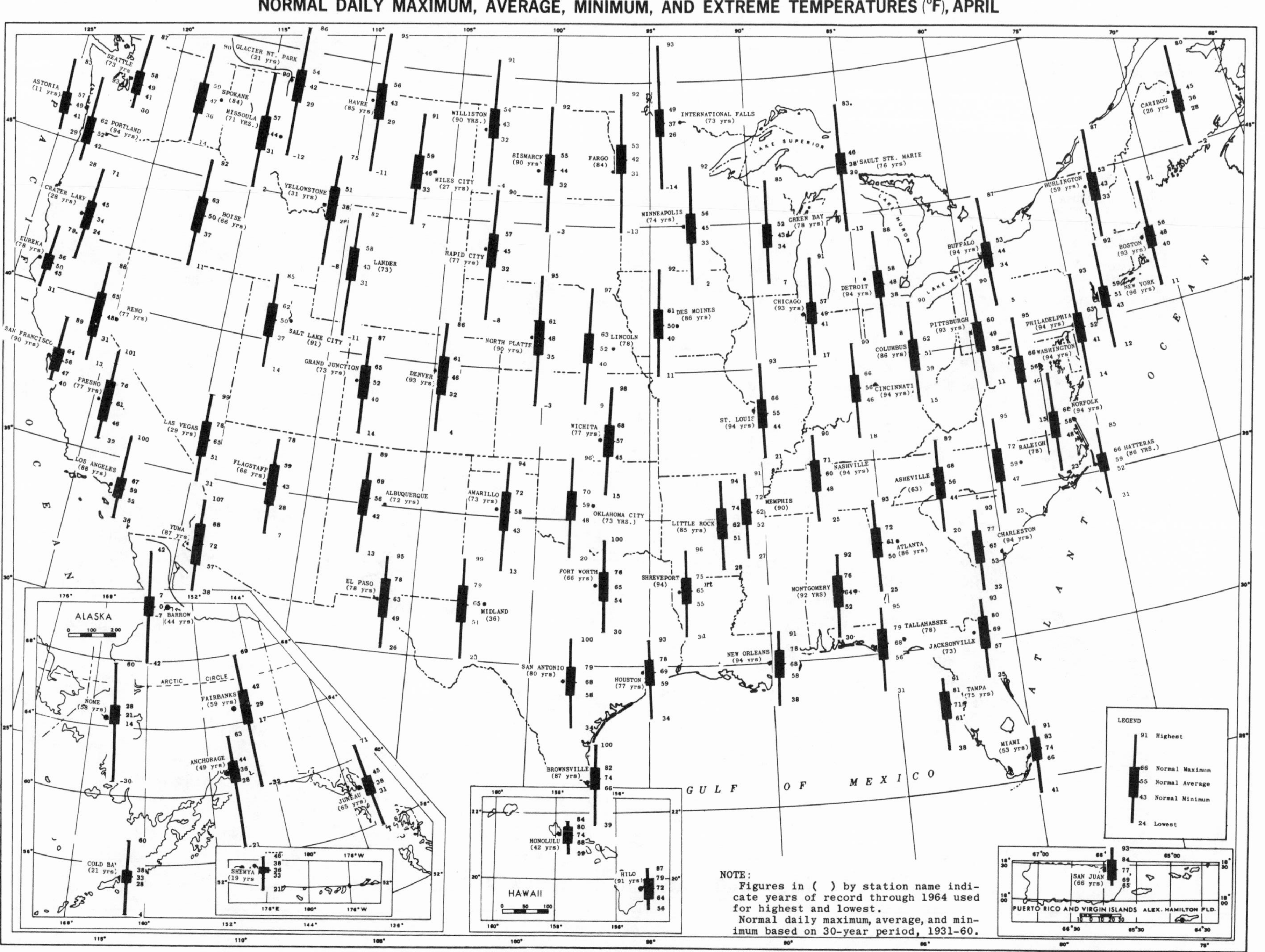

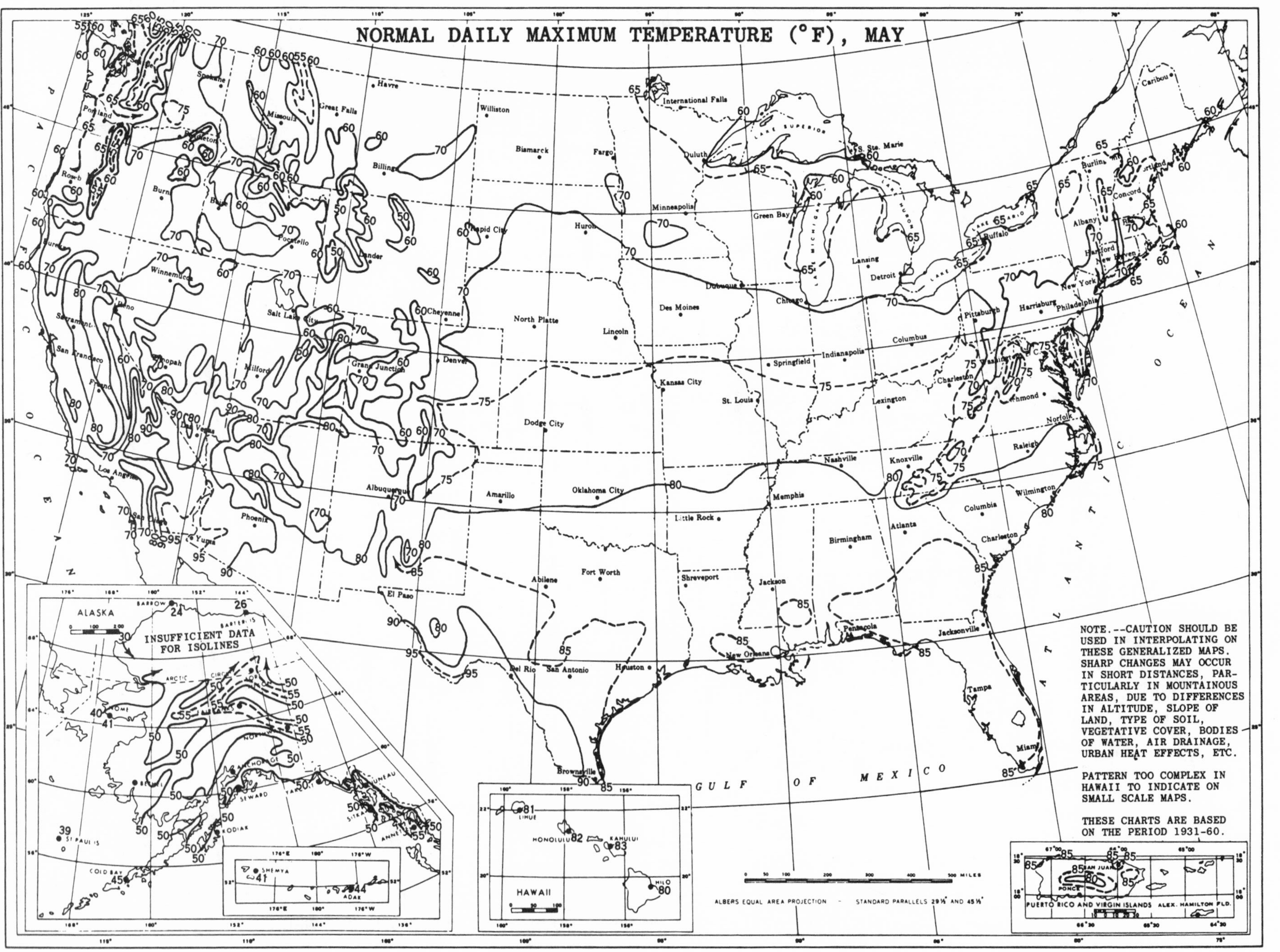
NORMAL DAILY MAXIMUM TEMPERATURE (°F), MAY
NOTE.--CAUTION SHOULD BE USED IN INTERPOLATING ON THESE GENERALIZED MAPS. SHARP CHANGES MAY OCCUR IN SHORT DISTANCES, PARTICULARLY IN MOUNTAINOUS AREAS, DUE TO DIFFERENCES IN ALTITUDE, SLOPE OF LAND, TYPE OF SOIL, VEGETATIVE COVER, BODIES OF WATER, AIR DRAINAGE, URBAN HEAT EFFECTS, ETC.
PATTERN TOO COMPLEX IN HAWAII TO INDICATE ON SMALL SCALE MAPS.
THESE CHARTS ARE BASED ON THE PERIOD 1931-60.
INSUFFICIENT DATA FOR ISOLINES
ALASKA
HAWAII
PUERTO RICO AND VIRGIN ISLANDS
ALBERS EQUAL AREA PROJECTION - STANDARD PARALLELS 29½° AND 45½°
PACIFIC OCEAN
ATLANTIC OCEAN
GULF OF MEXICO

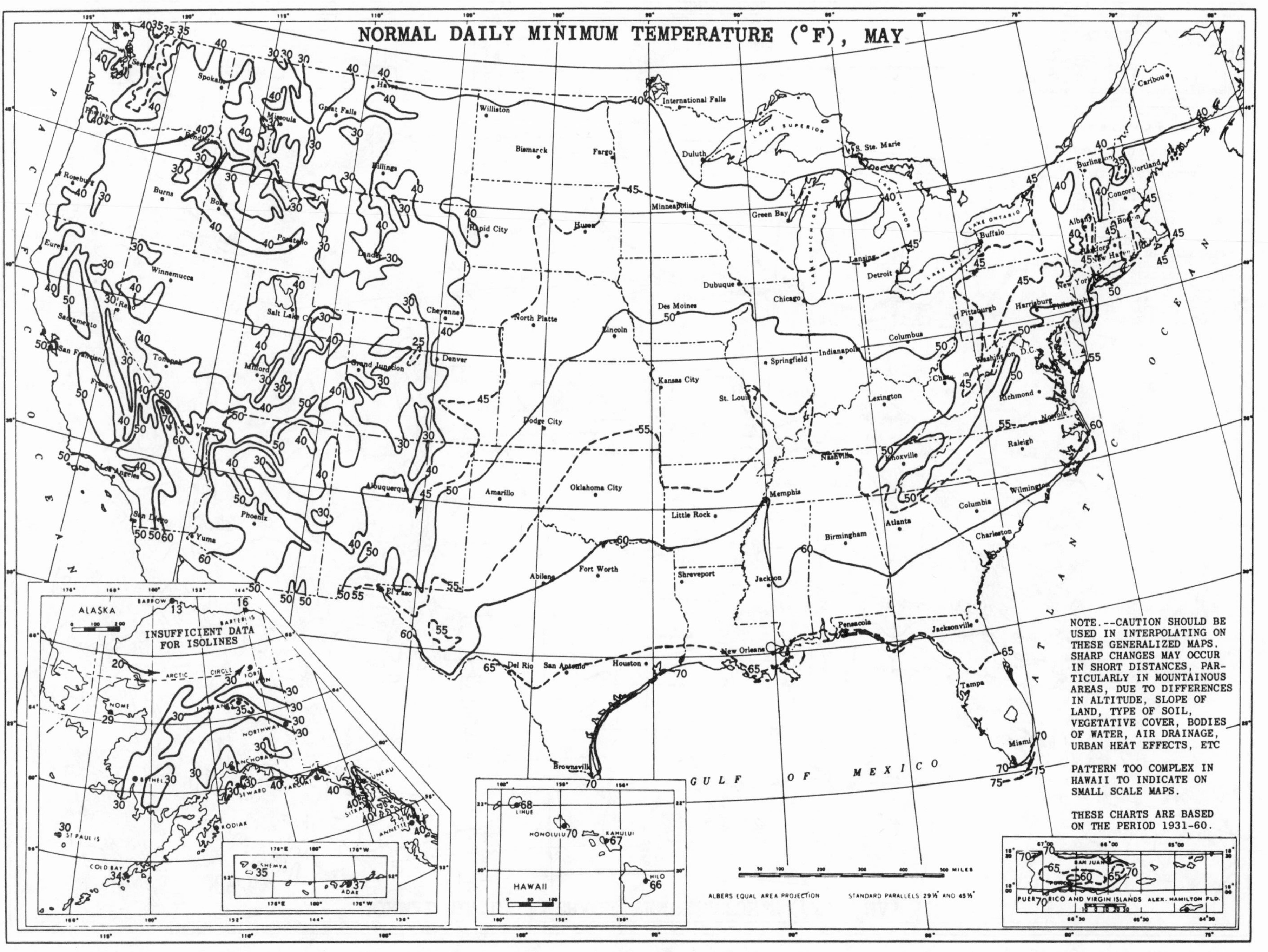
NORMAL DAILY MINIMUM TEMPERATURE (°F), MAY
NOTE.--CAUTION SHOULD BE USED IN INTERPOLATING ON THESE GENERALIZED MAPS. SHARP CHANGES MAY OCCUR IN SHORT DISTANCES, PARTICULARLY IN MOUNTAINOUS AREAS, DUE TO DIFFERENCES IN ALTITUDE, SLOPE OF LAND, TYPE OF SOIL, VEGETATIVE COVER, BODIES OF WATER, AIR DRAINAGE, URBAN HEAT EFFECTS, ETC
PATTERN TOO COMPLEX IN HAWAII TO INDICATE ON SMALL SCALE MAPS.
THESE CHARTS ARE BASED ON THE PERIOD 1931-60.
PACIFIC OCEAN
ATLANTIC OCEAN
GULF OF MEXICO
ALASKA
INSUFFICIENT DATA FOR ISOLINES
HAWAII
PUERTO RICO AND VIRGIN ISLANDS
ALBERS EQUAL AREA PROJECTION
STANDARD PARALLELS 29½° AND 45½°
MILES

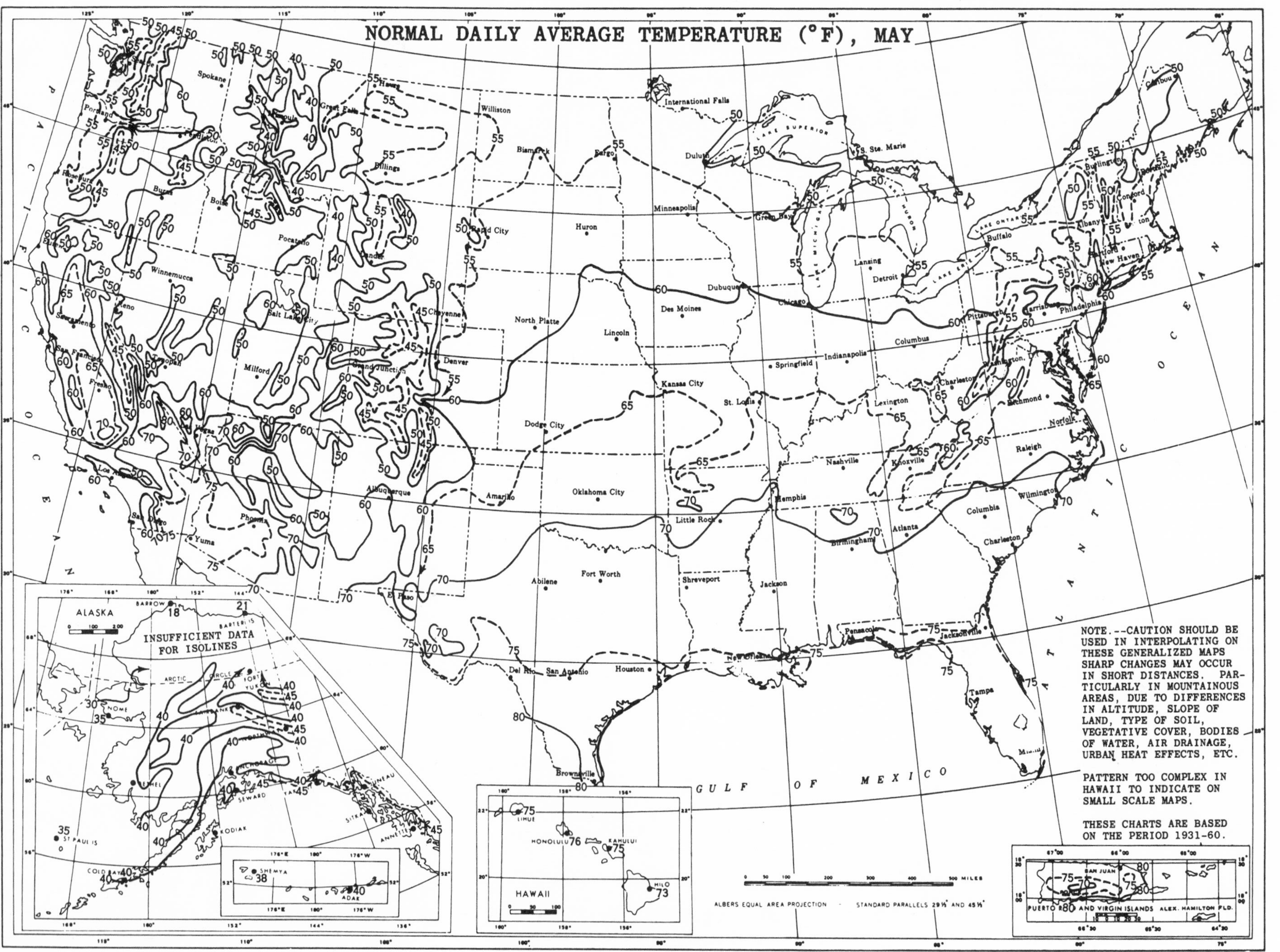
NORMAL DAILY AVERAGE TEMPERATURE (°F), MAY
NOTE.--CAUTION SHOULD BE USED IN INTERPOLATING ON THESE GENERALIZED MAPS SHARP CHANGES MAY OCCUR IN SHORT DISTANCES. PARTICULARLY IN MOUNTAINOUS AREAS, DUE TO DIFFERENCES IN ALTITUDE, SLOPE OF LAND, TYPE OF SOIL, VEGETATIVE COVER, BODIES OF WATER, AIR DRAINAGE, URBAN HEAT EFFECTS, ETC.
PATTERN TOO COMPLEX IN HAWAII TO INDICATE ON SMALL SCALE MAPS.
THESE CHARTS ARE BASED ON THE PERIOD 1931-60.
ALASKA
INSUFFICIENT DATA FOR ISOLINES
HAWAII
PUERTO RICO AND VIRGIN ISLANDS
ALBERS EQUAL AREA PROJECTION - STANDARD PARALLELS 29½° AND 45½°
PACIFIC OCEAN
ATLANTIC OCEAN
GULF OF MEXICO

NORMAL DAILY RANGE OF TEMPERATURE (°F), MAY

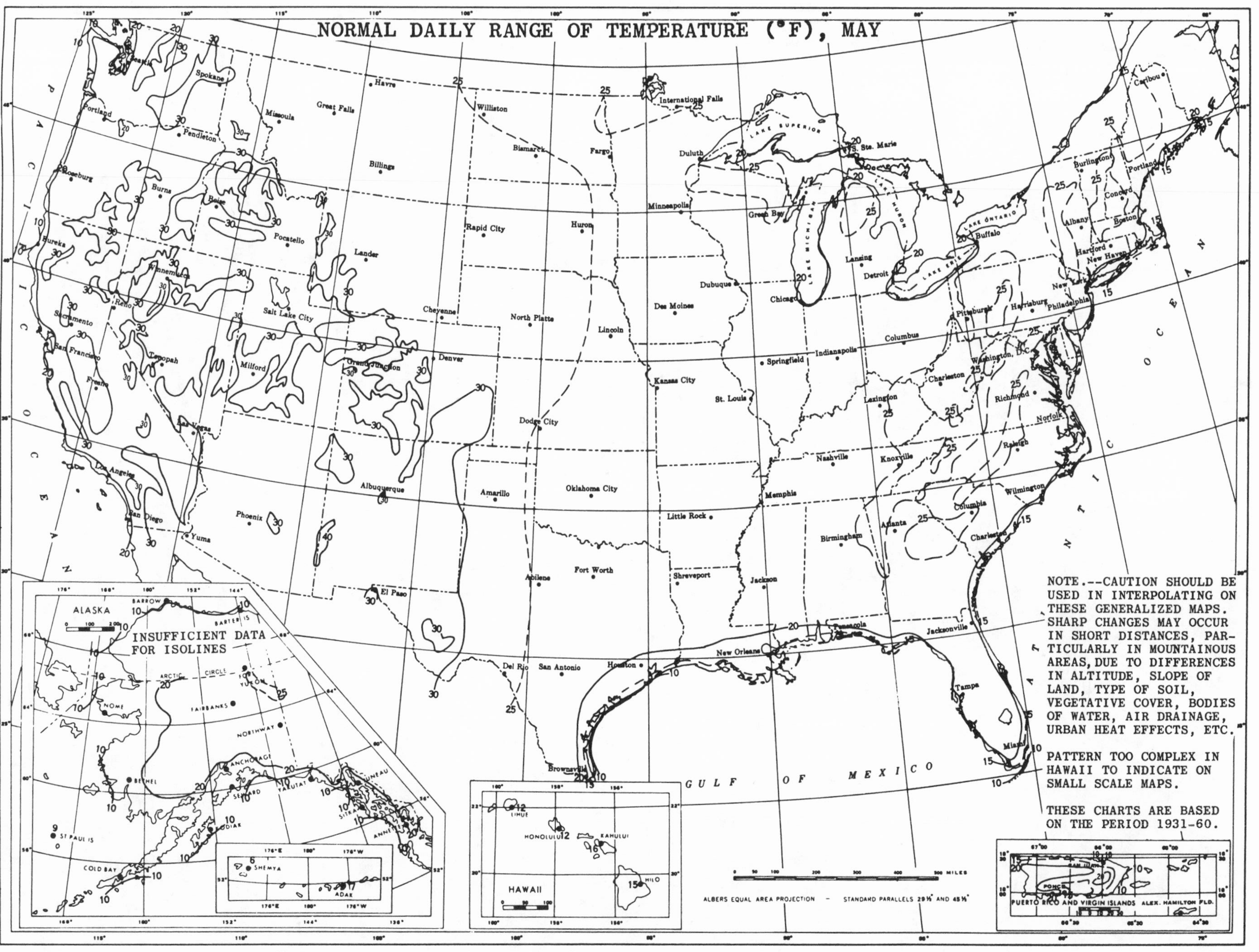

NOTE.--CAUTION SHOULD BE USED IN INTERPOLATING ON THESE GENERALIZED MAPS. SHARP CHANGES MAY OCCUR IN SHORT DISTANCES, PARTICULARLY IN MOUNTAINOUS AREAS, DUE TO DIFFERENCES IN ALTITUDE, SLOPE OF LAND, TYPE OF SOIL, VEGETATIVE COVER, BODIES OF WATER, AIR DRAINAGE, URBAN HEAT EFFECTS, ETC.

PATTERN TOO COMPLEX IN HAWAII TO INDICATE ON SMALL SCALE MAPS.

THESE CHARTS ARE BASED ON THE PERIOD 1931-60.

NORMAL DAILY MAXIMUM, AVERAGE, MINIMUM, AND EXTREME TEMPERATURES (°F), MAY

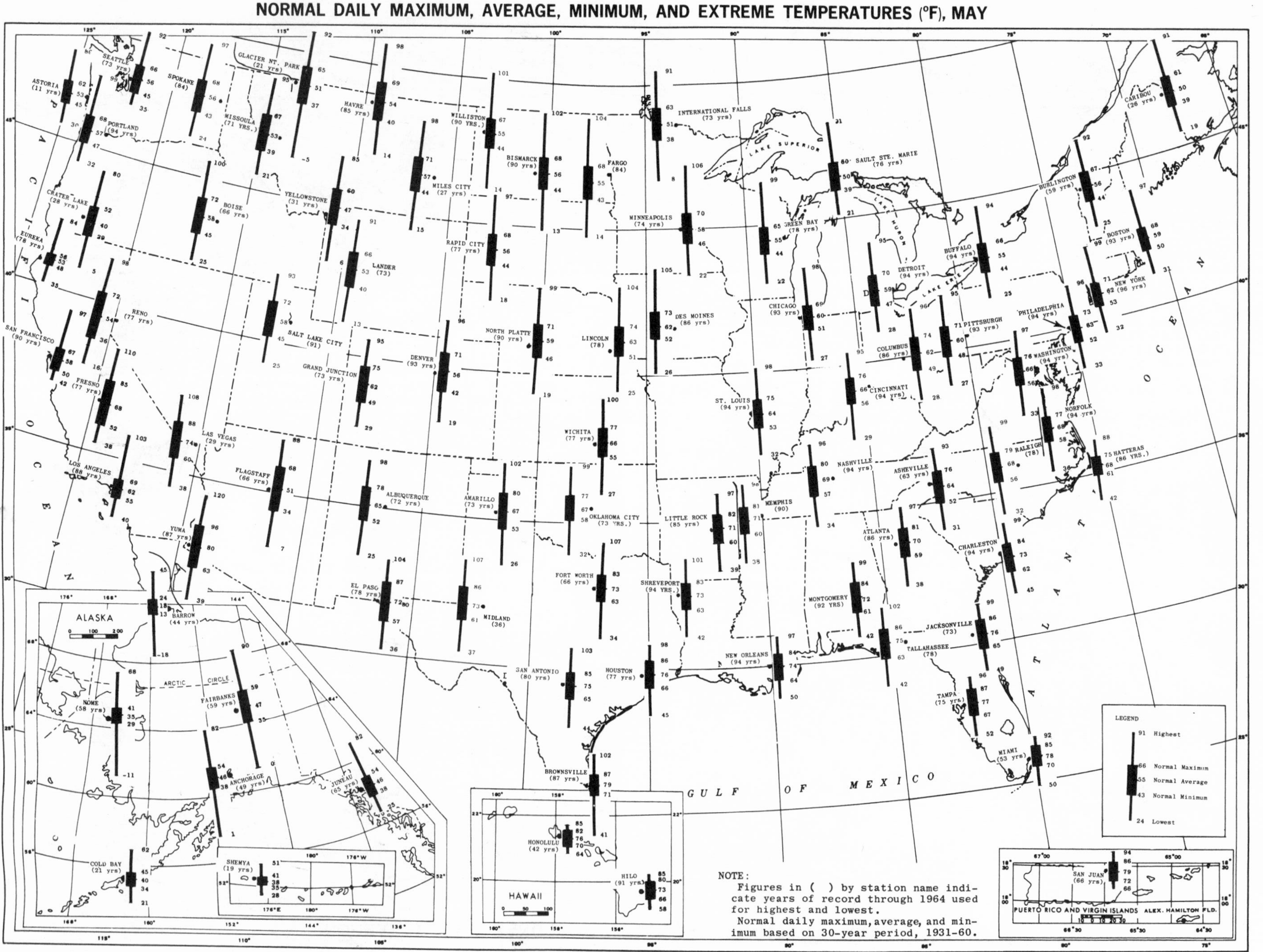

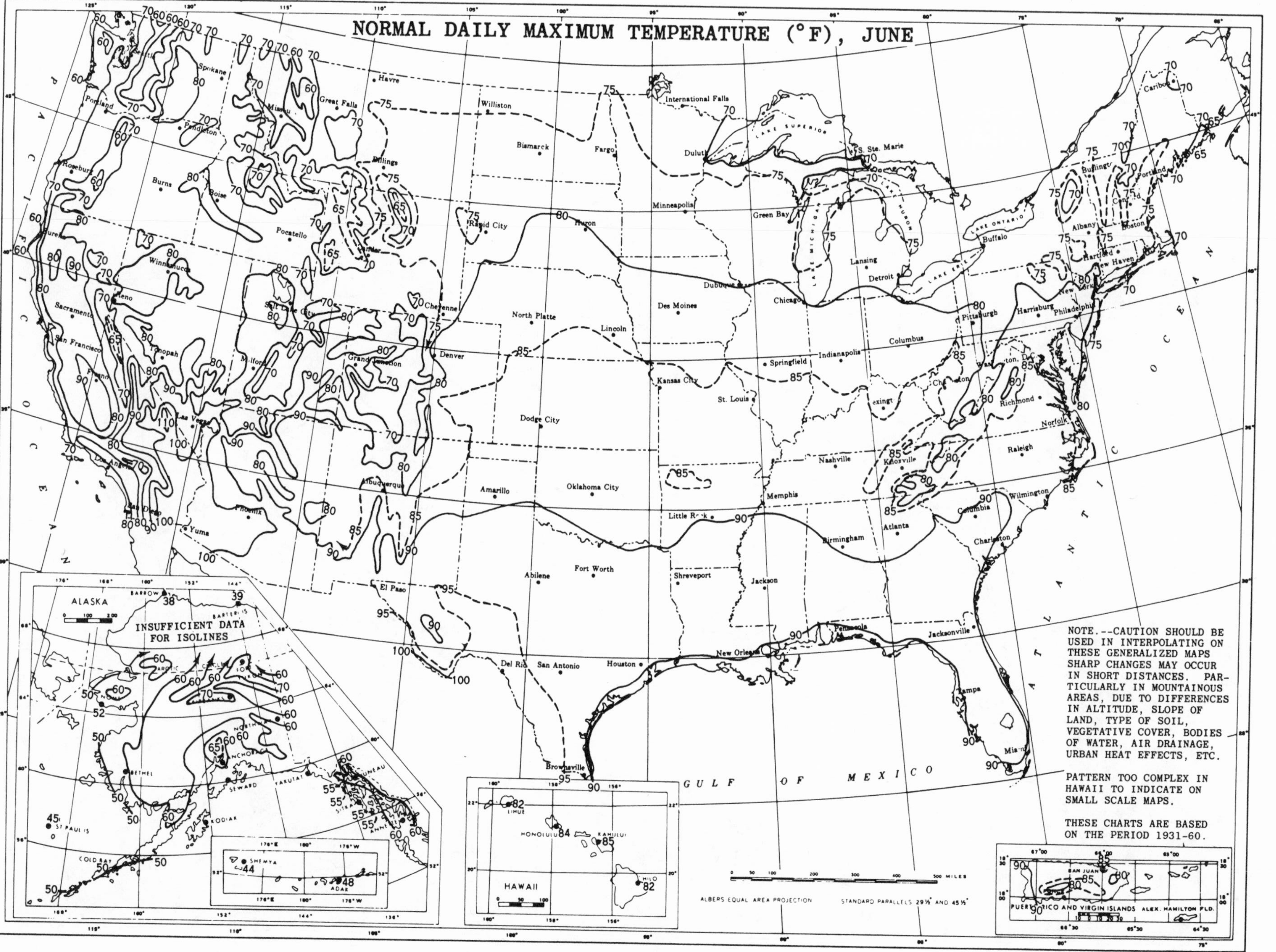
NORMAL DAILY MAXIMUM TEMPERATURE (°F), JUNE
NOTE.--CAUTION SHOULD BE USED IN INTERPOLATING ON THESE GENERALIZED MAPS SHARP CHANGES MAY OCCUR IN SHORT DISTANCES. PARTICULARLY IN MOUNTAINOUS AREAS, DUE TO DIFFERENCES IN ALTITUDE, SLOPE OF LAND, TYPE OF SOIL, VEGETATIVE COVER, BODIES OF WATER, AIR DRAINAGE, URBAN HEAT EFFECTS, ETC.
PATTERN TOO COMPLEX IN HAWAII TO INDICATE ON SMALL SCALE MAPS.
THESE CHARTS ARE BASED ON THE PERIOD 1931-60.
PACIFIC OCEAN
ATLANTIC OCEAN
GULF OF MEXICO
ALASKA
INSUFFICIENT DATA FOR ISOLINES
HAWAII
PUERTO RICO AND VIRGIN ISLANDS
ALBERS EQUAL AREA PROJECTION
STANDARD PARALLELS 29½° AND 45½°
Seattle
Spokane
Portland
Pendleton
Roseburg
Eureka
Burns
Boise
Reno
Winnemucca
Sacramento
San Francisco
Fresno
Tonopah
Los Angeles
San Diego
Yuma
Phoenix
Las Vegas
Milford
Salt Lake City
Pocatello
Lander
Missoula
Great Falls
Havre
Billings
Cheyenne
Denver
Grand Junction
Albuquerque
El Paso
Williston
Bismarck
Rapid City
North Platte
Dodge City
Amarillo
Abilene
Del Rio
San Antonio
Houston
Brownsville
Fort Worth
Oklahoma City
Lincoln
Huron
Fargo
International Falls
Duluth
Minneapolis
Des Moines
Kansas City
St. Louis
Springfield
Little Rock
Shreveport
Jackson
Memphis
Birmingham
New Orleans
Pensacola
Jacksonville
Tampa
Miami
Atlanta
Columbia
Charleston
Wilmington
Raleigh
Norfolk
Richmond
Nashville
Knoxville
Chicago
Dubuque
Green Bay
S. Ste. Marie
Lansing
Detroit
Indianapolis
Columbus
Pittsburgh
Buffalo
Harrisburg
Philadelphia
New York
New Haven
Hartford
Albany
Boston
Burlington
Concord
Portland
Caribou
LAKE SUPERIOR
LAKE MICHIGAN
LAKE HURON
LAKE ERIE
LAKE ONTARIO
BARROW
BARTER IS
NOME
BETHEL
ANCHORAGE
SEWARD
YAKUTAT
JUNEAU
SITKA
KODIAK
COLD BAY
ST PAUL IS
SHEMYA
ADAK
LIHUE
HONOLULU
KAHULUI
HILO
SAN JUAN
ALEX. HAMILTON FLD.

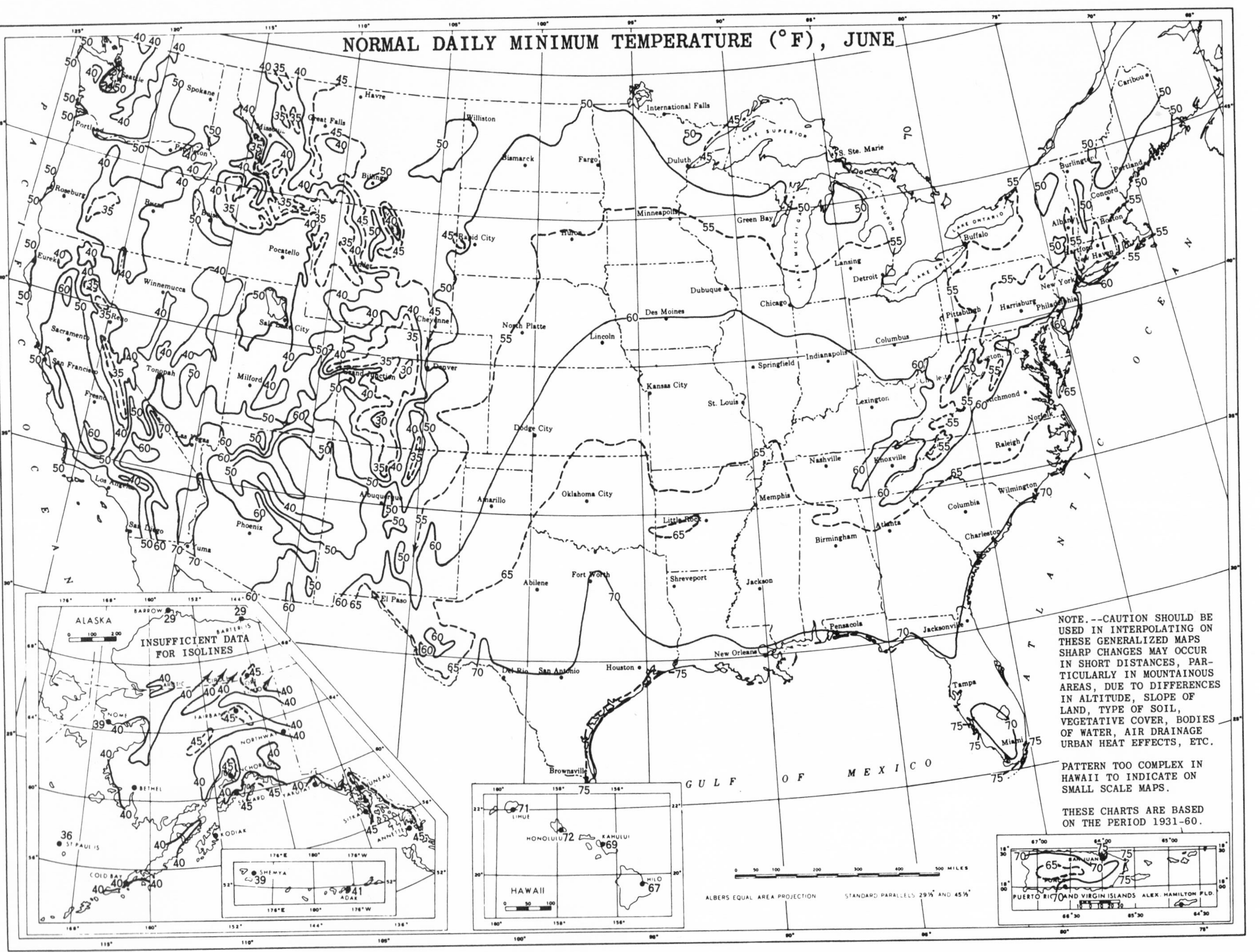

NORMAL DAILY MINIMUM TEMPERATURE (°F), JUNE
NOTE.--CAUTION SHOULD BE USED IN INTERPOLATING ON THESE GENERALIZED MAPS SHARP CHANGES MAY OCCUR IN SHORT DISTANCES, PARTICULARLY IN MOUNTAINOUS AREAS, DUE TO DIFFERENCES IN ALTITUDE, SLOPE OF LAND, TYPE OF SOIL, VEGETATIVE COVER, BODIES OF WATER, AIR DRAINAGE URBAN HEAT EFFECTS, ETC.
PATTERN TOO COMPLEX IN HAWAII TO INDICATE ON SMALL SCALE MAPS.
THESE CHARTS ARE BASED ON THE PERIOD 1931-60.
ALBERS EQUAL AREA PROJECTION
STANDARD PARALLELS 29½° AND 45½°
0 50 100 200 300 400 500 MILES
PACIFIC OCEAN
ATLANTIC OCEAN
GULF OF MEXICO
LAKE SUPERIOR
LAKE MICHIGAN
LAKE HURON
LAKE ERIE
LAKE ONTARIO
Seattle
Spokane
Portland
Pendleton
Roseburg
Eureka
Boise
Winnemucca
Reno
Sacramento
San Francisco
Fresno
Tonopah
Las Vegas
Los Angeles
San Diego
Yuma
Phoenix
Missoula
Great Falls
Havre
Billings
Pocatello
Salt Lake City
Milford
Grand Junction
Lander
Cheyenne
Denver
Albuquerque
El Paso
Williston
Bismarck
Rapid City
North Platte
Dodge City
Amarillo
Fargo
Huron
Lincoln
Oklahoma City
Abilene
Fort Worth
Del Rio
San Antonio
Houston
Brownsville
International Falls
Duluth
Minneapolis
Des Moines
Kansas City
Shreveport
Green Bay
Dubuque
St. Louis
Little Rock
Memphis
Jackson
New Orleans
Chicago
Springfield
Indianapolis
Nashville
Birmingham
Pensacola
Lansing
Detroit
Columbus
Lexington
Knoxville
Atlanta
Jacksonville
Tampa
Miami
S. Ste. Marie
Buffalo
Pittsburgh
Harrisburg
Richmond
Norfolk
Raleigh
Wilmington
Columbia
Charleston
Caribou
Portland
Burlington
Concord
Albany
Boston
Hartford
New Haven
New York
Philadelphia
ALASKA
0 100 200
INSUFFICIENT DATA FOR ISOLINES
BARROW 29
BARTER IS 29
NOME 39
BETHEL
KODIAK
ST PAUL IS 36
COLD BAY
SITKA
ANNETTE
JUNEAU
SHEMYA 39
ADAK 41
HAWAII
0 50 100
LIHUE 71
HONOLULU 72
KAHULUI 69
HILO 67
PUERTO RICO AND VIRGIN ISLANDS
SAN JUAN
ALEX. HAMILTON FLD.

NORMAL DAILY AVERAGE TEMPERATURE (°F), JUNE

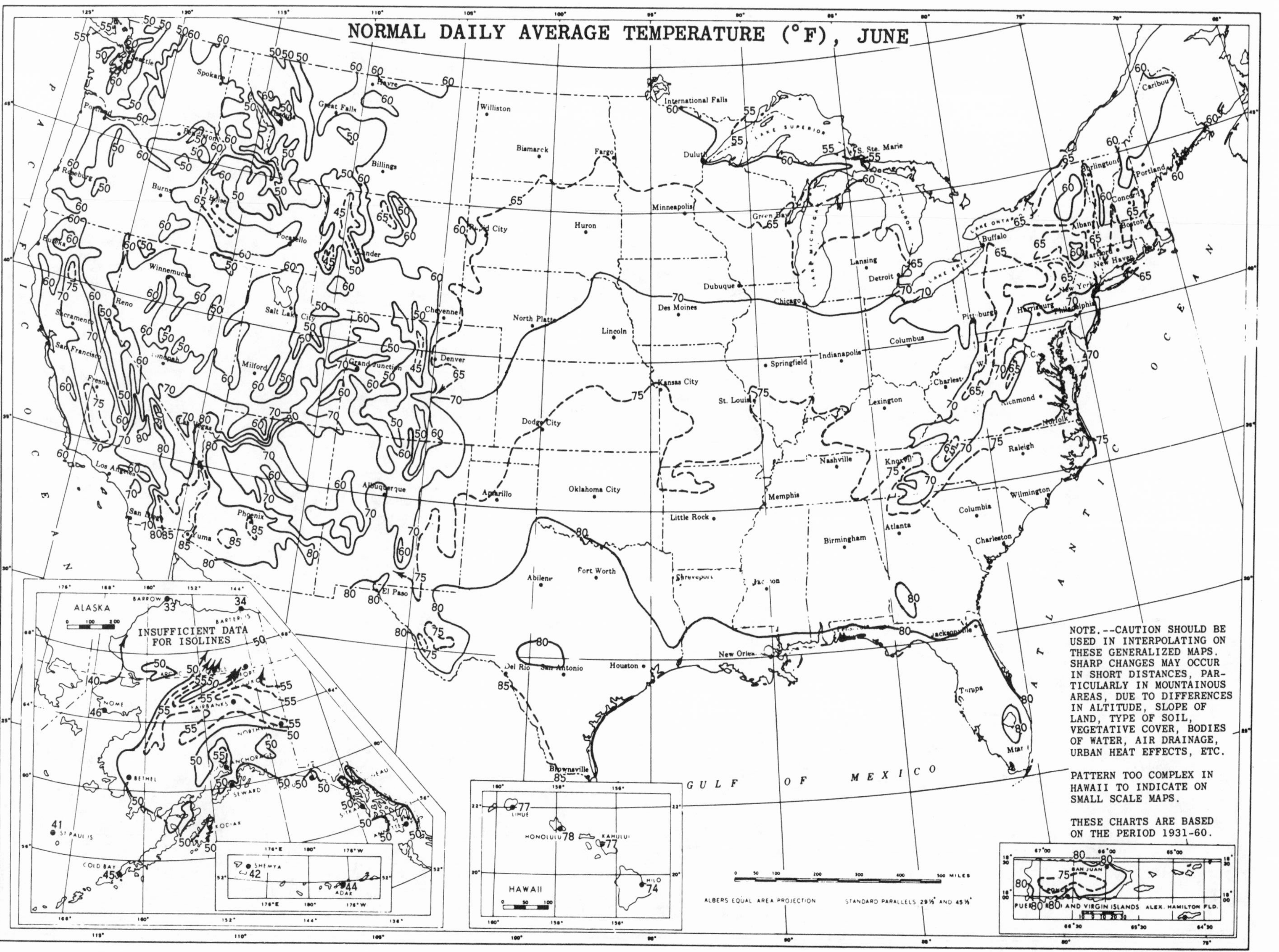
NOTE.--CAUTION SHOULD BE USED IN INTERPOLATING ON THESE GENERALIZED MAPS. SHARP CHANGES MAY OCCUR IN SHORT DISTANCES, PARTICULARLY IN MOUNTAINOUS AREAS, DUE TO DIFFERENCES IN ALTITUDE, SLOPE OF LAND, TYPE OF SOIL, VEGETATIVE COVER, BODIES OF WATER, AIR DRAINAGE, URBAN HEAT EFFECTS, ETC.
PATTERN TOO COMPLEX IN HAWAII TO INDICATE ON SMALL SCALE MAPS.
THESE CHARTS ARE BASED ON THE PERIOD 1931-60.
PACIFIC OCEAN
ATLANTIC OCEAN
GULF OF MEXICO
ALASKA
INSUFFICIENT DATA FOR ISOLINES
HAWAII
PUERTO RICO AND VIRGIN ISLANDS
ALBERS EQUAL AREA PROJECTION
STANDARD PARALLELS 29½° AND 45½°
500 MILES

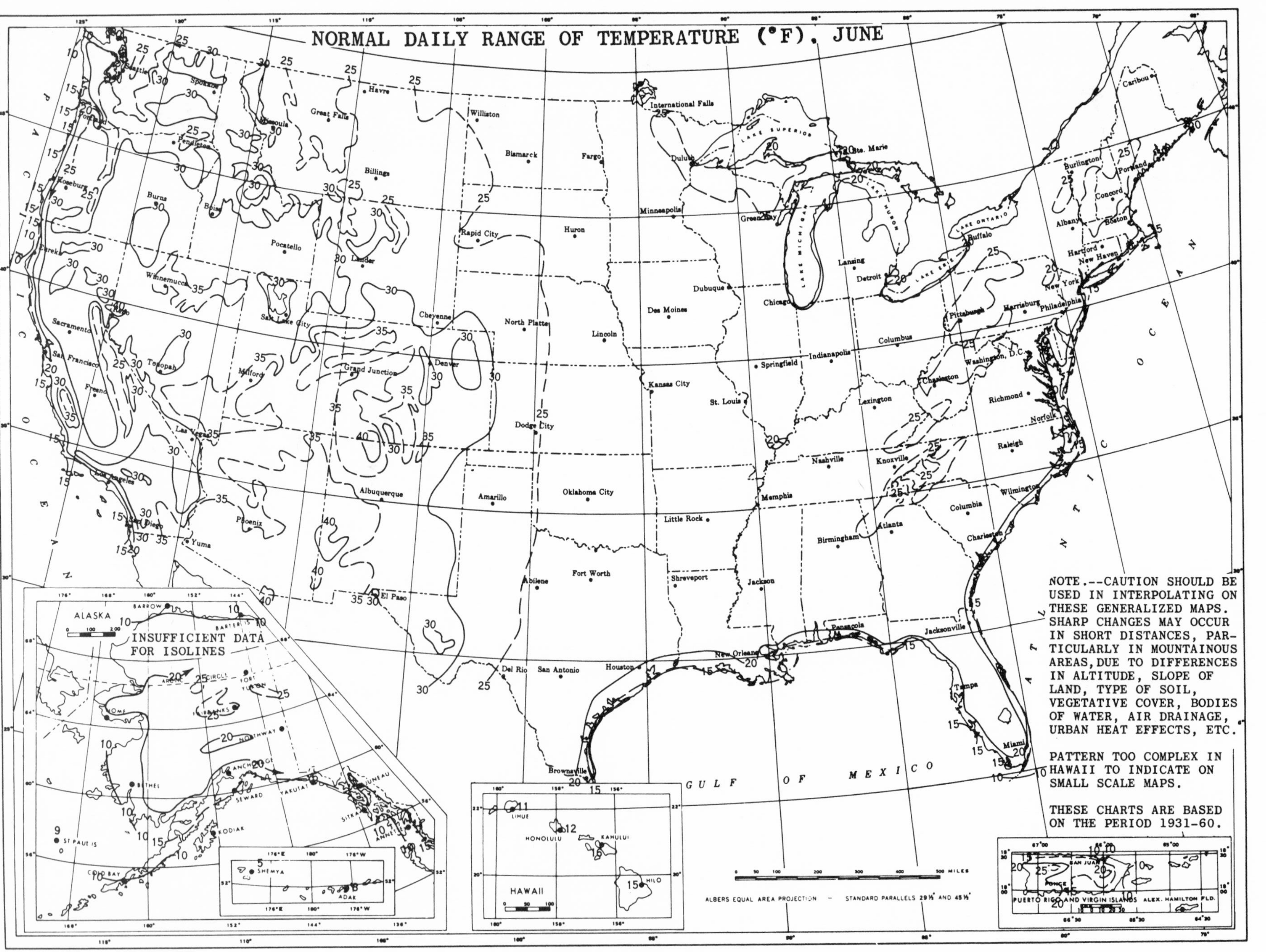
NORMAL DAILY RANGE OF TEMPERATURE (°F), JUNE
NOTE.--CAUTION SHOULD BE USED IN INTERPOLATING ON THESE GENERALIZED MAPS. SHARP CHANGES MAY OCCUR IN SHORT DISTANCES, PARTICULARLY IN MOUNTAINOUS AREAS, DUE TO DIFFERENCES IN ALTITUDE, SLOPE OF LAND, TYPE OF SOIL, VEGETATIVE COVER, BODIES OF WATER, AIR DRAINAGE, URBAN HEAT EFFECTS, ETC.
PATTERN TOO COMPLEX IN HAWAII TO INDICATE ON SMALL SCALE MAPS.
THESE CHARTS ARE BASED ON THE PERIOD 1931-60.
PACIFIC OCEAN
ATLANTIC OCEAN
GULF OF MEXICO
ALASKA
INSUFFICIENT DATA FOR ISOLINES
HAWAII
PUERTO RICO AND VIRGIN ISLANDS
ALEX. HAMILTON FLD.
ALBERS EQUAL AREA PROJECTION — STANDARD PARALLELS 29½° AND 45½°
0 50 100 200 300 400 500 MILES

NORMAL DAILY MAXIMUM, AVERAGE, MINIMUM, AND EXTREME TEMPERATURES (°F), JUNE

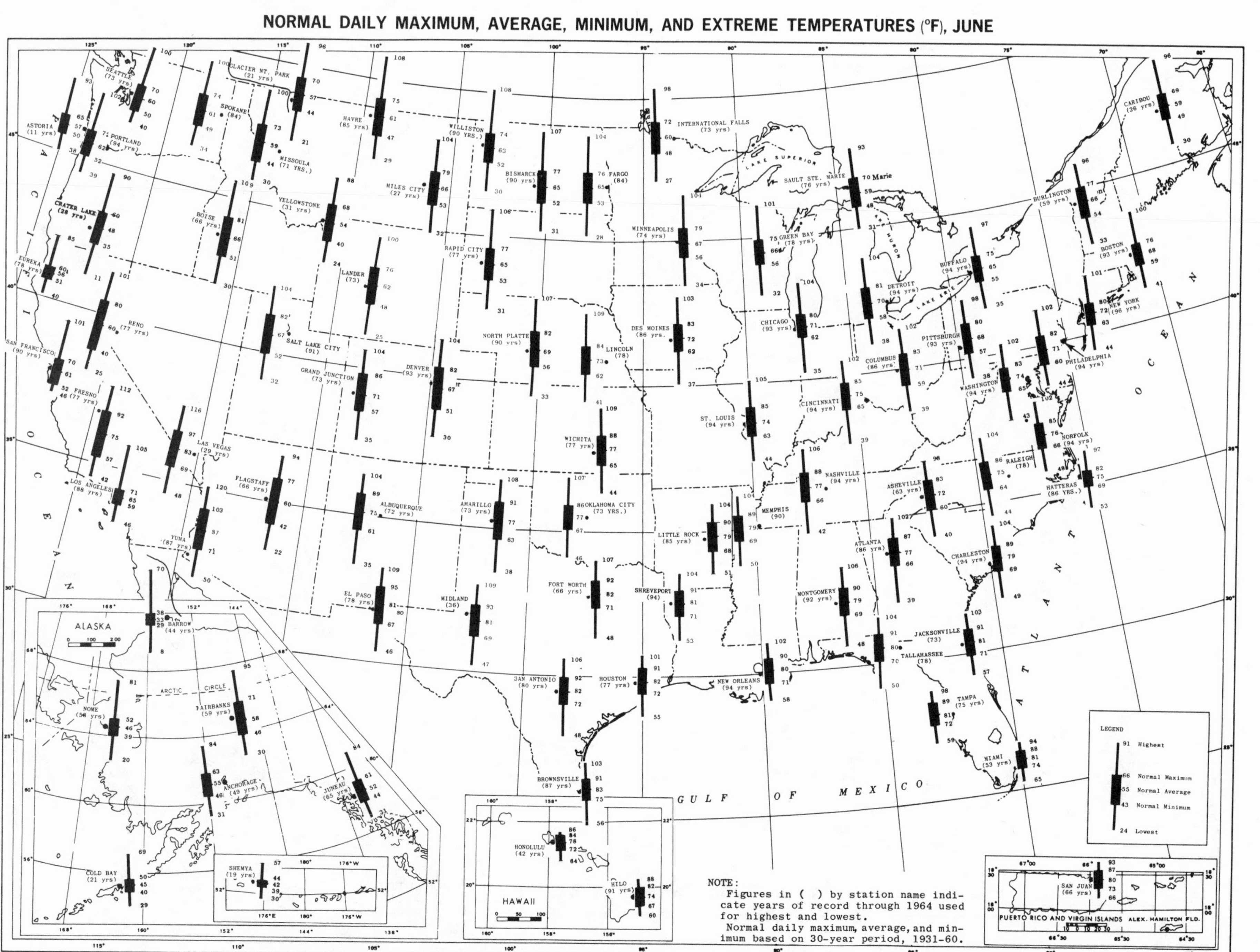

NOTE:
Figures in ( ) by station name indicate years of record through 1964 used for highest and lowest.
Normal daily maximum, average, and minimum based on 30-year period, 1931-60.

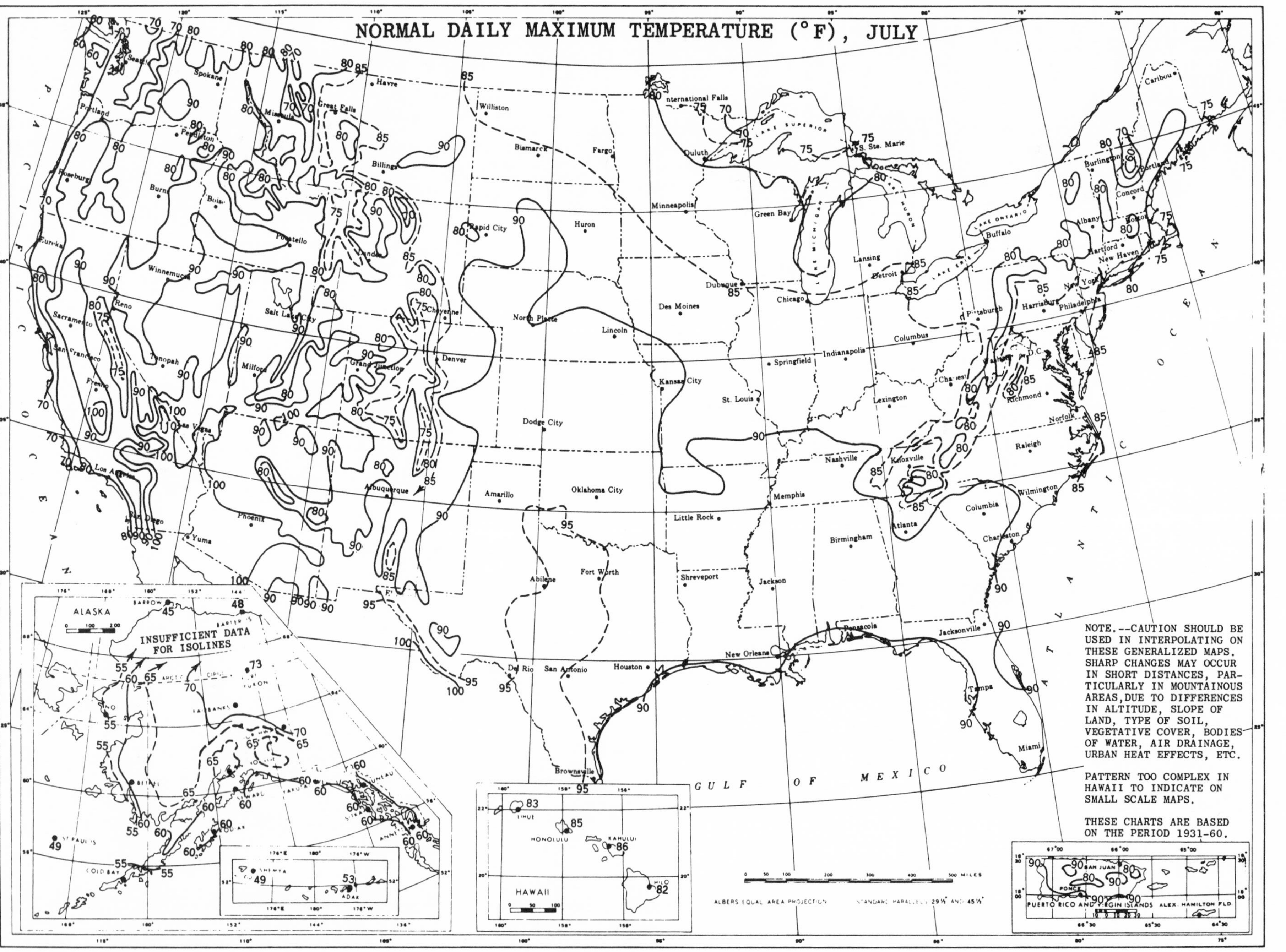

NORMAL DAILY MAXIMUM TEMPERATURE (° F), JULY
NOTE.--CAUTION SHOULD BE USED IN INTERPOLATING ON THESE GENERALIZED MAPS. SHARP CHANGES MAY OCCUR IN SHORT DISTANCES, PARTICULARLY IN MOUNTAINOUS AREAS, DUE TO DIFFERENCES IN ALTITUDE, SLOPE OF LAND, TYPE OF SOIL, VEGETATIVE COVER, BODIES OF WATER, AIR DRAINAGE, URBAN HEAT EFFECTS, ETC.
PATTERN TOO COMPLEX IN HAWAII TO INDICATE ON SMALL SCALE MAPS.
THESE CHARTS ARE BASED ON THE PERIOD 1931-60.
ALASKA
INSUFFICIENT DATA FOR ISOLINES
HAWAII
PUERTO RICO AND VIRGIN ISLANDS
ALBERS EQUAL AREA PROJECTION
GULF OF MEXICO
PACIFIC OCEAN
ATLANTIC OCEAN

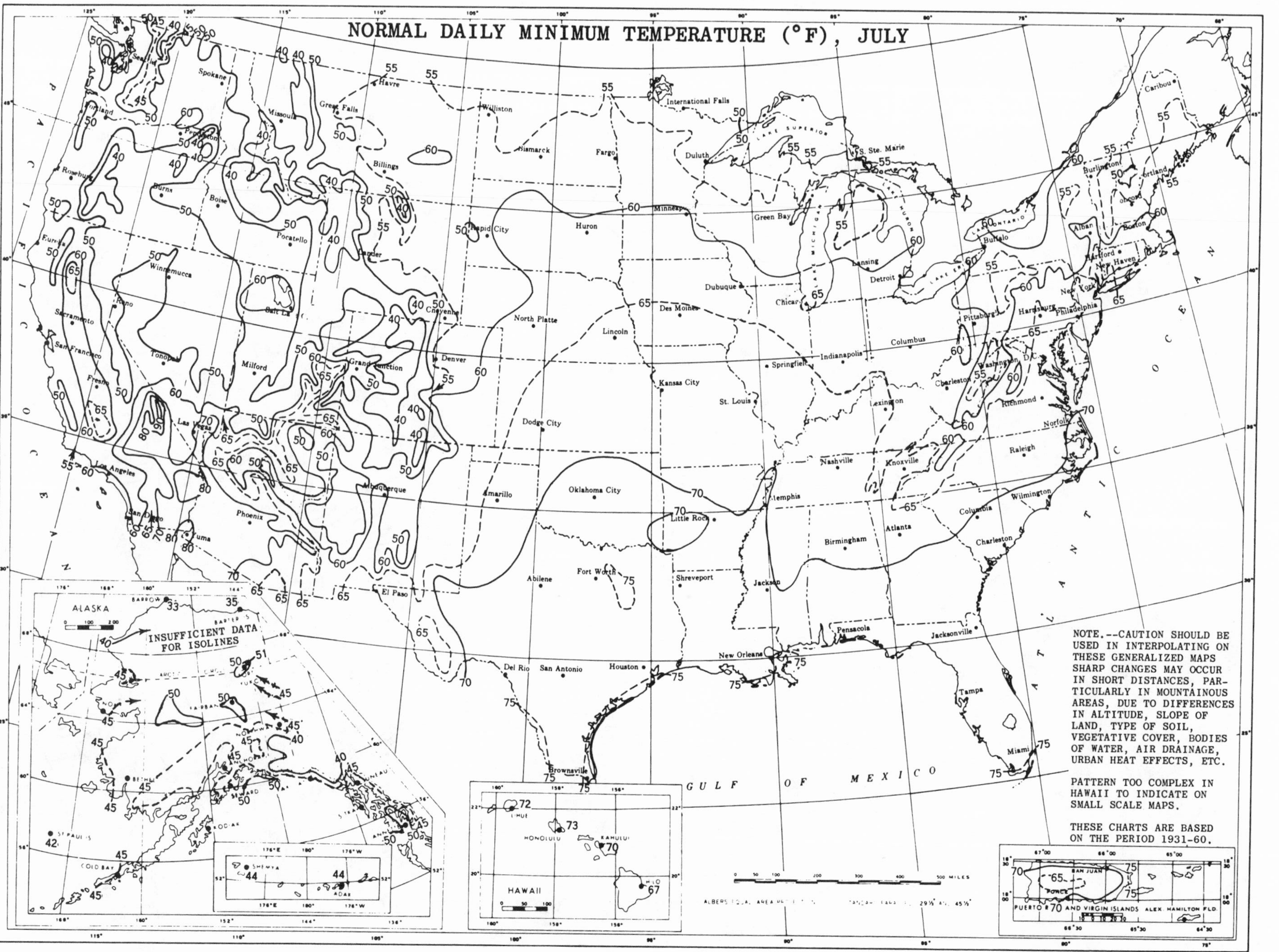
NORMAL DAILY MINIMUM TEMPERATURE (°F), JULY
PACIFIC OCEAN
ATLANTIC OCEAN
GULF OF MEXICO
Seattle
Spokane
Portland
Pendleton
Roseburg
Burns
Boise
Eureka
Winnemucca
Reno
Sacramento
San Francisco
Fresno
Tonopah
Las Vegas
Los Angeles
San Diego
Yuma
Phoenix
Missoula
Great Falls
Havre
Billings
Pocatello
Lander
Salt Lake City
Milford
Grand Junction
Cheyenne
Denver
Albuquerque
El Paso
Williston
Bismarck
Rapid City
Huron
Fargo
North Platte
Lincoln
Dodge City
Amarillo
Oklahoma City
Abilene
Fort Worth
Del Rio
San Antonio
Houston
Brownsville
International Falls
Duluth
Minneapolis
Des Moines
Kansas City
Dubuque
Green Bay
Lake Superior
Lake Michigan
Lake Huron
Lake Ontario
Lake Erie
S. Ste. Marie
Lansing
Detroit
Chicago
Springfield
St. Louis
Indianapolis
Columbus
Shreveport
Little Rock
Memphis
Jackson
Nashville
Lexington
Knoxville
Birmingham
Atlanta
New Orleans
Pensacola
Jacksonville
Tampa
Miami
Columbia
Charleston
Wilmington
Raleigh
Norfolk
Richmond
Washington, D.C.
Charleston
Pittsburgh
Harrisburg
Philadelphia
New York
Buffalo
Albany
Hartford
New Haven
Boston
Concord
Burlington
Portland
Caribou
ALASKA
INSUFFICIENT DATA FOR ISOLINES
Barrow
St. Paul Is.
Cold Bay
Kodiak
Bethel
Shemya
Adak
HAWAII
Lihue
Honolulu
Kahului
Hilo
PUERTO RICO AND VIRGIN ISLANDS
San Juan
Ponce
Alex. Hamilton Fld.
0 50 100 200 300 400 500 MILES
NOTE.--CAUTION SHOULD BE USED IN INTERPOLATING ON THESE GENERALIZED MAPS SHARP CHANGES MAY OCCUR IN SHORT DISTANCES, PARTICULARLY IN MOUNTAINOUS AREAS, DUE TO DIFFERENCES IN ALTITUDE, SLOPE OF LAND, TYPE OF SOIL, VEGETATIVE COVER, BODIES OF WATER, AIR DRAINAGE, URBAN HEAT EFFECTS, ETC.
PATTERN TOO COMPLEX IN HAWAII TO INDICATE ON SMALL SCALE MAPS.
THESE CHARTS ARE BASED ON THE PERIOD 1931-60.

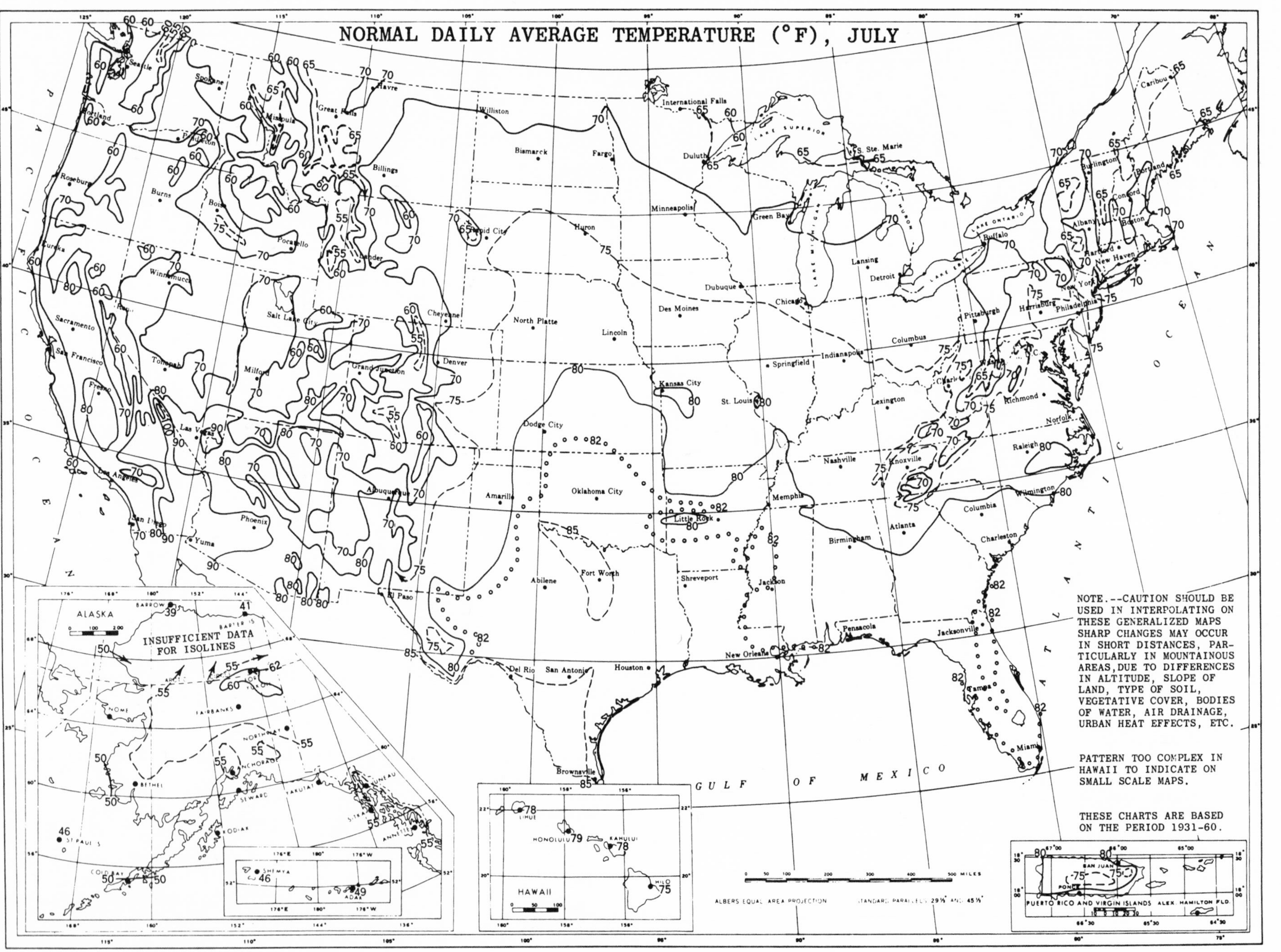
NORMAL DAILY AVERAGE TEMPERATURE (°F), JULY
NOTE.--CAUTION SHOULD BE USED IN INTERPOLATING ON THESE GENERALIZED MAPS SHARP CHANGES MAY OCCUR IN SHORT DISTANCES, PARTICULARLY IN MOUNTAINOUS AREAS, DUE TO DIFFERENCES IN ALTITUDE, SLOPE OF LAND, TYPE OF SOIL, VEGETATIVE COVER, BODIES OF WATER, AIR DRAINAGE, URBAN HEAT EFFECTS, ETC.
PATTERN TOO COMPLEX IN HAWAII TO INDICATE ON SMALL SCALE MAPS.
THESE CHARTS ARE BASED ON THE PERIOD 1931-60.
PACIFIC OCEAN
ATLANTIC OCEAN
GULF OF MEXICO
ALASKA
INSUFFICIENT DATA FOR ISOLINES
HAWAII
PUERTO RICO AND VIRGIN ISLANDS
ALBERS EQUAL AREA PROJECTION
STANDARD PARALLELS 29½° AND 45½°
Seattle
Portland
Spokane
Great Falls
Havre
Billings
Williston
Bismarck
Fargo
International Falls
Duluth
Minneapolis
Huron
Rapid City
North Platte
Lincoln
Des Moines
Dubuque
Chicago
Green Bay
S. Ste. Marie
Lansing
Detroit
Buffalo
Caribou
Portland
Burlington
Concord
Albany
Boston
Hartford
New Haven
New York
Philadelphia
Harrisburg
Pittsburgh
Columbus
Indianapolis
Springfield
St. Louis
Kansas City
Dodge City
Denver
Cheyenne
Lander
Pocatello
Boise
Burns
Roseburg
Eureka
Sacramento
San Francisco
Fresno
Los Angeles
San Diego
Yuma
Phoenix
Las Vegas
Tonopah
Winnemucca
Reno
Salt Lake City
Milford
Grand Junction
Albuquerque
El Paso
Amarillo
Oklahoma City
Abilene
Fort Worth
Del Rio
San Antonio
Houston
Brownsville
Shreveport
Little Rock
Memphis
Jackson
New Orleans
Nashville
Birmingham
Atlanta
Knoxville
Lexington
Charleston
Richmond
Norfolk
Raleigh
Wilmington
Columbia
Charleston
Jacksonville
Pensacola
Tampa
Miami
Barrow
Barter Is
Nome
Fairbanks
Northway
Anchorage
Seward
Yakutat
Juneau
Sitka
Annette
Kodiak
Bethel
St Paul Is
Cold Bay
Shemya
Adak
Lihue
Honolulu
Kahului
Hilo
San Juan
Ponce
Alex Hamilton Fld.

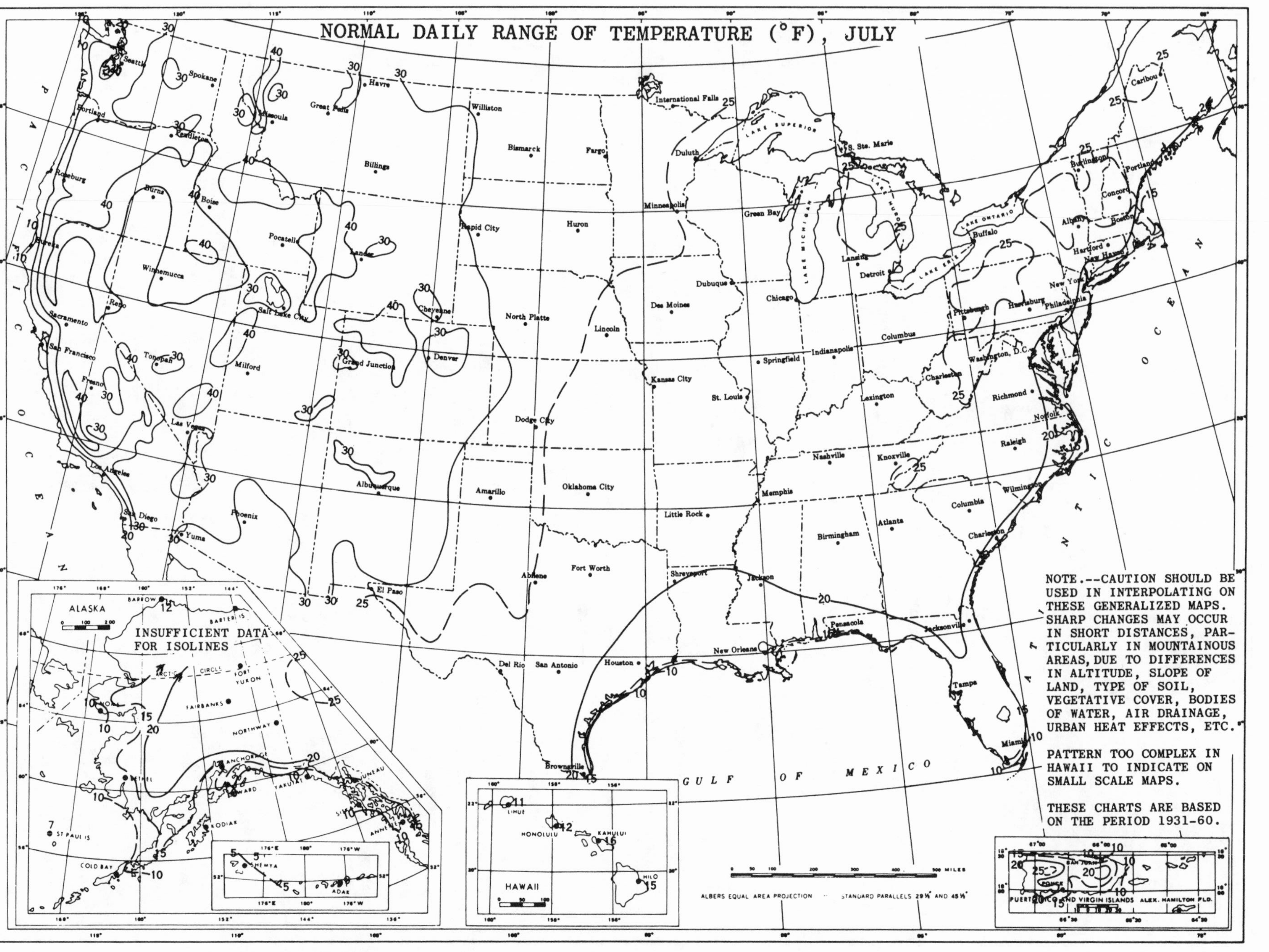
NORMAL DAILY RANGE OF TEMPERATURE (°F), JULY
NOTE.--CAUTION SHOULD BE USED IN INTERPOLATING ON THESE GENERALIZED MAPS. SHARP CHANGES MAY OCCUR IN SHORT DISTANCES, PARTICULARLY IN MOUNTAINOUS AREAS, DUE TO DIFFERENCES IN ALTITUDE, SLOPE OF LAND, TYPE OF SOIL, VEGETATIVE COVER, BODIES OF WATER, AIR DRAINAGE, URBAN HEAT EFFECTS, ETC.
PATTERN TOO COMPLEX IN HAWAII TO INDICATE ON SMALL SCALE MAPS.
THESE CHARTS ARE BASED ON THE PERIOD 1931-60.
ALASKA
INSUFFICIENT DATA FOR ISOLINES
HAWAII
PUERTO RICO AND VIRGIN ISLANDS
ALBERS EQUAL AREA PROJECTION
STANDARD PARALLELS 29½° AND 45½°
PACIFIC OCEAN
ATLANTIC OCEAN
GULF OF MEXICO
Seattle
Spokane
Portland
Roseburg
Eureka
Sacramento
San Francisco
Fresno
Los Angeles
San Diego
Yuma
Phoenix
Las Vegas
Tonopah
Reno
Winnemucca
Burns
Boise
Pendleton
Pocatello
Salt Lake City
Milford
Missoula
Great Falls
Havre
Billings
Lander
Grand Junction
Cheyenne
Denver
Albuquerque
El Paso
Williston
Bismarck
Rapid City
North Platte
Dodge City
Amarillo
Abilene
Del Rio
San Antonio
Brownsville
Houston
Fort Worth
Oklahoma City
Kansas City
Lincoln
Huron
Fargo
International Falls
Duluth
Minneapolis
Des Moines
Dubuque
St. Louis
Little Rock
Shreveport
New Orleans
Jackson
Memphis
Nashville
Birmingham
Atlanta
Knoxville
Lexington
Springfield
Indianapolis
Chicago
Green Bay
Sault Ste. Marie
Lansing
Detroit
Columbus
Charleston
Pittsburgh
Buffalo
Harrisburg
Philadelphia
New York
Washington, D.C.
Richmond
Norfolk
Raleigh
Wilmington
Columbia
Jacksonville
Pensacola
Tampa
Miami
Albany
Hartford
New Haven
Boston
Concord
Burlington
Portland
Caribou
Lake Superior
Lake Michigan
Lake Huron
Lake Erie
Lake Ontario
Barrow
Barter Is.
Fort Yukon
Fairbanks
Northway
Nome
Bethel
Anchorage
Seward
Kodiak
Cold Bay
St Paul Is
Juneau
Annette
Sitka
Yakutat
Shemya
Adak
Arctic Circle
Lihue
Honolulu
Kahului
Hilo
San Juan
Ponce
Alex. Hamilton Fld.

NORMAL DAILY MAXIMUM, AVERAGE, MINIMUM, AND EXTREME TEMPERATURES (°F), JULY

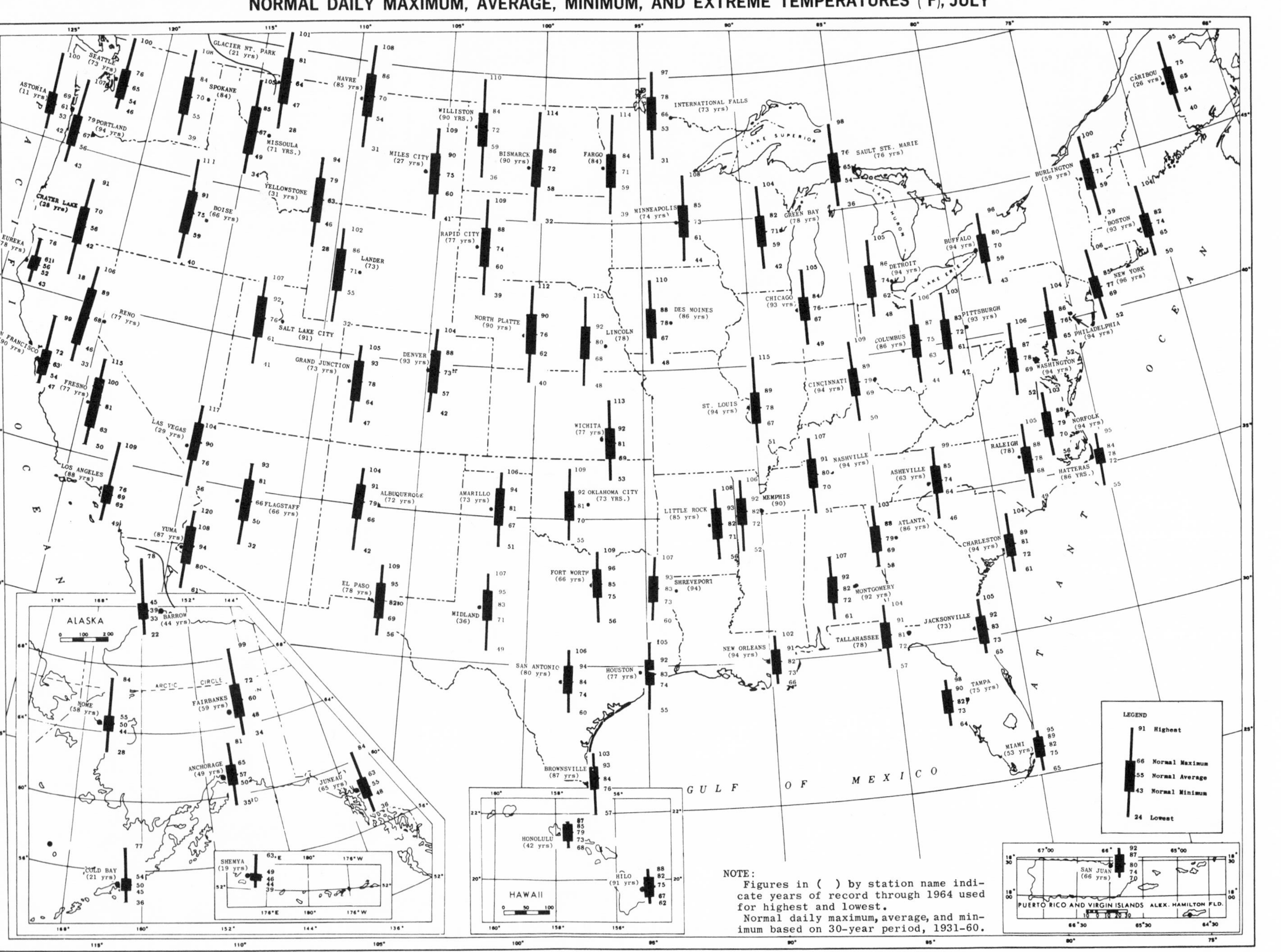

NOTE:
Figures in ( ) by station name indicate years of record through 1964 used for highest and lowest.
Normal daily maximum, average, and minimum based on 30-year period, 1931-60.

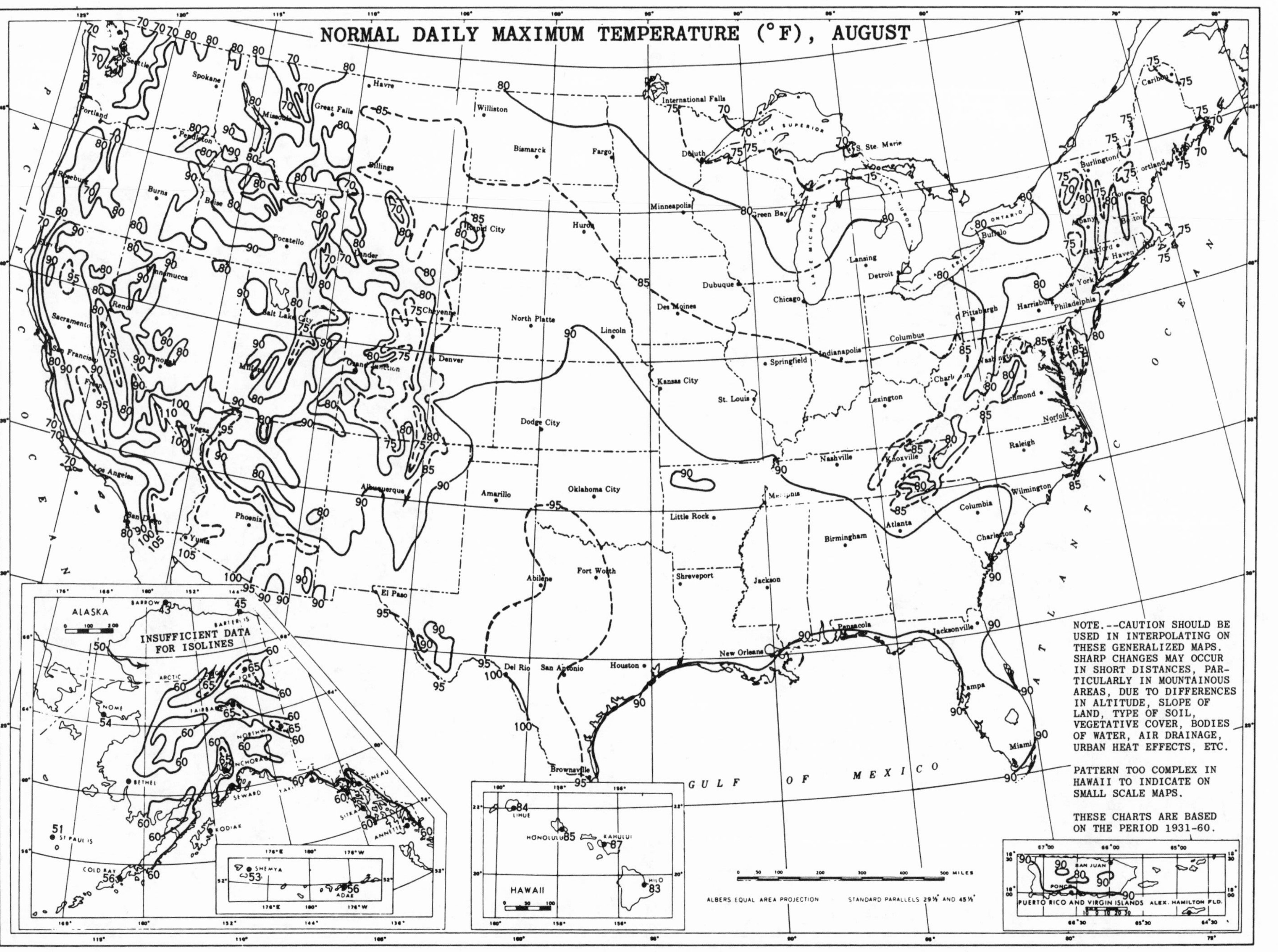
NORMAL DAILY MAXIMUM TEMPERATURE (°F), AUGUST
NOTE.--CAUTION SHOULD BE USED IN INTERPOLATING ON THESE GENERALIZED MAPS. SHARP CHANGES MAY OCCUR IN SHORT DISTANCES, PARTICULARLY IN MOUNTAINOUS AREAS, DUE TO DIFFERENCES IN ALTITUDE, SLOPE OF LAND, TYPE OF SOIL, VEGETATIVE COVER, BODIES OF WATER, AIR DRAINAGE, URBAN HEAT EFFECTS, ETC.
PATTERN TOO COMPLEX IN HAWAII TO INDICATE ON SMALL SCALE MAPS.
THESE CHARTS ARE BASED ON THE PERIOD 1931-60.
ALASKA
INSUFFICIENT DATA FOR ISOLINES
HAWAII
PUERTO RICO AND VIRGIN ISLANDS
ALBERS EQUAL AREA PROJECTION
STANDARD PARALLELS 29½° AND 45½°
PACIFIC OCEAN
ATLANTIC OCEAN
GULF OF MEXICO

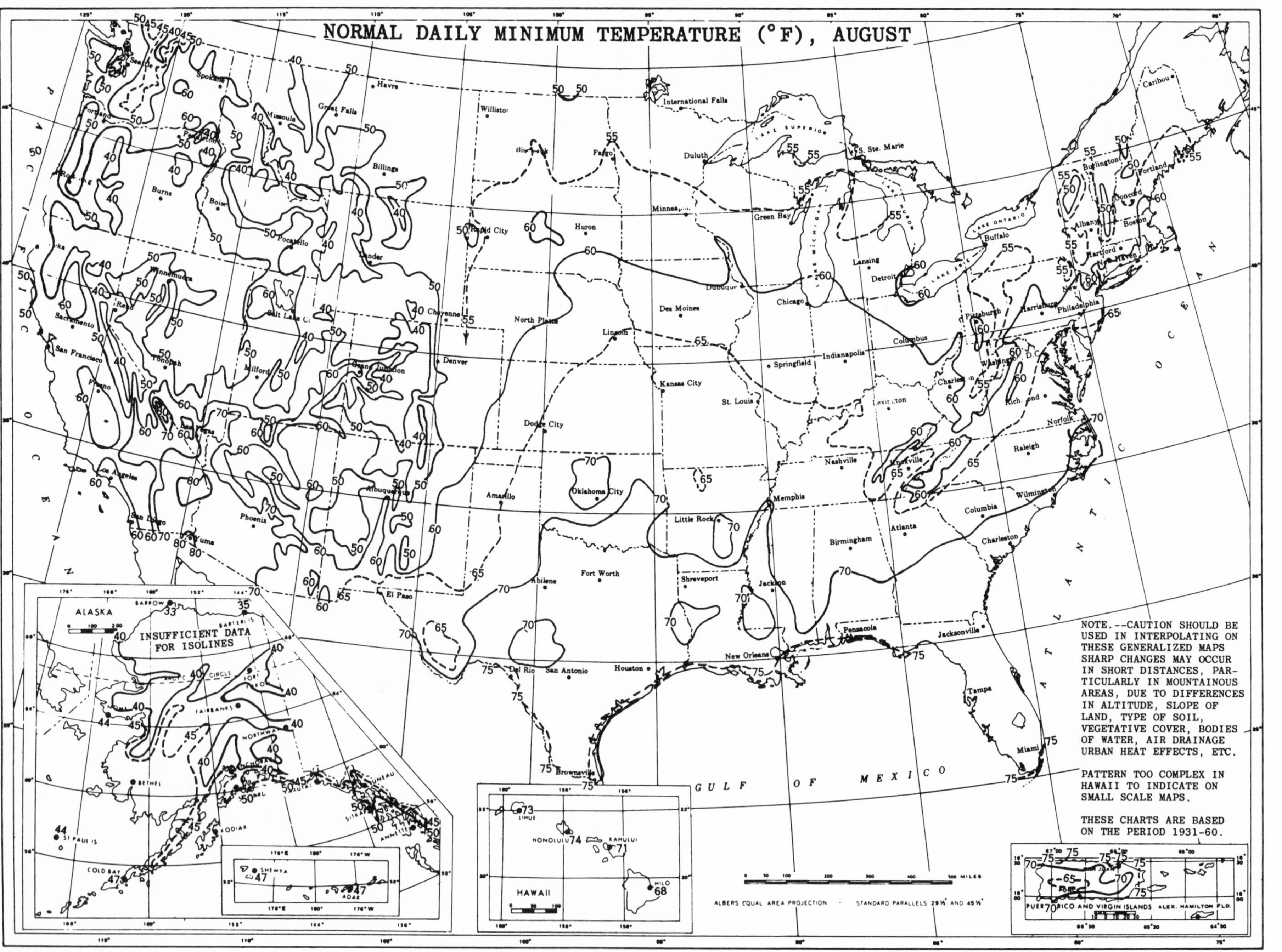
NORMAL DAILY MINIMUM TEMPERATURE (°F), AUGUST
NOTE.--CAUTION SHOULD BE USED IN INTERPOLATING ON THESE GENERALIZED MAPS SHARP CHANGES MAY OCCUR IN SHORT DISTANCES, PARTICULARLY IN MOUNTAINOUS AREAS, DUE TO DIFFERENCES IN ALTITUDE, SLOPE OF LAND, TYPE OF SOIL, VEGETATIVE COVER, BODIES OF WATER, AIR DRAINAGE URBAN HEAT EFFECTS, ETC.
PATTERN TOO COMPLEX IN HAWAII TO INDICATE ON SMALL SCALE MAPS.
THESE CHARTS ARE BASED ON THE PERIOD 1931-60.
ALBERS EQUAL AREA PROJECTION - STANDARD PARALLELS 29½° AND 45½°
PACIFIC OCEAN
ATLANTIC OCEAN
GULF OF MEXICO
ALASKA
INSUFFICIENT DATA FOR ISOLINES
HAWAII
PUERTO RICO AND VIRGIN ISLANDS
Havre
Great Falls
Missoula
Spokane
Seattle
Portland
Billings
Williston
Bismarck
Fargo
International Falls
Duluth
Sault Ste. Marie
Caribou
Burlington
Portland
Concord
Albany
Boston
Hartford
New Haven
Buffalo
Green Bay
Minneapolis
Huron
Rapid City
Pocatello
Boise
Burns
Lander
Cheyenne
Denver
North Platte
Lincoln
Des Moines
Dubuque
Chicago
Lansing
Detroit
Pittsburgh
Harrisburg
Philadelphia
New York
Columbus
Indianapolis
Springfield
Kansas City
St. Louis
Lexington
Charleston
Washington, D.C.
Richmond
Norfolk
Raleigh
Nashville
Knoxville
Memphis
Little Rock
Oklahoma City
Dodge City
Amarillo
Albuquerque
Grand Junction
Salt Lake City
Winnemucca
Reno
Sacramento
San Francisco
Fresno
Tonopah
Milford
Las Vegas
Los Angeles
San Diego
Yuma
Phoenix
El Paso
Abilene
Fort Worth
Shreveport
Jackson
Birmingham
Atlanta
Columbia
Wilmington
Charleston
Jacksonville
Pensacola
New Orleans
Houston
San Antonio
Del Rio
Brownsville
Tampa
Miami
Barrow
Barter Is
Fort Yukon
Fairbanks
Nome
Bethel
Northway
Anchorage
Juneau
Sitka
Annette
Kodiak
St Paul Is
Cold Bay
Shemya
Adak
Lihue
Honolulu
Kahului
Hilo
Alex. Hamilton Fld.

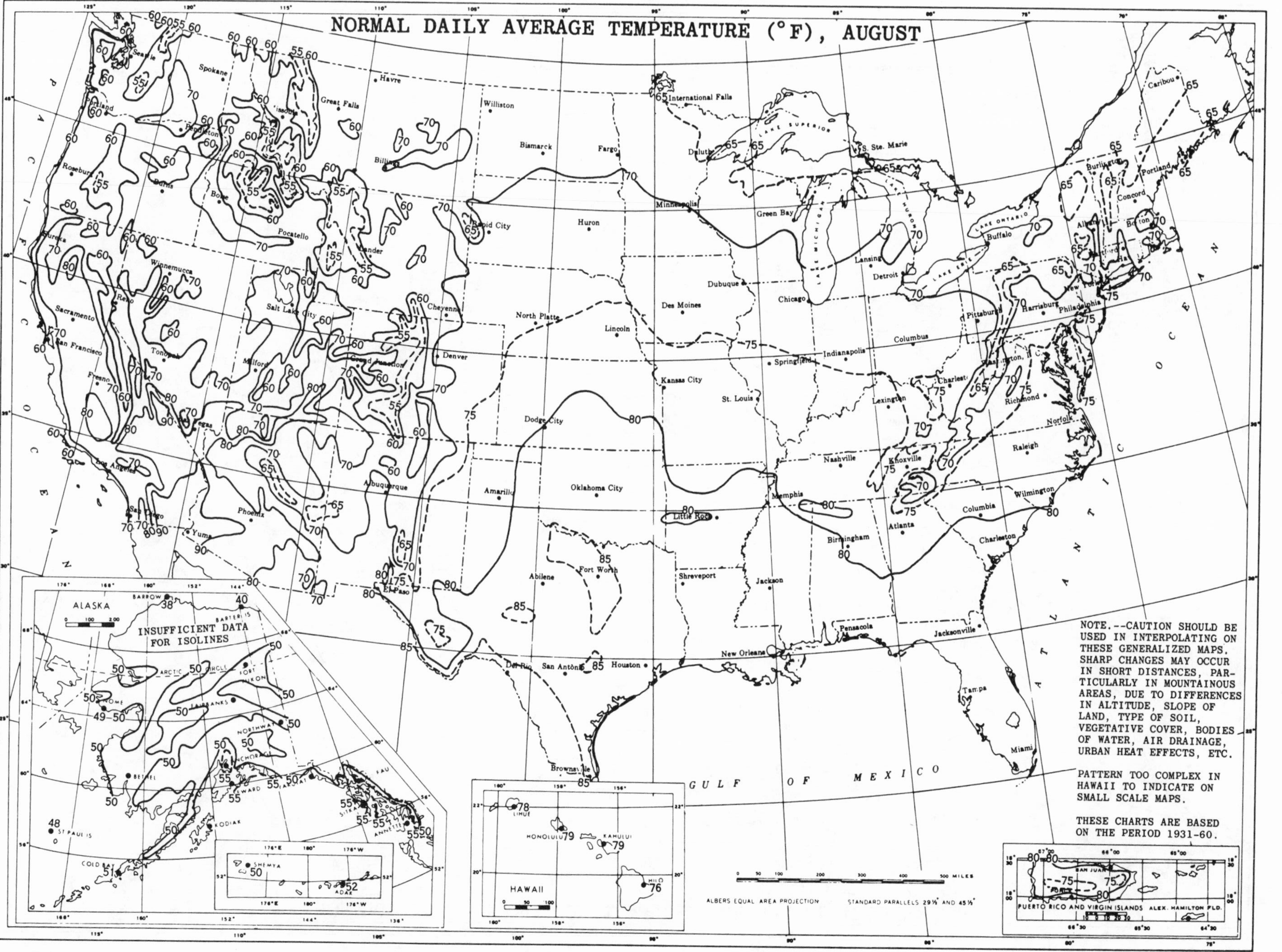
NORMAL DAILY AVERAGE TEMPERATURE (°F), AUGUST
NOTE.--CAUTION SHOULD BE USED IN INTERPOLATING ON THESE GENERALIZED MAPS. SHARP CHANGES MAY OCCUR IN SHORT DISTANCES, PARTICULARLY IN MOUNTAINOUS AREAS, DUE TO DIFFERENCES IN ALTITUDE, SLOPE OF LAND, TYPE OF SOIL, VEGETATIVE COVER, BODIES OF WATER, AIR DRAINAGE, URBAN HEAT EFFECTS, ETC.
PATTERN TOO COMPLEX IN HAWAII TO INDICATE ON SMALL SCALE MAPS.
THESE CHARTS ARE BASED ON THE PERIOD 1931-60.
PACIFIC OCEAN
ATLANTIC OCEAN
GULF OF MEXICO
ALASKA
INSUFFICIENT DATA FOR ISOLINES
HAWAII
PUERTO RICO AND VIRGIN ISLANDS
ALBERS EQUAL AREA PROJECTION
STANDARD PARALLELS 29½° AND 45½°
500 MILES

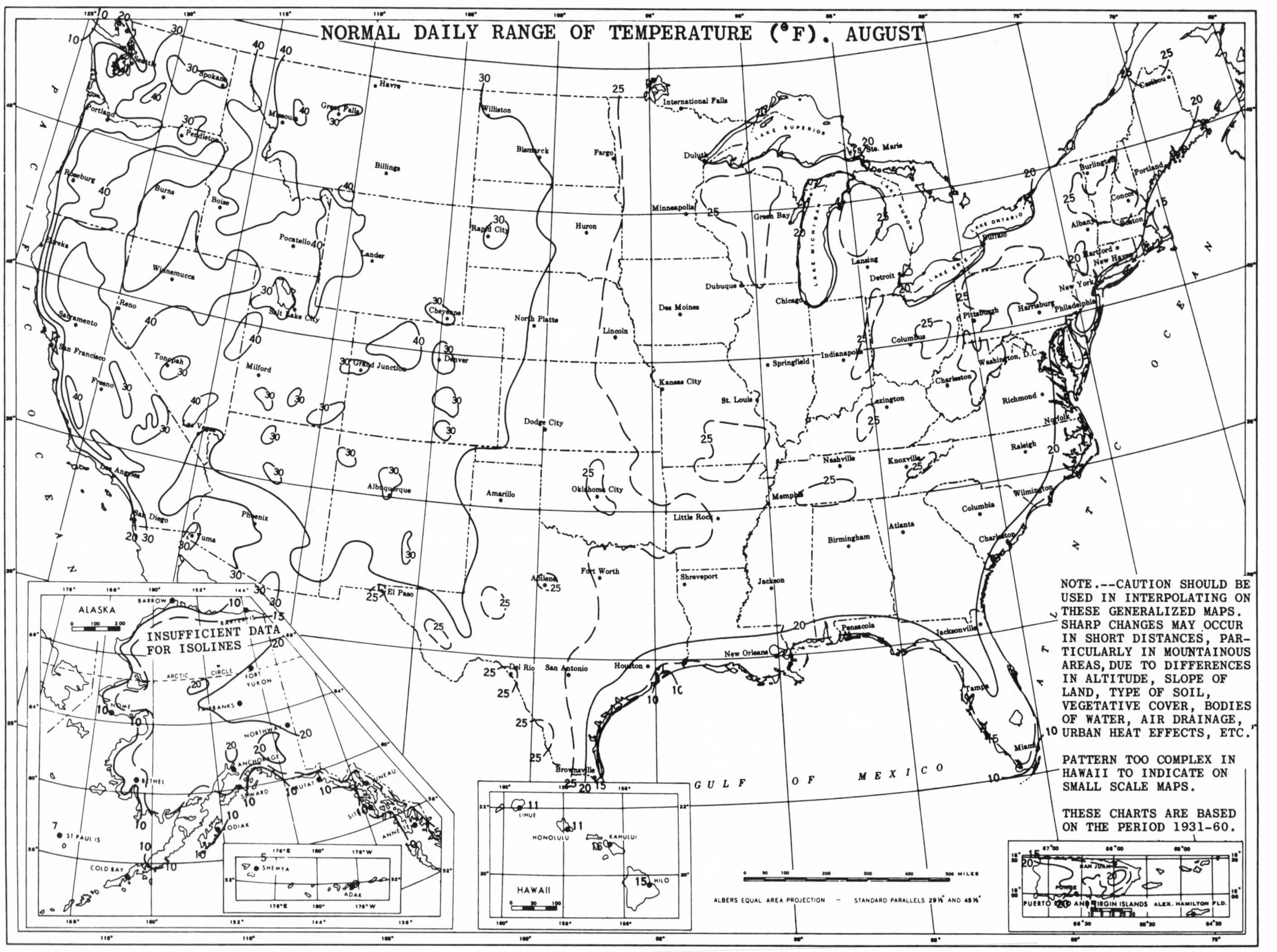
NORMAL DAILY RANGE OF TEMPERATURE (°F). AUGUST
NOTE.--CAUTION SHOULD BE USED IN INTERPOLATING ON THESE GENERALIZED MAPS. SHARP CHANGES MAY OCCUR IN SHORT DISTANCES, PARTICULARLY IN MOUNTAINOUS AREAS, DUE TO DIFFERENCES IN ALTITUDE, SLOPE OF LAND, TYPE OF SOIL, VEGETATIVE COVER, BODIES OF WATER, AIR DRAINAGE, URBAN HEAT EFFECTS, ETC.
PATTERN TOO COMPLEX IN HAWAII TO INDICATE ON SMALL SCALE MAPS.
THESE CHARTS ARE BASED ON THE PERIOD 1931-60.
INSUFFICIENT DATA FOR ISOLINES
ALASKA
HAWAII
PUERTO RICO AND VIRGIN ISLANDS
PACIFIC OCEAN
ATLANTIC OCEAN
GULF OF MEXICO
ALBERS EQUAL AREA PROJECTION — STANDARD PARALLELS 29½° AND 45½°

NORMAL DAILY MAXIMUM, AVERAGE, MINIMUM, AND EXTREME TEMPERATURES (°F), AUGUST

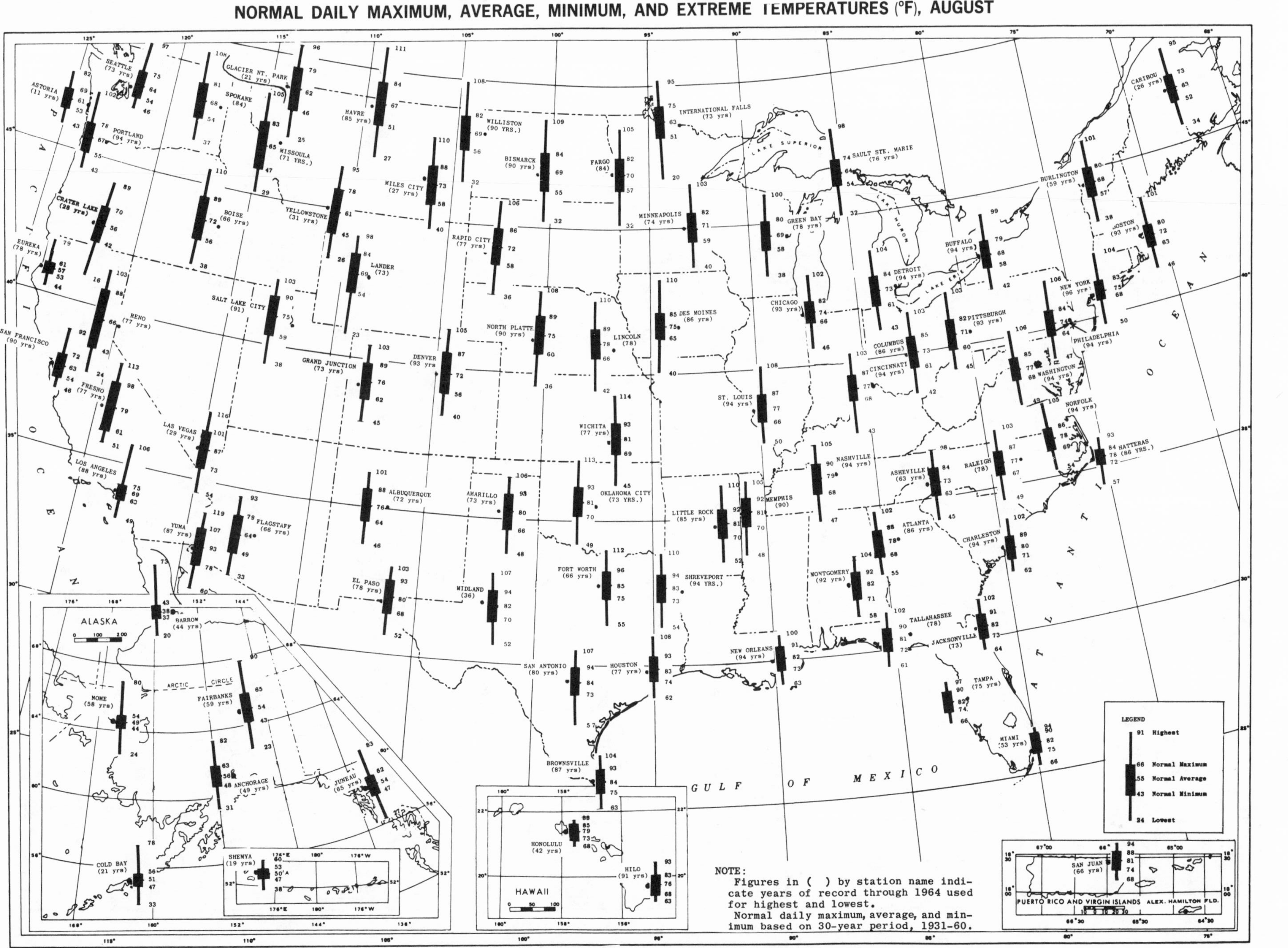

NOTE:
Figures in ( ) by station name indicate years of record through 1964 used for highest and lowest.
Normal daily maximum, average, and minimum based on 30-year period, 1931-60.

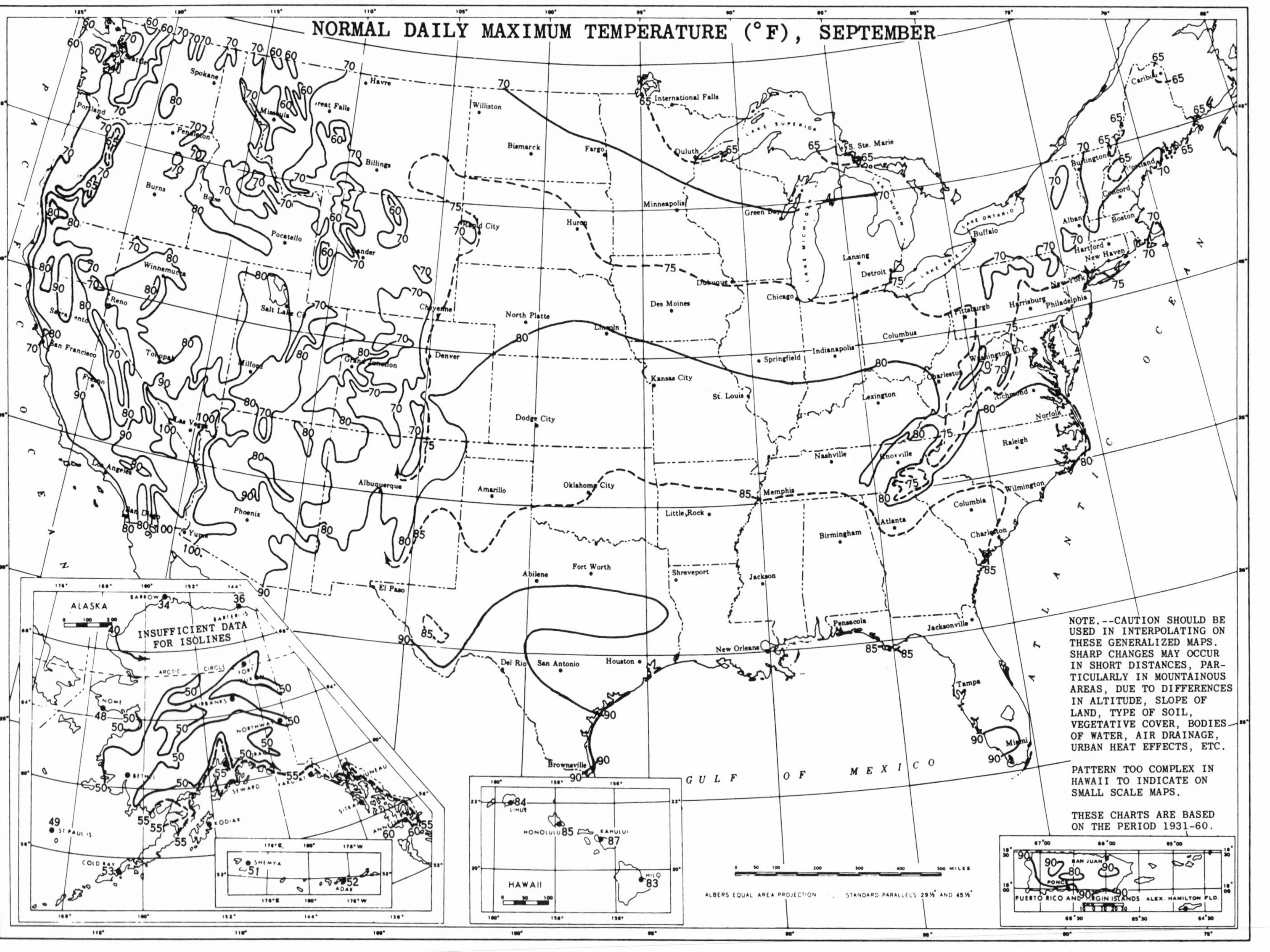
NORMAL DAILY MAXIMUM TEMPERATURE (°F), SEPTEMBER
NOTE.--CAUTION SHOULD BE USED IN INTERPOLATING ON THESE GENERALIZED MAPS. SHARP CHANGES MAY OCCUR IN SHORT DISTANCES, PARTICULARLY IN MOUNTAINOUS AREAS, DUE TO DIFFERENCES IN ALTITUDE, SLOPE OF LAND, TYPE OF SOIL, VEGETATIVE COVER, BODIES OF WATER, AIR DRAINAGE, URBAN HEAT EFFECTS, ETC.
PATTERN TOO COMPLEX IN HAWAII TO INDICATE ON SMALL SCALE MAPS.
THESE CHARTS ARE BASED ON THE PERIOD 1931-60.
INSUFFICIENT DATA FOR ISOLINES
ALASKA
HAWAII
PUERTO RICO AND VIRGIN ISLANDS
PACIFIC OCEAN
ATLANTIC OCEAN
GULF OF MEXICO
ALBERS EQUAL AREA PROJECTION
STANDARD PARALLELS 29½° AND 45½°

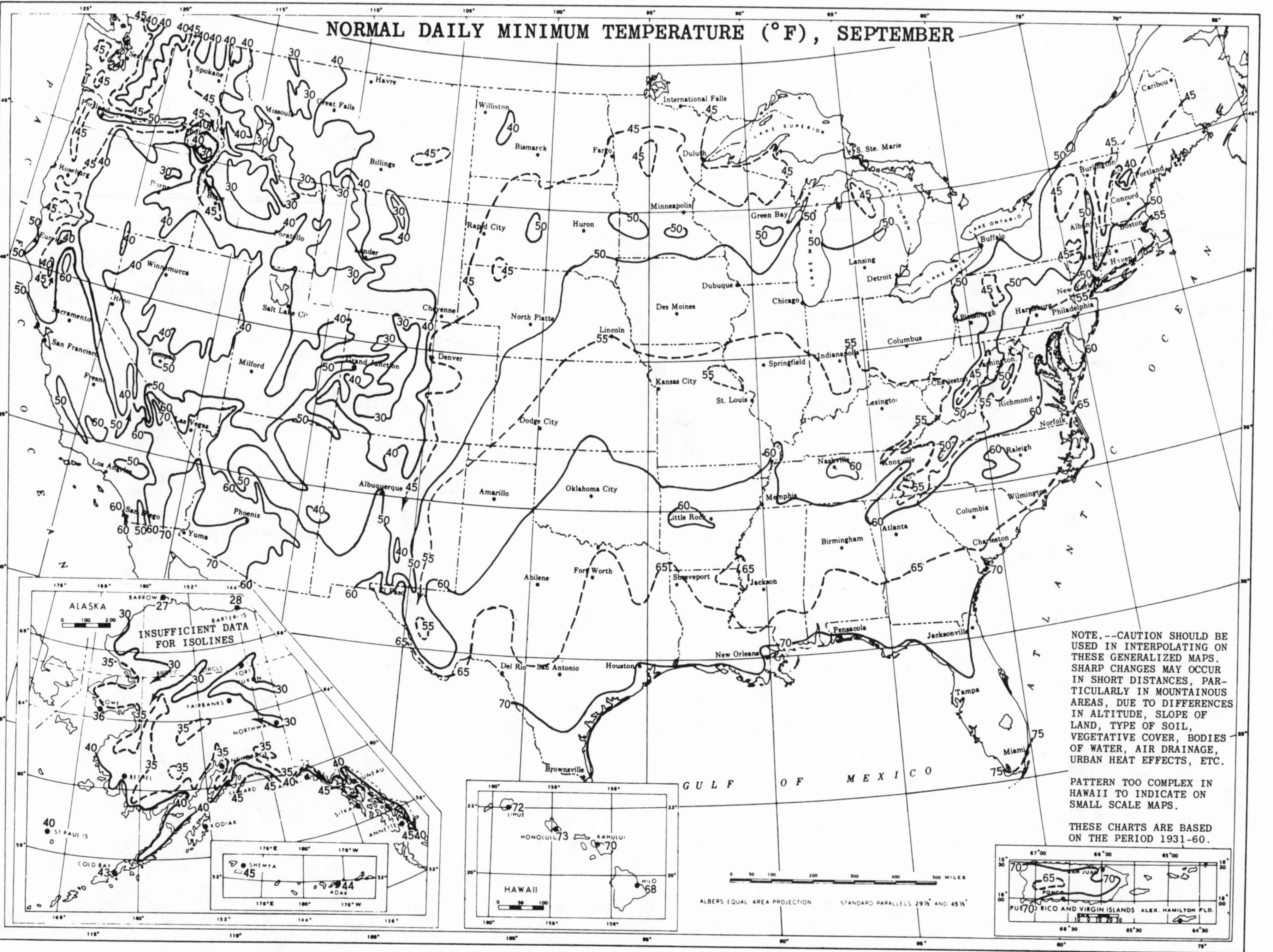
NORMAL DAILY MINIMUM TEMPERATURE (°F), SEPTEMBER
NOTE.--CAUTION SHOULD BE USED IN INTERPOLATING ON THESE GENERALIZED MAPS. SHARP CHANGES MAY OCCUR IN SHORT DISTANCES, PARTICULARLY IN MOUNTAINOUS AREAS, DUE TO DIFFERENCES IN ALTITUDE, SLOPE OF LAND, TYPE OF SOIL, VEGETATIVE COVER, BODIES OF WATER, AIR DRAINAGE, URBAN HEAT EFFECTS, ETC.
PATTERN TOO COMPLEX IN HAWAII TO INDICATE ON SMALL SCALE MAPS.
THESE CHARTS ARE BASED ON THE PERIOD 1931-60.
ALBERS EQUAL AREA PROJECTION
STANDARD PARALLELS 29½° AND 45½°
0 50 100 200 300 400 500 MILES
PACIFIC OCEAN
ATLANTIC OCEAN
GULF OF MEXICO
ALASKA
INSUFFICIENT DATA FOR ISOLINES
HAWAII
PUERTO RICO AND VIRGIN ISLANDS
Spokane
Havre
Great Falls
Missoula
Williston
Bismarck
Billings
International Falls
Fargo
Duluth
S. Ste. Marie
Caribou
Portland
Roseburg
Rapid City
Huron
Minneapolis
Green Bay
Lansing
Detroit
Buffalo
Burlington
Concord
Albany
Boston
Winnemucca
Reno
Salt Lake City
Cheyenne
North Platte
Des Moines
Dubuque
Chicago
Sacramento
San Francisco
Fresno
Tonopah
Milford
Grand Junction
Denver
Lincoln
Kansas City
Springfield
Indianapolis
Columbus
Pittsburgh
Harrisburg
Philadelphia
New York
St. Louis
Lexington
Charleston
Richmond
Norfolk
Las Vegas
Los Angeles
San Diego
Yuma
Phoenix
Albuquerque
Amarillo
Dodge City
Oklahoma City
Little Rock
Memphis
Nashville
Knoxville
Raleigh
Wilmington
Atlanta
Columbia
Birmingham
Abilene
Fort Worth
Shreveport
Jackson
Del Rio
San Antonio
Houston
New Orleans
Pensacola
Jacksonville
Tampa
Miami
Brownsville
El Paso
Barrow
Barter Is
Fort Yukon
Fairbanks
Nome
Northway
Bethel
Kodiak
Cold Bay
St Paul Is
Juneau
Sitka
Annette
Shemya
Adak
Lihue
Honolulu
Kahului
Hilo
San Juan
Ponce
Alex. Hamilton Fld.

NORMAL DAILY AVERAGE TEMPERATURE (°F), SEPTEMBER

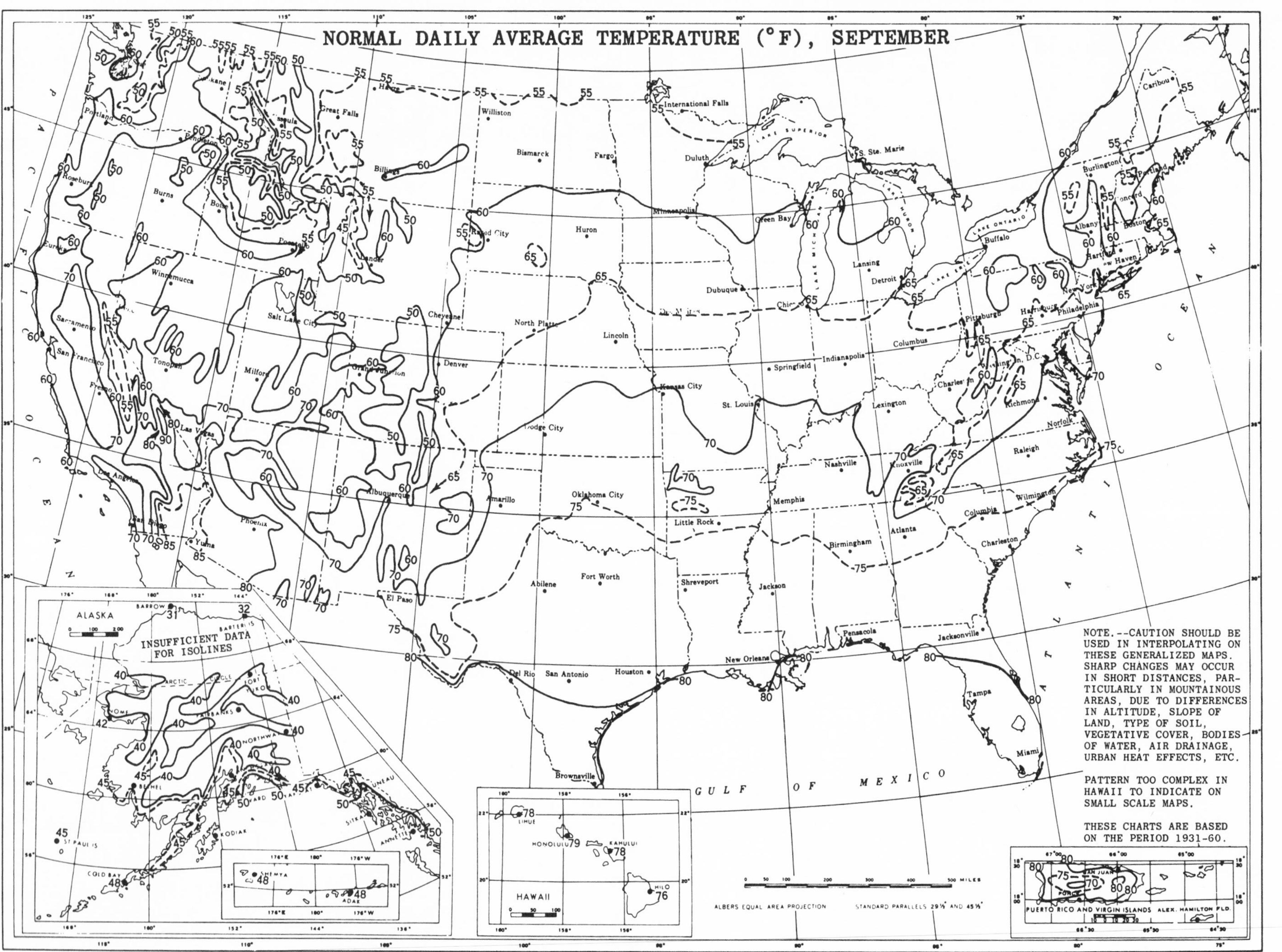
NOTE.--CAUTION SHOULD BE USED IN INTERPOLATING ON THESE GENERALIZED MAPS. SHARP CHANGES MAY OCCUR IN SHORT DISTANCES, PARTICULARLY IN MOUNTAINOUS AREAS, DUE TO DIFFERENCES IN ALTITUDE, SLOPE OF LAND, TYPE OF SOIL, VEGETATIVE COVER, BODIES OF WATER, AIR DRAINAGE, URBAN HEAT EFFECTS, ETC.
PATTERN TOO COMPLEX IN HAWAII TO INDICATE ON SMALL SCALE MAPS.
THESE CHARTS ARE BASED ON THE PERIOD 1931-60.
PACIFIC OCEAN
ATLANTIC OCEAN
GULF OF MEXICO
ALASKA
INSUFFICIENT DATA FOR ISOLINES
HAWAII
PUERTO RICO AND VIRGIN ISLANDS
ALBERS EQUAL AREA PROJECTION
STANDARD PARALLELS 29½° AND 45½°
MILES

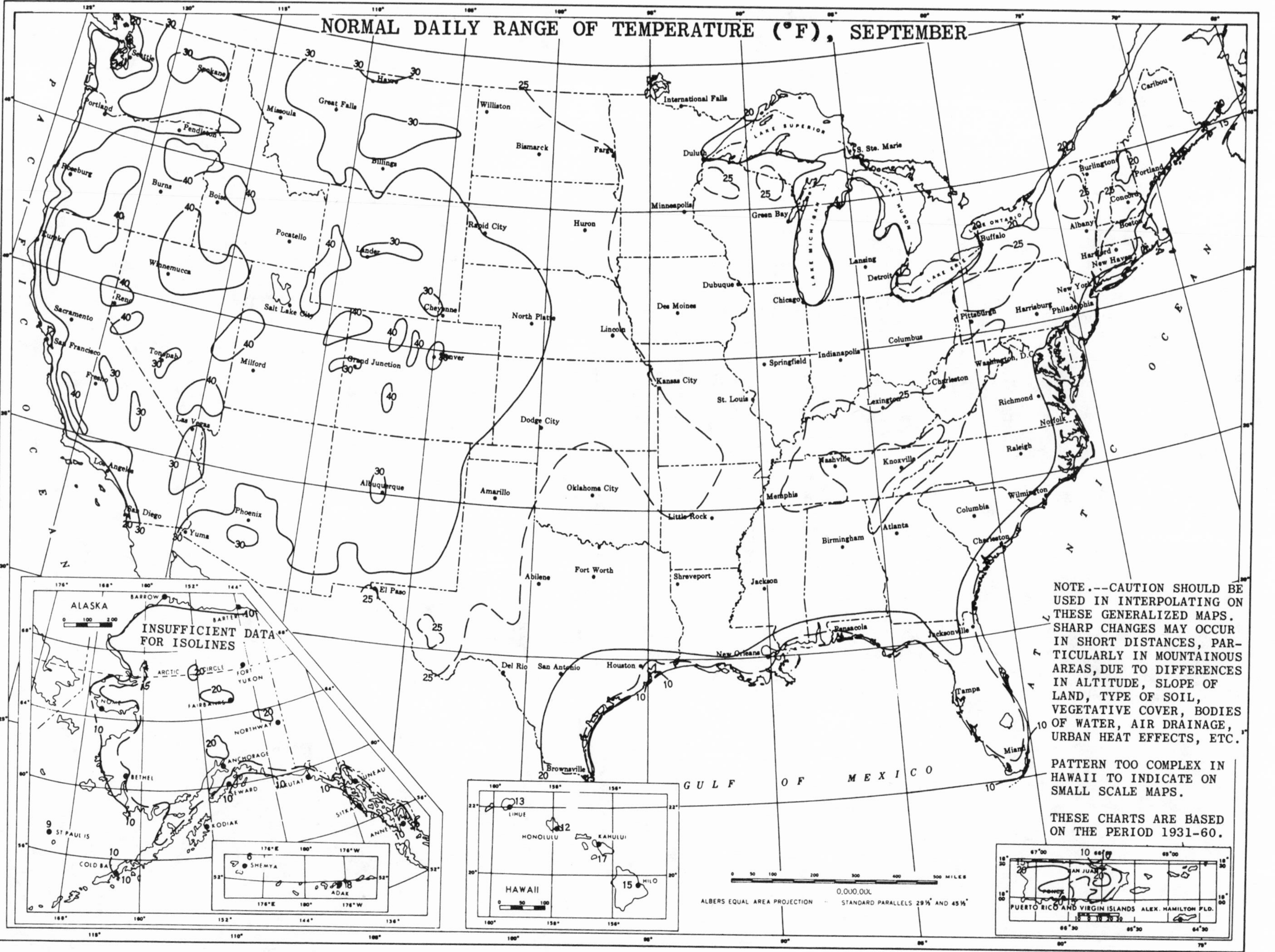
NORMAL DAILY RANGE OF TEMPERATURE (°F), SEPTEMBER
NOTE.--CAUTION SHOULD BE USED IN INTERPOLATING ON THESE GENERALIZED MAPS. SHARP CHANGES MAY OCCUR IN SHORT DISTANCES, PARTICULARLY IN MOUNTAINOUS AREAS, DUE TO DIFFERENCES IN ALTITUDE, SLOPE OF LAND, TYPE OF SOIL, VEGETATIVE COVER, BODIES OF WATER, AIR DRAINAGE, URBAN HEAT EFFECTS, ETC.
PATTERN TOO COMPLEX IN HAWAII TO INDICATE ON SMALL SCALE MAPS.
THESE CHARTS ARE BASED ON THE PERIOD 1931-60.
ALASKA
INSUFFICIENT DATA FOR ISOLINES
HAWAII
PUERTO RICO AND VIRGIN ISLANDS
ALBERS EQUAL AREA PROJECTION
STANDARD PARALLELS 29½° AND 45½°
PACIFIC OCEAN
ATLANTIC OCEAN
GULF OF MEXICO

NORMAL DAILY MAXIMUM, AVERAGE, MINIMUM, AND EXTREME TEMPERATURES (°F), SEPTEMBER

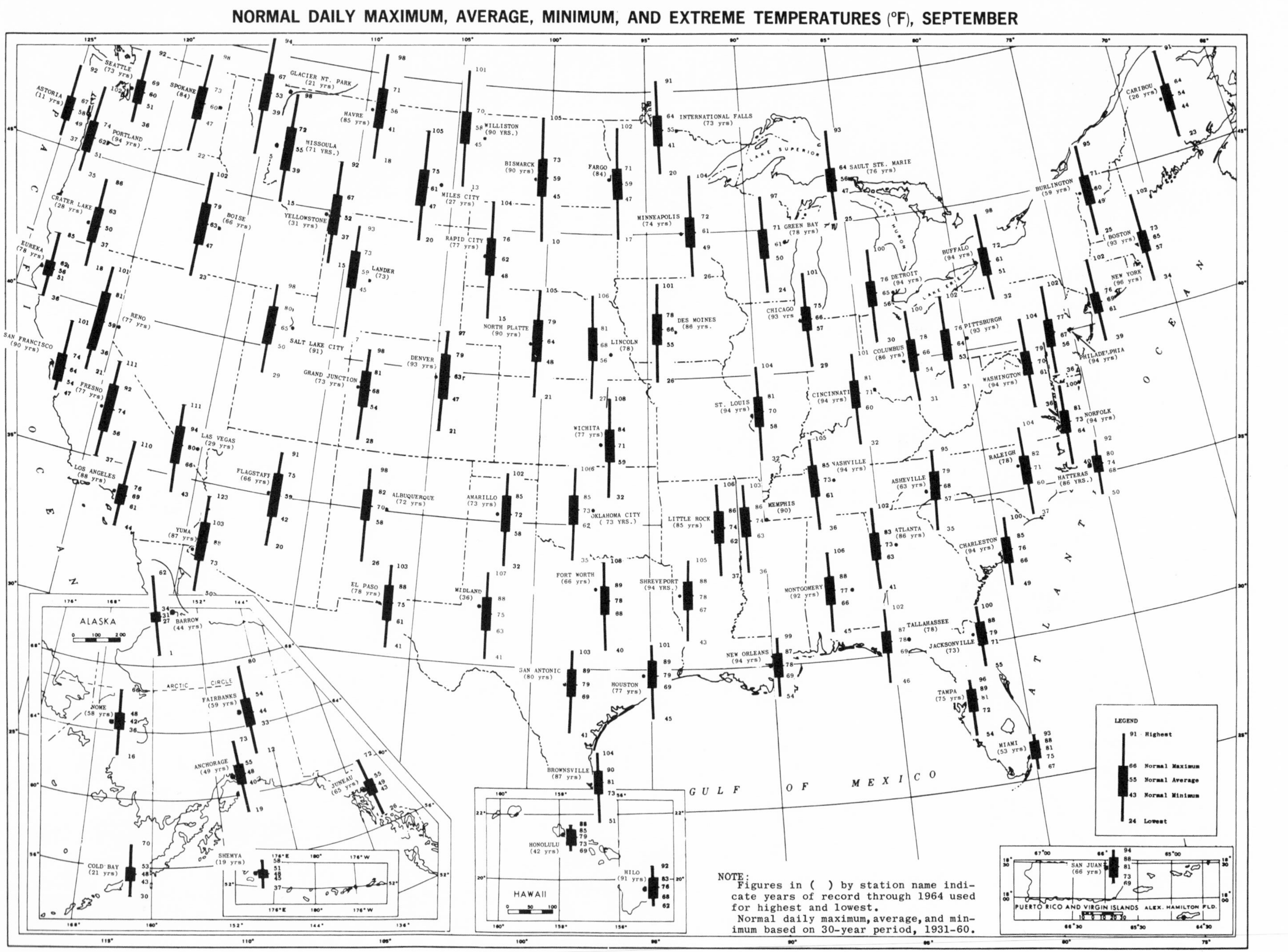

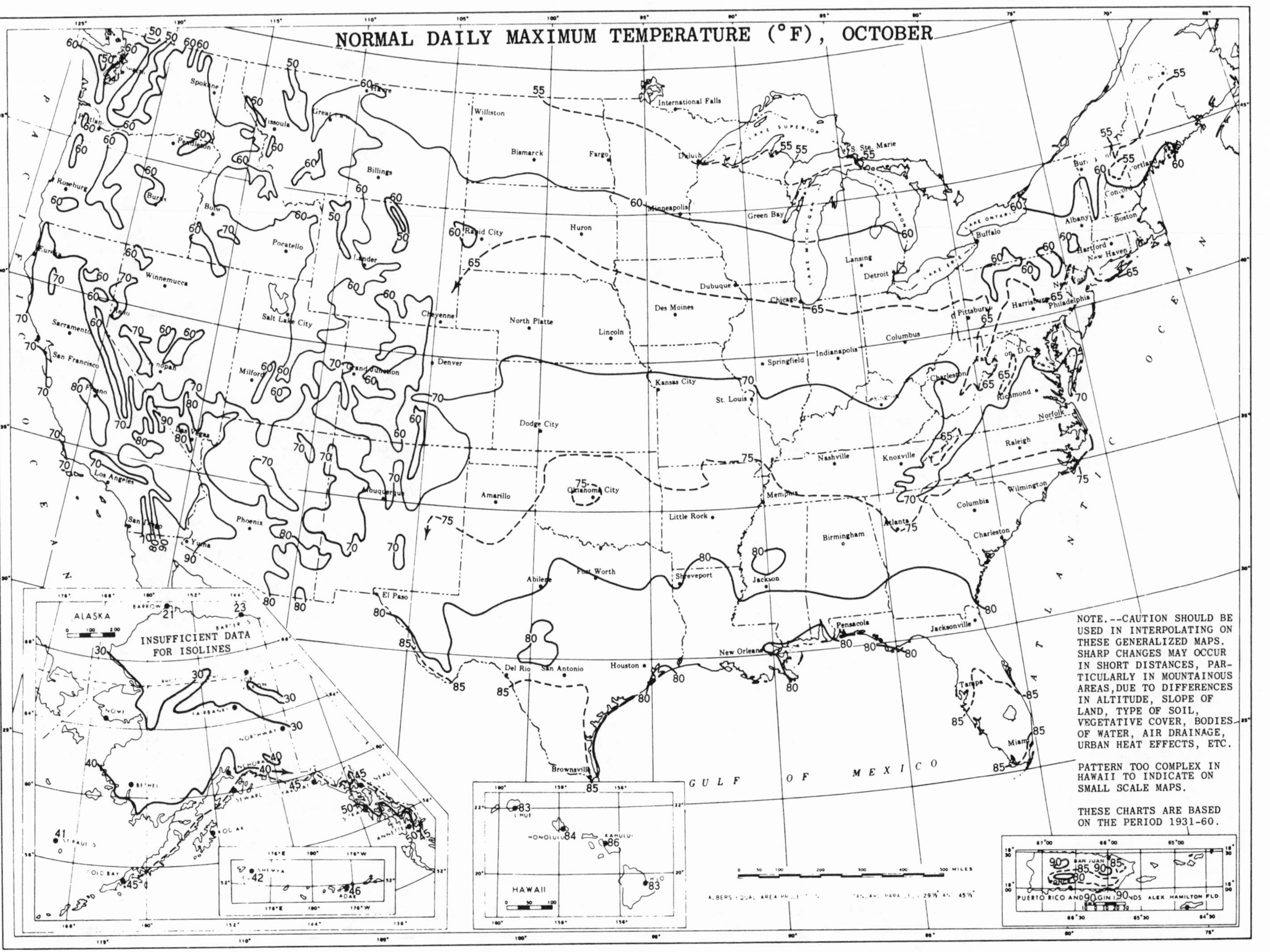
NORMAL DAILY MAXIMUM TEMPERATURE (°F), OCTOBER
NOTE.--CAUTION SHOULD BE USED IN INTERPOLATING ON THESE GENERALIZED MAPS. SHARP CHANGES MAY OCCUR IN SHORT DISTANCES, PARTICULARLY IN MOUNTAINOUS AREAS, DUE TO DIFFERENCES IN ALTITUDE, SLOPE OF LAND, TYPE OF SOIL, VEGETATIVE COVER, BODIES OF WATER, AIR DRAINAGE, URBAN HEAT EFFECTS, ETC.
PATTERN TOO COMPLEX IN HAWAII TO INDICATE ON SMALL SCALE MAPS.
THESE CHARTS ARE BASED ON THE PERIOD 1931-60.
ALASKA
INSUFFICIENT DATA FOR ISOLINES
HAWAII
PUERTO RICO AND VIRGIN ISLANDS
PACIFIC OCEAN
ATLANTIC OCEAN
GULF OF MEXICO
Spokane
Williston
International Falls
Bismarck
Fargo
Duluth
Minneapolis
Green Bay
Huron
Rapid City
Billings
Pocatello
Lander
Winnemucca
Roseburg
Eureka
Sacramento
San Francisco
Fresno
Salt Lake City
Milford
Grand Junction
Cheyenne
Denver
North Platte
Lincoln
Des Moines
Dubuque
Chicago
Lansing
Detroit
Buffalo
Albany
Boston
Concord
Hartford
New Haven
Harrisburg
Philadelphia
Pittsburgh
Columbus
Indianapolis
Springfield
Kansas City
St. Louis
Dodge City
Amarillo
Oklahoma City
Little Rock
Memphis
Nashville
Knoxville
Charleston
Richmond
Norfolk
Raleigh
Wilmington
Columbia
Atlanta
Birmingham
Jackson
Shreveport
Fort Worth
Abilene
El Paso
Albuquerque
Phoenix
Yuma
Los Angeles
Las Vegas
Del Rio
San Antonio
Houston
New Orleans
Pensacola
Jacksonville
Tampa
Miami
Brownsville

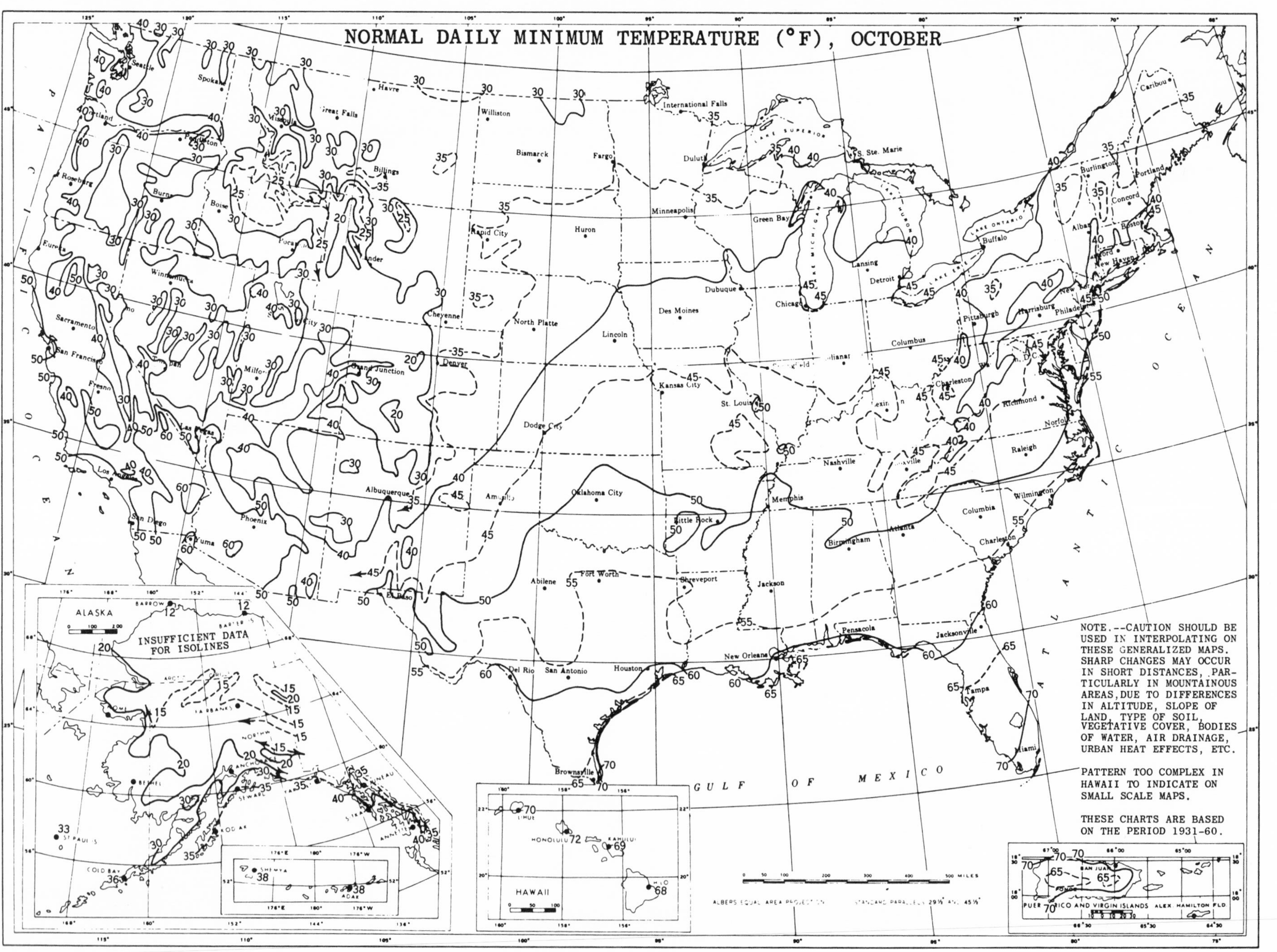
NORMAL DAILY MINIMUM TEMPERATURE (°F), OCTOBER
NOTE.--CAUTION SHOULD BE USED IN INTERPOLATING ON THESE GENERALIZED MAPS. SHARP CHANGES MAY OCCUR IN SHORT DISTANCES, PARTICULARLY IN MOUNTAINOUS AREAS, DUE TO DIFFERENCES IN ALTITUDE, SLOPE OF LAND, TYPE OF SOIL, VEGETATIVE COVER, BODIES OF WATER, AIR DRAINAGE, URBAN HEAT EFFECTS, ETC.
PATTERN TOO COMPLEX IN HAWAII TO INDICATE ON SMALL SCALE MAPS.
THESE CHARTS ARE BASED ON THE PERIOD 1931-60.
ALASKA
INSUFFICIENT DATA FOR ISOLINES
HAWAII
PUERTO RICO AND VIRGIN ISLANDS
PACIFIC OCEAN
ATLANTIC OCEAN
GULF OF MEXICO
ALBERS EQUAL AREA PROJECTION

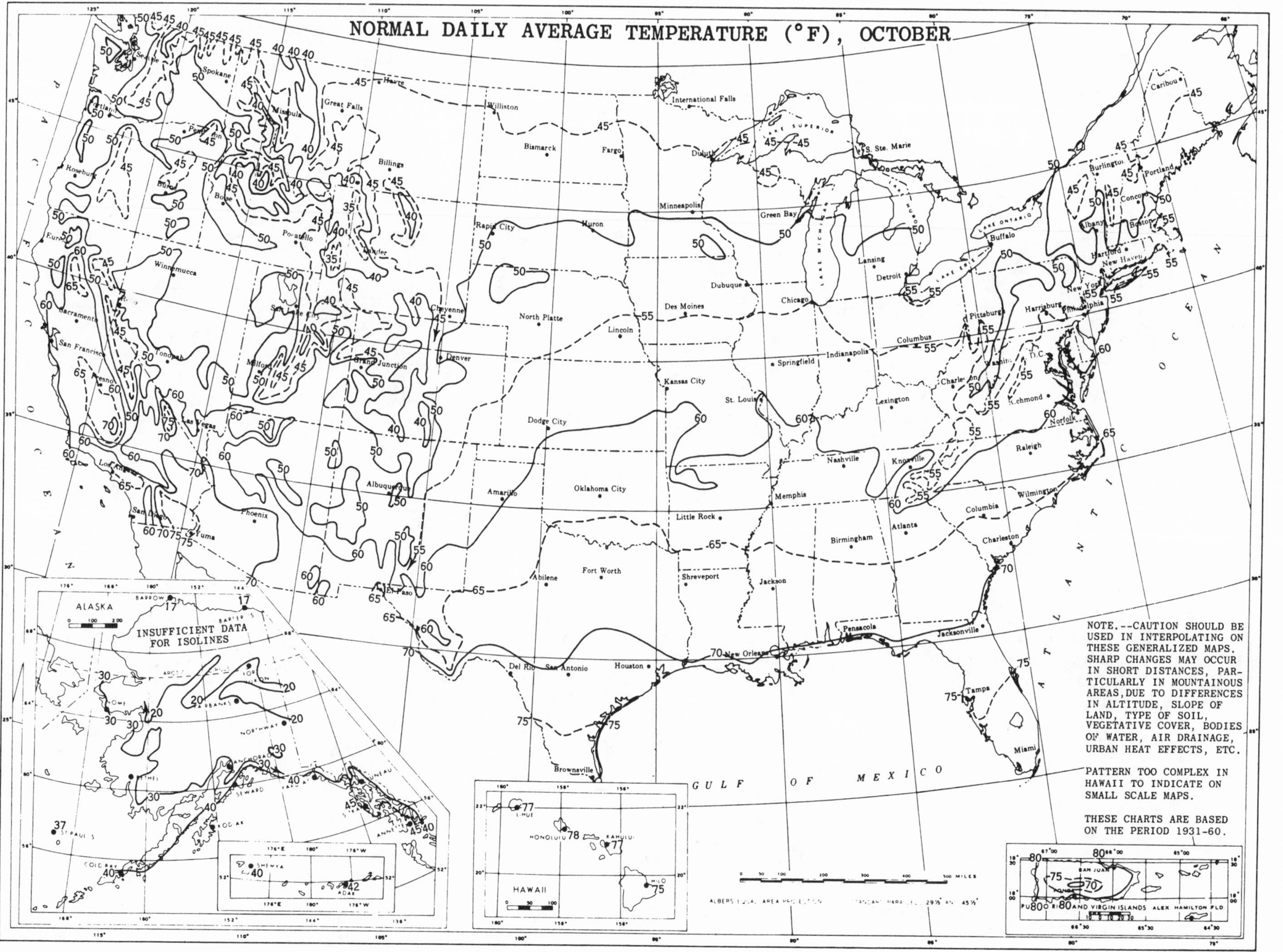
NORMAL DAILY AVERAGE TEMPERATURE (°F), OCTOBER
NOTE.--CAUTION SHOULD BE USED IN INTERPOLATING ON THESE GENERALIZED MAPS. SHARP CHANGES MAY OCCUR IN SHORT DISTANCES, PARTICULARLY IN MOUNTAINOUS AREAS, DUE TO DIFFERENCES IN ALTITUDE, SLOPE OF LAND, TYPE OF SOIL, VEGETATIVE COVER, BODIES OF WATER, AIR DRAINAGE, URBAN HEAT EFFECTS, ETC.
PATTERN TOO COMPLEX IN HAWAII TO INDICATE ON SMALL SCALE MAPS.
THESE CHARTS ARE BASED ON THE PERIOD 1931-60.
ALASKA
INSUFFICIENT DATA FOR ISOLINES
HAWAII
PUERTO RICO AND VIRGIN ISLANDS
PACIFIC OCEAN
ATLANTIC OCEAN
GULF OF MEXICO
Seattle
Spokane
Portland
Pendleton
Boise
Roseburg
Eureka
Winnemucca
Reno
Sacramento
San Francisco
Fresno
Tonopah
Las Vegas
Los Angeles
San Diego
Yuma
Phoenix
Missoula
Great Falls
Havre
Billings
Pocatello
Lander
Salt Lake City
Milford
Grand Junction
Cheyenne
Denver
Albuquerque
El Paso
Williston
Bismarck
Fargo
Rapid City
Huron
North Platte
Lincoln
Dodge City
Amarillo
Oklahoma City
Abilene
Fort Worth
Del Rio
San Antonio
Houston
Brownsville
International Falls
Duluth
Minneapolis
Des Moines
Dubuque
Kansas City
St. Louis
Springfield
Little Rock
Shreveport
Jackson
New Orleans
Memphis
Nashville
Birmingham
Atlanta
Pensacola
Jacksonville
Tampa
Miami
Green Bay
Chicago
Indianapolis
Lansing
Detroit
Sault Ste. Marie
Columbus
Lexington
Knoxville
Charleston
Columbia
Wilmington
Raleigh
Norfolk
Richmond
Washington, D.C.
Pittsburgh
Harrisburg
Philadelphia
New York
Buffalo
Albany
Hartford
New Haven
Boston
Concord
Burlington
Portland
Caribou
LAKE SUPERIOR
LAKE MICHIGAN
LAKE HURON
LAKE ERIE
LAKE ONTARIO
Barrow
Barter Is.
Nome
Bethel
Fairbanks
Northway
Anchorage
Seward
Kodiak
Juneau
Annette
Cold Bay
St. Paul Is.
Shemya
Adak
Lihue
Honolulu
Kahului
Hilo
San Juan
0 50 100 200 300 400 500 MILES
ALBERS EQUAL AREA PROJECTION STANDARD PARALLELS 29½° AND 45½°

NORMAL DAILY RANGE OF TEMPERATURE (°F), OCTOBER

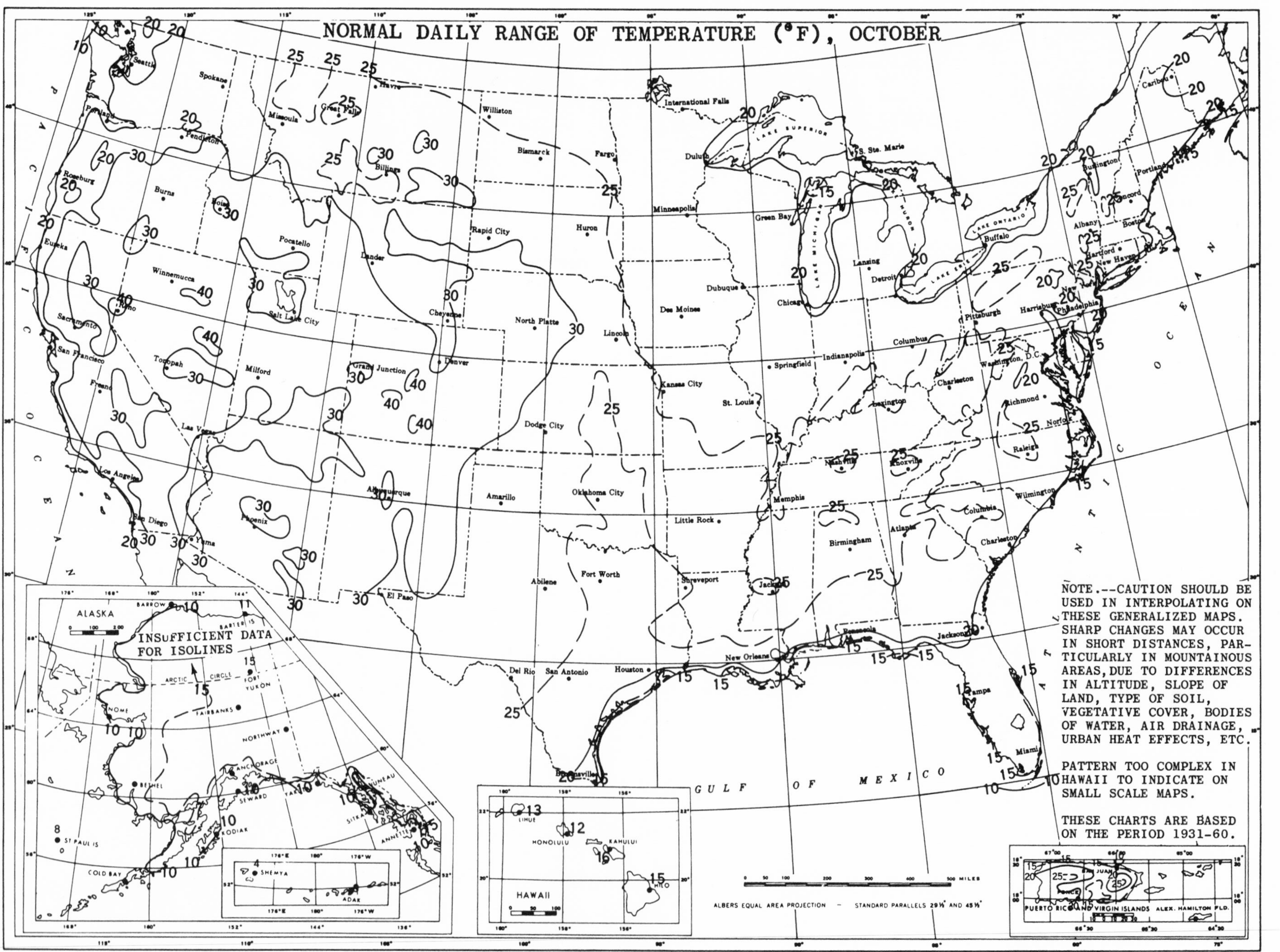

NOTE.--CAUTION SHOULD BE USED IN INTERPOLATING ON THESE GENERALIZED MAPS. SHARP CHANGES MAY OCCUR IN SHORT DISTANCES, PARTICULARLY IN MOUNTAINOUS AREAS, DUE TO DIFFERENCES IN ALTITUDE, SLOPE OF LAND, TYPE OF SOIL, VEGETATIVE COVER, BODIES OF WATER, AIR DRAINAGE, URBAN HEAT EFFECTS, ETC.
PATTERN TOO COMPLEX IN HAWAII TO INDICATE ON SMALL SCALE MAPS.
THESE CHARTS ARE BASED ON THE PERIOD 1931-60.
ALASKA
INSUFFICIENT DATA FOR ISOLINES
HAWAII
PUERTO RICO AND VIRGIN ISLANDS
ALBERS EQUAL AREA PROJECTION — STANDARD PARALLELS 29½° AND 45½°
PACIFIC OCEAN
ATLANTIC OCEAN
GULF OF MEXICO

NORMAL DAILY MAXIMUM, AVERAGE, MINIMUM, AND EXTREME TEMPERATURES (°F), OCTOBER

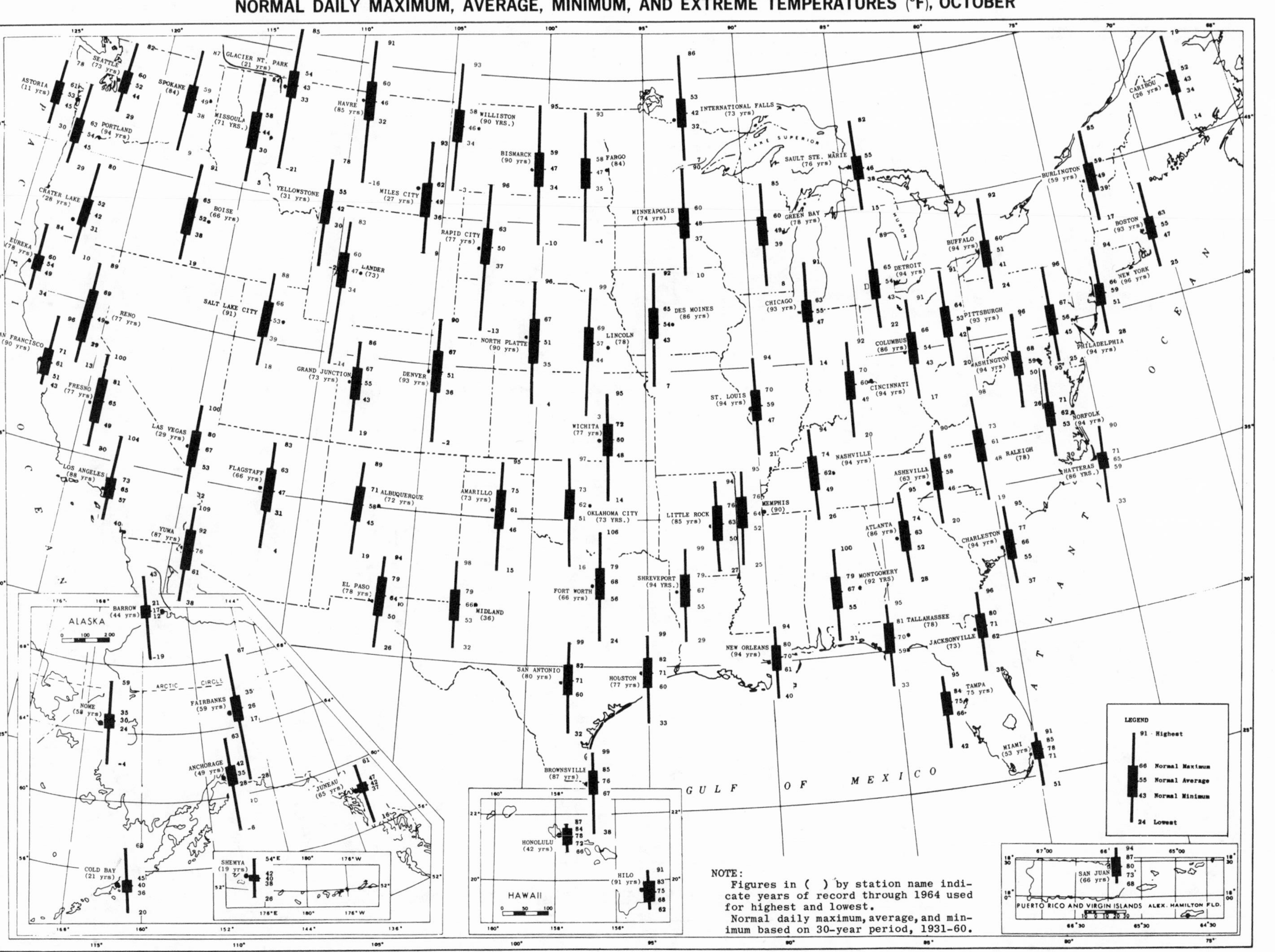

NOTE:
Figures in ( ) by station name indicate years of record through 1964 used for highest and lowest.
Normal daily maximum, average, and minimum based on 30-year period, 1931-60.

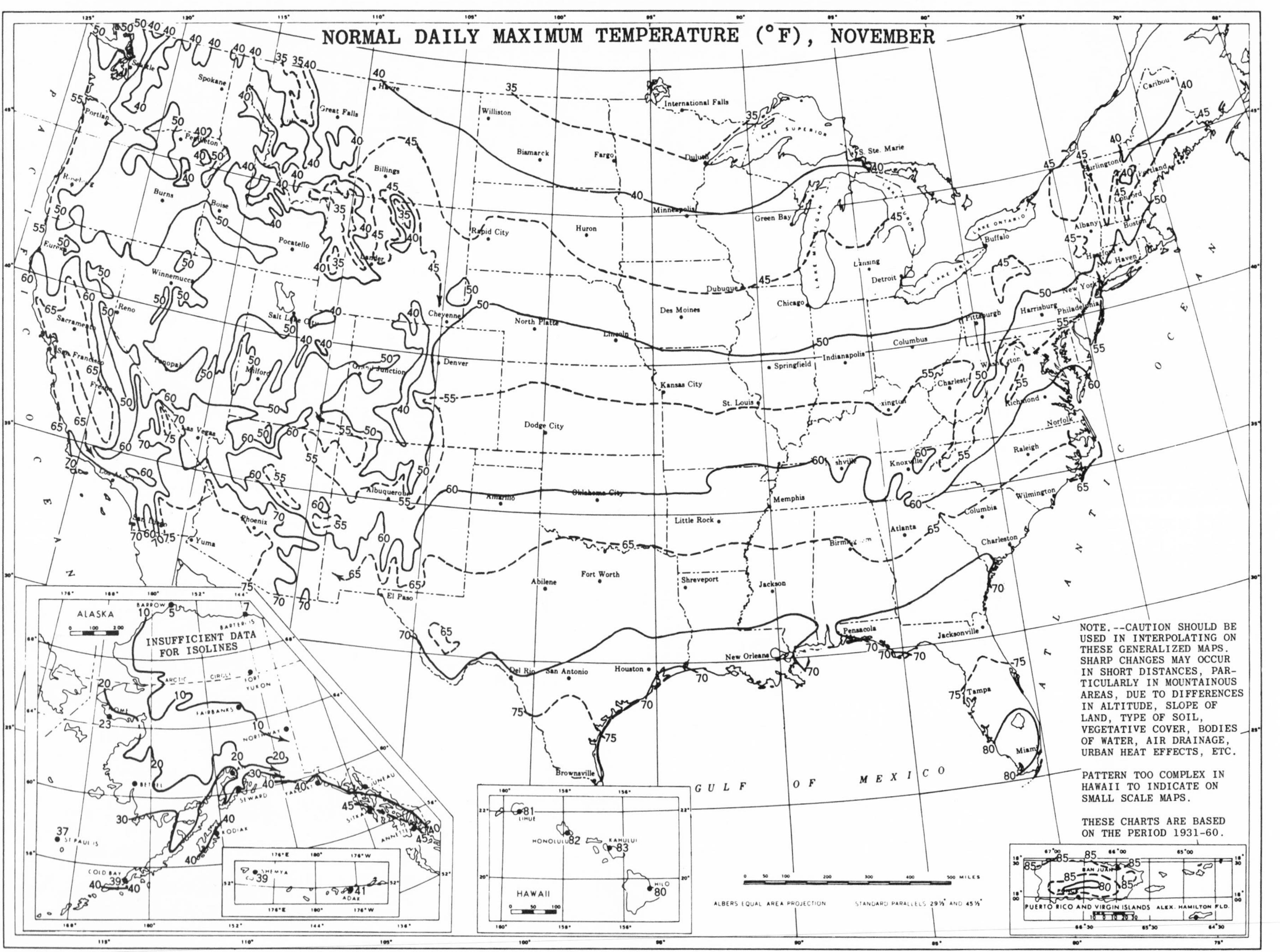
NORMAL DAILY MAXIMUM TEMPERATURE (°F), NOVEMBER
NOTE.--CAUTION SHOULD BE USED IN INTERPOLATING ON THESE GENERALIZED MAPS. SHARP CHANGES MAY OCCUR IN SHORT DISTANCES, PARTICULARLY IN MOUNTAINOUS AREAS, DUE TO DIFFERENCES IN ALTITUDE, SLOPE OF LAND, TYPE OF SOIL, VEGETATIVE COVER, BODIES OF WATER, AIR DRAINAGE, URBAN HEAT EFFECTS, ETC.
PATTERN TOO COMPLEX IN HAWAII TO INDICATE ON SMALL SCALE MAPS.
THESE CHARTS ARE BASED ON THE PERIOD 1931-60.
ALBERS EQUAL AREA PROJECTION
STANDARD PARALLELS 29½° AND 45½°
0 50 100 200 300 400 500 MILES
PACIFIC OCEAN
ATLANTIC OCEAN
GULF OF MEXICO
LAKE SUPERIOR
LAKE MICHIGAN
LAKE HURON
LAKE ERIE
LAKE ONTARIO
Seattle
Spokane
Portland
Pendleton
Roseburg
Burns
Boise
Eureka
Pocatello
Winnemucca
Reno
Sacramento
San Francisco
Fresno
Tonopah
Milford
Salt Lake City
Grand Junction
Las Vegas
Los Angeles
San Diego
Yuma
Phoenix
Albuquerque
El Paso
Great Falls
Havre
Billings
Lander
Cheyenne
Denver
Williston
Bismarck
Rapid City
Fargo
Huron
North Platte
Dodge City
Amarillo
Abilene
Fort Worth
Del Rio
San Antonio
Brownsville
Houston
Oklahoma City
Lincoln
Kansas City
Des Moines
International Falls
Duluth
Minneapolis
Green Bay
Dubuque
Chicago
Springfield
St. Louis
Little Rock
Shreveport
Memphis
Jackson
New Orleans
Sault Ste. Marie
Lansing
Detroit
Indianapolis
Columbus
Pittsburgh
Buffalo
Charleston
Washington
Lexington
Nashville
Knoxville
Birmingham
Atlanta
Pensacola
Jacksonville
Tampa
Miami
Columbia
Charleston
Wilmington
Raleigh
Norfolk
Richmond
Harrisburg
Philadelphia
New York
New Haven
Hartford
Albany
Boston
Concord
Burlington
Portland
Caribou
ALASKA
INSUFFICIENT DATA FOR ISOLINES
BARROW
BARTER IS.
ARCTIC CIRCLE
FORT YUKON
FAIRBANKS
NORTHWAY
NOME
BETHEL
SEWARD
YAKUTAT
JUNEAU
SITKA
ANNETTE
KODIAK
COLD BAY
ST PAUL IS
SHEMYA
ADAK
HAWAII
LIHUE
HONOLULU
KAHULUI
HILO
PUERTO RICO AND VIRGIN ISLANDS
SAN JUAN
ALEX. HAMILTON FLD.

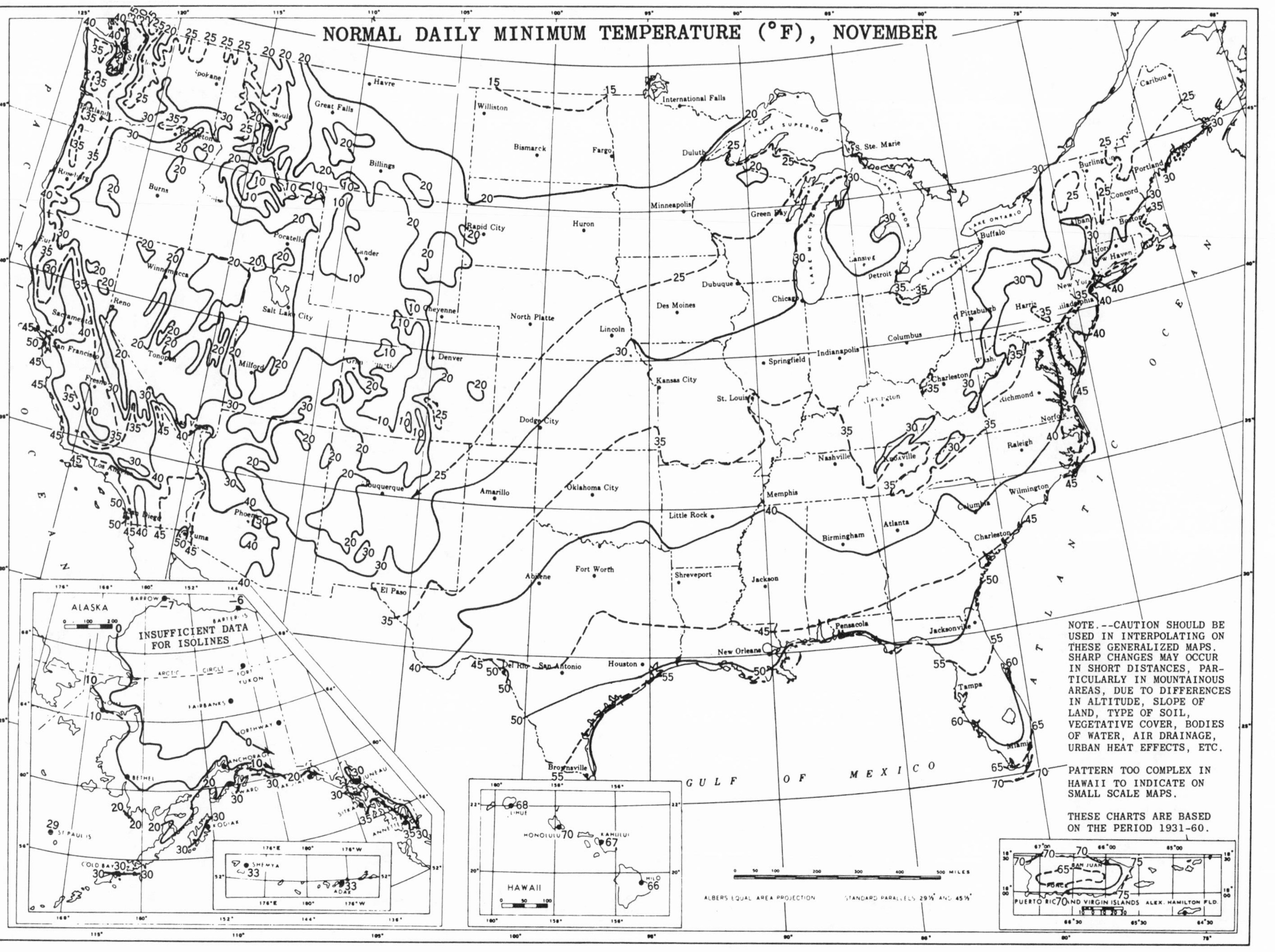
NORMAL DAILY MINIMUM TEMPERATURE (°F), NOVEMBER
NOTE.--CAUTION SHOULD BE USED IN INTERPOLATING ON THESE GENERALIZED MAPS. SHARP CHANGES MAY OCCUR IN SHORT DISTANCES, PARTICULARLY IN MOUNTAINOUS AREAS, DUE TO DIFFERENCES IN ALTITUDE, SLOPE OF LAND, TYPE OF SOIL, VEGETATIVE COVER, BODIES OF WATER, AIR DRAINAGE, URBAN HEAT EFFECTS, ETC.
PATTERN TOO COMPLEX IN HAWAII TO INDICATE ON SMALL SCALE MAPS.
THESE CHARTS ARE BASED ON THE PERIOD 1931-60.
PACIFIC OCEAN
ATLANTIC OCEAN
GULF OF MEXICO
ALASKA
INSUFFICIENT DATA FOR ISOLINES
HAWAII
PUERTO RICO AND VIRGIN ISLANDS
ALEX. HAMILTON FLD.
ALBERS EQUAL AREA PROJECTION
STANDARD PARALLELS 29½° AND 45½°
MILES
Spokane
Havre
Great Falls
Billings
Williston
Bismarck
Fargo
International Falls
Duluth
S. Ste. Marie
Caribou
Burns
Winnemucca
Reno
Sacramento
San Francisco
Fresno
Tonopah
Salt Lake City
Pocatello
Lander
Rapid City
Huron
Minneapolis
Green Bay
Milford
Cheyenne
Denver
North Platte
Lincoln
Des Moines
Dubuque
Chicago
Detroit
Buffalo
Albany
Burlington
Concord
Portland
Boston
Hartford
New Haven
New York
Philadelphia
Harrisburg
Pittsburgh
Columbus
Indianapolis
Springfield
Kansas City
St. Louis
Charleston
Richmond
Norfolk
Raleigh
Nashville
Knoxville
Dodge City
Albuquerque
Amarillo
Oklahoma City
Memphis
Little Rock
Birmingham
Atlanta
Wilmington
Columbia
Phoenix
Yuma
San Diego
Los Angeles
Las Vegas
El Paso
Abilene
Fort Worth
Shreveport
Jackson
Del Rio
San Antonio
Houston
New Orleans
Pensacola
Jacksonville
Tampa
Miami
Brownsville
Barrow
Barter Is.
Fort Yukon
Fairbanks
Northway
Arctic Circle
Bethel
Anchorage
Juneau
Sitka
Kodiak
Annette
St. Paul Is.
Cold Bay
Shemya
Adak
Lihue
Honolulu
Kahului
Hilo
San Juan
Ponce

NORMAL DAILY AVERAGE TEMPERATURE (°F), NOVEMBER

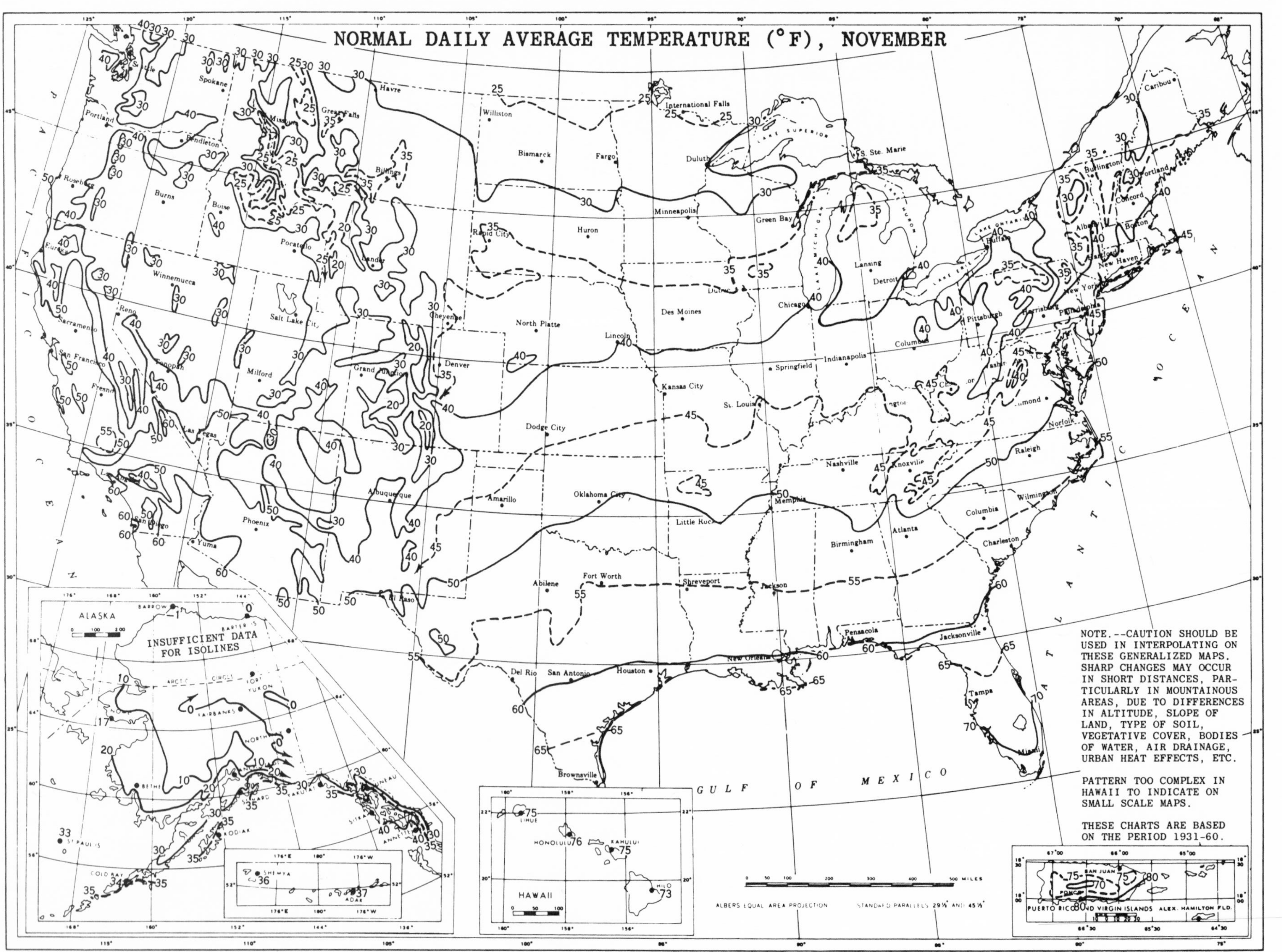

PACIFIC OCEAN
ATLANTIC OCEAN
GULF OF MEXICO
ALASKA
INSUFFICIENT DATA FOR ISOLINES
HAWAII
PUERTO RICO AND VIRGIN ISLANDS
ALBERS EQUAL AREA PROJECTION
STANDARD PARALLELS 29½° AND 45½°
NOTE.--CAUTION SHOULD BE USED IN INTERPOLATING ON THESE GENERALIZED MAPS. SHARP CHANGES MAY OCCUR IN SHORT DISTANCES, PARTICULARLY IN MOUNTAINOUS AREAS, DUE TO DIFFERENCES IN ALTITUDE, SLOPE OF LAND, TYPE OF SOIL, VEGETATIVE COVER, BODIES OF WATER, AIR DRAINAGE, URBAN HEAT EFFECTS, ETC.
PATTERN TOO COMPLEX IN HAWAII TO INDICATE ON SMALL SCALE MAPS.
THESE CHARTS ARE BASED ON THE PERIOD 1931-60.

NORMAL DAILY RANGE OF TEMPERATURE (°F), NOVEMBER

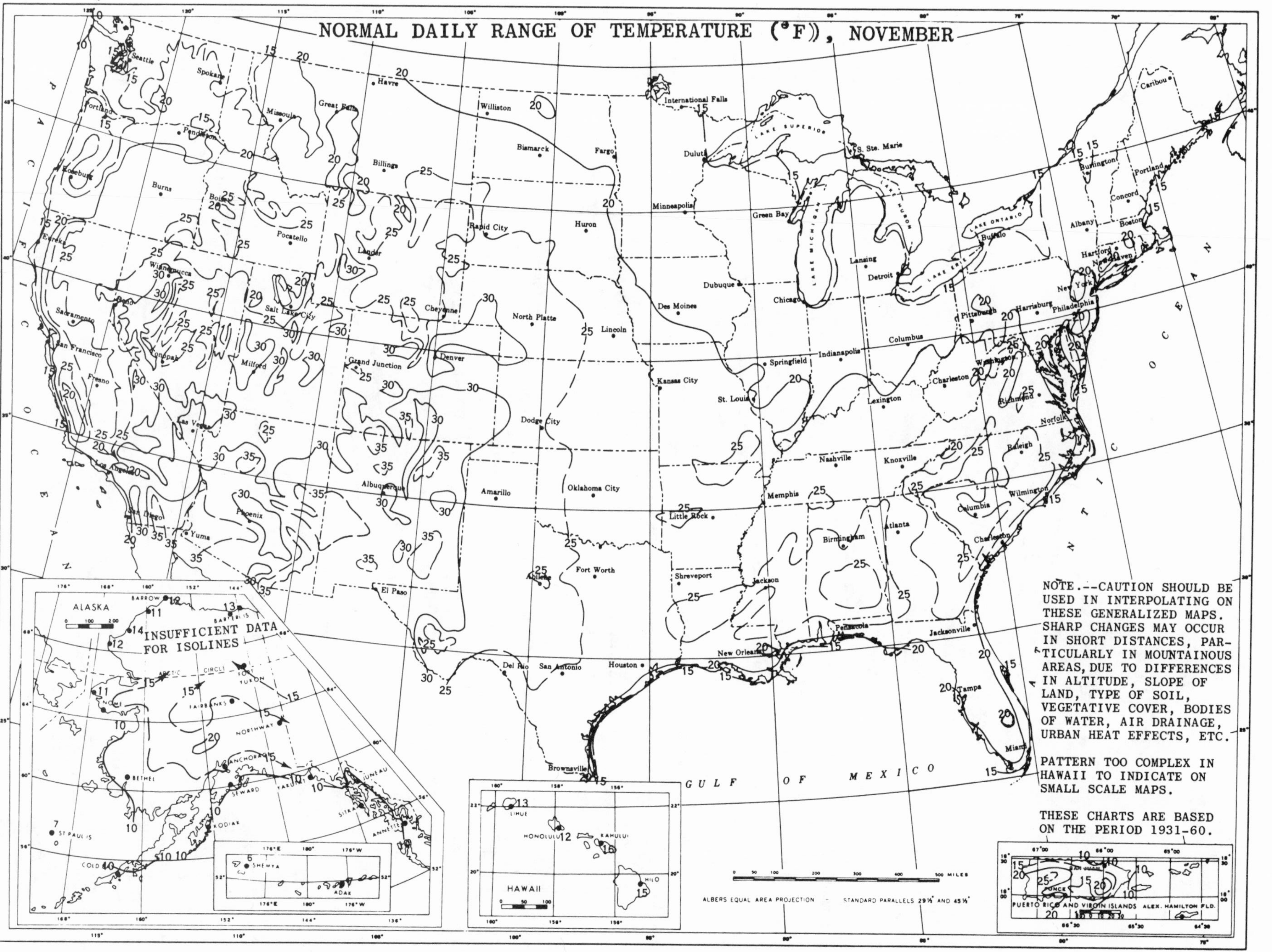

NORMAL DAILY RANGE OF TEMPERATURE (°F), NOVEMBER
NOTE.--CAUTION SHOULD BE USED IN INTERPOLATING ON THESE GENERALIZED MAPS. SHARP CHANGES MAY OCCUR IN SHORT DISTANCES, PARTICULARLY IN MOUNTAINOUS AREAS, DUE TO DIFFERENCES IN ALTITUDE, SLOPE OF LAND, TYPE OF SOIL, VEGETATIVE COVER, BODIES OF WATER, AIR DRAINAGE, URBAN HEAT EFFECTS, ETC.
PATTERN TOO COMPLEX IN HAWAII TO INDICATE ON SMALL SCALE MAPS.
THESE CHARTS ARE BASED ON THE PERIOD 1931-60.
PACIFIC OCEAN
ATLANTIC OCEAN
GULF OF MEXICO
ALASKA
INSUFFICIENT DATA FOR ISOLINES
HAWAII
PUERTO RICO AND VIRGIN ISLANDS
ALEX. HAMILTON FLD.
ALBERS EQUAL AREA PROJECTION - STANDARD PARALLELS 29½ AND 45½
MILES

NORMAL DAILY MAXIMUM, AVERAGE, MINIMUM, AND EXTREME TEMPERATURES (°F), NOVEMBER

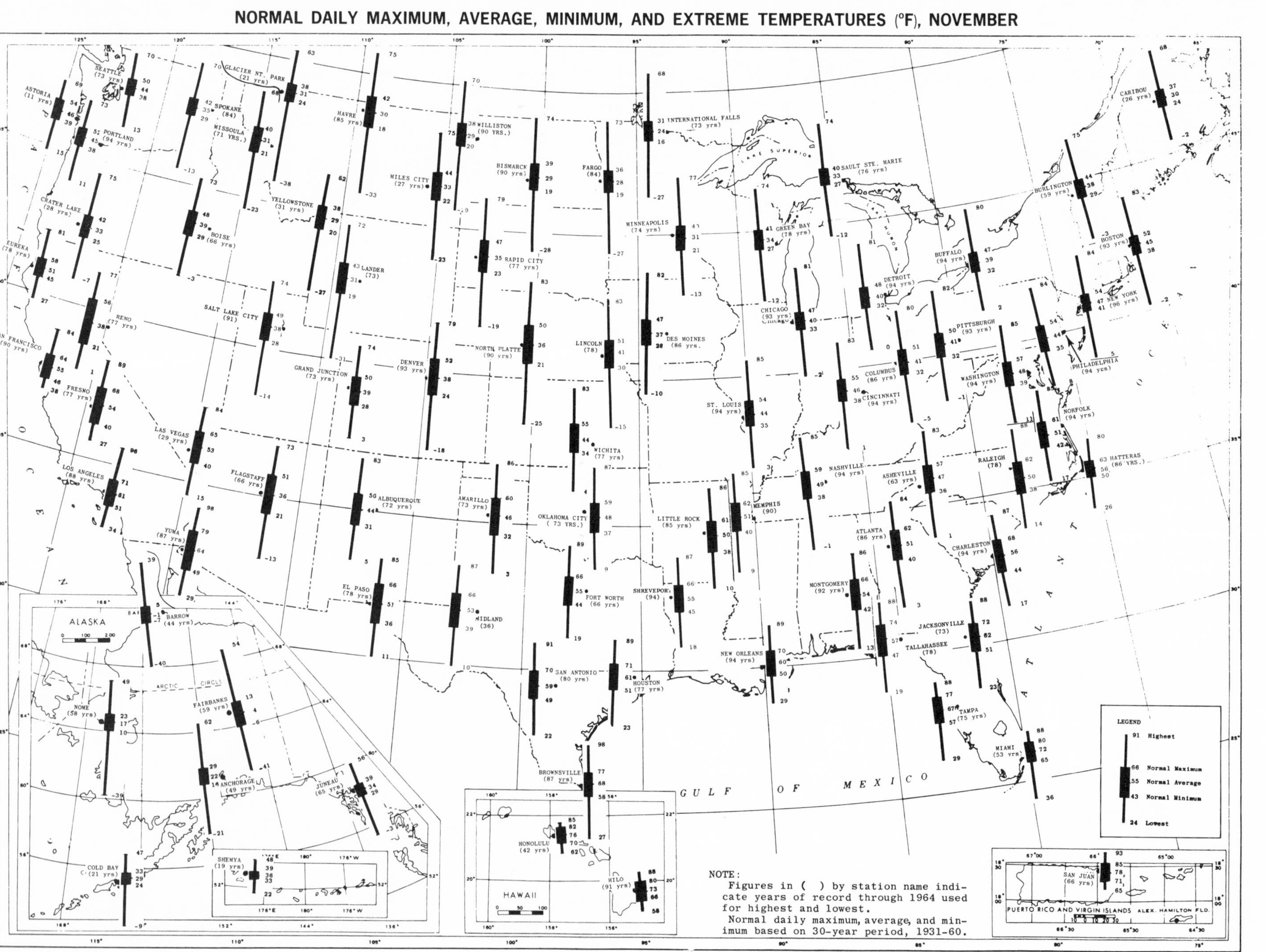

NOTE:
Figures in ( ) by station name indicate years of record through 1964 used for highest and lowest.
Normal daily maximum, average, and minimum based on 30-year period, 1931-60.

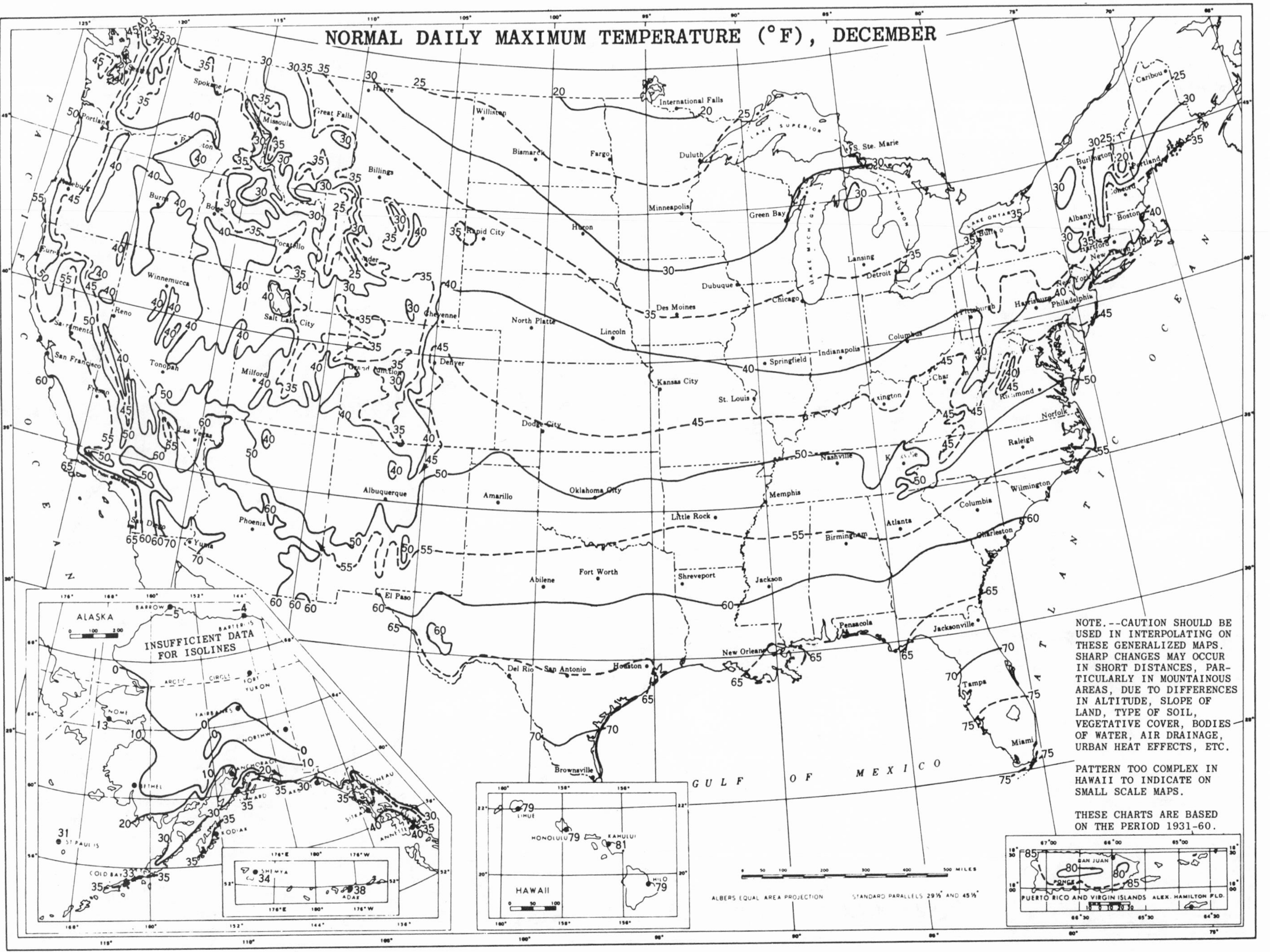
NORMAL DAILY MAXIMUM TEMPERATURE (°F), DECEMBER
NOTE.--CAUTION SHOULD BE USED IN INTERPOLATING ON THESE GENERALIZED MAPS. SHARP CHANGES MAY OCCUR IN SHORT DISTANCES, PARTICULARLY IN MOUNTAINOUS AREAS, DUE TO DIFFERENCES IN ALTITUDE, SLOPE OF LAND, TYPE OF SOIL, VEGETATIVE COVER, BODIES OF WATER, AIR DRAINAGE, URBAN HEAT EFFECTS, ETC.
PATTERN TOO COMPLEX IN HAWAII TO INDICATE ON SMALL SCALE MAPS.
THESE CHARTS ARE BASED ON THE PERIOD 1931-60.
PACIFIC OCEAN
ATLANTIC OCEAN
GULF OF MEXICO
ALASKA
INSUFFICIENT DATA FOR ISOLINES
HAWAII
PUERTO RICO AND VIRGIN ISLANDS
ALBERS EQUAL AREA PROJECTION
STANDARD PARALLELS 29½° AND 45½°

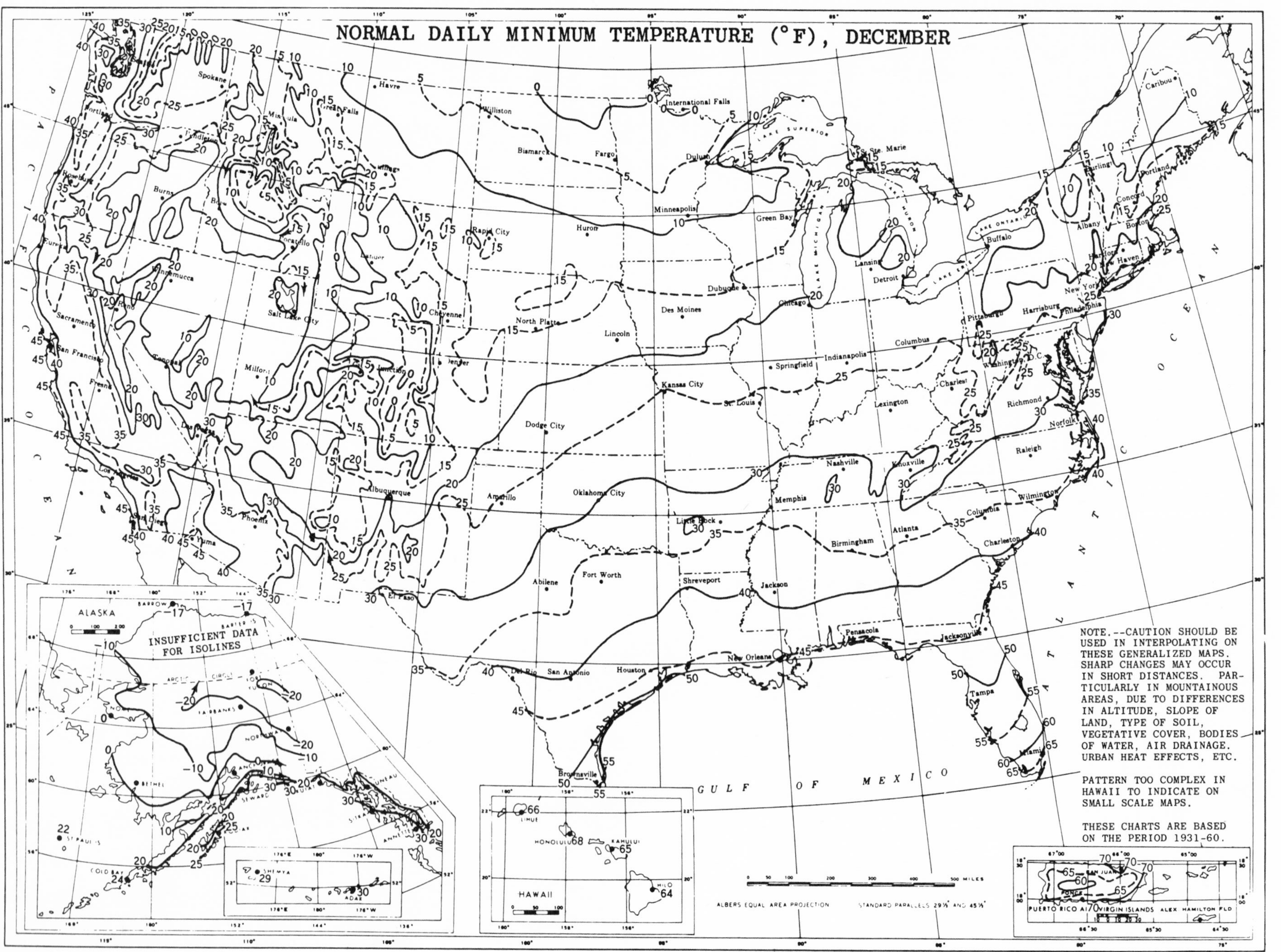
NORMAL DAILY MINIMUM TEMPERATURE (°F), DECEMBER
NOTE.--CAUTION SHOULD BE USED IN INTERPOLATING ON THESE GENERALIZED MAPS. SHARP CHANGES MAY OCCUR IN SHORT DISTANCES. PARTICULARLY IN MOUNTAINOUS AREAS, DUE TO DIFFERENCES IN ALTITUDE, SLOPE OF LAND, TYPE OF SOIL, VEGETATIVE COVER, BODIES OF WATER, AIR DRAINAGE, URBAN HEAT EFFECTS, ETC.
PATTERN TOO COMPLEX IN HAWAII TO INDICATE ON SMALL SCALE MAPS.
THESE CHARTS ARE BASED ON THE PERIOD 1931-60.
ALASKA
INSUFFICIENT DATA FOR ISOLINES
HAWAII
PUERTO RICO AND VIRGIN ISLANDS
ALBERS EQUAL AREA PROJECTION
STANDARD PARALLELS 29½° AND 45½°
GULF OF MEXICO
PACIFIC OCEAN
ATLANTIC OCEAN
Spokane
Portland
Havre
Williston
Bismarck
Fargo
International Falls
Duluth
Sault Ste. Marie
Caribou
Burlington
Portland
Concord
Albany
Boston
Hartford
New Haven
New York
Philadelphia
Harrisburg
Buffalo
Pittsburgh
Washington, D.C.
Richmond
Norfolk
Raleigh
Wilmington
Columbia
Charleston
Jacksonville
Tampa
Miami
Atlanta
Birmingham
Pensacola
New Orleans
Jackson
Memphis
Nashville
Knoxville
Lexington
Charleston
Columbus
Indianapolis
Springfield
St. Louis
Chicago
Detroit
Lansing
Green Bay
Minneapolis
Dubuque
Des Moines
Kansas City
Lincoln
Huron
North Platte
Rapid City
Cheyenne
Denver
Dodge City
Amarillo
Oklahoma City
Little Rock
Shreveport
Fort Worth
Abilene
El Paso
Del Rio
San Antonio
Houston
Brownsville
Albuquerque
Phoenix
Yuma
San Diego
Los Angeles
Fresno
San Francisco
Sacramento
Eureka
Roseburg
Reno
Winnemucca
Burns
Boise
Pendleton
Missoula
Great Falls
Billings
Pocatello
Salt Lake City
Milford
Tonopah
Grand Junction
Lake Superior
Lake Michigan
Lake Huron
Lake Erie
Lake Ontario
Barrow
Barter Is.
Arctic Circle
Fairbanks
Nome
Bethel
Anchorage
Seward
Juneau
Sitka
Annette
Cold Bay
St. Paul Is.
Shemya
Adak
Lihue
Honolulu
Kahului
Hilo
San Juan

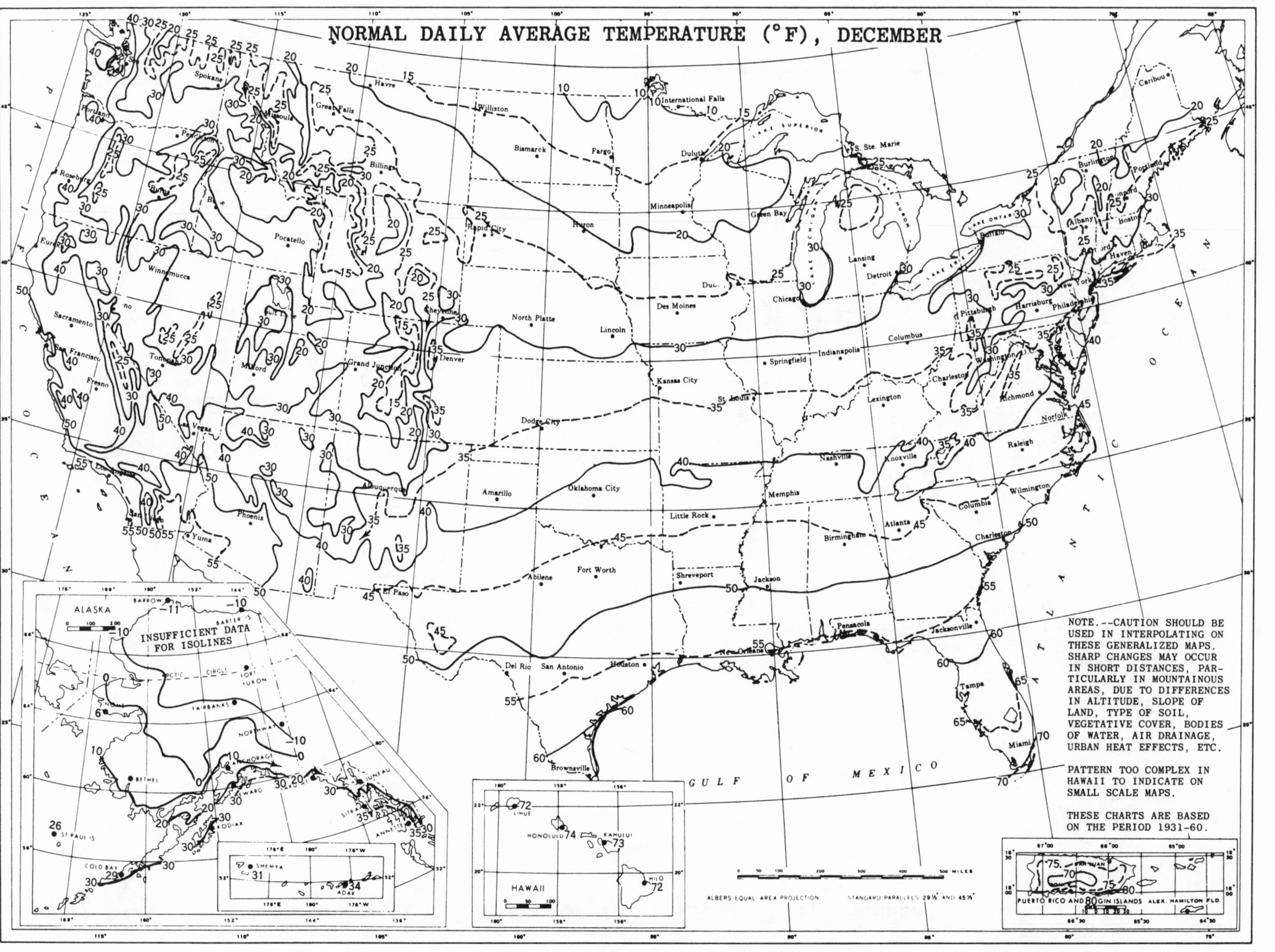
NORMAL DAILY AVERAGE TEMPERATURE (°F), DECEMBER
NOTE.--CAUTION SHOULD BE USED IN INTERPOLATING ON THESE GENERALIZED MAPS. SHARP CHANGES MAY OCCUR IN SHORT DISTANCES, PARTICULARLY IN MOUNTAINOUS AREAS, DUE TO DIFFERENCES IN ALTITUDE, SLOPE OF LAND, TYPE OF SOIL, VEGETATIVE COVER, BODIES OF WATER, AIR DRAINAGE, URBAN HEAT EFFECTS, ETC.
PATTERN TOO COMPLEX IN HAWAII TO INDICATE ON SMALL SCALE MAPS.
THESE CHARTS ARE BASED ON THE PERIOD 1931-60.
ALASKA
INSUFFICIENT DATA FOR ISOLINES
HAWAII
PUERTO RICO AND VIRGIN ISLANDS
ALBERS EQUAL AREA PROJECTION
PACIFIC OCEAN
ATLANTIC OCEAN
GULF OF MEXICO

NORMAL DAILY RANGE OF TEMPERATURE (°F), DECEMBER

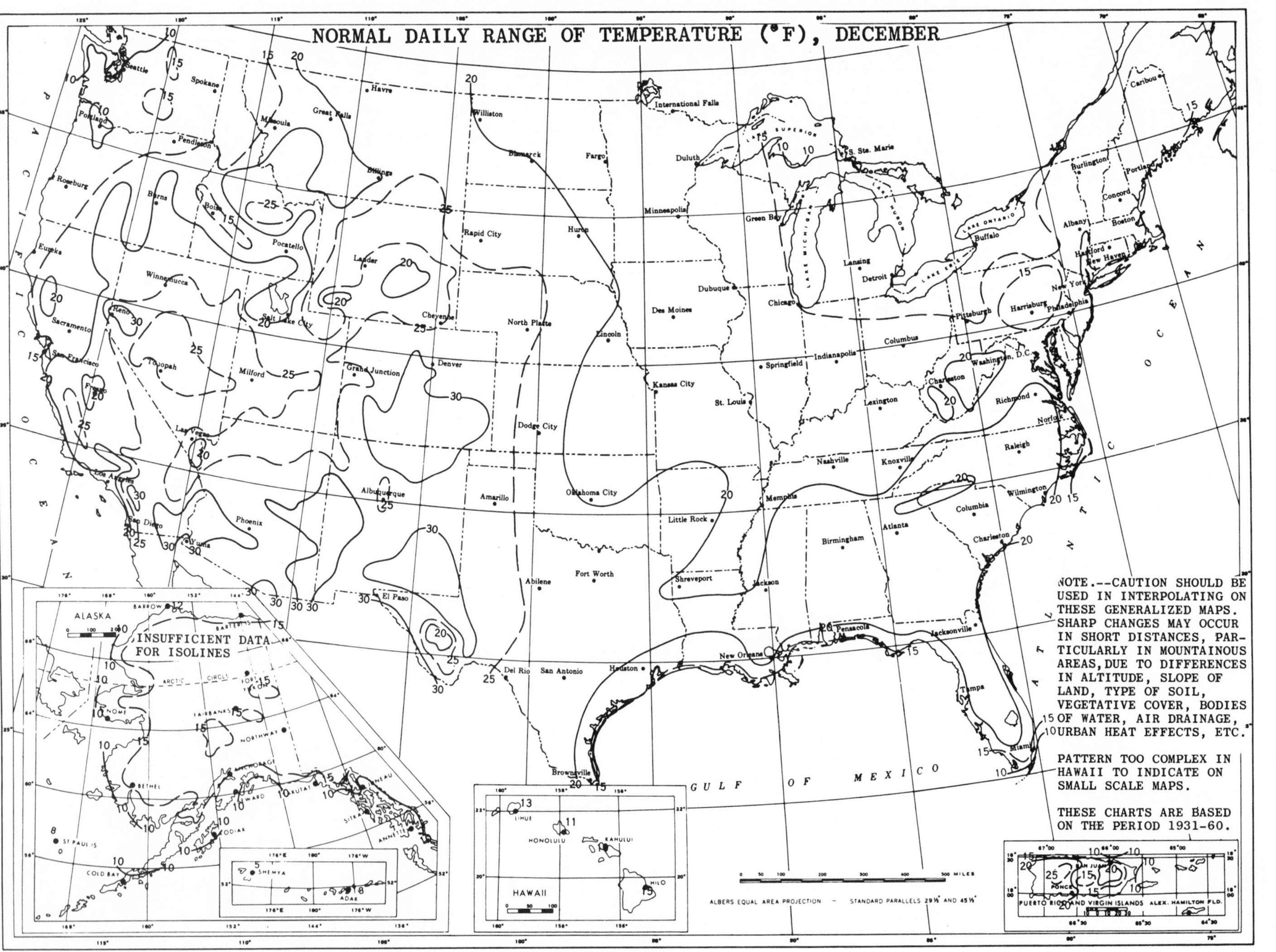
NOTE.--CAUTION SHOULD BE USED IN INTERPOLATING ON THESE GENERALIZED MAPS. SHARP CHANGES MAY OCCUR IN SHORT DISTANCES, PARTICULARLY IN MOUNTAINOUS AREAS, DUE TO DIFFERENCES IN ALTITUDE, SLOPE OF LAND, TYPE OF SOIL, VEGETATIVE COVER, BODIES OF WATER, AIR DRAINAGE, URBAN HEAT EFFECTS, ETC.
PATTERN TOO COMPLEX IN HAWAII TO INDICATE ON SMALL SCALE MAPS.
THESE CHARTS ARE BASED ON THE PERIOD 1931-60.
ALASKA
INSUFFICIENT DATA FOR ISOLINES
HAWAII
PUERTO RICO AND VIRGIN ISLANDS
ALBERS EQUAL AREA PROJECTION - STANDARD PARALLELS 29½° AND 45½°
PACIFIC OCEAN
ATLANTIC OCEAN
GULF OF MEXICO

NORMAL DAILY MAXIMUM, AVERAGE, MINIMUM, AND EXTREME TEMPERATURES (°F), DECEMBER

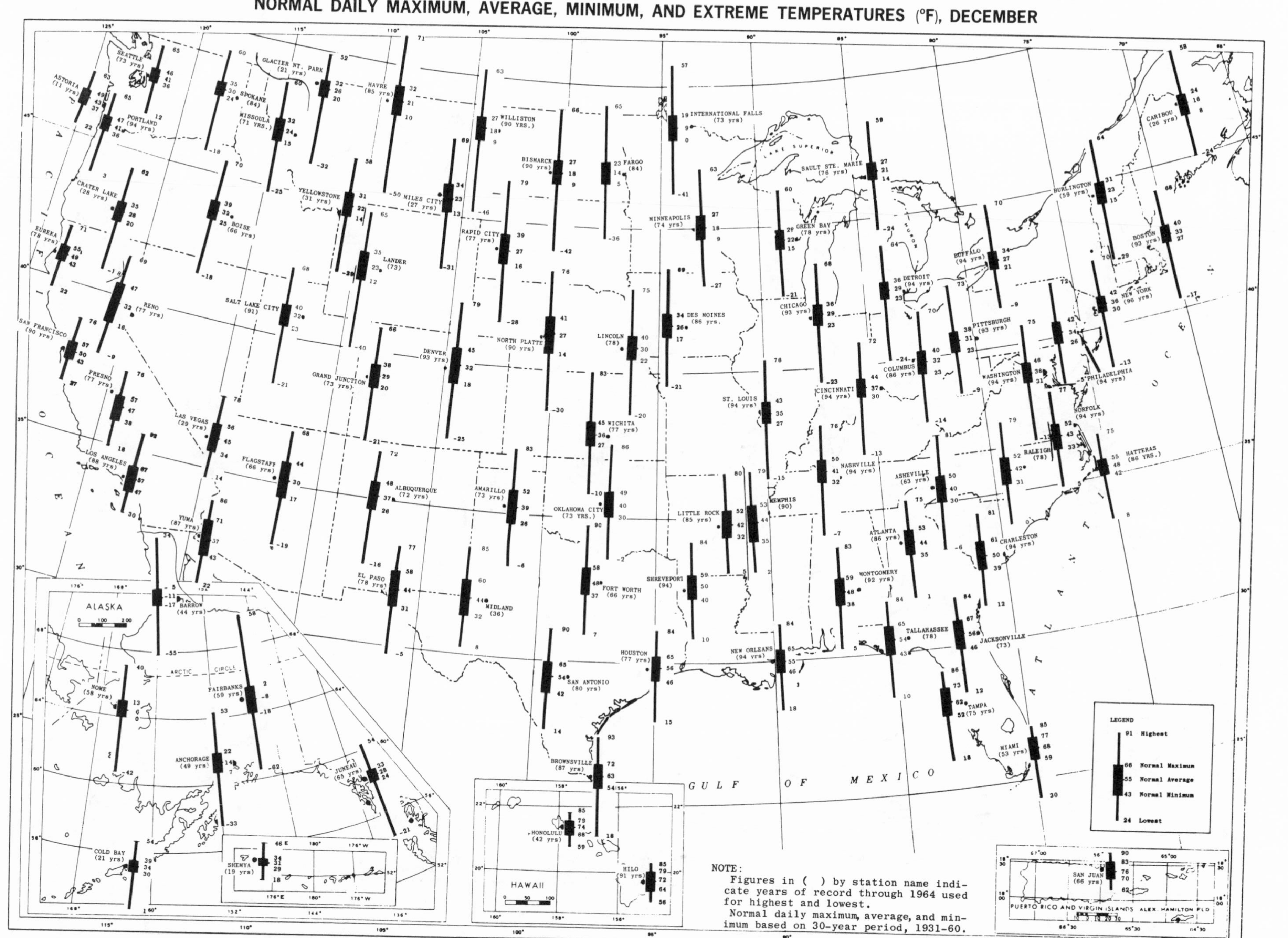

NOTE:
Figures in ( ) by station name indicate years of record through 1964 used for highest and lowest.
Normal daily maximum, average, and minimum based on 30-year period, 1931-60.

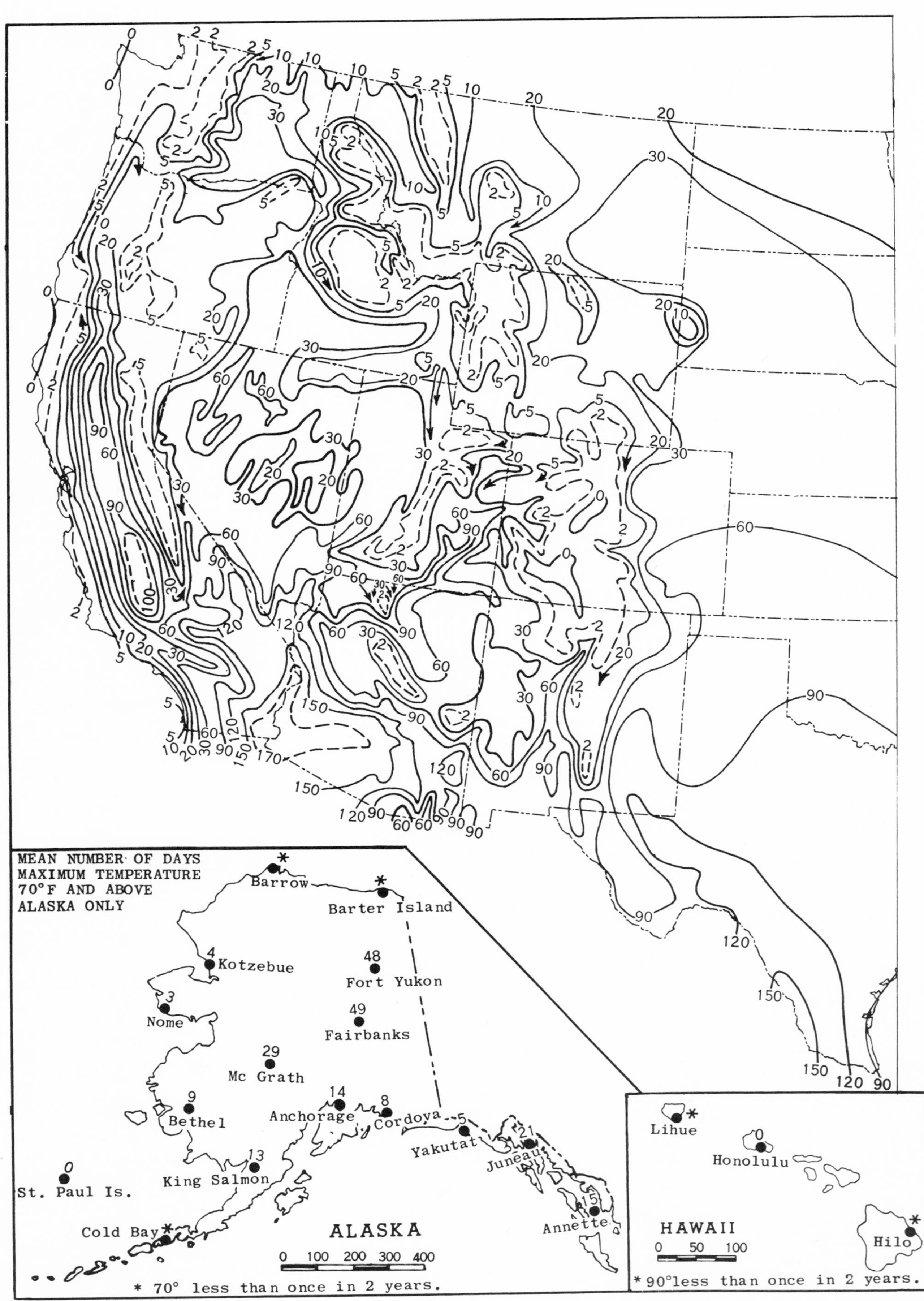
MEAN NUMBER OF DAYS
MAXIMUM TEMPERATURE
70°F AND ABOVE
ALASKA ONLY
Barrow
Barter Island
4 Kotzebue
48 Fort Yukon
3 Nome
49 Fairbanks
29 Mc Grath
9 Bethel
14 Anchorage
8 Cordova
5 Yakutat
21 Juneau
13 King Salmon
0 St. Paul Is.
15 Annette
Cold Bay
ALASKA
0 100 200 300 400
* 70° less than once in 2 years.
Lihue
0 Honolulu
Hilo
HAWAII
0 50 100
* 90° less than once in 2 years.

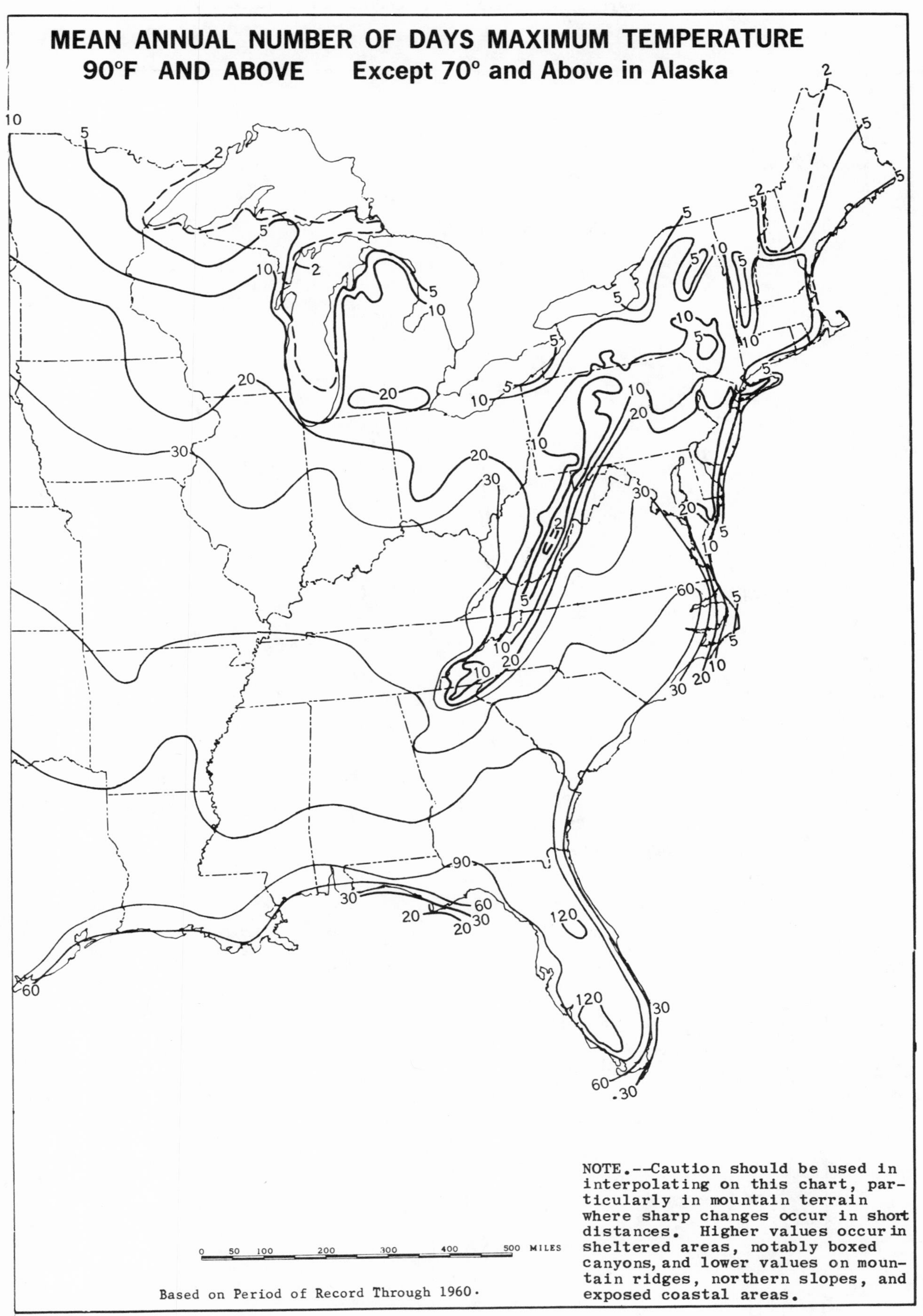

MEAN ANNUAL NUMBER OF DAYS MAXIMUM TEMPERATURE 90°F AND ABOVE Except 70° and Above in Alaska

Based on Period of Record Through 1960.

NOTE.--Caution should be used in interpolating on this chart, particularly in mountain terrain where sharp changes occur in short distances. Higher values occur in sheltered areas, notably boxed canyons, and lower values on mountain ridges, northern slopes, and exposed coastal areas.

MEAN MONTHLY NUMBER OF DAYS MAXIMUM TEMPERATURE
For Selected

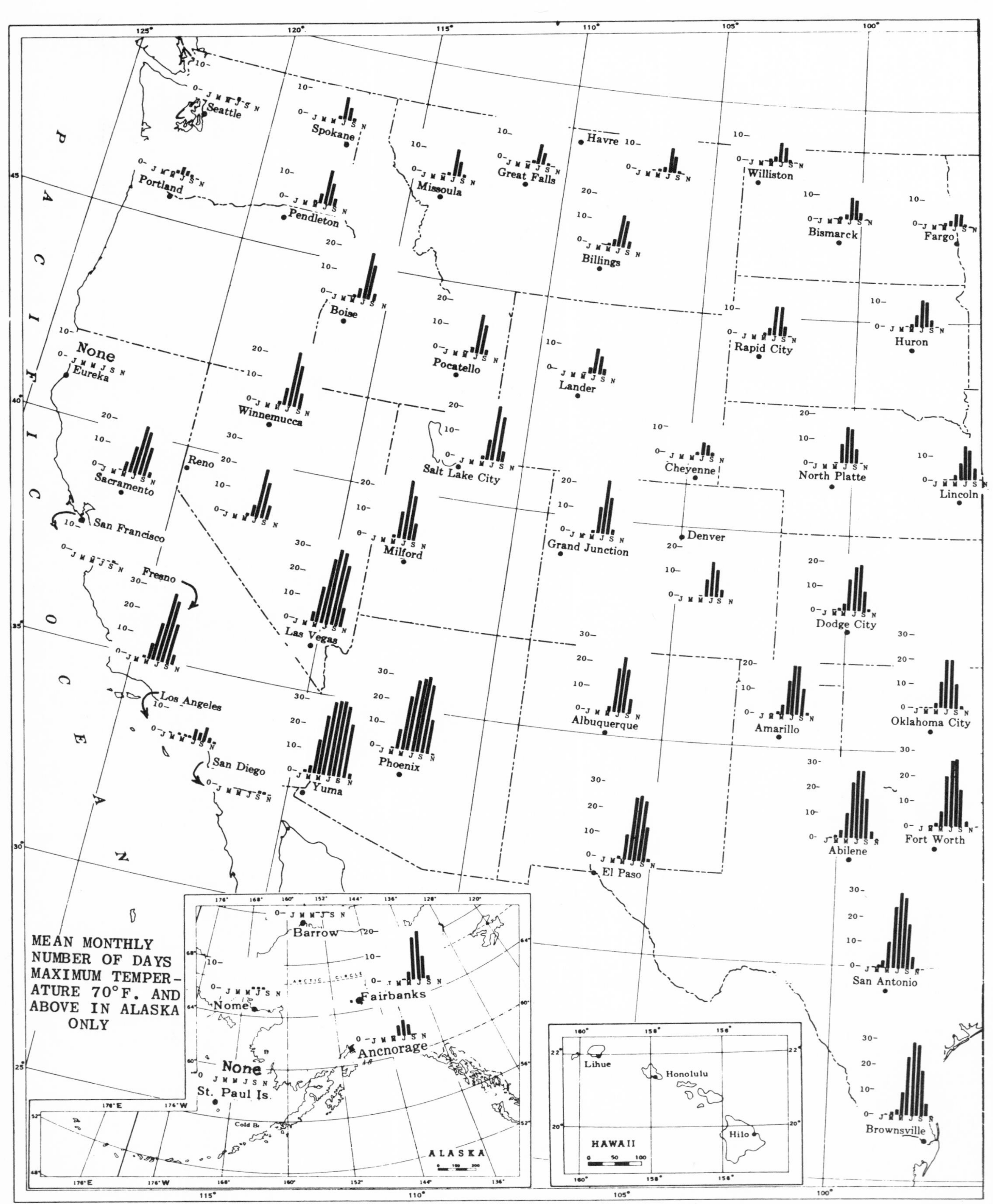

90°F. AND ABOVE EXCEPT 70°F. AND ABOVE IN ALASKA

Stations

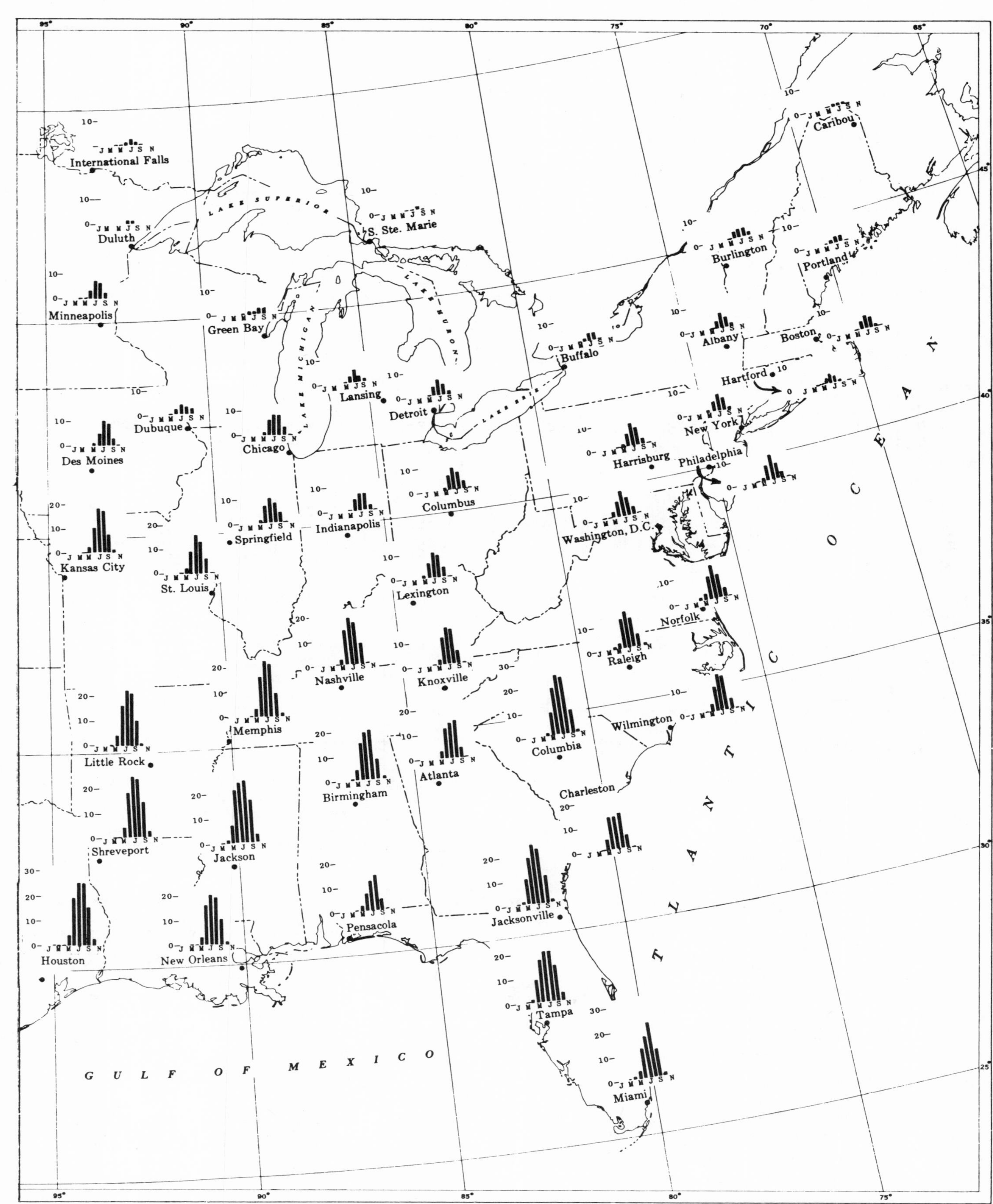

MEAN NUMBER OF DAYS MAXIMUM TEMPERATURE 90°F. AND ABOVE EXCEPT 70°F. AND ABOVE IN ALASKA

States and Stations	Yrs.	Jan.	Feb.	Mar.	Apr.	May	June	July	Aug.	Sept.	Oct.	Nov.	Dec.	Annual
ALA. BIRMINGHAM	17	0	0	0	*	4	15	19	20	8	1	0	0	67
MOBILE	19	0	0	*	*	5	17	18	20	9	1	0	0	70
MONTGOMERY	16	0	0	0	*	8	19	22	24	13	1	0	0	87
ARIZ. FLAGSTAFF U	43	0	0	0	0	0	1	1	*	*	0	0	0	2
PHOENIX	21	0	0	*	8	22	29	30	31	28	14	*	0	163
PRESCOTT	18	0	0	0	0	1	10	17	11	8	*	0	0	48
TUCSON	20	0	*	*	5	18	28	29	29	26	10	*	0	146
WINSLOW	29	0	0	0	*	2	18	25	18	9	*	0	0	72
YUMA	10	0	1	3	14	24	30	31	31	29	22	2	0	187
ARK. FORT SMITH	15	0	0	*	*	4	17	24	24	14	1	0	0	84
LITTLE ROCK	19	0	0	0	*	4	16	22	21	10	1	0	0	74
TEXARKANA	18	0	0	*	*	3	19	25	24	13	1	0	0	85
CALIF. BAKERSFIELD	23	0	0	0	2	9	19	30	27	18	5	*	0	110
BISHOP	13	0	0	0	1	5	19	30	28	17	1	0	0	101
BLUE CANYON	17	0	0	0	0	0	*	*	*	*	0	0	0	1
BURBANK	18	0	*	0	1	2	3	12	12	13	4	1	*	47
EUREKA	50	0	0	0	0	0	0	0	0	0	0	0	0	0
FRESNO	21	0	0	0	1	7	16	29	26	17	4	0	0	100
LONG BEACH	17	0	0	0	*	1	*	2	2	3	2	1	*	11
LOS ANGELES U	20	0	*	0	1	1	1	4	3	6	2	1	0	19
MT. SHASTA	18	0	0	0	0	*	1	8	6	4	*	0	0	20
OAKLAND	30	0	0	0	0	*	1	*	*	2	*	0	0	4
RED BLUFF	16	0	0	*	2	7	15	28	25	17	4	*	0	98
SACRAMENTO	27	0	0	0	*	4	11	20	18	12	2	0	0	67
SANDBERG	28	0	0	0	0	*	2	8	7	4	0	0	0	21
SAN DIEGO	20	0	0	0	*	*	*	*	*	1	1	*	0	3
SAN FRANCISCO	24	0	0	0	0	*	*	*	*	1	*	0	0	1
SANTA MARIA	18	0	0	0	*	*	*	*	*	1	1	1	*	4
COLO. ALAMOSA	15	0	0	0	0	0	*	1	0	0	0	0	0	1
COLORADO SPRINGS	12	0	0	0	0	0	4	8	5	2	0	0	0	19
DENVER	26	0	0	0	0	*	7	14	11	3	0	0	0	35
GRAND JUNCTION U	64	0	0	0	0	1	11	22	15	2	0	0	0	51
PUEBLO	20	0	0	0	0	1	12	19	17	7	*	0	0	57
CONN. BRIDGEPORT	12	0	0	0	0	0	2	4	2	*	0	0	0	8
HARTFORD	51	0	0	0	*	*	2	4	3	1	*	0	0	10
NEW HAVEN	17	0	0	0	0	0	1	1	1	*	0	0	0	3
DEL. WILMINGTON	13	0	0	0	0	1	5	9	5	1	*	0	0	21
D.C. WASHINGTON U	88	0	0	*	*	2	6	10	7	3	*	0	0	28
FLA. APALACHICOLA	31	0	0	0	*	1	4	5	5	2	*	0	0	17
DAYTONA BEACH	17	0	0	*	1	6	13	17	16	8	1	0	0	62
EVERGLADES	26	0	*	1	5	15	25	27	28	22	9	1	0	133
FORT MYERS	20	0	*	1	4	16	22	25	27	21	5	*	0	121
JACKSONVILLE	19	0	0	*	1	10	19	24	22	11	1	0	0	88
KEY WEST	12	0	0	0	0	1	10	19	20	10	1	0	0	61
KEY WEST U	87	0	0	0	*	1	5	10	12	5	*	*	0	33
LAKELAND U	20	0	0	*	1	9	18	21	22	12	2	0	0	85
MIAMI	18	0	0	*	1	3	12	17	22	11	1	0	0	67
MIAMI BEACH	17	0	0	*	*	1	2	2	6	2	*	0	0	13
ORLANDO	18	0	0	1	3	13	21	24	25	17	3	0	0	107
PENSACOLA	21	0	0	0	*	2	7	12	14	4	*	0	0	39
TALLAHASSEE	21	0	0	0	1	9	18	20	21	11	1	0	0	81
TAMPA	14	0	0	*	1	9	17	20	20	14	3	0	0	84
WEST PALM BEACH	20	0	*	1	2	5	16	23	24	15	3	*	*	89
GA. ATHENS	17	0	0	0	*	4	14	16	17	6	*	0	0	57
ATLANTA	12	0	0	0	0	3	12	14	15	4	*	0	0	48
AUGUSTA	10	0	0	0	1	9	19	24	25	9	1	0	0	88
COLUMBUS	15	0	0	0	*	8	18	21	22	10	1	0	0	80
MACON	12	0	0	0	1	12	21	24	25	12	2	0	0	97
ROME	15	0	0	0	*	4	15	20	21	8	1	0	0	69
SAVANNAH	10	0	0	*	1	7	15	20	21	6	1	0	0	71
THOMASVILLE	36	0	0	*	*	7	17	19	20	12	1	0	0	76
HAWAII HILO	15	0	0	0	0	0	0	0	*	*	*	0	0	*
HONOLULU	38	0	0	0	0	0	0	0	0	0	0	0	0	0
LIHUE	11	0	0	0	0	0	0	0	0	0	*	0	0	*
IDAHO BOISE	21	0	0	0	*	1	4	19	14	3	0	0	0	41
IDAHO FALLS	11	0	0	0	0	*	2	13	10	2	0	0	0	27
LEWISTON	14	0	0	0	0	1	3	16	12	4	0	0	0	36
POCATELLO	22	0	0	0	0	*	2	16	12	2	0	0	0	32
ILL. CAIRO U	18	0	0	0	0	2	13	18	17	5	*	0	0	55
CHICAGO	18	0	0	0	0	1	6	8	8	3	*	0	0	26
CHICAGO U	85	0	0	0	*	1	3	5	3	1	*	0	0	13
MOLINE	26	0	0	0	*	1	6	10	8	3	*	0	0	28
PEORIA	21	0	0	0	0	1	6	8	8	3	*	0	0	26
ROCKFORD	10	0	0	0	0	1	5	7	6	3	0	0	0	22
SPRINGFIELD	13	0	0	0	0	1	7	10	8	4	*	0	0	30

States and Stations	Yrs.	Jan.	Feb.	Mar.	Apr.	May	June	July	Aug.	Sept.	Oct.	Nov.	Dec.	Annual
IND. EVANSVILLE	20	0	0	0	0	1	10	14	13	5	*	0	0	43
FORT WAYNE	14	0	0	0	0	*	4	6	6	2	*	0	0	18
INDIANAPOLIS	20	0	0	0	0	*	5	7	7	2	*	0	0	21
SOUTH BEND	20	0	0	0	*	*	4	6	6	2	0	0	0	18
IOWA BURLINGTON U	62	0	0	0	*	1	7	12	10	4	*	0	0	34
DES MOINES	27	0	0	0	0	1	5	10	9	3	*	0	0	28
DUBUQUE	10	0	0	0	0	*	2	4	3	2	0	0	0	11
SIOUX CITY	20	0	0	0	*	1	7	12	10	4	*	0	0	34
WATERLOO	12	0	0	0	*	*	3	6	6	2	*	0	0	17
KANS. CONCORDIA U	76	0	0	*	*	1	9	16	15	7	*	0	0	48
DODGE CITY	18	0	0	*	1	3	13	18	19	8	1	0	0	63
GOODLAND U	40	0	0	0	*	2	11	20	18	8	1	0	0	60
TOPEKA	14	0	0	0	1	1	10	15	17	8	1	0	0	53
WICHITA	7	0	0	0	1	2	12	21	23	10	1	0	0	70
KY. LEXINGTON	16	0	0	0	0	1	6	10	9	4	*	0	0	30
LOUISVILLE	13	0	0	0	*	2	10	16	14	6	1	0	0	49
LA. BATON ROUGE	9	0	0	0	1	7	22	26	24	15	2	0	0	97
LAKE CHARLES	22	0	0	*	*	3	19	23	24	15	2	0	0	86
NEW ORLEANS	45	0	0	*	*	3	16	20	19	10	1	0	0	69
SHREVEPORT U	80	0	0	*	*	4	18	24	23	14	2	0	0	85
MAINE CARIBOU	21	0	0	0	0	*	1	1	1	*	0	0	0	3
PORTLAND	20	0	0	0	0	*	1	2	2	*	0	0	0	5
MD. BALTIMORE U	88	0	0	*	*	1	5	10	6	2	*	0	0	24
FREDERICK	11	0	0	0	*	1	7	10	8	2	*	0	0	28
MASS. BLUE HILL OBS.	75	0	0	0	0	*	1	2	1	*	0	0	0	4
BOSTON	9	0	0	0	0	*	3	5	4	1	*	0	0	13
NANTUCKET	14	0	0	0	0	0	0	*	*	0	0	0	0	*
PITTSFIELD	22	0	0	0	0	0	*	1	1	*	0	0	0	2
WORCESTER U	10	0	0	0	0	0	2	2	2	*	0	0	0	6
MICH. ALPENA U	45	0	0	0	0	*	1	1	1	*	0	0	0	3
DETROIT	27	0	0	0	0	*	3	6	4	1	0	0	0	14
ESCANABA U	51	0	0	0	0	*	*	*	*	*	0	0	0	*
FLINT	19	0	0	0	0	*	3	4	4	1	0	0	0	12
GRAND RAPIDS	21	0	0	0	0	*	2	5	5	1	0	0	0	13
LANSING	43	0	0	0	0	*	2	5	2	1	0	0	0	10
MARQUETTE	23	0	0	0	0	*	1	2	2	1	0	0	0	6
MUSKEGON	21	0	0	0	0	0	1	2	2	1	0	0	0	6
SAULT STE. MARIE	19	0	0	0	0	0	*	*	1	*	0	0	0	1
MINN. DULUTH	19	0	0	0	0	0	*	1	1	0	0	0	0	2
INTERNATIONAL FALLS	21	0	0	0	*	*	1	2	1	*	0	0	0	4
MINNEAPOLIS	22	0	0	0	*	*	3	7	6	2	0	0	0	18
ROCHESTER	10	0	0	0	0	*	3	5	5	1	0	0	0	14
ST. CLOUD	21	0	0	0	0	*	2	4	4	1	*	0	0	11
MISS. JACKSON U	58	0	0	*	1	7	21	24	25	17	3	0	0	98
MERIDIAN	15	0	0	0	1	8	21	24	24	12	2	0	0	92
VICKSBURG	23	0	0	0	*	2	15	20	21	9	1	0	0	68
MO. COLUMBIA	21	0	0	0	*	1	8	14	14	6	*	0	0	43
KANSAS CITY	27	0	0	0	*	2	10	18	17	7	1	0	0	55
ST. JOSEPH U	43	0	0	0	*	1	8	16	13	5	*	0	0	43
ST. LOUIS U	23	0	0	0	*	2	9	16	13	6	*	0	0	46
SPRINGFIELD	15	0	0	0	*	*	8	13	14	6	*	0	0	41
MONT. BILLINGS	26	0	0	0	*	1	3	13	11	3	0	0	0	31
BUTTE	29	0	0	0	0	0	*	3	1	*	0	0	0	4
GLASGOW	17	0	0	0	*	1	2	13	9	2	*	0	0	28
GREAT FALLS	23	0	0	0	0	*	1	8	5	1	*	0	0	15
HAVRE U	57	0	0	0	*	1	2	10	7	1	*	0	0	21
HELENA	20	0	0	0	0	*	1	8	5	1	0	0	0	15
KALISPELL	11	0	0	0	0	0	*	5	3	1	0	0	0	9
MILES CITY	16	0	0	0	*	1	3	16	14	4	*	0	0	38
MISSOULA	16	0	0	0	0	*	1	11	6	1	0	0	0	19
NEBR. GRAND ISLAND	22	0	0	0	*	1	8	16	14	6	0	0	0	45
LINCOLN U	66	0	0	*	*	2	7	14	12	5	*	0	0	40
NORFOLK	15	0	0	0	*	1	7	11	11	5	*	0	0	35
NORTH PLATTE	9	0	0	0	0	*	8	15	14	6	*	0	0	43
OMAHA	25	0	0	0	*	1	8	14	12	5	*	0	0	40
SCOTTSBLUFF	17	0	0	0	0	1	6	15	13	5	*	0	0	40
VALENTINE U	64	0	0	0	*	1	5	12	10	4	*	0	0	32
NEV. ELKO	30	0	0	0	0	*	4	20	16	3	0	0	0	44
ELY	22	0	0	0	0	0	1	9	5	1	0	0	0	16
LAS VEGAS	12	0	0	0	4	15	27	31	30	25	8	0	0	140
RENO	18	0	0	0	0	1	5	20	15	6	0	0	0	47
WINNEMUCCA	11	0	0	0	0	1	7	22	17	6	0	0	0	53
N. H. CONCORD	19	0	0	0	0	1	4	5	3	1	0	0	0	14
MT. WASHINGTON	28	0	0	0	0	0	0	0	0	0	0	0	0	0

MEAN NUMBER OF DAYS MAXIMUM TEMPERATURE 90°F. AND ABOVE EXCEPT 70°F. AND ABOVE IN ALASKA

States and Stations	Yrs.	Jan.	Feb.	Mar.	Apr.	May	June	July	Aug.	Sept.	Oct.	Nov.	Dec.	Annual
N. J. ATLANTIC CITY U	79	0	0	0	0	*	1	1	1	*	*	0	0	3
NEWARK	19	0	0	0	*	1	6	9	7	2	*	0	0	25
TRENTON	28	0	0	0	*	1	4	7	5	1	*	0	0	18
N. MEX. ALBUQUERQUE	21	0	0	0	0	2	18	23	18	6	0	0	0	67
CLAYTON	15	0	0	0	*	1	11	14	11	4	*	0	0	41
RATON	15	0	0	0	0	0	5	7	4	1	0	0	0	17
ROSWELL	13	0	0	0	3	11	25	26	26	15	2	0	0	108
N. Y. ALBANY	14	0	0	0	0	*	3	6	4	1	0	0	0	14
BINGHAMTON U	64	0	0	0	*	*	2	4	3	1	*	0	0	10
BUFFALO	17	0	0	0	0	*	1	3	3	1	0	0	0	8
NEW YORK U	44	0	0	0	*	1	3	6	4	1	*	0	0	15
NEW YORK	20	0	0	0	0	1	4	7	5	1	*	0	0	18
ROCHESTER	20	0	0	0	0	*	3	5	4	2	*	0	0	14
SYRACUSE	11	0	0	0	0	0	2	4	3	1	0	0	0	10
N. C. ASHEVILLE	30	0	0	0	0	*	4	6	3	1	*	0	0	14
CAPE HATERAS U	78	0	0	0	0	0	*	1	*	*	*	0	0	1
CHARLOTTE	21	0	0	*	*	4	14	16	15	6	1	0	0	56
GREENSBORO	32	0	0	*	*	2	10	12	9	4	*	0	0	37
RALEIGH	16	0	0	*	1	2	12	15	12	5	1	0	0	48
WILMINGTON	9	0	0	0	*	3	10	15	14	4	*	0	0	46
WINSTON-SALEM U	57	0	0	*	*	3	10	13	10	5	*	0	0	41
N. DAK. BISMARCK	21	0	0	0	*	1	2	9	8	3	*	0	0	23
DEVILS LAKE U	56	0	0	0	*	*	1	4	4	1	0	0	0	10
FARGO	19	0	0	0	*	1	2	5	5	1	*	0	0	14
WILLISTON U	44	0	0	0	*	1	2	8	6	1	*	0	0	18
OHIO AKRON-CANTON	12	0	0	0	0	0	2	3	3	2	0	0	0	10
CINCINNATI (ABBE.)	45	0	0	0	*	1	5	11	8	4	*	0	0	29
CLEVELAND	19	0	0	0	0	*	5	6	5	2	*	0	0	18
COLUMBUS	21	0	0	0	0	1	6	9	7	3	*	0	0	26
DAYTON	17	0	0	0	0	*	4	6	6	2	0	0	0	18
SANDUSKY	33	0	0	0	*	1	4	7	5	2	*	0	0	19
TOLEDO U	81	0	0	0	0	*	3	5	3	1	0	0	0	12
YOUNGSTOWN	17	0	0	0	0	*	2	4	4	1	0	0	0	11
OKLA. OKLAHOMA CITY U	62	0	*	*	*	2	11	20	20	10	1	0	0	64
TULSA	22	0	0	*	*	2	13	21	22	11	2	0	0	71
OREG. ASTORIA	7	0	0	0	0	0	*	*	0	*	0	0	0	1
BURNS	10	0	0	0	0	*	1	11	7	2	0	0	0	21
EUGENE	18	0	0	0	0	*	1	6	4	3	0	0	0	14
MEACHAM	16	0	0	0	0	0	0	2	1	*	0	0	0	3
MEDFORD	31	0	0	0	*	2	5	16	15	8	1	0	0	47
PENDLETON	25	0	0	0	0	1	4	14	9	3	0	0	0	31
PORTLAND U	58	0	0	0	*	*	1	3	2	1	*	0	0	8
ROSEBURG	8	0	0	0	*	1	2	7	6	3	*	0	0	20
SALEM	23	0	0	0	0	*	2	6	5	3	0	0	0	16
SEXTON SUMMIT	16	0	0	0	0	0	*	1	1	*	0	0	0	2
PA. ALLENTOWN	17	0	0	0	0	*	5	8	4	1	*	0	0	18
ERIE U	77	0	0	0	0	*	1	2	1	*	0	0	0	4
HARRISBURG	22	0	0	0	*	1	5	9	7	2	*	0	0	24
PHILADELPHIA	20	0	0	0	*	1	6	10	6	2	*	0	0	25
PITTSBURGH	8	0	0	0	0	0	1	3	3	2	0	0	0	9
READING	20	0	0	0	*	1	6	9	6	2	*	0	0	24
WILLIAMSPORT	16	0	0	0	*	*	5	8	4	1	*	0	0	18
R. I. BLOCK ISLAND	10	0	0	0	0	0	*	*	0	0	0	0	0	*
PROVIDENCE	7	0	0	0	0	*	2	2	2	*	0	0	0	6
S. C. CHARLESTON	18	0	0	0	*	4	13	13	14	5	*	0	0	49
COLUMBIA	13	0	0	0	1	9	19	24	23	9	1	0	0	86
FLORENCE	12	0	0	0	1	7	16	21	21	7	1	0	0	74
GREENVILLE	7	0	0	0	*	3	11	17	18	6	0	0	0	55
SPARTANBURG	17	0	0	0	*	3	13	15	13	4	1	0	0	49
S. DAK. HURON	21	0	0	0	*	1	5	11	10	3	*	0	0	30
RAPID CITY	18	0	0	0	0	1	3	12	12	4	*	0	0	32
Sioux Falls	15	0	0	0	*	1	4	10	9	3	*	0	0	27
TENN. BRISTOL	15	0	0	0	0	*	5	5	7	2	*	0	0	19
CHATTANOOGA	20	0	0	0	*	4	14	18	17	6	1	0	0	60
KNOXVILLE	18	0	0	0	*	2	11	15	14	5	*	0	0	47
MEMPHIS	19	0	0	0	*	3	16	22	21	9	1	0	0	72
NASHVILLE	19	0	0	0	*	2	14	19	17	8	*	0	0	60
OAK RIDGE	15	0	0	0	0	2	9	12	11	5	*	0	0	39
TEX. ABILENE	21	0	*	1	3	10	22	27	27	16	3	*	0	109
AMARILLO	20	0	0	*	1	4	16	20	20	9	1	0	0	71
AUSTIN	19	0	*	1	1	9	23	29	28	18	5	*	*	114
BROWNSVILLE	18	0	*	1	2	9	23	29	28	18	5	*	0	115
CORPUS CHRISTI	22	0	*	*	1	4	19	29	28	16	3	*	0	100
TEX. DALLAS	20	0	0	*	1	7	21	27	27	17	2	0	0	102
EL PASO	21	0	0	0	1	10	25	26	24	14	1	0	0	101
FT. WORTH	13	0	0	*	1	6	20	26	27	15	2	0	*	98
GALVESTON	10	0	0	0	*	1	4	13	14	3	*	0	0	35
HOUSTON	22	0	*	*	*	4	19	25	25	15	2	0	0	90
LAREDO	17	*	2	6	13	23	28	31	29	25	12	1	*	171
LUBBOCK	14	0	0	0	2	8	21	23	22	11	1	0	0	88
MIDLAND	13	0	0	*	4	12	25	27	27	15	3	0	0	113
PORT ARTHUR	7	0	0	0	*	3	19	24	22	13	2	0	0	83
SAN ANGELO	13	0	*	1	4	12	23	27	27	16	3	*	*	113
SAN ANTONIO	18	0	*	1	3	10	24	30	28	18	5	*	*	119
VICTORIA	14	0	*	*	1	7	25	30	28	19	5	*	0	115
WACO	18	0	*	*	1	7	23	28	28	18	4	*	*	109
WICHITA FALLS	17	0	0	1	3	8	21	26	27	17	3	0	0	106
UTAH MILFORD	12	0	0	0	0	1	11	24	18	7	0	0	0	61
SALT LAKE CITY	32	0	0	0	0	1	8	22	18	4	0	0	0	53
WENDOVER	11	0	0	0	0	1	8	22	19	4	0	0	0	54
VT. BURLINGTON	17	0	0	0	0	*	2	3	3	1	0	0	0	9
VA. LYNCHBURG	16	0	0	0	*	1	7	10	8	3	*	0	0	29
NORFOLK	12	0	0	0	1	2	9	13	9	3	*	0	0	37
RICHMOND	31	0	0	*	1	3	10	14	11	5	1	0	0	45
ROANOKE	13	0	0	0	2	2	9	14	11	4	*	0	0	42
WASH. OLYMPIA	19	0	0	0	0	*	1	3	2	*	0	0	0	6
SEATTLE U	27	0	0	0	0	*	*	1	*	*	0	0	0	2
SPOKANE	13	0	0	0	0	*	1	9	5	1	0	0	0	16
STAMPEDE PASS	17	0	0	0	0	0	0	0	*	0	0	0	0	*
TATOOSH ISLAND U	58	0	0	0	0	0	0	0	0	0	0	0	0	0
WALLA WALLA U	46	0	0	0	*	1	4	16	13	3	0	0	0	37
YAKIMA	14	0	0	0	0	2	4	15	10	4	0	0	0	35
W.I. SAN JUAN P. R.	62	*	*	*	1	2	1	*	1	2	2	*	*	9
W. VA. CHARLESTON	13	0	0	0	*	1	5	9	7	3	*	0	0	25
HUNTINGTON	11	0	0	0	1	3	10	14	12	6	1	0	0	47
PARKERSBURG	72	0	0	0	*	1	5	8	6	3	*	0	0	23
WIS. GREEN BAY	11	0	0	0	0	*	1	1	2	2	0	0	0	6
LA CROSSE	10	0	0	0	*	*	3	5	4	2	0	0	0	14
MADISON	21	0	0	0	*	*	4	7	6	2	0	0	0	19
MILWAUKEE	20	0	0	0	0	*	2	3	4	1	0	0	0	10
WYO. CASPER	21	0	0	0	0	0	3	12	8	2	0	0	0	25
CHEYENNE	25	0	0	0	0	0	1	5	4	1	0	0	0	11
LANDER	14	0	0	0	0	*	2	10	7	2	0	0	0	21
SHERIDAN	20	0	0	0	0	*	2	12	11	3	0	0	0	28
YELLOWSTONE	30	0	0	0	0	0	0	1	*	*	0	0	0	1

MEAN NUMBER OF DAYS MAXIMUM TEMPERATURE 70°F. AND ABOVE

States and Stations	Yrs.	Jan.	Feb.	Mar.	Apr.	May	June	July	Aug.	Sept.	Oct.	Nov.	Dec.	Annual
ALASKA ANCHORAGE	35	0	0	0	0	*	4	6	4	*	0	0	0	14
ANNETTE	13	0	0	0	*	1	3	4	5	2	0	0	0	15
BARROW	40	0	0	0	0	0	*	*	*	0	0	0	0	*
BARTER ISLAND	13	0	0	0	0	0	0	*	*	0	0	0	0	*
Bethel	18	0	0	0	0	*	3	4	1	*	0	0	0	9
COLD BAY	16	0	0	0	0	0	0	*	*	*	0	0	0	*
CORDOVA	15	0	0	0	0	*	2	3	3	*	*	0	0	8
FAIRBANKS	31	0	0	0	*	3	17	19	9	1	0	0	0	49
JUNEAU	17	0	0	0	*	2	7	7	5	*	0	0	0	21
KING SALMON	15	0	0	0	0	*	3	6	3	*	0	0	0	13
KOTZEBUE	17	0	0	0	0	*	1	2	1	0	0	0	0	4
MCGRATH	18	0	0	0	0	1	11	13	4	*	0	0	0	29
NOME	14	0	0	0	0	0	1	1	1	0	0	0	0	3
ST. PAUL ISLAND	43	0	0	0	0	0	0	0	0	0	0	0	0	0
SHEMYA	15	0	0	0	0	0	0	0	0	0	0	0	0	0
YAKUTAT	14	0	0	0	0	1	1	2	1	*	0	0	0	5

*LESS THAN ONCE IN 2 YEARS

DATA FROM AIRPORT, EXCEPT THOSE MARKED WITH U FOR URBAN.

THESE CHARTS AND TABULATIONS WERE DERIVED FROM "NORMALS, MEANS, AND EXTREMES" TABLE IN U. S. WEATHER BUREAU PUBLICATION LOCAL CLIMATOLOGICAL DATA 1960.

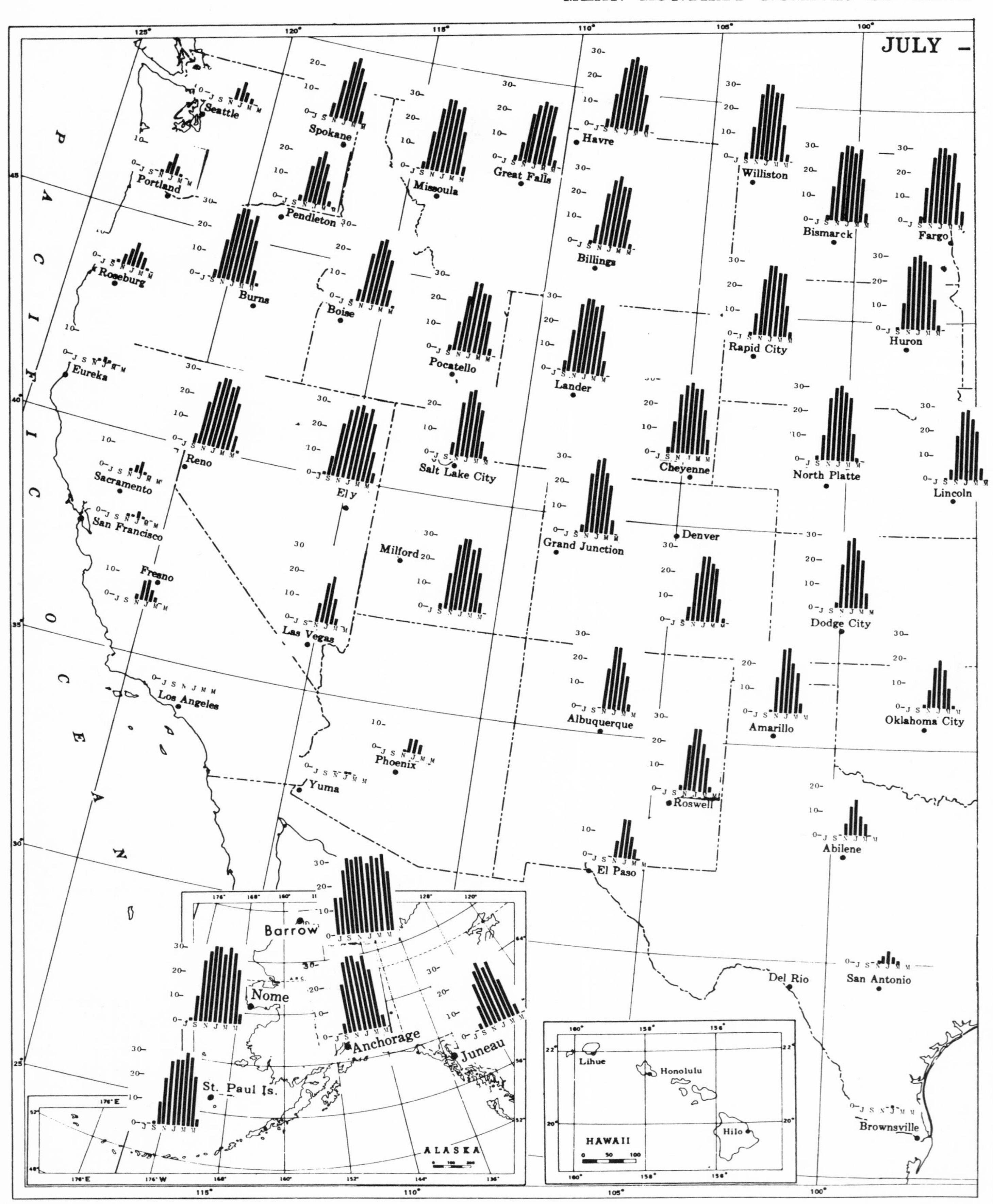
JULY –
Seattle
Spokane
Missoula
Great Falls
Havre
Williston
Bismarck
Fargo
Portland
Pendleton
Billings
Roseburg
Burns
Boise
Pocatello
Rapid City
Huron
Eureka
Lander
Reno
Ely
Salt Lake City
Cheyenne
North Platte
Lincoln
Sacramento
San Francisco
Grand Junction
Denver
Milford
Fresno
Las Vegas
Dodge City
Los Angeles
Albuquerque
Amarillo
Oklahoma City
Phoenix
Yuma
Roswell
Abilene
El Paso
Barrow
Nome
Anchorage
Juneau
St. Paul Is.
ALASKA
Del Rio
San Antonio
Lihue
Honolulu
Hilo
HAWAII
Brownsville
P A C I F I C O C E A N

MINIMUM TEMPERATURE 32°F AND BELOW

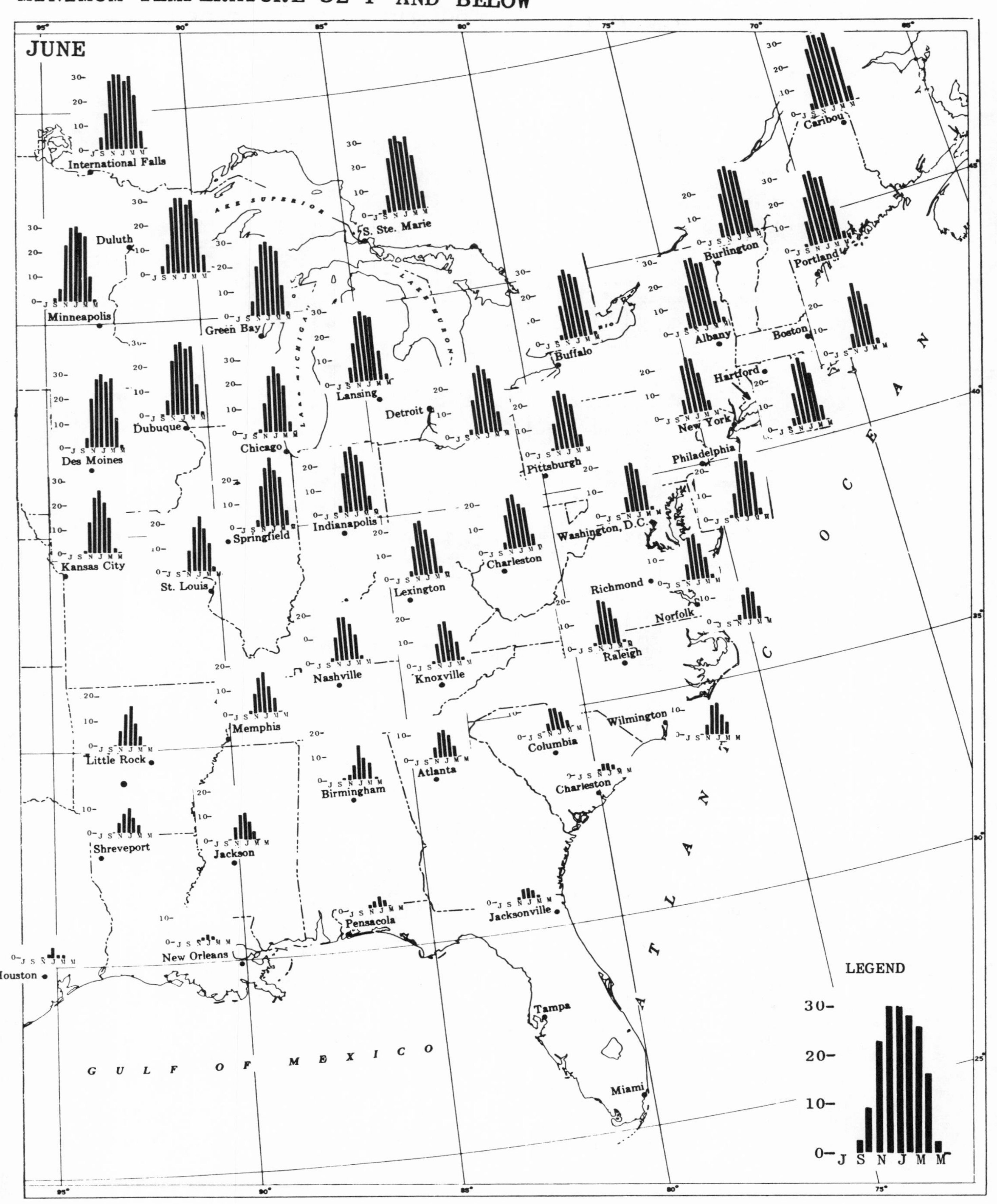

MEAN NUMBER OF DAYS MINIMUM TEMPERATURE 32°F AND BELOW

State and Stations	Yrs.	Jan.	Feb.	Mar.	Apr.	May	June	July	Aug.	Sept.	Oct.	Nov.	Dec.	Annual
ALA. BIRMINGHAM	18	14	9	7	1	0	0	0	0	0	1	8	14	53
MOBILE	20	7	4	1	0	0	0	0	0	0	*	2	5	18
MONTGOMERY U	83	7	4	1	*	0	0	0	0	0	*	2	6	20
ALASKA ANCHORAGE	35	31	28	30	24	5	*	0	*	4	20	29	31	201
ANNETTE	13	20	16	15	5	*	0	0	0	0	1	7	15	79
BARROW	41	31	28	31	30	31	24	15	15	26	31	30	31	323
BARTER ISLAND	14	31	28	31	30	31	23	8	9	24	31	30	31	307
BETHEL	18	31	28	31	28	14	*	*	*	6	26	29	31	225
COLD BAY	18	26	25	28	24	7	*	0	0	*	7	20	27	163
CORDOVA	16	29	26	29	25	10	1	0	*	6	17	22	27	191
FAIRBANKS	32	31	28	31	28	9	*	0	1	13	29	30	31	232
FT. YUKON	30	31	28	31	29	17	1	0	2	15	30	30	31	245
JUNEAU	18	26	23	25	16	4	*	0	*	2	9	18	23	146
KING SALMON	16	29	27	29	26	10	1	0	*	6	21	26	30	205
KOTZEBUE	18	31	28	31	30	26	7	0	*	9	29	30	31	252
MCGRATH	18	31	28	31	29	11	*	0	1	10	28	30	31	231
NOME	39	31	28	31	29	22	4	0	1	10	25	29	31	241
ST. PAUL ISLAND	44	27	27	30	28	20	3	*	*	1	9	19	26	189
SHEMYA	16	23	24	25	14	2	0	0	0	0	1	12	22	124
YAKUTAT	15	28	26	28	23	8	*	0	*	3	12	19	25	171
ARIZ. FLAGSTAFF	12	31	27	30	24	15	3	0	0	3	18	28	30	211
PHOENIX	21	6	4	0	0	0	0	0	0	0	0	1	6	17
PRESCOTT	19	27	24	21	7	1	0	0	0	0	3	20	27	131
TUCSON	21	7	5	1	*	0	0	0	0	0	0	2	6	21
WINSLOW	29	28	23	20	7	1	0	0	0	0	3	22	28	132
YUMA	11	1	1	*	0	0	0	0	0	0	0	*	*	2
ARK. FT. SMITH	15	20	14	7	1	0	0	0	0	0	1	10	18	71
LITTLE ROCK	17	16	9	4	*	0	0	0	0	0	*	6	13	48
TEXARKANA	19	13	7	3	0	0	0	0	0	0	*	5	11	39
CALIF. BAKERSFIELD	24	7	2	*	0	0	0	0	0	0	0	1	5	15
BISHOP	14	30	25	22	7	1	*	0	0	*	6	24	30	146
BLUE CANYON	18	20	18	20	10	5	1	0	0	*	3	11	15	102
BURBANK	19	2	1	*	0	0	0	0	0	0	0	*	*	4
EUREKA	51	2	1	*	*	0	0	0	0	0	0	*	1	4
FRESNO	21	9	5	2	*	0	0	0	0	0	0	2	8	26
LONG BEACH	17	1	*	0	0	0	0	0	0	0	0	0	0	1
LOS ANGELES U	21	*	0	0	0	0	0	0	0	0	0	0	*	*
MT. SHASTA	19	26	22	21	11	4	*	*	0	1	6	19	26	135
OAKLAND	31	4	1	*	0	0	0	0	0	0	0	*	2	7
RED BLUFF	17	10	3	1	*	0	0	0	0	0	*	1	6	23
SACRAMENTO	28	5	1	*	0	0	0	0	0	0	0	1	3	9
SANDBERG	29	13	11	10	4	1	*	0	0	0	1	5	9	54
SAN DIEGO	21	*	0	0	0	0	0	0	0	0	0	0	0	*
SAN FRANCISCO	34	3	1	*	*	0	0	0	0	0	0	1	1	6
SANTA MARIA	19	7	3	1	*	0	0	0	0	0	*	1	3	15
COLO. ALAMOSA	16	31	28	31	26	13	2	0	*	10	26	30	31	226
COLORADO SPRINGS	12	30	27	27	16	2	*	0	0	*	9	25	30	166
DENVER U	84	27	24	22	10	2	*	0	0	*	6	20	27	139
GRAND JUNCTION	15	31	25	20	6	*	0	0	0	0	3	24	30	138
PUEBLO	21	30	27	26	10	1	0	0	0	*	8	26	30	158
CONN. BRIDGEPORT	13	25	22	18	3	0	0	0	0	0	*	9	23	101
HARTFORD U	51	27	25	20	6	*	0	0	0	*	3	13	25	119
NEW HAVEN	18	26	24	20	4	*	0	0	0	0	2	11	25	112
DEL. WILMINGTON	14	25	22	17	3	0	0	0	0	0	1	12	23	102
D.C. WASHINGTON	20	20	17	10	1	0	0	0	0	0	*	6	19	73
FLA. APALACHICOLA	31	2	1	*	0	0	0	0	0	0	0	*	1	4
DAYTONA BEACH	18	2	1	*	0	0	0	0	0	0	0	*	2	5
EVERGLADES	26	*	0	0	0	0	0	0	0	0	0	0	*	*
FT. MYERS	20	*	*	0	0	0	0	0	0	0	0	0	*	*
JACKSONVILLE	20	4	3	1	0	0	0	0	0	0	0	1	4	11
KEY WEST	13	0	0	0	0	0	0	0	0	0	0	0	0	0
KEY WEST U	87	0	0	0	0	0	0	0	0	0	0	0	0	0
LAKELAND U	21	*	*	*	0	0	0	0	0	0	0	*	*	1
MIAMI	19	0	*	0	0	0	0	0	0	0	0	0	0	*
MIAMI BEACH	20	0	0	0	0	0	0	0	0	0	0	0	0	0
ORLANDO	18	1	1	*	0	0	0	0	0	0	0	*	1	3
PENSACOLA	22	4	2	*	0	0	0	0	0	0	0	1	2	9
TALLAHASSEE	22	7	4	1	*	0	0	0	0	0	0	2	6	20
TAMPA	15	1	*	0	0	0	0	0	0	0	0	*	*	2
WEST PALM BEACH	21	*	0	0	0	0	0	0	0	0	0	0	*	*
GA. ATHENS	18	13	9	6	1	0	0	0	0	0	*	6	14	49
ATLANTA U	77	11	9	4	*	0	0	0	0	0	*	4	10	38
AUGUSTA	11	15	9	5	*	0	0	0	0	0	1	9	14	54
COLUMBUS	16	12	8	5	*	0	0	0	0	0	*	6	12	44
MACON	13	10	6	3	*	0	0	0	0	0	*	5	10	34
ROME	16	17	14	11	2	0	0	0	0	0	2	13	19	79
SAVANNAH	11	10	5	3	0	0	0	0	0	0	*	3	9	30
THOMASVILLE	39	5	3	1	*	0	0	0	0	0	0	1	4	14
HAWAII Hilo	16	0	0	0	0	0	0	0	0	0	0	0	0	0
HONOLULU	39	0	0	0	0	0	0	0	0	0	0	0	0	0
LIHUE	12	0	0	0	0	0	0	0	0	0	0	0	0	0
IDAHO BOISE	21	27	23	19	7	1	0	0	0	*	5	20	26	128
IDAHO FALLS	12	31	28	31	22	10	2	0	*	7	23	29	31	213
LEWSITON	15	24	18	14	4	*	0	0	0	*	4	15	21	101
POCATELLO	23	29	25	25	14	3	*	0	0	2	12	23	29	161
ILL. CAIRO U	19	19	13	8	*	0	0	0	0	0	*	7	17	64
CHICAGO	19	29	24	20	6	*	0	0	0	0	2	16	26	123
CHICAGO U	85	27	24	19	4	*	0	0	0	*	1	12	23	110
MOLINE	21	31	26	22	8	*	0	0	0	*	5	19	28	139
PEORIA	19	29	25	21	5	*	0	0	0	*	2	18	27	127
ROCKFORD	11	30	27	27	9	1	0	0	0	*	6	21	29	150
SPRINGFIELD	12	28	23	20	6	*	0	0	0	0	3	17	24	121
IND. EVANSVILLE	20	24	20	15	2	0	0	0	0	*	2	15	23	101
FT. WAYNE	10	28	25	23	8	*	0	0	0	*	5	18	27	134
INDIANAPOLIS	18	27	23	20	6	*	0	0	0	*	3	16	25	120
SOUTH BEND	21	29	26	23	10	1	0	0	0	*	4	17	27	137
IOWA BURLINGTON U	61	29	25	20	6	*	0	0	0	*	4	17	27	128
DES MOINES	21	30	27	23	7	1	0	0	0	*	4	20	28	140
DUBUQUE	11	30	27	28	12	1	0	0	0	*	6	22	29	155
SIOUX CITY	20	31	27	24	10	1	0	0	0	*	6	23	30	152
WATERLOO	12	31	27	26	12	1	0	0	0	1	7	23	30	158

MEAN NUMBER OF DAYS MINIMUM TEMPERATURE 32°F AND BELOW

State and Stations	Yrs.	Jan.	Feb.	Mar.	Apr.	May	June	July	Aug.	Sept.	Oct.	Nov.	Dec.	Annual
KANS. CONCORDIA U	76	28	23	18	4	*	0	0	0	*	3	16	26	118
DODGE CITY	19	29	24	20	6	*	0	0	0	0	2	18	28	127
GOODLAND	41	30	26	25	11	1	*	0	0	1	7	25	30	156
TOPEKA	15	29	25	20	6	*	0	0	0	0	3	19	27	129
WICHITA	8	29	23	16	3	0	0	0	0	0	1	14	26	112
KY. LEXINGTON	17	23	19	15	3	*	0	0	0	0	2	13	22	97
LOUISVILLE	13	23	17	15	2	0	0	0	0	0	2	13	20	92
LA. BATON ROUGE	9	6	3	1	0	0	0	0	0	0	*	2	6	18
LAKE CHARLES	22	5	2	1	0	0	0	0	0	0	0	1	3	12
NEW ORLEANS	46	2	1	*	0	0	0	0	0	0	0	*	1	4
SHREVEPORT	9	10	6	3	0	0	0	0	0	0	*	4	8	31
MAINE CARIBOU	22	31	28	30	23	6	*	0	0	3	15	25	30	192
PORTLAND	21	30	27	27	14	3	0	0	0	1	10	20	29	160
MD. BALTIMORE U	89	20	18	10	1	0	0	0	0	0	*	5	17	71
FREDERICK	12	27	22	20	6	*	0	0	0	*	3	17	25	119
MASS. BLUE HILL OBS.	76	29	26	24	10	*	0	0	0	*	3	14	26	132
BOSTON	10	26	22	17	2	0	0	0	0	0	*	6	21	94
NANTUCKER	15	24	22	18	4	*	0	0	0	0	*	5	19	93
PITTSFIELD	23	30	25	28	15	2	0	0	0	2	10	20	28	160
WORCESTER	6	30	27	28	12	1	0	0	0	*	6	15	29	148
MICH. ALPENA U	45	30	28	28	16	2	0	0	0	*	5	18	28	155
DETROIT	28	28	26	22	8	*	0	0	0	*	2	14	25	125
ESCANABA U	52	31	28	28	17	3	*	0	0	1	6	20	29	163
FLINT	19	30	27	25	13	2	0	0	0	1	6	18	27	149
GRAND RAPIDS	22	30	27	25	12	2	*	0	0	1	6	19	28	150
LANSING	43	29	27	26	12	2	0	0	0	*	6	16	27	145
MARQUETTE	24	31	28	28	17	4	0	0	0	*	4	20	29	161
MUSKEGON	21	30	27	26	12	1	0	0	0	*	4	17	27	144
SAULT STE. MARIE	20	31	28	30	22	7	*	0	*	1	7	22	30	178
MINN. DULUTH	19	31	28	30	22	7	*	0	0	3	12	27	31	191
INTERNATIONAL FALLS	22	31	28	30	23	8	*	0	*	5	15	28	31	199
MINNEAPOLIS	22	31	28	27	10	1	0	0	0	1	5	23	30	156
ROCHESTER	10	30	28	26	16	2	*	C	0	2	9	23	30	166
ST. CLOUD	22	31	28	29	17	3	0	0	0	2	11	26	31	178
MISS. JACKSON U	59	11	7	3	*	0	0	0	0	0	*	5	10	36
MERIDIAN	16	13	8	5	1	0	0	0	0	0	1	7	13	47
VICKSBURG	24	8	4	1	0	0	0	0	0	0	*	2	5	21
MO. COLUMBIA	22	27	23	17	3	0	0	0	0	*	1	15	24	110
KANSAS CITY	27	26	21	15	2	*	0	0	0	0	1	13	23	101
ST. JOSEPH	13	30	25	20	5	*	0	0	0	0	2	18	27	127
ST. LOUIS U	23	23	18	12	2	0	0	0	0	0	*	9	19	83
SPRINGFIELD	15	25	20	16	3	*	0	0	0	0	2	15	22	103
MONT. BILLINGS	26	29	25	25	12	2	*	0	0	1	8	21	27	149
BUTTE	29	31	28	30	26	15	3	*	1	11	24	29	30	228
GLASGOW	18	31	28	29	17	3	*	0	0	2	14	28	31	184
GREAT FALLS	24	26	25	25	13	3	*	0	0	2	8	20	24	146
MONT. HAVRE U	57	30	27	27	15	3	*	0	*	3	14	25	29	172
HELENA	21	30	27	28	17	5	*	0	*	4	16	26	29	182
KALISPELL	11	29	27	28	18	4	*	0	*	5	18	25	29	182
MILES CITY	17	31	28	27	16	3	*	0	0	1	11	27	30	174
MISSOULA	16	30	27	28	18	4	*	0	*	3	16	26	30	181
NEBR. GRAND ISLAND	22	31	27	24	9	*	0	0	0	0	6	23	30	150
LINCOLN U	68	29	25	20	5	*	0	0	0	*	4	18	28	129
NORFOLK	16	31	27	25	11	1	0	0	0	1	7	25	30	158
NORTH PLATTE	10	31	28	28	15	2	0	0	0	1	14	28	31	178
OMAHA	25	30	26	22	6	*	0	0	0	*	3	20	29	136
SCOTTSBLUFF	18	31	28	29	17	3	*	0	0	1	12	28	31	180
VALENTINE U	67	31	27	26	12	2	0	0	0	1	10	26	30	164
NEV. ELKO	31	30	27	29	22	11	3	*	1	9	23	28	30	214
ELY	23	31	28	30	24	13	4	*	1	8	22	28	30	218
LAS VEGAS	19	20	11	3	*	0	0	0	0	0	*	8	17	59
RENO	19	29	26	27	20	7	1	0	*	4	18	27	29	188
WINNEMUCCA	12	28	25	27	20	9	2	*	*	7	21	26	28	195
N. H. CONCORD	20	30	28	27	15	4	0	0	0	2	12	21	29	168
MT. WASHINGTON	29	31	28	31	29	19	6	1	3	12	23	28	31	242
N. J. ATLANTIC CITY U	85	20	18	12	2	0	0	0	0	0	*	6	17	75
NEWARK	20	25	22	15	2	0	0	0	0	0	*	9	22	94
TRENTON	29	24	21	15	2	0	0	0	0	0	*	7	21	89
N. MEX. ALBUQUERQUE	21	26	20	15	3	0	0	0	0	0	*	17	26	107
CLAYTON	16	29	25	24	10	1	0	0	0	0	5	22	28	144
RATON	16	31	28	29	18	3	0	0	0	1	16	29	31	185
ROSWELL	51	25	19	11	2	*	0	0	0	0	2	15	25	99
N. Y. ALBANY U	82	28	26	23	8	*	0	0	0	*	3	15	26	129
BINGHAMTON U	65	28	25	23	10	1	0	0	0	*	5	16	26	134
BUFFALO	17	28	26	24	10	1	0	0	0	*	2	14	26	131
NEW YORK U	89	23	22	16	4	0	0	0	0	0	*	6	20	91
NEW YORK	21	23	19	12	1	0	0	0	0	0	0	4	18	77
ROCHESTER	21	29	26	25	10	1	0	0	0	*	3	15	26	136
SYRACUSE	12	29	26	26	8	1	0	0	0	0	3	15	27	135
N. C. ASHEVILLE	31	18	16	12	2	0	0	0	0	0	2	13	19	83
CAPE HATTERAS R	81	5	4	1	*	0	0	0	0	0	0	*	3	13
CHARLOTTE	21	17	13	9	1	0	0	0	0	0	*	8	17	65
GREENSBORO	33	20	17	12	2	0	0	0	0	0	1	12	21	85
RALEIGH	17	18	15	10	1	*	0	0	0	0	1	9	19	73
WILMINGTON U	81	8	7	3	*	0	0	0	0	0	*	2	7	27
WINSTON-SALEM U	56	23	18	11	3	*	0	0	0	0	2	12	21	90
N. DAK. BISMARCK	19	31	28	30	18	4	0	0	0	2	14	28	31	186
DEVILS LAKE U	57	31	28	29	19	6	*	0	*	3	15	28	30	189
FARGO	19	31	28	29	17	5	0	0	0	2	12	27	31	182
WILLISTON U	45	31	28	28	15	3	*	0	0	2	13	27	31	178

MEAN NUMBER OF DAYS MINIMUM TEMPERATURE 32°F AND BELOW

State and Stations	Yrs.	Jan.	Feb.	Mar.	Apr.	May	June	July	Aug.	Sept.	Oct.	Nov.	Dec.	Annual
OHIO AKRON CANTON	13	28	25	24	11	1	0	0	0	*	3	18	26	136
CINCINNATI (ABBE)	46	23	20	16	3	*	0	0	0	0	2	11	22	98
CLEVELAND	19	27	24	21	8	*	0	0	0	*	1	13	25	119
COLUMBUS	21	26	23	20	6	*	0	0	0	*	3	15	24	117
DAYTON	18	27	23	19	6	*	0	0	0	0	2	15	24	116
SANDUSKY	34	26	24	20	5	0	0	0	0	0	1	12	23	111
TOLEDO U	82	27	25	21	7	*	0	0	0	*	2	14	24	121
YOUNGSTOWN	18	28	26	23	11	1	0	0	0	*	4	17	26	136
OKLA. OKLAHOMA CITY U	63	20	16	8	1	0	0	0	0	0	1	7	17	70
TULSA	22	21	16	10	1	0	0	0	0	0	1	9	18	76
OREG. ASTORIA	8	8	6	8	2	*	0	0	0	0	*	6	6	36
BURNS	11	30	26	28	18	6	1	0	*	3	16	27	30	184
EUGENE	19	15	10	8	3	*	0	0	0	*	2	8	12	59
MEACHAM	17	28	26	27	20	7	1	*	0	10	10	22	27	170
MEDFORD	31	19	15	10	3	*	*	0	0	*	3	13	16	79
PENDLETON	26	22	17	10	2	*	0	0	0	0	2	15	19	86
PORTLAND U	59	9	4	1	*	0	0	0	0	0	*	1	5	22
ROSEBURG U	77	10	7	5	1	*	0	0	0	0	*	1	4	7	35
SALEM	24	15	11	9	3	*	0	0	0	*	1	9	11	60
SEXTON-SUMMIT	17	20	17	22	14	6	*	0	0	*	4	10	18	112
PA. ALLENTOWN	18	28	26	22	7	*	0	0	0	*	4	16	27	129
ERIE U	84	27	25	23	8	*	0	0	0	0	1	10	23	117
HARRISBURG	23	26	22	18	3	*	0	0	0	*	1	12	23	106
PHILADELPHIA	22	26	23	18	3	*	0	0	0	*	1	12	23	106
PITTSBURGH U	81	24	23	18	5	0	0	0	0	0	1	11	22	103
READING	20	24	21	14	2	0	0	0	0	0	1	8	22	92
WILLIAMSPORT	17	28	25	22	7	1	0	0	0	*	4	16	27	130
R. I. BLOCK ISLAND U	75	22	22	16	2	0	0	0	0	0	*	5	18	85
PROVIDENCE U	51	26	24	18	4	*	0	0	0	0	1	10	23	106
S. C. CHARLESTON	19	10	7	3	*	0	0	0	0	0	*	4	9	33
CHARLESTON U	85	3	2	*	0	0	0	0	0	0	0	1	3	9
COLUMBIA U	68	9	7	3	*	0	0	0	0	0	*	3	9	31
FLORENCE U	56	11	11	4	*	0	0	0	0	0	*	2	11	39
GREENVILLE U	38	12	9	5	*	0	0	0	0	0	*	5	11	42
SPARTANBURG	18	14	10	6	1	0	0	0	0	0	*	5	16	52
S. DAK. HURON U	74	31	28	27	13	2	*	0	0	1	11	26	30	169
RAPID CITY U	56	29	26	26	13	2	*	0	0	1	9	23	29	158
SIOUX FALLS	16	31	28	28	14	2	0	0	0	1	9	26	30	169
TENN. BRISTOL	16	20	17	14	3	*	0	0	0	0	3	14	22	94
CHATTANOOGA U	77	14	11	6	1	0	0	0	0	0	*	6	13	51
KNOXVILLE U	85	17	13	8	1	0	0	0	0	0	1	8	16	64
MEMPHIS	20	17	11	6	*	0	0	0	0	0	1	8	15	57
NASHVILLE	20	18	15	10	1	0	0	0	0	0	1	10	18	74
OAK RIDGE	13	17	14	12	2	0	0	0	0	0	1	14	19	80
TEX. ABILENE	21	15	8	5	*	0	0	0	0	0	*	5	12	45
AMARILLO	20	27	21	17	4	*	0	0	0	0	1	15	26	111
AUSTIN	19	7	4	2	0	0	0	0	0	0	*	2	4	19
BROWNSVILLE	19	1	*	*	0	0	0	0	0	0	0	0	*	1
CORPUS CHRISTI	22	3	1	*	0	0	0	0	0	0	0	*	1	5
DALLAS	20	12	6	3	*	0	0	0	0	0	*	4	8	33
EL PASO	21	16	9	4	*	0	0	0	0	0	*	7	16	52
FT. WORTH U	57	11	8	3	*	0	0	0	0	0	*	2	8	32
GALVESTON U	84	2	1	*	0	0	0	0	0	0	0	*	1	4
HOUSTON	23	4	1	1	0	0	0	0	0	0	0	*	1	7
LAREDO	18	2	1	*	0	0	0	0	0	0	0	*	1	4
LUBBOCK	15	26	19	13	3	0	0	0	0	0	1	15	25	102
MIDLAND	26	18	11	6	1	0	0	0	0	0	*	8	20	64
PORT ARTHUR	11	4	2	1	0	0	0	0	0	0	0	1	3	11
SAN ANGELO	13	13	8	4	*	0	0	0	0	0	*	5	12	42
SAN ANTONIO U	70	5	3	1	0	0	0	0	0	0	*	1	3	13
VICTORIA U	50	4	2	1	0	0	0	0	0	0	*	1	3	11
WACO	25	9	5	2	*	0	0	0	0	0	0	3	7	26
WICHITA FALLS	17	18	12	7	1	0	0	0	0	0	*	7	16	61
UTAH MILFORD	13	30	26	27	17	4	*	0	0	2	15	27	30	179
SALT LAKE CITY	32	28	23	20	7	1	0	0	0	*	5	22	27	134
WENDOVER	12	29	24	18	4	*	0	0	0	0	3	22	29	129
VT. BURLINGTON	64	29	27	26	13	1	0	0	0	*	6	18	28	148
VA. LYNCHBURG U	81	20	17	10	2	*	0	0	0	0	1	9	19	78
NORFOLK U	85	13	11	5	*	0	0	0	0	0	*	2	10	41
RICHMOND U	58	19	17	9	1	0	0	0	0	0	*	7	18	71
ROANOKE	14	20	17	14	2	0	0	0	0	0	1	12	20	86
WASH. OLYMPIA	19	18	14	15	7	1	0	0	0	*	4	10	14	84
SEATTLE U	64	8	4	2	*	0	0	0	0	0	*	1	5	20
SPOKANE U	74	26	22	17	5	*	0	0	0	*	6	16	24	116
STAMPEDE PASS	18	31	27	30	23	11	2	*	0	1	11	26	30	191
TATOOSH ISLAND R	59	4	1	1	0	0	0	0	0	0	0	*	1	7
WALLA WALLA U	47	20	12	5	1	*	0	0	0	*	1	10	17	66
YAKIMA U	46	28	22	17	6	1	*	0	0	1	7	21	27	130
W.I. SAN JUAN P. R.	62	0	0	0	0	0	0	0	0	0	0	0	0	0
W. VA. CHARLESTON	14	22	18	16	5	*	0	0	0	0	3	14	21	97
HUNTINGTON U	12	21	17	14	3	0	0	0	0	0	1	14	20	91
PARKERSBURG	73	22	21	15	4	*	0	0	0	*	2	12	21	97
WIS. GREEN BAY U	69	30	28	26	12	1	0	0	0	*	6	20	29	152
LA CROSSE	11	31	27	28	10	*	0	0	0	*	5	22	30	153
MADISON U	77	30	27	25	9	*	0	0	0	*	4	19	29	143
MILWAUKEE U	85	29	26	22	7	*	0	0	0	*	3	16	26	129
WYO. CASPER	22	30	27	27	18	5	*	0	0	3	13	24	28	174
CHEYENNE U	83	29	27	27	18	6	*	0	*	2	13	24	28	174
LANDER U	64	31	28	29	18	6	*	*	*	4	17	28	31	192
SHERIDAN	21	31	28	29	17	3	*	0	0	2	14	27	30	182
YELLOWSTONE MAM.H.S.	30	31	27	30	23	12	3	*	*	7	19	28	30	210

* LESS THAN ONCE IN 2 YEARS.

DATA FROM AIRPORT, EXCEPT THOSE MARKED WITH U FOR URBAN AND R FOR RURAL.

THESE CHARTS AND TABULATIONS WERE DERIVED FROM "NORMALS, MEANS, AND EXTREMES" TABLE IN U. S. WEATHER BUREAU PUBLICATION LOCAL CLIMATOLOGICAL DATA (THROUGH 1961 USUALLY).

DUE TO ROUNDING TO WHOLE NUMBERS THE SUM OF THE MONTHLY VALUES MAY NOT EQUAL THE ANNUAL.

MEAN ANNUAL NUMBER OF DAYS MINIMUM TEMPERATURE 32°F AND BELOW

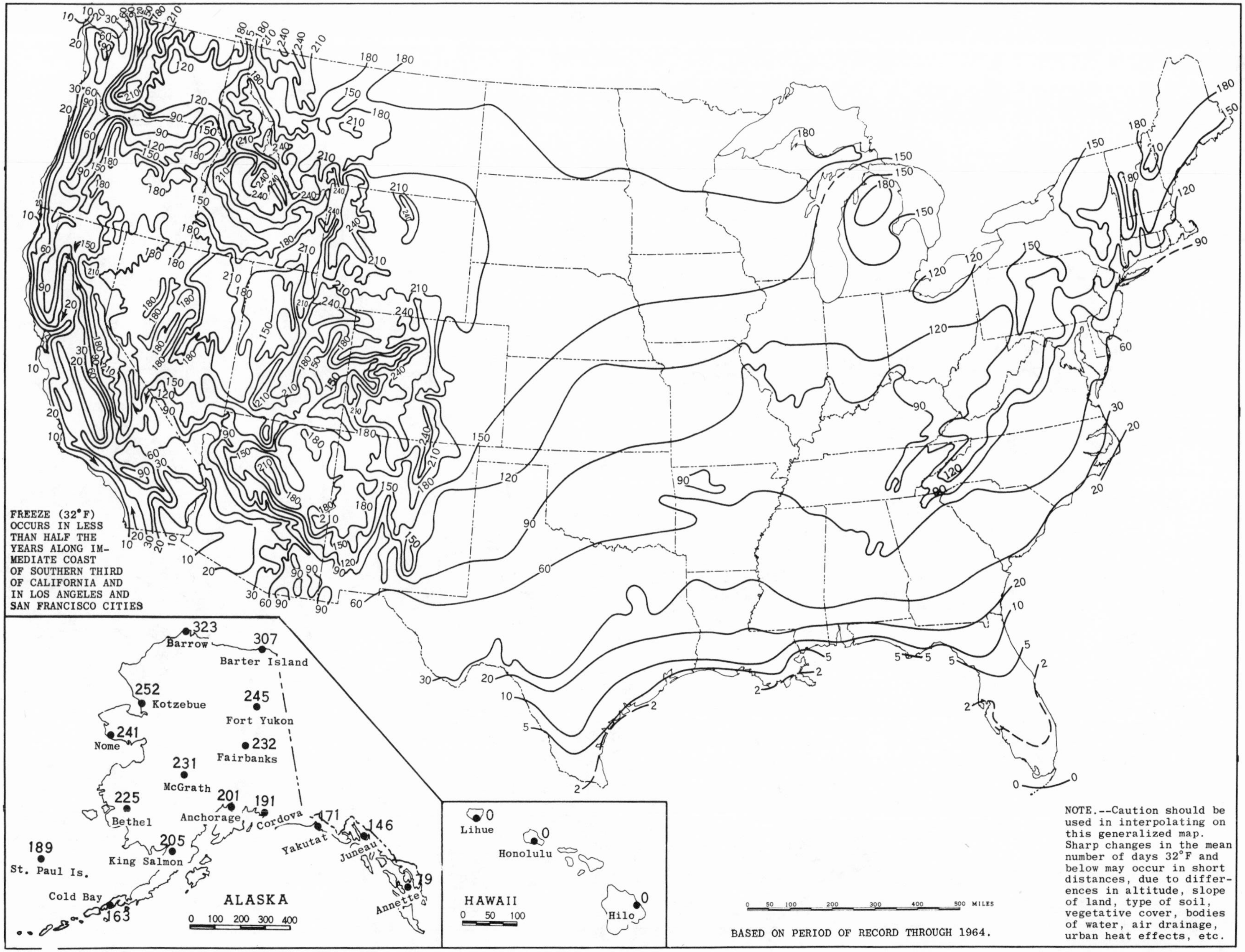

FREEZE (32°F) OCCURS IN LESS THAN HALF THE YEARS ALONG IMMEDIATE COAST OF SOUTHERN THIRD OF CALIFORNIA AND IN LOS ANGELES AND SAN FRANCISCO CITIES

ALASKA

323 Barrow
307 Barter Island
252 Kotzebue
245 Fort Yukon
241 Nome
232 Fairbanks
231 McGrath
225 Bethel
201 Anchorage
191 Cordova
171 Yakutat
146 Juneau
79 Annette
205 King Salmon
189 St. Paul Is.
163 Cold Bay

0 100 200 300 400

HAWAII

0 Lihue
0 Honolulu
0 Hilo

0 50 100

0 50 100 200 300 400 500 MILES

BASED ON PERIOD OF RECORD THROUGH 1964.

NOTE.--Caution should be used in interpolating on this generalized map. Sharp changes in the mean number of days 32°F and below may occur in short distances, due to differences in altitude, slope of land, type of soil, vegetative cover, bodies of water, air drainage, urban heat effects, etc.

MEAN DATE OF LAST 32°(F.)

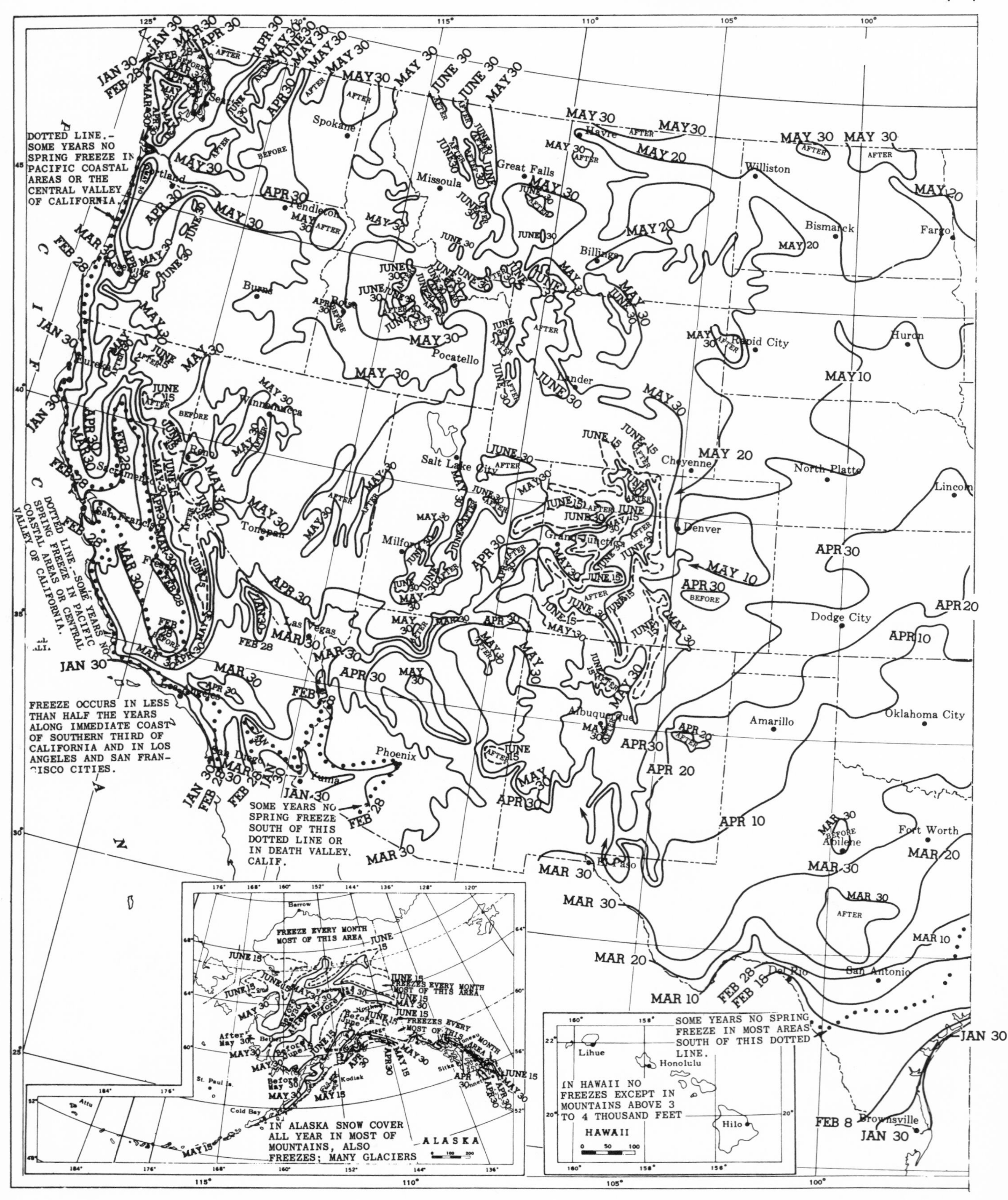

TEMPERATURE IN SPRING

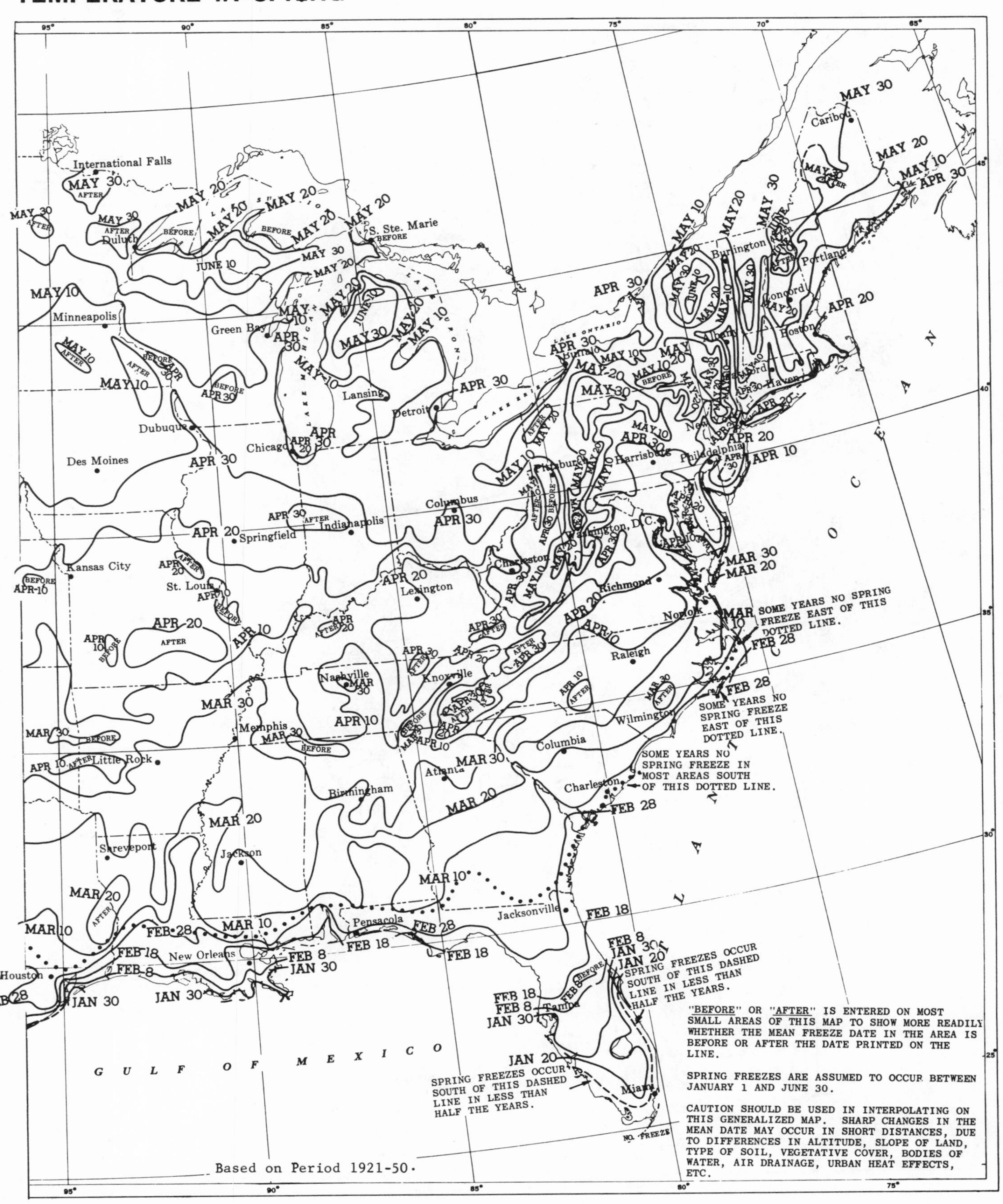

International Falls
Duluth
S. Ste. Marie
Minneapolis
Green Bay
Lansing
Detroit
Dubuque
Chicago
Des Moines
Springfield
Indianapolis
Columbus
Kansas City
St. Louis
Lexington
Charleston
Pittsburgh
Harrisburg
Philadelphia
Washington, D.C.
Richmond
Norfolk
Raleigh
Nashville
Knoxville
Memphis
Little Rock
Wilmington
Columbia
Atlanta
Birmingham
Charleston
Shreveport
Jackson
Jacksonville
Pensacola
New Orleans
Houston
Tampa
Miami
Buffalo
Burlington
Concord
Portland
Boston
Caribou
LAKE ONTARIO
LAKE ERIE
LAKE HURON
LAKE MICHIGAN
GULF OF MEXICO
ATLANTIC OCEAN
SOME YEARS NO SPRING FREEZE EAST OF THIS DOTTED LINE.
SOME YEARS NO SPRING FREEZE EAST OF THIS DOTTED LINE.
SOME YEARS NO SPRING FREEZE IN MOST AREAS SOUTH OF THIS DOTTED LINE.
SPRING FREEZES OCCUR SOUTH OF THIS DASHED LINE IN LESS THAN HALF THE YEARS.
SPRING FREEZES OCCUR SOUTH OF THIS DASHED LINE IN LESS THAN HALF THE YEARS.
NO FREEZE
"BEFORE" OR "AFTER" IS ENTERED ON MOST SMALL AREAS OF THIS MAP TO SHOW MORE READILY WHETHER THE MEAN FREEZE DATE IN THE AREA IS BEFORE OR AFTER THE DATE PRINTED ON THE LINE.
SPRING FREEZES ARE ASSUMED TO OCCUR BETWEEN JANUARY 1 AND JUNE 30.
CAUTION SHOULD BE USED IN INTERPOLATING ON THIS GENERALIZED MAP. SHARP CHANGES IN THE MEAN DATE MAY OCCUR IN SHORT DISTANCES, DUE TO DIFFERENCES IN ALTITUDE, SLOPE OF LAND, TYPE OF SOIL, VEGETATIVE COVER, BODIES OF WATER, AIR DRAINAGE, URBAN HEAT EFFECTS, ETC.
Based on Period 1921-50.

MEAN DATE OF FIRST 32°(F.)

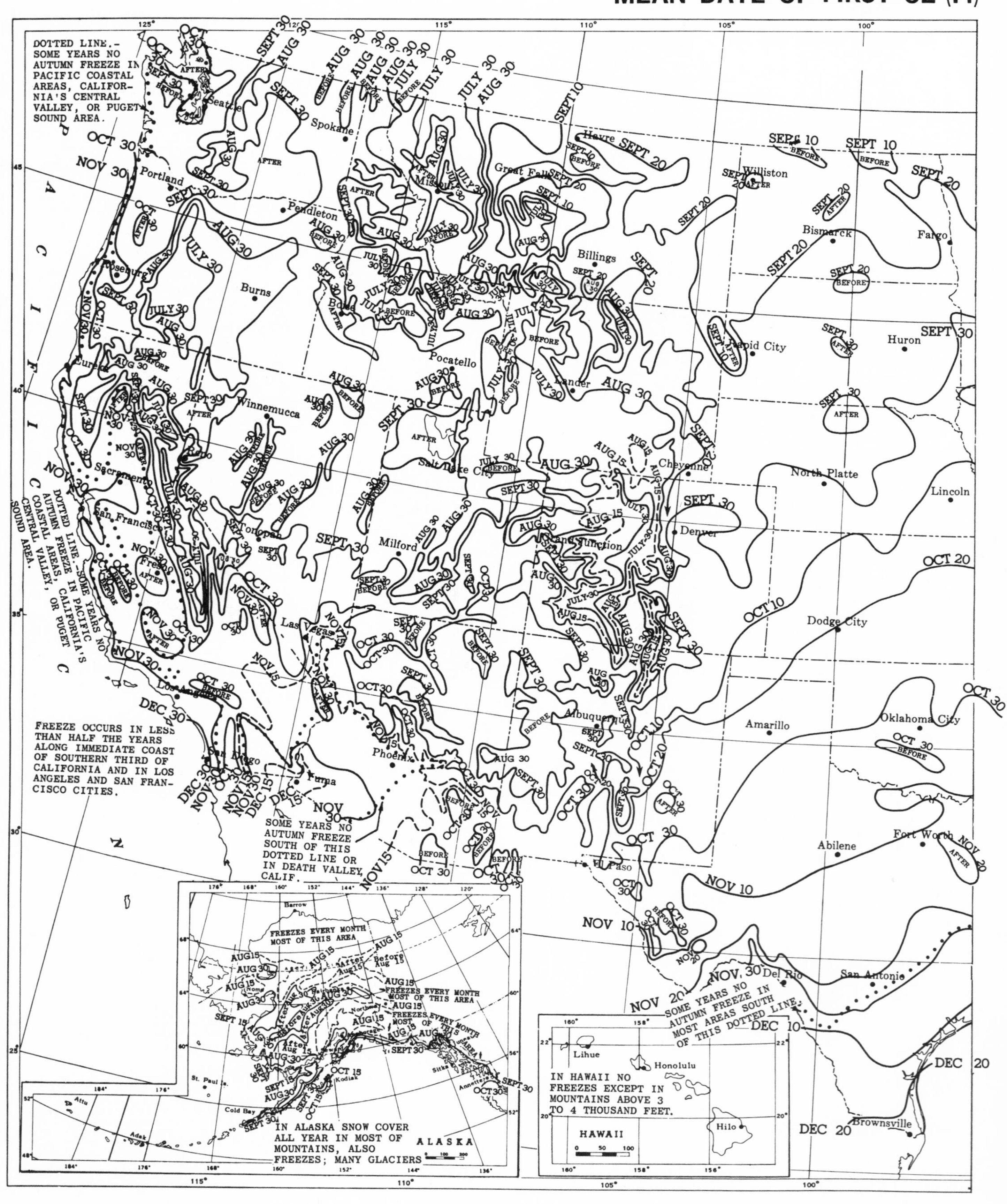

TEMPERATURE IN AUTUMN

"BEFORE" OR "AFTER IS ENTERED ON MOST SMALL AREAS OF THIS MAP TO SHOW MORE READILY WHETHER THE MEAN FREEZE DATE IN THE AREA IS BEFORE OR AFTER THE DATE PRINTED ON THE LINE.
AUTUMN (FALL) FREEZES ARE ASSUMED TO OCCUR BETWEEN JULY 1 AND DECEMBER 31.
CAUTION SHOULD BE USED IN INTERPOLATING ON THIS GENERALIZED MAP. SHARP CHANGES IN THE MEAN DATE MAY OCCUR IN SHORT DISTANCES, DUE TO DIFFERENCES IN ALTITUDE, SLOPE OF LAND, TYPE OF SOIL, VEGETATIVE COVER, BODIES OF WATER, AIR DRAINAGE, URBAN HEAT EFFECTS, ETC.
Based on Period 1921-50.
SOME YEARS NO AUTUMN FREEZE EAST OF THIS DOTTED LINE.
SOME YEARS NO AUTUMN FREEZE IN MOST AREAS SOUTH OF THIS DOTTED LINE.
FALL FREEZES OCCUR SOUTH OF THIS DASHED LINE IN LESS THAN HALF THE YEARS.
NO FREEZE
GULF OF MEXICO
ATLANTIC OCEAN
International Falls
Duluth
S. Ste. Marie
Caribou
Minneapolis
Green Bay
Lansing
Detroit
Buffalo
Burlington
Concord
Portland
Albany
Boston
Hartford
New Haven
New York
Dubuque
Chicago
Des Moines
Harrisburg
Philadelphia
Pittsburgh
Columbus
Indianapolis
Springfield
Kansas City
St. Louis
Washington, D.C.
Charleston
Lexington
Richmond
Norfolk
Raleigh
Nashville
Knoxville
Memphis
Wilmington
Columbia
Little Rock
Atlanta
Birmingham
Charleston
Shreveport
Jackson
Jacksonville
Pensacola
New Orleans
Houston
Tampa
Miami

MEAN LENGTH OF FREEZE-FREE
Temperature in Spring and First

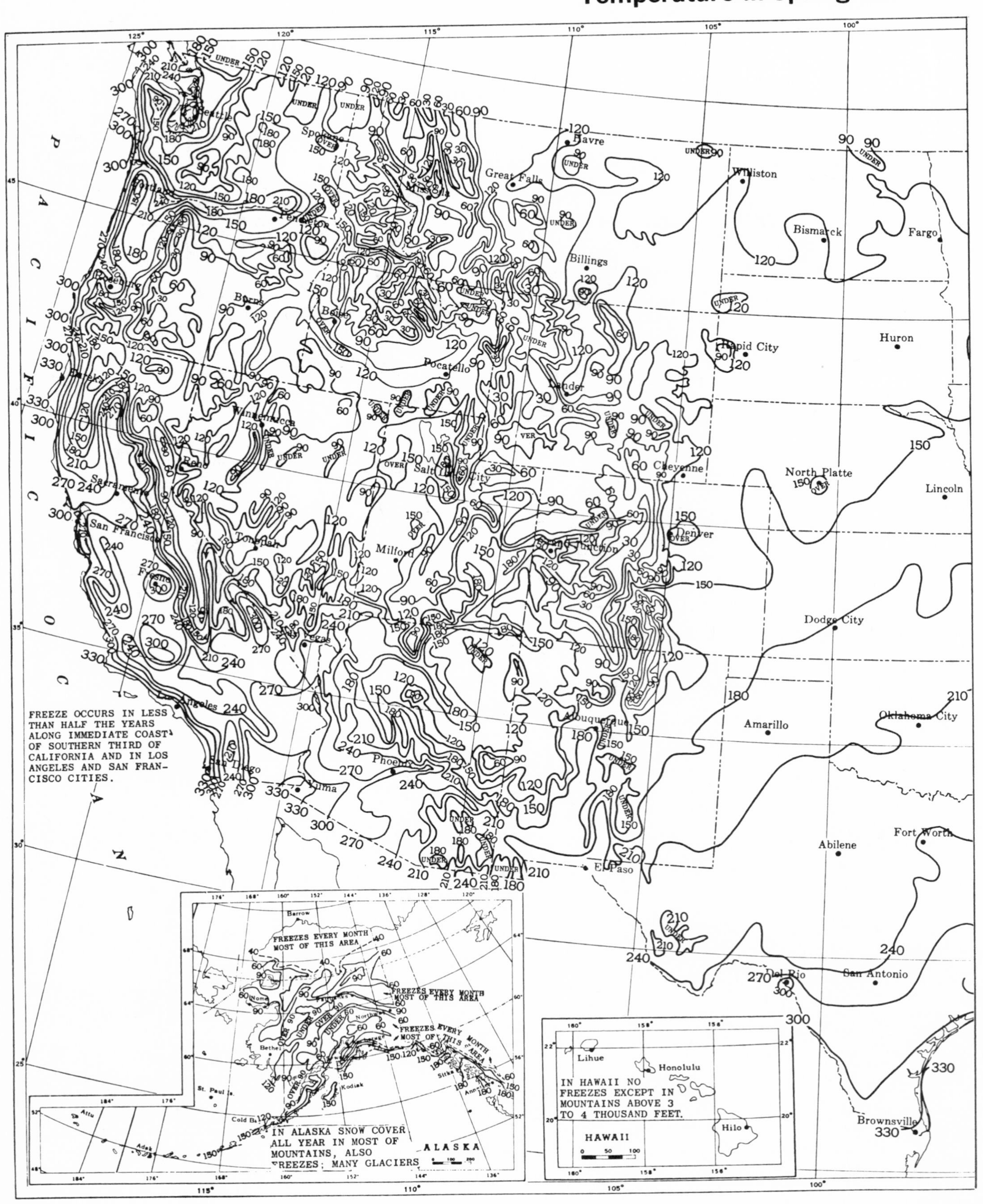

PERIOD (Days) Between Last 32°(F.)
32°(F.) Temperature in Autumn

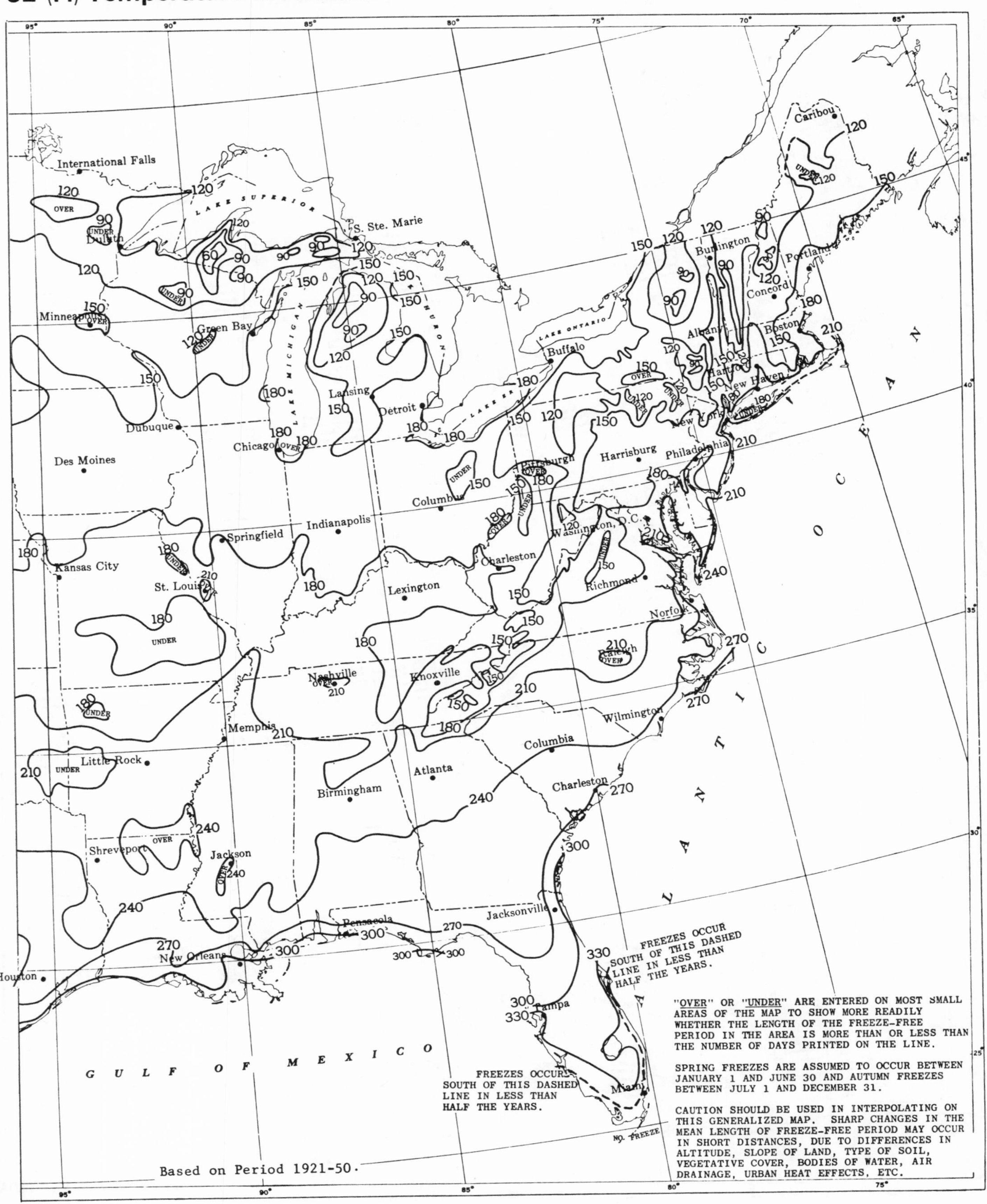

"OVER" OR "UNDER" ARE ENTERED ON MOST SMALL AREAS OF THE MAP TO SHOW MORE READILY WHETHER THE LENGTH OF THE FREEZE-FREE PERIOD IN THE AREA IS MORE THAN OR LESS THAN THE NUMBER OF DAYS PRINTED ON THE LINE.
SPRING FREEZES ARE ASSUMED TO OCCUR BETWEEN JANUARY 1 AND JUNE 30 AND AUTUMN FREEZES BETWEEN JULY 1 AND DECEMBER 31.
CAUTION SHOULD BE USED IN INTERPOLATING ON THIS GENERALIZED MAP. SHARP CHANGES IN THE MEAN LENGTH OF FREEZE-FREE PERIOD MAY OCCUR IN SHORT DISTANCES, DUE TO DIFFERENCES IN ALTITUDE, SLOPE OF LAND, TYPE OF SOIL, VEGETATIVE COVER, BODIES OF WATER, AIR DRAINAGE, URBAN HEAT EFFECTS, ETC.
FREEZES OCCUR SOUTH OF THIS DASHED LINE IN LESS THAN HALF THE YEARS.
FREEZES OCCUR SOUTH OF THIS DASHED LINE IN LESS THAN HALF THE YEARS.
NO. FREEZE
Based on Period 1921-50.
GULF OF MEXICO
ATLANTIC OCEAN
LAKE SUPERIOR
LAKE MICHIGAN
LAKE HURON
LAKE ERIE
LAKE ONTARIO
International Falls
Duluth
S. Ste. Marie
Minneapolis
Green Bay
Caribou
Burlington
Portland
Concord
Boston
Albany
Buffalo
Hartford
New Haven
New York
Lansing
Detroit
Dubuque
Chicago
Des Moines
Pittsburgh
Harrisburg
Philadelphia
Columbus
Indianapolis
Springfield
Washington, D.C.
Kansas City
St. Louis
Charleston
Richmond
Lexington
Norfolk
Raleigh
Nashville
Knoxville
Memphis
Wilmington
Columbia
Little Rock
Atlanta
Birmingham
Charleston
Shreveport
Jackson
Jacksonville
Pensacola
New Orleans
Houston
Tampa
Miami

MEAN LENGTH OF PERIOD BETWEEN LAST IN SPRING AND FIRST IN
For Selected

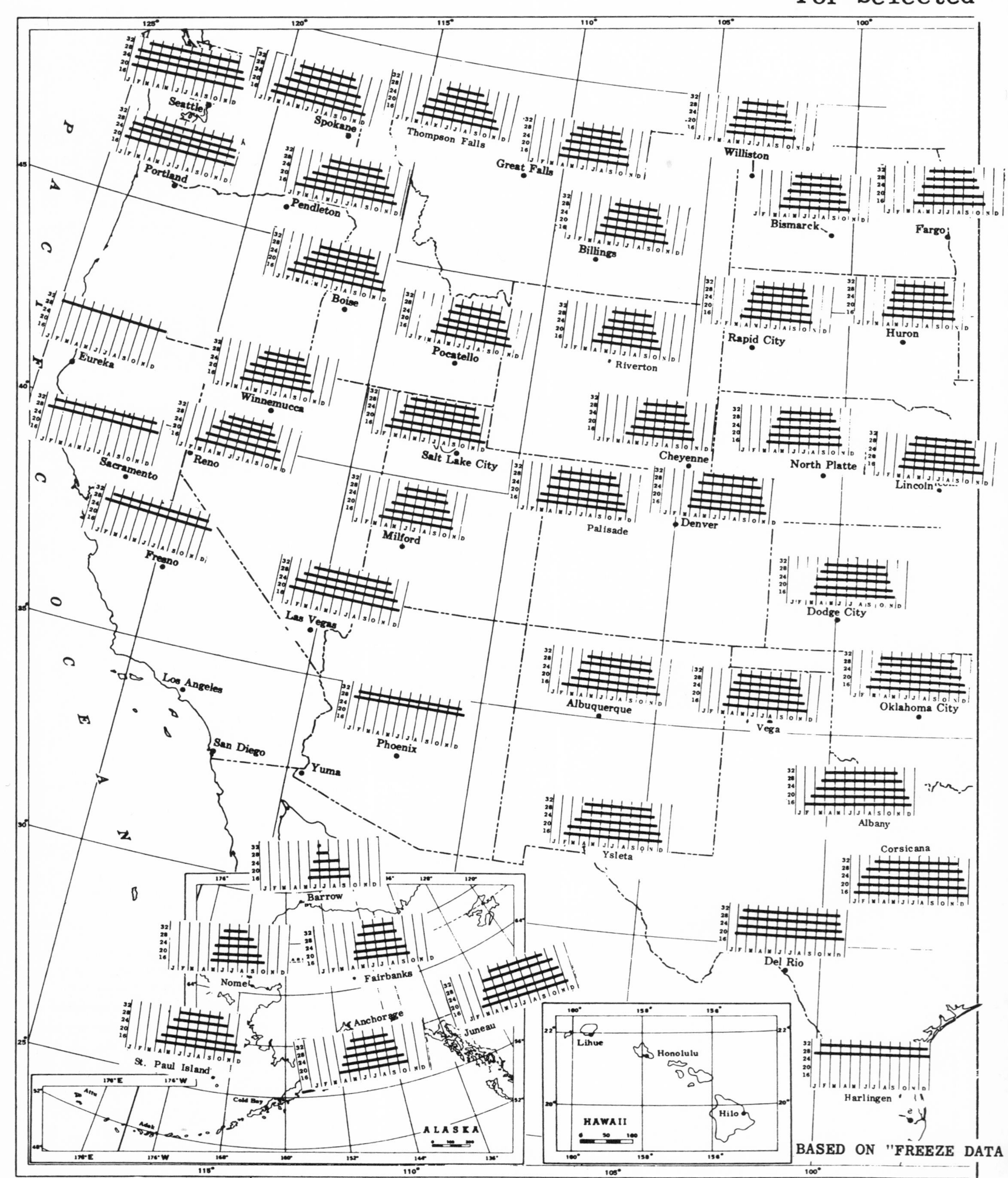

AUTUMN OF 32°F, 28°F, 24°F, 20°F, and 16°F, RESPECTIVELY
Stations

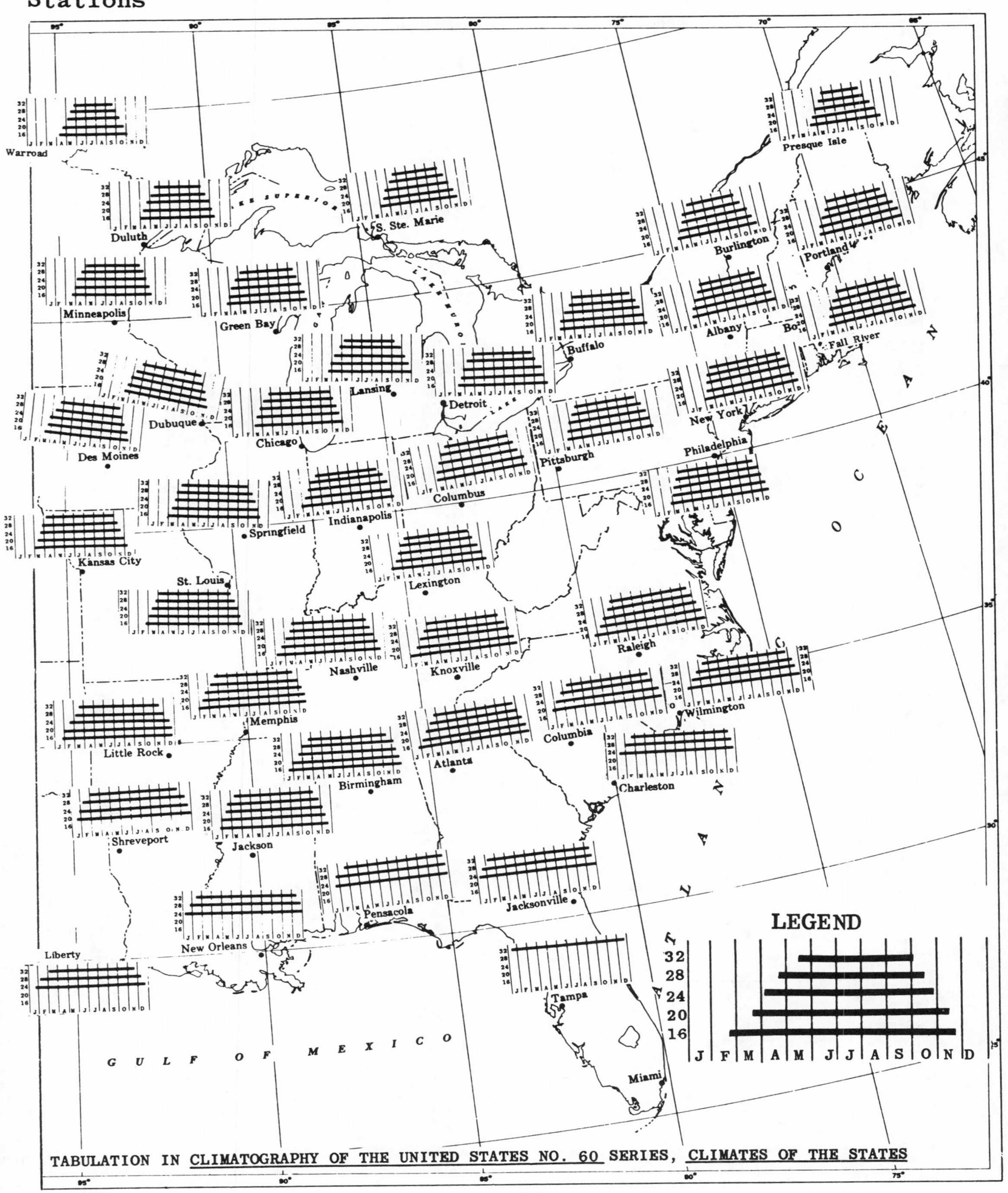

TABULATION IN CLIMATOGRAPHY OF THE UNITED STATES NO. 60 SERIES, CLIMATES OF THE STATES

MEAN DATE OF LAST 32°F. TEMPERATURE IN SPRING, FIRST 32°F. IN AUTUMN, AND MEAN LENGTH OF FREEZE-FREE PERIOD (Days)

State and Station	Mean date last 32°F. in spring	Mean date first 32°F. in fall	Mean freeze-free period (no. days)
ALA.Birmingham	Mar. 19	Nov. 14	241
Mobile U	Feb. 17	Dec. 12	298
Montgomery U	Feb. 27	Dec. 3	279
ALASKA.Anchorage	May 18	Sept. 13	118
Barrow	June 27	July 5	8
Cordova	May 10	Oct. 2	145
Fairbanks	May 24	Aug. 29	97
Juneau	Apr. 27	Oct. 19	176
Nome	June 12	Aug. 24	73
ARIZ.Flagstaff	June 8	Oct. 2	116
Phoenix	Jan. 27	Dec. 11	317
Tucson	Mar. 6	Nov. 23	261
Winslow	Apr. 28	Oct. 21	176
Yuma U	Jan. 11	Dec. 27	350
ARK.Fort Smith	Mar. 23	Nov. 9	231
Little Rock	Mar. 16	Nov. 15	244
CALIF.Bakersfield	Feb. 14	Nov. 28	287
Eureka U	Jan. 24	Dec. 25	335
Fresno	Feb. 3	Dec. 3	303
Los Angeles U.	*	*	*
Red Bluff	Feb. 25	Nov. 29	277
Sacramento	Jan. 24	Dec. 11	321
San Diego	*	*	*
San Francisco U.	*	*	*
COLO.Denver U.	May 2	Oct. 14	165
Palisades	Apr. 22	Oct. 17	178
Pueblo	Apr. 28	Oct. 12	167
CONN.Hartford	Apr. 22	Oct. 19	180
New Haven	Apr. 15	Oct. 27	195
D.C.Washington U	Apr. 10	Oct. 28	200
FLA.Apalachicola U	Feb. 2	Dec. 21	322
Fort Myers	*	*	*
Jacksonville U	Feb. 6	Dec. 16	313
Key West	*	*	*
Lakeland	Jan. 10	Dec. 25	349
Miami	*	*	*
Orlando	Jan. 31	Dec. 17	319
Pensacola U.	Feb. 18	Dec. 15	300
Tallahassee	Feb. 26	Dec. 3	280
Tampa	Jan. 10	Dec. 26	349
GA.Atlanta U	Mar. 20	Nov. 19	244
Augusta	Mar. 7	Nov. 22	260
Macon	Mar. 12	Nov. 19	252
Savannah	Feb. 21	Dec. 9	291
IDAHO.Boise	Apr. 29	Oct. 16	171
Pocatello	May 8	Sept. 30	145
Salmon	June 4	Sept. 6	94
ILL.Cairo U.	Mar. 23	Nov. 11	233
Chicago U.	Apr. 19	Oct. 28	192
Freeport	May 8	Oct. 4	149
Peoria	Apr. 22	Oct. 16	177
Springfield U.	Apr. 8	Oct. 30	205
IND.Evansville	Apr. 2	Nov. 4	216
Fort Wayne	Apr. 24	Oct. 20	179
Indianapolis U	Apr. 17	Oct. 27	193
South Bend	May 3	Oct. 16	165
IOWA.Des Moines U.	Apr. 20	Oct. 19	183
Dubuque U.	Apr. 19	Oct. 19	184
Keokuk	Apr. 12	Oct. 26	197
Sioux City	Apr. 28	Oct. 12	167
KANS.Concordia U	Apr. 16	Oct. 24	191
Dodge City	Apr. 22	Oct. 24	184
Goodland	May 5	Oct. 9	157
Topeka U	Apr. 9	Oct. 26	200
Wichita	Apr. 5	Nov. 1	210
KY.Lexington	Apr. 13	Oct. 28	198
Louisville U	Apr. 1	Nov. 7	220
LA.Lake Charles	Feb. 18	Dec. 6	291
New Orleans	Feb. 13	Dec. 12	302
Shreveport	Mar. 1	Nov. 27	272
MAINE.Greenville	May 27	Sept. 20	116
Portland	Apr. 29	Oct. 15	169
MD.Annapolis	Mar. 4	Nov. 15	225
Baltimore U.	Mar. 28	Nov. 17	234
Frederick	Mar. 24	Oct. 17	176
MASS.Boston	Apr. 16	Oct. 25	192
Nantucket	Apr. 12	Nov. 16	219
MICH.Alpena U.	May 6	Oct. 9	156
Detroit	Apr. 25	Oct. 23	181
Escanaba U	May 14	Oct. 6	145
Grand Rapids U	Apr. 25	Oct. 27	185
Marquette U.	May 14	Oct. 17	156
S. Ste. Marie	May 18	Oct. 3	138
MINN.Albert Lee	May 3	Oct. 6	156
Big Falls R.S.	June 4	Sept. 7	95
Brainerd	May 16	Sept. 24	131
Duluth	May 22	Sept. 24	125
Minneapolis	Apr. 30	Oct. 13	166
St. Cloud	May 9	Sept. 29	144
MISS.Jackson	Mar. 10	Nov. 13	248
Meridian	Mar. 13	Nov. 14	246
Vicksburg U.	Mar. 8	Nov. 15	252
MO.Columbia	Apr. 9	Oct. 24	198
Kansas City	Apr. 5	Oct. 31	210
St. Louis U.	Apr. 2	Nov. 8	220
Springfield	Apr. 10	Oct. 31	203
MONT.Billings	May 15	Sept. 24	132
Glasgow U.	May 19	Sept. 20	124
Great Falls	May 14	Sept. 26	135
Havre U.	May 9	Sept. 23	138
Helena	May 12	Sept. 23	134
Kalispell	May 12	Sept. 23	135
Miles City	May 5	Oct. 3	150
Superior	June 5	Aug. 30	85
NEBR.Grand Island	Apr. 29	Oct. 6	160
Lincoln	Apr. 20	Oct. 17	180
Norfolk	May 4	Oct. 3	152
North Platte	Apr. 30	Oct. 7	160
Omaha	Apr. 14	Oct. 20	189
Valentine Lakes	May 7	Sept. 30	146
NEV.Elko	June 6	Sept. 3	89
Las Vegas	Mar. 13	Nov. 13	245
Reno	May 14	Oct. 2	141
Winnemucca	May 18	Sept. 21	125
N.H.Concord	May 11	Sept. 30	142
N.J.Cape May	Apr. 4	Nov. 15	225
Trenton U	Apr. 8	Nov. 5	211
N.MEX.Albuquerque	Apr. 16	Oct. 29	196
Roswell	Apr. 9	Nov. 2	208
N.Y.Albany	Apr. 27	Oct. 13	169
Binghamton U.	May 4	Oct. 6	154
Buffalo	Apr. 30	Oct. 25	179
New York U.	Apr. 7	Nov. 12	219
Rochester	Apr. 28	Oct. 21	176
Syracuse	Apr. 30	Oct. 15	168
N.C.Asheville U	Apr. 12	Oct. 24	195
Charlotte U	Mar. 21	Nov. 15	239
Greenville	Mar. 28	Nov. 5	222
Hatteras	Feb. 25	Dec. 18	296
Raleigh U	Mar. 24	Nov. 16	237
Wilmington U.	Mar. 8	Nov. 24	262
N.DAK.Bismarck	May 11	Sept. 24	136
Devils Lake U	May 18	Sept. 22	127
Fargo	May 13	Sept. 27	137
Williston U	May 14	Sept. 23	132
OHIO.Akron-Canton	Apr. 29	Oct. 20	173
Cincinnati (Abbe)	Apr. 15	Oct. 25	192
Cleveland	Apr. 21	Nov. 2	195
Columbus U.	Apr. 17	Oct. 30	196
Dayton	Apr. 20	Oct. 21	184
Toledo	Apr. 24	Oct. 25	184
OKLA.Okla.City U.	Mar. 28	Nov. 7	223
Tulsa	Mar. 31	Nov. 2	216
OREG.Astoria	Mar. 18	Nov. 24	251
Bend	June 17	Aug. 17	62
Medford	Apr. 25	Oct. 20	178
Pendleton	Apr. 27	Oct. 8	163
Portland U.	Feb. 25	Dec. 1	279
Salem	Apr. 14	Oct. 27	197
PA.Allentown	Apr. 20	Oct. 16	180
Harrisburg	Apr. 10	Oct. 28	201
Philadelphia U.	Mar. 30	Nov. 17	232
Pittsburgh	Apr. 20	Oct. 23	187
Scranton U.	Apr. 24	Oct. 14	174
R.I.Providence U.	Apr. 13	Oct. 27	197
S.C.Charleston U.	Feb. 19	Dec. 10	294
Columbia U.	Mar. 14	Nov. 21	252
Greenville	Mar. 23	Nov. 17	239
S.DAK.Huron U	May 4	Sept. 30	149
Rapid City U.	May 7	Oct. 4	150
Sioux Falls U	May 5	Oct. 3	152
TENN.Chattanooga U.	Mar. 26	Nov. 10	229
Knoxville U	Mar. 31	Nov. 6	220
Memphis U	Mar. 20	Nov. 12	237
Nashville U	Mar. 28	Nov. 7	224
TEX.Albany	Mar. 30	Nov. 9	224
Balmorhea	Apr. 1	Nov. 12	226
Beeville	Feb. 21	Dec. 6	288
College Station	Mar. 1	Dec. 1	275
Corsicana	Mar. 13	Nov. 27	259
Dalhart Exp. Sta.	Apr. 23	Oct. 18	178
Dallas	Mar. 18	Nov. 22	249
Del Rio	Feb. 12	Dec. 9	300
Encinal	Feb. 15	Dec. 12	301
Houston	Feb. 5	Dec. 11	309
Lampasas	Apr. 1	Nov. 10	223
Matagorda	Feb. 12	Dec. 17	308
Midland	Apr. 3	Nov. 6	218
Mission	Jan. 30	Dec. 21	325
Mount Pleasant	Mar. 23	Nov. 12	233
Nacogdoches	Mar. 15	Nov. 13	243
Plainview	Apr. 10	Nov. 6	211
Presidio	Mar. 20	Nov. 13	238
Quanah	Mar. 31	Nov. 7	221
San Angelo	Mar. 25	Nov. 15	235
Ysleta	Apr. 6	Oct. 30	207
UTAH.Blanding	May 18	Oct. 14	148
Salt Lake City	Apr. 12	Nov. 1	202
VT.Burlington	May 8	Oct. 3	148
VA.Lynchburg	Apr. 6	Oct. 27	205
Norfolk U	Mar. 18	Nov. 27	254
Richmond U.	Apr. 2	Nov. 8	220
Roanoke	Apr. 20	Oct. 24	187
WASH.Bumping Lake	June 17	Aug. 16	60
Seattle U	Feb. 23	Dec. 1	281
Spokane	Apr. 20	Oct. 12	175
Tatoosh Island	Jan. 25	Dec. 20	329
Walla Walla U	Mar. 28	Nov. 1	218
Yakima	Apr. 19	Oct. 15	179
W.VA.Charleston	Apr. 18	Oct. 28	193
Parkersburg	Apr. 16	Oct. 21	189
WIS.Green Bay	May 6	Oct. 13	161
La Crosse U	May 1	Oct. 8	161
Madison U	Apr. 26	Oct. 19	177
Milwaukee U	Apr. 20	Oct. 25	188
WYO.Casper	May 18	Sept. 25	130
Cheyenne	May 20	Sept. 27	130
Lander	May 15	Sept. 20	128
Sheridan	May 21	Sept. 21	123

* Occurs in less than 1 year in 10. No freeze of record in Key West, Fla.
U Indicates urban.

Charts and tabulation were derived from the Freeze Data tabulation in Climatography of the United States No. 60 - Climates of the States.

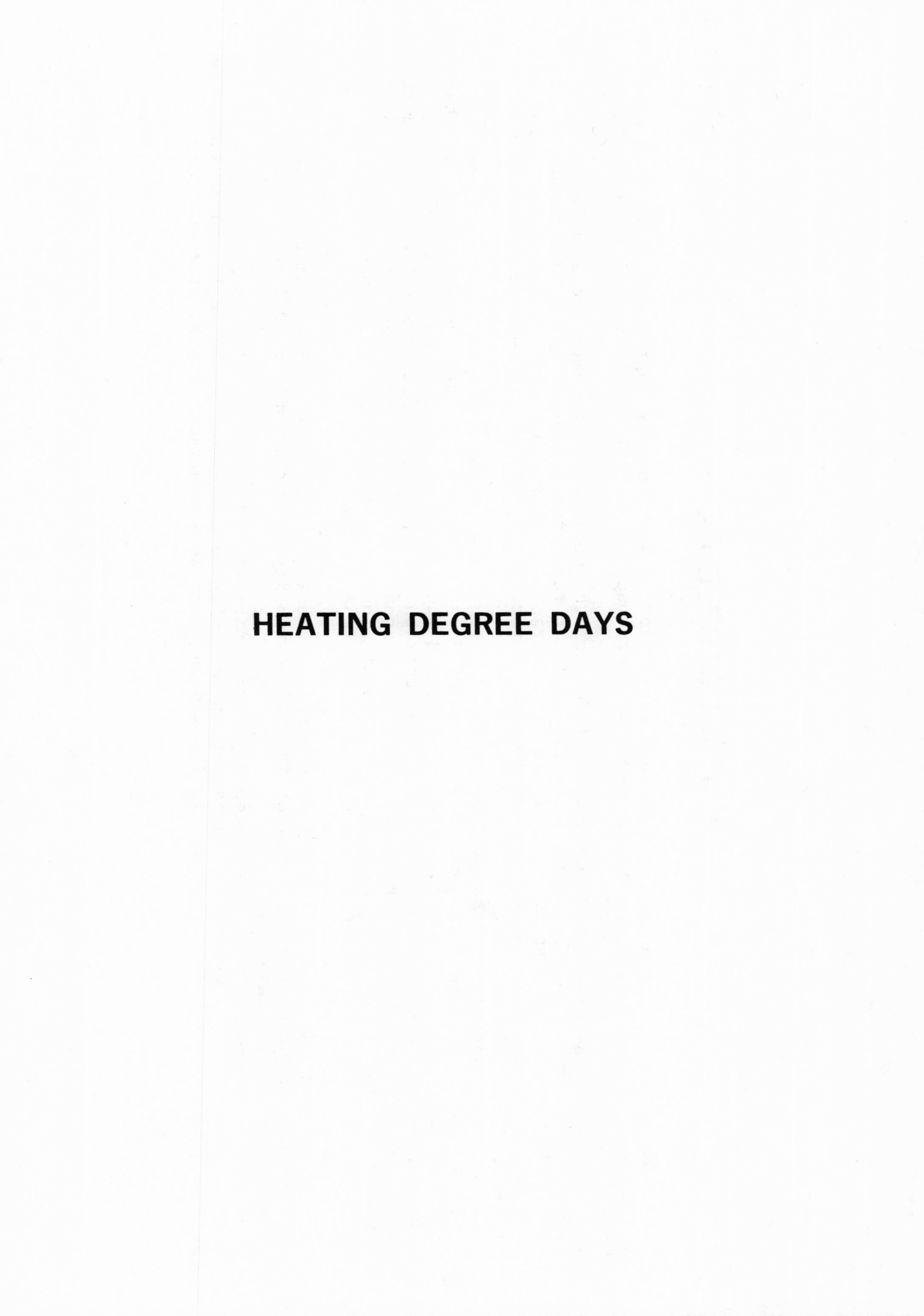

HEATING DEGREE DAYS

HEATING DEGREE DAYS

One of the most practical of weather statistics is the "heating degree day." First devised some 50 years ago, the degree day system has been in quite general use by the heating industry for more than 30 years.

Heating degree days are the number of degrees the daily average temperature is below 65°. Normally heating is not required in a building when the outdoor average daily temperature is 65°. Heating degree days are determined by substracting the average daily temperatures below 65° from the base 65°. A day with an average temperature of 50° has 15 heating degree days (65-50 = 15) while one with an average temperature of 65° or higher has none.

Several characteristics make the degree day figures especially useful. They are cumulative so that the degree day sum for a period of days represents the total heating load for that period. The relationship between degree days and fuel consumption is linear, i.e., doubling the degree days usually doubles the fuel consumption. Comparing normal seasonal degree days in different locations gives a rough estimate of seasonal fuel consumption. For example, it would require roughly 4-1/2 times as much fuel to heat a building in Chicago, Ill., where the mean annual total heating degree days are about 6,200 than to heat a similar building in New Orleans, La., where the annual total heating degree days are around 1,400. Using degree days has the advantage that the consumption ratios are fairly constant, i.e., the fuel consumed per 100 degree days is about the same whether the 100 degree days occur in only 3 or 4 days or are spread over 7 or 8 days.

The rapid adoption of the degree day system paralleled the spread of automatic fuel systems in the 1930's. Since oil and gas are more costly to store than solid fuels, this places a premium on the scheduling of deliveries and the precise evaluation of use rates and peak demands.

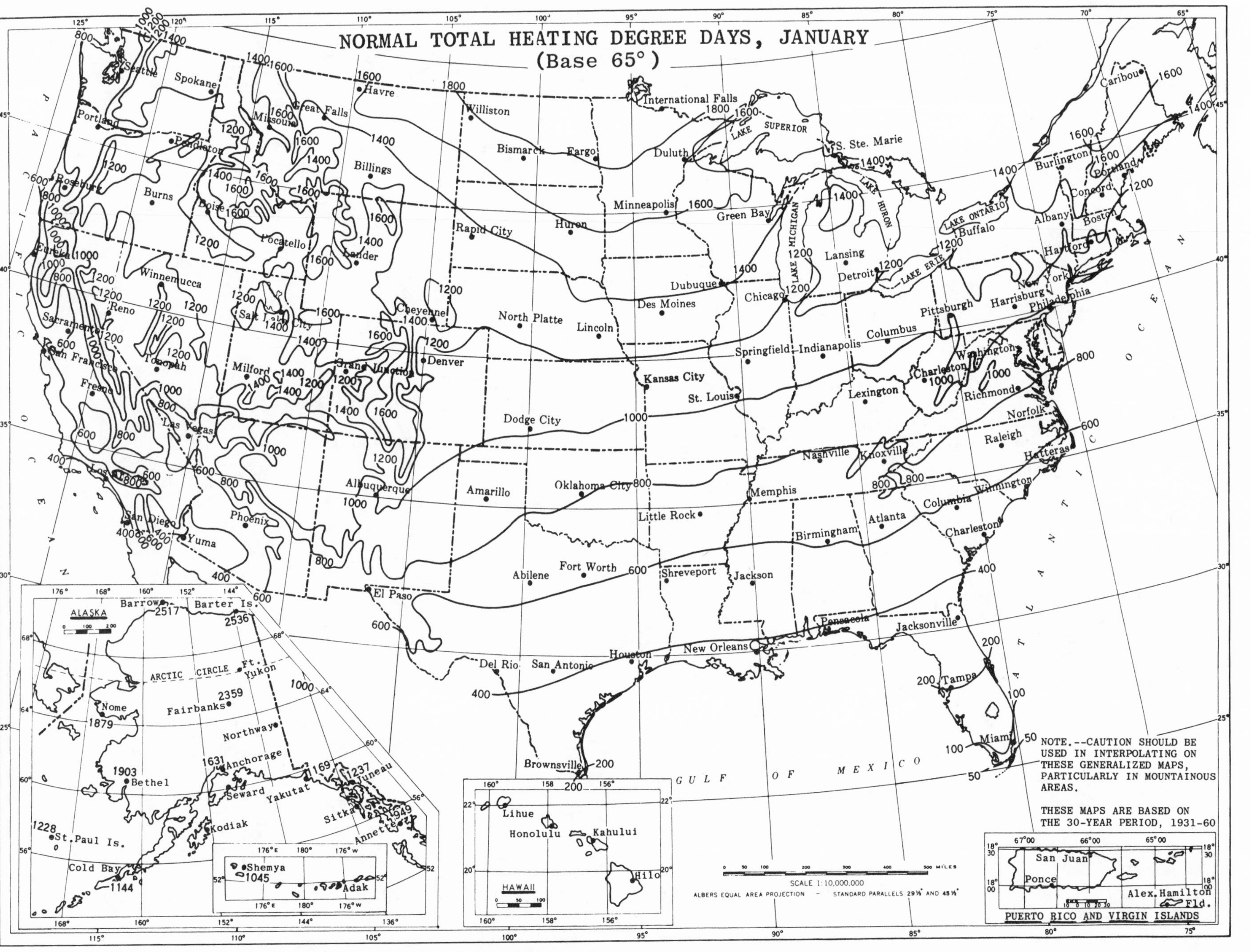
NORMAL TOTAL HEATING DEGREE DAYS, JANUARY
(Base 65°)
NOTE.--CAUTION SHOULD BE USED IN INTERPOLATING ON THESE GENERALIZED MAPS, PARTICULARLY IN MOUNTAINOUS AREAS.
THESE MAPS ARE BASED ON THE 30-YEAR PERIOD, 1931-60
SCALE 1:10,000,000
ALBERS EQUAL AREA PROJECTION - STANDARD PARALLELS 29½° AND 45½°
PUERTO RICO AND VIRGIN ISLANDS
San Juan
Ponce
Alex. Hamilton Fld.
HAWAII
Lihue
Honolulu
Kahului
Hilo
ALASKA
ARCTIC CIRCLE
Barrow 2517
Barter Is. 2536
Ft. Yukon
Fairbanks 2359
Northway
Nome 1879
Bethel 1903
Anchorage 1631
Seward
Yakutat 1169
Juneau 1237
Sitka
Annette 949
Kodiak
Cold Bay 1144
St. Paul Is. 1228
Shemya 1045
Adak
PACIFIC OCEAN
ATLANTIC OCEAN
GULF OF MEXICO
LAKE SUPERIOR
LAKE MICHIGAN
LAKE HURON
LAKE ERIE
LAKE ONTARIO
Seattle
Spokane
Portland
Pendleton
Burns
Roseburg
Eureka
Missoula
Great Falls
Havre
Billings
Boise
Pocatello
Lander
Winnemucca
Reno
Sacramento
San Francisco
Fresno
Tonopah
Las Vegas
Los Angeles
San Diego
Yuma
Phoenix
Milford
Salt Lake City
Grand Junction
Cheyenne
Denver
Albuquerque
El Paso
Williston
Bismarck
Rapid City
Fargo
Huron
North Platte
Dodge City
Amarillo
Abilene
Del Rio
San Antonio
Brownsville
Houston
Fort Worth
Oklahoma City
Lincoln
Des Moines
Kansas City
Minneapolis
Duluth
International Falls
Green Bay
Dubuque
St. Louis
Little Rock
Shreveport
New Orleans
Jackson
Memphis
Springfield
Chicago
Indianapolis
Lansing
Detroit
S. Ste. Marie
Nashville
Birmingham
Pensacola
Knoxville
Atlanta
Lexington
Columbus
Charleston
Pittsburgh
Harrisburg
Washington
Richmond
Norfolk
Raleigh
Hatteras
Wilmington
Columbia
Jacksonville
Tampa
Miami
Buffalo
Albany
Burlington
Concord
Portland
Boston
Hartford
New York
Philadelphia
Caribou

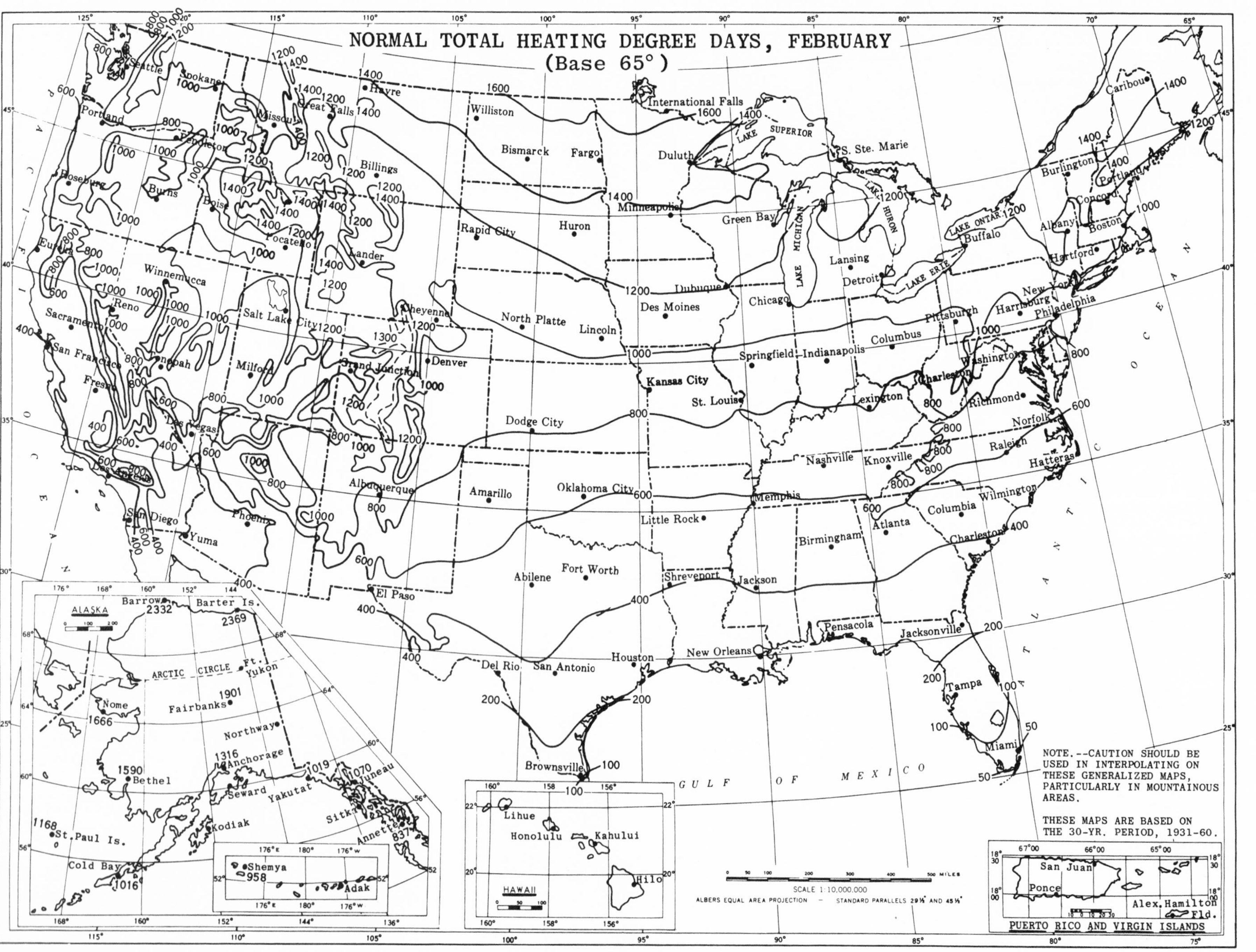
NORMAL TOTAL HEATING DEGREE DAYS, FEBRUARY
(Base 65°)
NOTE.--CAUTION SHOULD BE USED IN INTERPOLATING ON THESE GENERALIZED MAPS, PARTICULARLY IN MOUNTAINOUS AREAS.
THESE MAPS ARE BASED ON THE 30-YR. PERIOD, 1931-60.
SCALE 1:10,000,000
ALBERS EQUAL AREA PROJECTION — STANDARD PARALLELS 29½° AND 45½°
ALASKA
HAWAII
PUERTO RICO AND VIRGIN ISLANDS
GULF OF MEXICO
ATLANTIC OCEAN
PACIFIC OCEAN
Barrow 2332
Barter Is. 2369
Fairbanks 1901
Nome 1666
Bethel 1590
Anchorage 1316
St. Paul Is. 1168
Cold Bay 1016
Juneau 1070
Yakutat 1019
Annette 837
Shemya 958
Adak

NORMAL TOTAL HEATING DEGREE DAYS, MARCH
(Base 65°)
NOTE.--CAUTION SHOULD BE USED IN INTERPOLATING ON THESE GENERALIZED MAPS, PARTICULARLY IN MOUNTAINOUS AREAS.
THESE MAPS ARE BASED ON THE 30-YR. PERIOD, 1931-60.
ALASKA
HAWAII
PUERTO RICO AND VIRGIN ISLANDS
GULF OF MEXICO
ATLANTIC OCEAN
PACIFIC OCEAN
SCALE 1:10,000,000
ALBERS EQUAL AREA PROJECTION - STANDARD PARALLELS 29½° AND 45½°

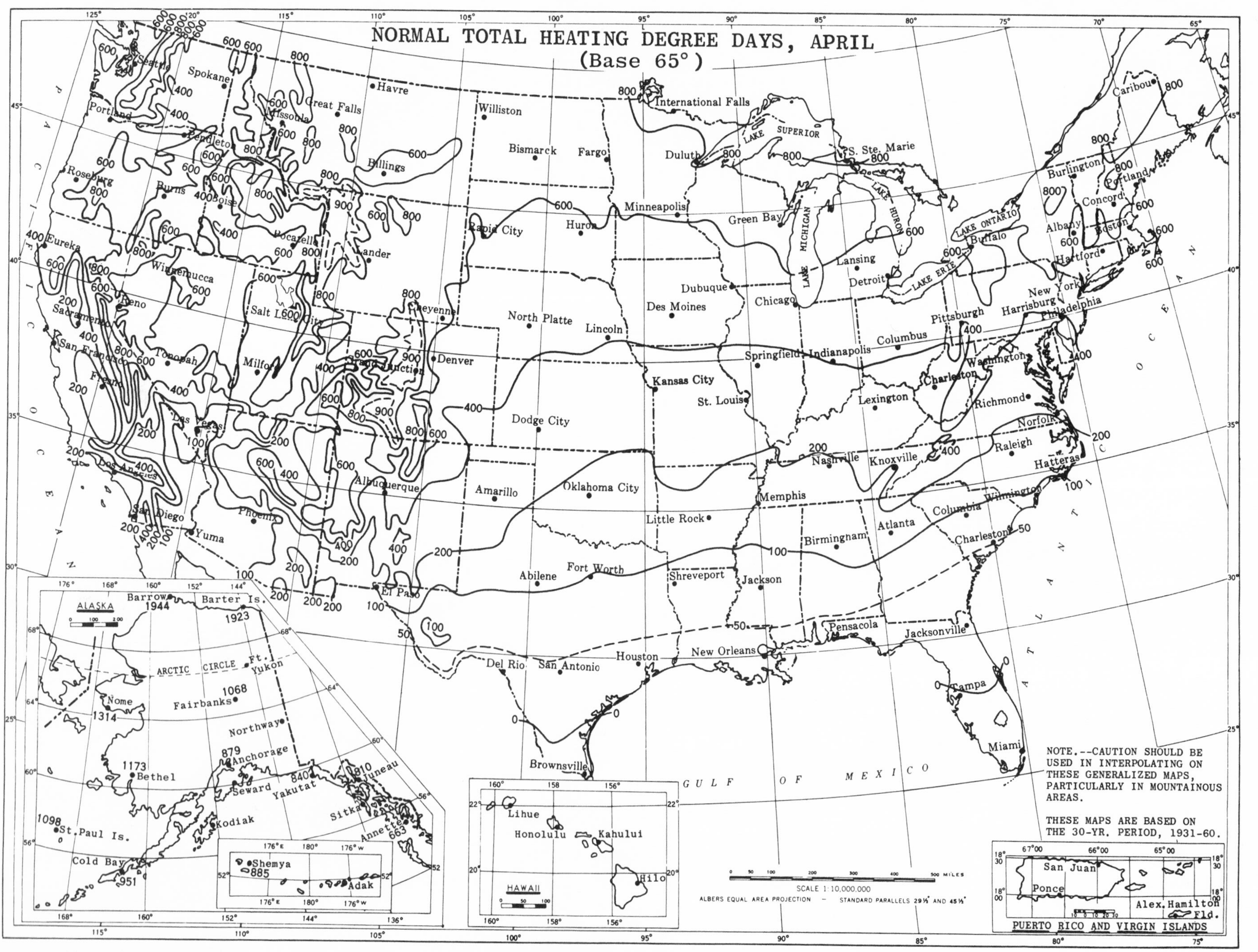
NORMAL TOTAL HEATING DEGREE DAYS, APRIL
(Base 65°)
NOTE.--CAUTION SHOULD BE USED IN INTERPOLATING ON THESE GENERALIZED MAPS, PARTICULARLY IN MOUNTAINOUS AREAS.
THESE MAPS ARE BASED ON THE 30-YR. PERIOD, 1931-60.
SCALE 1:10,000,000
ALBERS EQUAL AREA PROJECTION — STANDARD PARALLELS 29½° AND 45½°
PACIFIC OCEAN
ATLANTIC OCEAN
GULF OF MEXICO
LAKE SUPERIOR
LAKE MICHIGAN
LAKE HURON
LAKE ERIE
LAKE ONTARIO
Seattle
Spokane
Portland
Pendleton
Roseburg
Burns
Boise
Eureka
Winnemucca
Reno
Sacramento
San Francisco
Fresno
Tonopah
Las Vegas
Los Angeles
San Diego
Yuma
Phoenix
Missoula
Great Falls
Havre
Billings
Pocatello
Salt Lake City
Milford
Lander
Cheyenne
Denver
Grand Junction
Albuquerque
El Paso
Williston
Bismarck
Fargo
Rapid City
Huron
North Platte
Lincoln
Dodge City
Amarillo
Oklahoma City
Abilene
Fort Worth
Del Rio
San Antonio
Houston
Brownsville
International Falls
Duluth
Minneapolis
Des Moines
Kansas City
St. Louis
Little Rock
Shreveport
Green Bay
Dubuque
Chicago
Springfield
Indianapolis
Memphis
Jackson
New Orleans
S. Ste. Marie
Lansing
Detroit
Columbus
Lexington
Nashville
Knoxville
Birmingham
Atlanta
Pensacola
Jacksonville
Tampa
Miami
Pittsburgh
Charleston
Washington
Richmond
Norfolk
Raleigh
Hatteras
Wilmington
Columbia
Charleston
Buffalo
Harrisburg
Philadelphia
New York
Hartford
Albany
Boston
Concord
Burlington
Portland
Caribou
ALASKA
Barrow 1944
Barter Is. 1923
ARCTIC CIRCLE
Ft. Yukon
Fairbanks 1068
Nome 1314
Northway
Anchorage 879
Bethel 1173
Seward
Yakutat 840
Juneau 810
Sitka
Annette 663
Kodiak
St. Paul Is. 1098
Cold Bay 951
Shemya 885
Adak
HAWAII
Lihue
Honolulu
Kahului
Hilo
PUERTO RICO AND VIRGIN ISLANDS
San Juan
Ponce
Alex. Hamilton Fld.

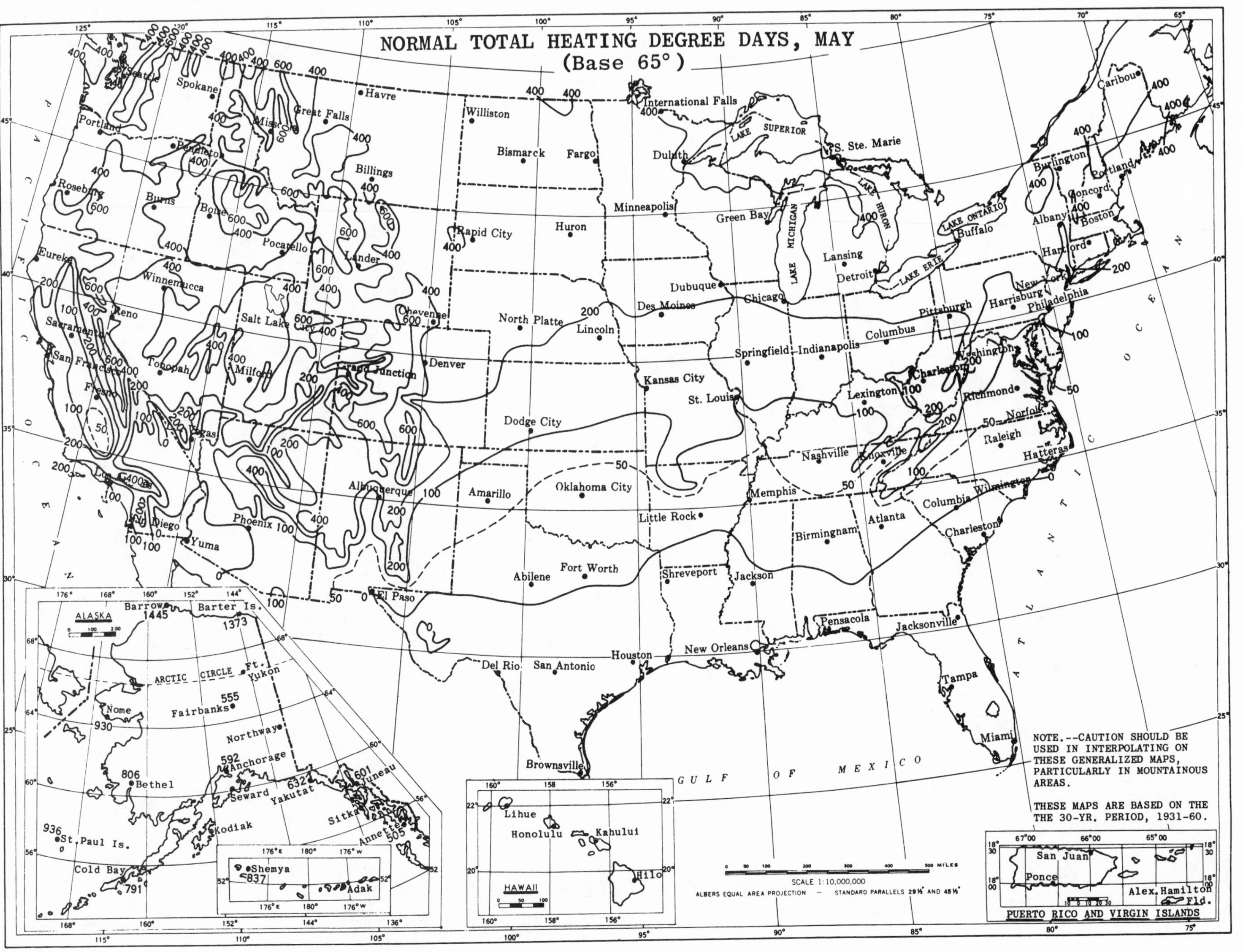
NORMAL TOTAL HEATING DEGREE DAYS, MAY
(Base 65°)
NOTE.--CAUTION SHOULD BE USED IN INTERPOLATING ON THESE GENERALIZED MAPS, PARTICULARLY IN MOUNTAINOUS AREAS.
THESE MAPS ARE BASED ON THE THE 30-YR. PERIOD, 1931-60.
SCALE 1:10,000,000
ALBERS EQUAL AREA PROJECTION — STANDARD PARALLELS 29½° AND 45½°
GULF OF MEXICO
ATLANTIC OCEAN
PACIFIC OCEAN
ALASKA
Barrow 1445
Barter Is. 1373
ARCTIC CIRCLE
Ft. Yukon
Fairbanks 555
Nome 930
Northway
Anchorage 592
Bethel 806
Seward
Yakutat 632
Juneau 601
Sitka
Annette 505
Kodiak
St. Paul Is. 936
Cold Bay 791
Shemya 837
Adak
HAWAII
Lihue
Honolulu
Kahului
Hilo
PUERTO RICO AND VIRGIN ISLANDS
San Juan
Ponce
Alex. Hamilton Fld.
LAKE SUPERIOR
LAKE MICHIGAN
LAKE HURON
LAKE ERIE
LAKE ONTARIO

NORMAL TOTAL HEATING DEGREE DAYS, JUNE
(Base 65°)
NOTE.--CAUTION SHOULD BE USED IN INTERPOLATING ON THESE GENERALIZED MAPS, PARTICULARLY IN MOUNTAINOUS AREAS.
THESE MAPS ARE BASED ON THE 30-YR. PERIOD, 1931-60.
SCALE 1:10,000,000
ALBERS EQUAL AREA PROJECTION — STANDARD PARALLELS 29½° AND 45½°
ALASKA
HAWAII
PUERTO RICO AND VIRGIN ISLANDS
PACIFIC OCEAN
ATLANTIC OCEAN
GULF OF MEXICO

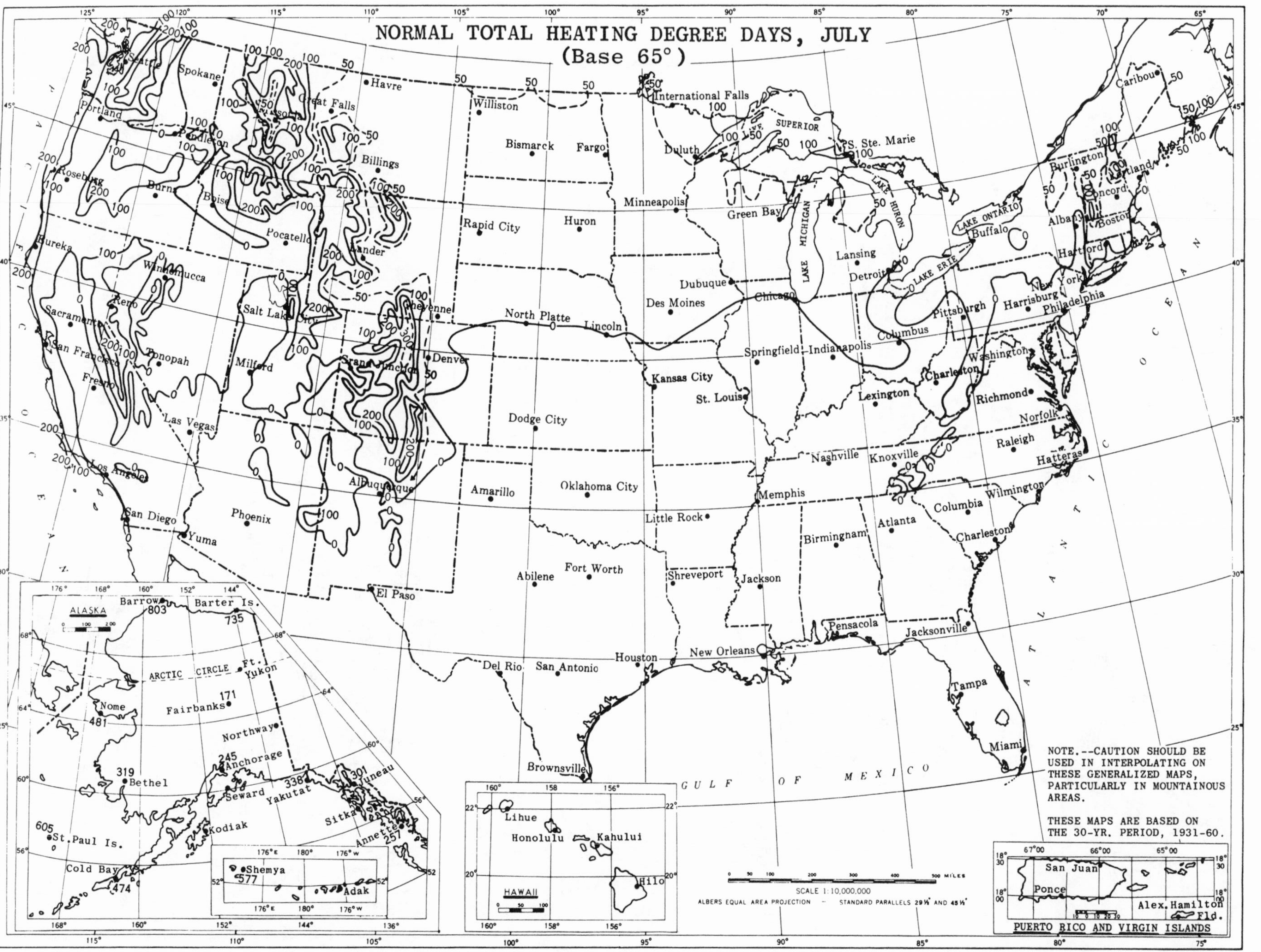
NORMAL TOTAL HEATING DEGREE DAYS, JULY
(Base 65°)
NOTE.--CAUTION SHOULD BE USED IN INTERPOLATING ON THESE GENERALIZED MAPS, PARTICULARLY IN MOUNTAINOUS AREAS.
THESE MAPS ARE BASED ON THE 30-YR. PERIOD, 1931-60.
SCALE 1:10,000,000
ALBERS EQUAL AREA PROJECTION — STANDARD PARALLELS 29½° AND 45½°
GULF OF MEXICO
ATLANTIC OCEAN
PACIFIC OCEAN
ALASKA
HAWAII
PUERTO RICO AND VIRGIN ISLANDS

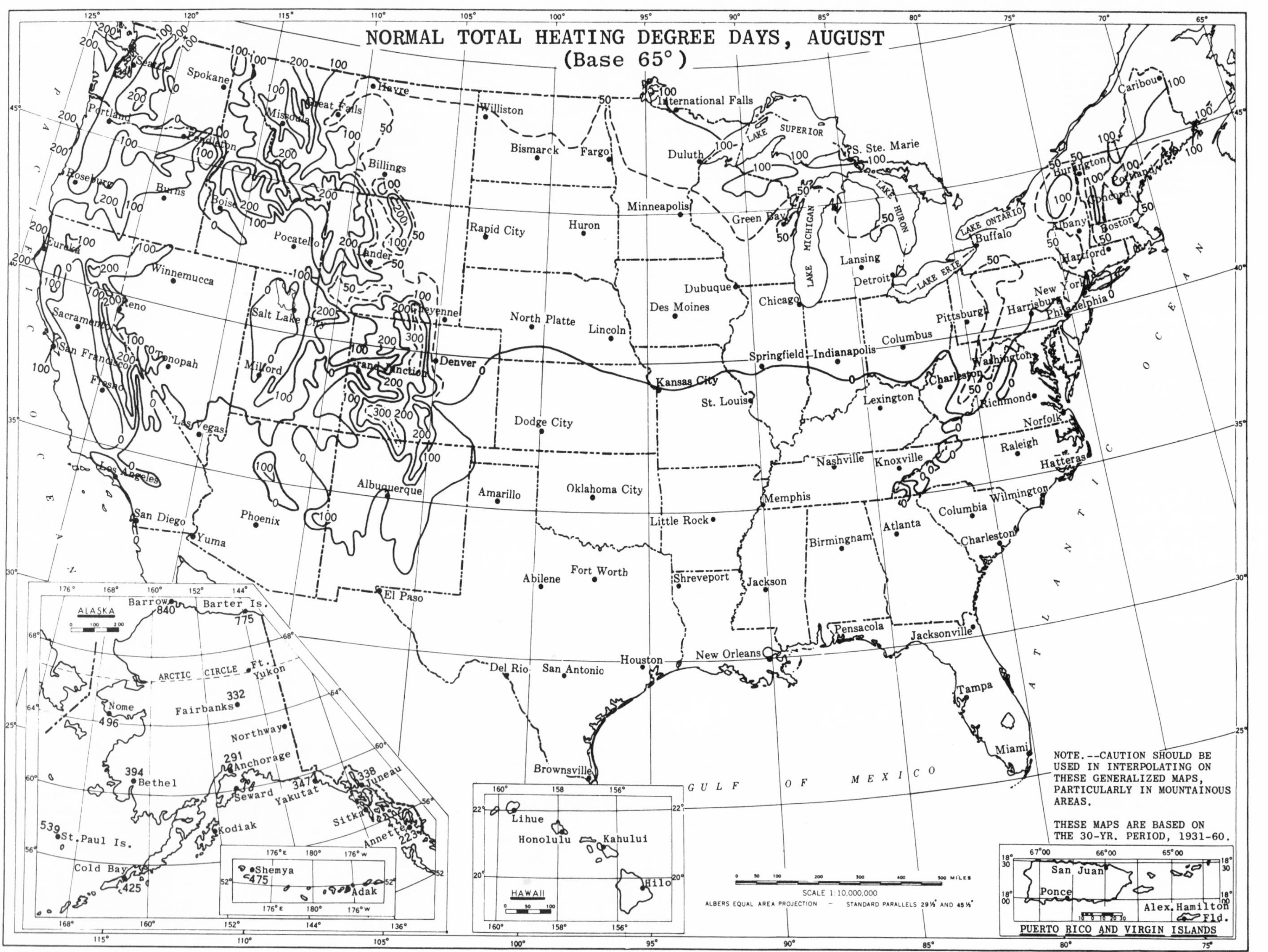
NORMAL TOTAL HEATING DEGREE DAYS, AUGUST
(Base 65°)
NOTE.--CAUTION SHOULD BE USED IN INTERPOLATING ON THESE GENERALIZED MAPS, PARTICULARLY IN MOUNTAINOUS AREAS.
THESE MAPS ARE BASED ON THE 30-YR. PERIOD, 1931-60.
SCALE 1:10,000,000
ALBERS EQUAL AREA PROJECTION - STANDARD PARALLELS 29½° AND 45½°
PACIFIC OCEAN
ATLANTIC OCEAN
GULF OF MEXICO
LAKE SUPERIOR
LAKE MICHIGAN
LAKE HURON
LAKE ERIE
LAKE ONTARIO
Seattle
Spokane
Portland
Pendleton
Roseburg
Burns
Boise
Eureka
Winnemucca
Reno
Sacramento
San Francisco
Tonopah
Fresno
Los Angeles
San Diego
Las Vegas
Yuma
Phoenix
Missoula
Great Falls
Havre
Billings
Pocatello
Lander
Salt Lake City
Milford
Cheyenne
Grand Junction
Denver
Albuquerque
El Paso
Williston
Bismarck
Fargo
Rapid City
Huron
North Platte
Lincoln
Dodge City
Amarillo
Oklahoma City
Abilene
Fort Worth
Del Rio
San Antonio
Houston
Brownsville
International Falls
Duluth
Minneapolis
Green Bay
Dubuque
Des Moines
Kansas City
St. Louis
Springfield
Chicago
Indianapolis
Little Rock
Shreveport
Memphis
Jackson
New Orleans
S. Ste. Marie
Lansing
Detroit
Columbus
Lexington
Nashville
Knoxville
Birmingham
Atlanta
Pensacola
Jacksonville
Tampa
Miami
Buffalo
Pittsburgh
Charleston
Washington
Richmond
Norfolk
Raleigh
Hatteras
Wilmington
Columbia
Charleston
Harrisburg
Philadelphia
New York
Hartford
Albany
Boston
Burlington
Concord
Portland
Caribou
ALASKA
Barrow 840
Barter Is. 775
ARCTIC CIRCLE
Ft. Yukon
Fairbanks 332
Nome 496
Northway
Anchorage 291
Bethel 394
Seward
Yakutat 347
Juneau 338
Sitka
Annette 223
Kodiak
St. Paul Is. 539
Cold Bay 425
Shemya 475
Adak
HAWAII
Lihue
Honolulu
Kahului
Hilo
San Juan
Ponce
Alex. Hamilton Fld.
PUERTO RICO AND VIRGIN ISLANDS

NORMAL TOTAL HEATING DEGREE DAYS, SEPTEMBER
(Base 65°)
NOTE.--CAUTION SHOULD BE USED IN INTERPOLATING ON THESE GENERALIZED MAPS, PARTICULARLY IN MOUNTAINOUS AREAS.
THESE MAPS ARE BASED ON THE 30-YR. PERIOD, 1931-60.
SCALE 1:10,000,000
ALBERS EQUAL AREA PROJECTION — STANDARD PARALLELS 29½° AND 45½°
ALASKA
HAWAII
PUERTO RICO AND VIRGIN ISLANDS
GULF OF MEXICO
ATLANTIC OCEAN
PACIFIC OCEAN
Seattle
Spokane
Portland
Pendleton
Roseburg
Burns
Boise
Pocatello
Eureka
Winnemucca
Reno
Sacramento
San Francisco
Fresno
Tonopah
Las Vegas
Los Angeles
San Diego
Yuma
Phoenix
Salt Lake City
Milford
Grand Junction
Denver
Cheyenne
Albuquerque
El Paso
Havre
Great Falls
Missoula
Billings
Lander
Williston
Bismarck
Fargo
Rapid City
Huron
North Platte
Lincoln
Dodge City
Amarillo
Oklahoma City
Abilene
Fort Worth
Del Rio
San Antonio
Houston
Brownsville
International Falls
Duluth
Minneapolis
Green Bay
Des Moines
Dubuque
Kansas City
St. Louis
Springfield
Chicago
Indianapolis
Lansing
Detroit
Columbus
Lexington
Nashville
Knoxville
Memphis
Little Rock
Shreveport
Jackson
Birmingham
Atlanta
New Orleans
Pensacola
Jacksonville
Tampa
Miami
Charleston
Columbia
Wilmington
Raleigh
Norfolk
Richmond
Hatteras
Washington
Pittsburgh
Harrisburg
Philadelphia
New York
Hartford
Albany
Boston
Concord
Portland
Burlington
Buffalo
Caribou
S. Ste. Marie
LAKE SUPERIOR
LAKE MICHIGAN
LAKE HURON
LAKE ERIE
LAKE ONTARIO
Barrow 1035
Barter Is. 987
ARCTIC CIRCLE
Ft. Yukon
Fairbanks 642
Nome 693
Northway
Anchorage 516
Bethel 612
Seward
Yakutat 474
Juneau 483
Sitka
Annette 342
Kodiak
St. Paul Is. 612
Cold Bay 525
Shemya 501
Adak
Lihue
Honolulu
Kahului
Hilo
San Juan
Ponce
Alex. Hamilton Fld.

NORMAL TOTAL HEATING DEGREE DAYS, OCTOBER
(Base 65°)
NOTE.--CAUTION SHOULD BE USED IN INTERPOLATING ON THESE GENERALIZED MAPS, PARTICULARLY IN MOUNTAINOUS AREAS.
THESE MAPS ARE BASED ON THE 30-YR. PERIOD, 1931-60.
SCALE 1:10,000,000
ALBERS EQUAL AREA PROJECTION — STANDARD PARALLELS 29½° AND 45½°
PUERTO RICO AND VIRGIN ISLANDS
HAWAII
ALASKA
GULF OF MEXICO
ATLANTIC OCEAN
PACIFIC OCEAN

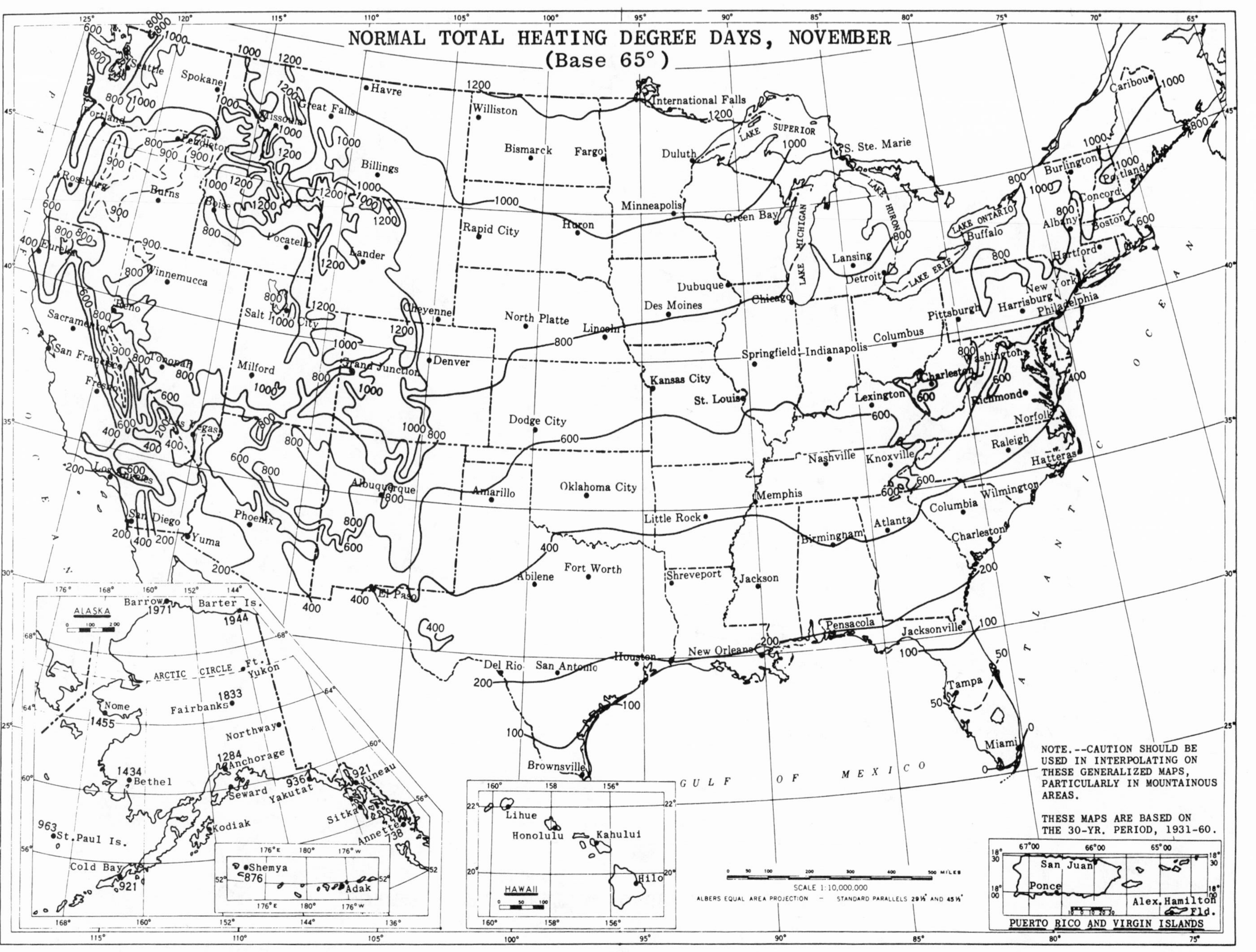
NORMAL TOTAL HEATING DEGREE DAYS, NOVEMBER
(Base 65°)
NOTE.--CAUTION SHOULD BE USED IN INTERPOLATING ON THESE GENERALIZED MAPS, PARTICULARLY IN MOUNTAINOUS AREAS.
THESE MAPS ARE BASED ON THE 30-YR. PERIOD, 1931-60.
SCALE 1:10,000,000
ALBERS EQUAL AREA PROJECTION — STANDARD PARALLELS 29½° AND 45½°
PUERTO RICO AND VIRGIN ISLANDS
San Juan
Ponce
Alex. Hamilton Fld.
HAWAII
Lihue
Honolulu
Kahului
Hilo
ALASKA
Barrow 1971
Barter Is. 1944
ARCTIC CIRCLE
Ft. Yukon
Nome 1455
Fairbanks 1833
Northway
Anchorage 1284
Bethel 1434
Seward
Yakutat 936
Juneau 921
Sitka
Annette 738
Kodiak
St. Paul Is. 963
Cold Bay 921
Shemya 876
Adak
PACIFIC OCEAN
ATLANTIC OCEAN
GULF OF MEXICO
LAKE SUPERIOR
LAKE MICHIGAN
LAKE HURON
LAKE ERIE
LAKE ONTARIO
Seattle
Spokane
Portland
Pendleton
Roseburg
Burns
Eureka
Boise
Pocatello
Winnemucca
Reno
Sacramento
San Francisco
Tonopah
Fresno
Las Vegas
Los Angeles
San Diego
Yuma
Phoenix
Milford
Salt Lake City
Great Falls
Missoula
Havre
Billings
Lander
Grand Junction
Cheyenne
Denver
Albuquerque
El Paso
Williston
Bismarck
Fargo
Rapid City
Huron
North Platte
Lincoln
Dodge City
Amarillo
Oklahoma City
Abilene
Fort Worth
Del Rio
San Antonio
Brownsville
Houston
International Falls
Duluth
Minneapolis
Green Bay
S. Ste. Marie
Dubuque
Des Moines
Chicago
Kansas City
St. Louis
Springfield
Indianapolis
Little Rock
Shreveport
Jackson
Memphis
New Orleans
Lansing
Detroit
Columbus
Lexington
Nashville
Knoxville
Birmingham
Atlanta
Pensacola
Jacksonville
Tampa
Miami
Buffalo
Pittsburgh
Harrisburg
Philadelphia
New York
Washington
Charleston
Richmond
Norfolk
Raleigh
Hatteras
Wilmington
Columbia
Charleston
Hartford
Albany
Boston
Concord
Burlington
Portland
Caribou

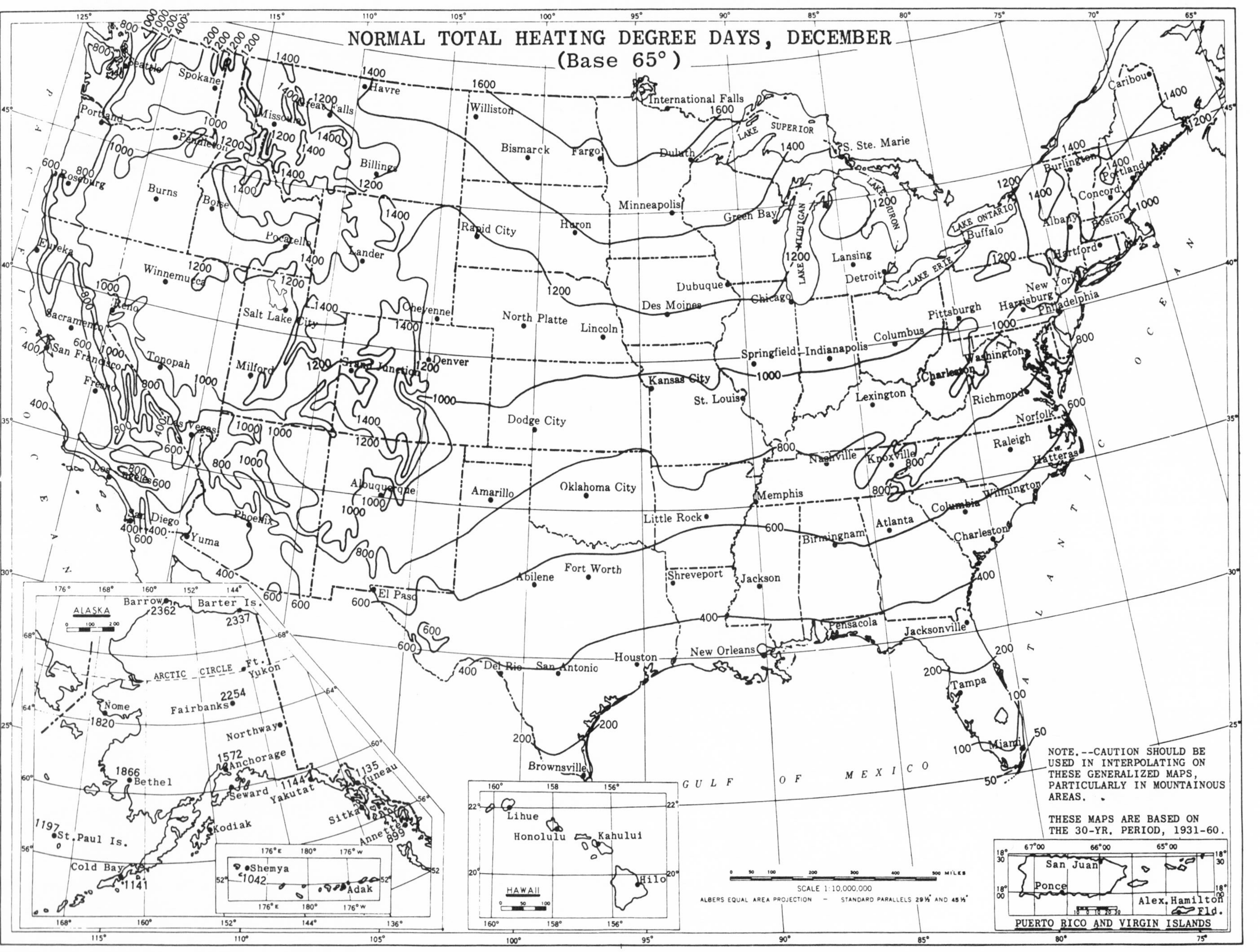
NORMAL TOTAL HEATING DEGREE DAYS, DECEMBER
(Base 65°)
NOTE.--CAUTION SHOULD BE USED IN INTERPOLATING ON THESE GENERALIZED MAPS, PARTICULARLY IN MOUNTAINOUS AREAS.
THESE MAPS ARE BASED ON THE 30-YR. PERIOD, 1931-60.
SCALE 1:10,000,000
ALBERS EQUAL AREA PROJECTION — STANDARD PARALLELS 29½° AND 45½°
PACIFIC OCEAN
ATLANTIC OCEAN
GULF OF MEXICO
LAKE SUPERIOR
LAKE MICHIGAN
LAKE HURON
LAKE ONTARIO
LAKE ERIE
Seattle
Spokane
Portland
Pendleton
Roseburg
Burns
Boise
Eureka
Winnemucca
Reno
Sacramento
San Francisco
Fresno
Tonopah
Las Vegas
Los Angeles
San Diego
Yuma
Phoenix
Havre
Great Falls
Missoula
Billings
Pocatello
Lander
Salt Lake City
Milford
Cheyenne
Denver
Grand Junction
Albuquerque
El Paso
Williston
Bismarck
Fargo
Huron
Rapid City
North Platte
Lincoln
Dodge City
Amarillo
Oklahoma City
Abilene
Fort Worth
Del Rio
San Antonio
Houston
Brownsville
International Falls
Duluth
Minneapolis
Green Bay
S. Ste. Marie
Dubuque
Des Moines
Chicago
Kansas City
St. Louis
Springfield
Indianapolis
Little Rock
Shreveport
Memphis
Jackson
New Orleans
Lansing
Detroit
Columbus
Lexington
Nashville
Knoxville
Birmingham
Atlanta
Pensacola
Jacksonville
Tampa
Miami
Buffalo
Pittsburgh
Harrisburg
Philadelphia
New York
Washington
Charleston
Richmond
Norfolk
Raleigh
Hatteras
Wilmington
Columbia
Charleston
Caribou
Burlington
Portland
Concord
Albany
Boston
Hartford
ALASKA
Barrow 2362
Barter Is. 2337
ARCTIC CIRCLE
Ft. Yukon
Fairbanks 2254
Nome 1820
Northway
Bethel 1866
1572 Anchorage
Seward
1144 Yakutat
1135 Juneau
Sitka
Annette 899
Kodiak
1197 St. Paul Is.
Cold Bay 1141
Shemya 1042
Adak
HAWAII
Lihue
Honolulu
Kahului
Hilo
PUERTO RICO AND VIRGIN ISLANDS
San Juan
Ponce
Alex. Hamilton Fld.

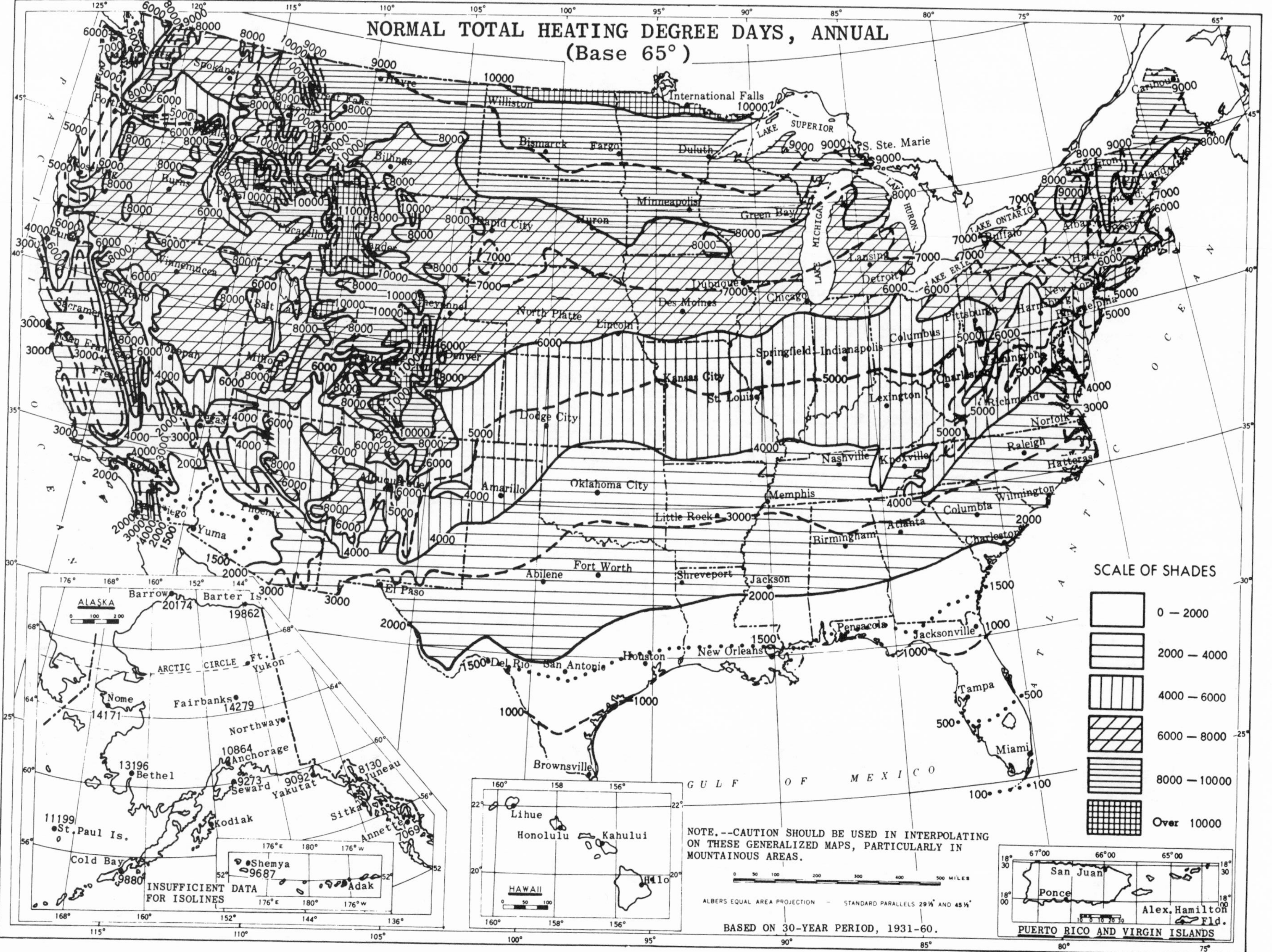

NORMAL TOTAL HEATING DEGREE DAYS, ANNUAL
(Base 65°)
SCALE OF SHADES
0 — 2000
2000 — 4000
4000 — 6000
6000 — 8000
8000 — 10000
Over 10000
NOTE.--CAUTION SHOULD BE USED IN INTERPOLATING ON THESE GENERALIZED MAPS, PARTICULARLY IN MOUNTAINOUS AREAS.
ALBERS EQUAL AREA PROJECTION — STANDARD PARALLELS 29½° AND 45½°
BASED ON 30-YEAR PERIOD, 1931-60.
GULF OF MEXICO
ATLANTIC OCEAN
PACIFIC OCEAN
LAKE SUPERIOR
LAKE MICHIGAN
LAKE HURON
LAKE ONTARIO
LAKE ERIE
International Falls
Williston
Bismarck
Fargo
Duluth
S. Ste. Marie
Minneapolis
Green Bay
Huron
Rapid City
Havre
Spokane
Billings
Great Falls
Lansing
Detroit
Buffalo
Albany
Burlington
Portland
Caribou
Dubuque
Chicago
Des Moines
North Platte
Lincoln
Cheyenne
Denver
Salt Lake
Winnemucca
Pocatello
Lander
Burns
Boise
Reno
Sacramento
San Francisco
Fresno
Tonopah
Milford
Grand Junction
Las Vegas
Los Angeles
San Diego
Yuma
Phoenix
Albuquerque
Amarillo
El Paso
Dodge City
Kansas City
St. Louis
Springfield
Indianapolis
Columbus
Pittsburgh
Harrisburg
New York
Philadelphia
Hartford
Washington
Charleston
Lexington
Richmond
Norfolk
Raleigh
Hatteras
Nashville
Knoxville
Memphis
Oklahoma City
Little Rock
Birmingham
Atlanta
Columbia
Wilmington
Charleston
Abilene
Fort Worth
Shreveport
Jackson
Pensacola
Jacksonville
Del Rio
San Antonio
Houston
New Orleans
Brownsville
Tampa
Miami
ALASKA
Barrow 20174
Barter Is. 19862
ARCTIC CIRCLE
Ft. Yukon
Nome 14171
Fairbanks 14279
Northway
10864 Anchorage
13196 Bethel
9273 Seward
9092 Yakutat
8130 Juneau
Sitka
Annette 7069
Kodiak
11199 St. Paul Is.
Cold Bay 9880
Shemya 9687
Adak
INSUFFICIENT DATA FOR ISOLINES
HAWAII
Lihue
Honolulu
Kahului
Hilo
San Juan
Ponce
Alex. Hamilton Fld.
PUERTO RICO AND VIRGIN ISLANDS

NORMAL TOTAL HEATING DEGREE DAYS
(Base 65°)

STATE AND STATION	JULY	AUG.	SEP.	OCT.	NOV.	DEC.	JAN.	FEB.	MAR.	APR.	MAY	JUNE	ANNUAL
ALA. BIRMINGHAM	0	0	6	93	363	555	592	462	363	108	9	0	2551
HUNTSVILLE	0	0	12	127	426	663	694	557	434	138	19	0	3070
MOBILE	0	0	0	22	213	357	415	300	211	42	0	0	1560
MONTGOMERY	0	0	0	68	330	527	543	417	316	90	0	0	2291
ALASKA ANCHORAGE	245	291	516	930	1284	1572	1631	1316	1293	879	592	315	10864
ANNETTE	242	208	327	567	738	899	949	837	843	648	490	321	7069
BARROW	803	840	1035	1500	1971	2362	2517	2332	2468	1944	1445	957	20174
BARTER IS.	735	775	987	1482	1944	2337	2536	2369	2477	1923	1373	924	19862
BETHEL	319	394	612	1042	1434	1866	1903	1590	1655	1173	806	402	13196
COLD BAY	474	425	525	772	918	1122	1153	1036	1122	951	791	591	9880
CORDOVA	366	391	522	781	1017	1221	1299	1086	1113	864	660	444	9764
FAIRBANKS	171	332	642	1203	1833	2254	2359	1901	1739	1068	555	222	14279
JUNEAU	301	338	483	725	921	1135	1237	1070	1073	810	601	381	9075
KING SALMON	313	322	513	908	1290	1606	1600	1333	1411	966	673	408	11343
KOTZEBUE	381	446	723	1249	1728	2127	2192	1932	2080	1554	1057	636	16105
MCGRATH	208	338	633	1184	1791	2232	2294	1817	1758	1122	648	258	14283
NOME	481	496	693	1094	1455	1820	1879	1666	1770	1314	930	573	14171
SAINT PAUL	605	539	612	862	963	1197	1228	1168	1265	1098	936	726	11199
SHEMYA	577	475	501	784	876	1042	1045	958	1011	885	837	696	9687
YAKUTAT	338	347	474	716	936	1144	1169	1019	1042	840	632	435	9092
ARIZ. FLAGSTAFF	46	68	201	558	867	1073	1169	991	911	651	437	180	7152
PHOENIX	0	0	0	22	234	415	474	328	217	75	0	0	1765
PRESCOTT	0	0	27	245	579	797	865	711	605	360	158	15	4362
TUCSON	0	0	0	25	231	406	471	344	242	75	6	0	1800
WINSLOW	0	0	6	245	711	1008	1054	770	601	291	96	0	4782
YUMA	0	0	0	0	148	319	363	228	130	29	0	0	1217
ARK. FORT SMITH	0	0	12	127	450	704	781	596	456	144	22	0	3292
LITTLE ROCK	0	0	9	127	465	716	756	577	434	126	9	0	3219
TEXARKANA	0	0	0	78	345	561	626	468	350	105	0	0	2533
CALIF. BAKERSFIELD	0	0	0	37	282	502	546	364	267	105	19	0	2122
BISHOP	0	0	42	248	576	797	874	666	539	306	143	36	4227
BLUE CANYON	34	50	120	347	579	766	865	781	791	582	397	195	5507
BURBANK	0	0	6	43	177	301	366	277	239	138	81	18	1646
EUREKA	270	257	258	329	414	499	546	470	505	438	372	285	4643
FRESNO	0	0	0	78	339	558	586	406	319	150	56	0	2492
LONG BEACH	0	0	12	40	156	288	375	297	267	168	90	18	1711
LOS ANGELES	28	22	42	78	180	291	372	302	288	219	158	81	2061
MT. SHASTA	25	34	123	406	696	902	983	784	738	525	347	159	5722
OAKLAND	53	50	45	127	309	481	527	400	353	255	180	90	2870
POINT ARGUELLO	202	186	162	205	291	400	474	392	403	339	298	243	3595
RED BLUFF	0	0	0	53	318	555	605	428	341	168	47	0	2515
SACRAMENTO	0	0	12	81	363	577	614	442	360	216	102	6	2773
SANDBERG	0	0	30	202	480	691	778	661	620	426	264	57	4209
SAN DIEGO	6	0	15	37	123	251	313	249	202	123	84	36	1439
SAN FRANCISCO	81	78	60	143	306	462	508	395	363	279	214	126	3015
SANTA CATALINA	16	0	9	50	165	279	353	308	326	249	192	105	2052
SANTA MARIA	99	93	96	146	270	391	459	370	363	282	233	165	2967
COLO. ALAMOSA	65	99	279	639	1065	1420	1476	1162	1020	696	440	168	8529
COLORADO SPRINGS	9	25	132	456	825	1032	1128	938	893	582	319	84	6423
DENVER	6	9	117	428	819	1035	1132	938	887	558	288	66	6283
GRAND JUNCTION	0	0	30	313	786	1113	1209	907	729	387	146	21	5641
PUEBLO	0	0	54	326	750	986	1085	871	772	429	174	15	5462
CONN. BRIDGEPORT	0	0	66	307	615	986	1079	966	853	510	208	27	5617
HARDFORT	0	6	99	372	711	1119	1209	1061	899	495	177	24	6172
NEW HAVEN	0	12	87	347	648	1011	1097	991	871	543	245	45	5897
DEL. WILMINGTON	0	0	51	270	588	927	980	874	735	387	112	6	4930
FLA. APALACHICOLA	0	0	0	16	153	319	347	260	180	33	0	0	1308
DAYTONA BEACH	0	0	0	0	75	211	248	190	140	15	0	0	879
FORT MYERS	0	0	0	0	24	109	146	101	62	0	0	0	442
JACKSONVILLE	0	0	0	12	144	310	332	246	174	21	0	0	1239
KEY WEST	0	0	0	0	0	28	40	31	9	0	0	0	108
LAKELAND	0	0	0	0	57	164	195	146	99	0	0	0	661
MIAMI BEACH	0	0	0	0	0	40	56	36	9	0	0	0	141
ORLANDO	0	0	0	0	72	198	220	165	105	6	0	0	766
PENSACOLA	0	0	0	19	195	353	400	277	183	36	0	0	1463
TALLAHASSEE	0	0	0	28	198	360	375	286	202	36	0	0	1485
TAMPA	0	0	0	0	60	171	202	148	102	0	0	0	683
WEST PALM BEACH	0	0	0	0	6	65	87	64	31	0	0	0	253
GA. ATHENS	0	0	12	115	405	632	642	529	431	141	22	0	2929
ATLANTA	0	0	18	127	414	626	639	529	437	168	25	0	2983
AUGUSTA	0	0	0	78	333	552	549	445	350	90	0	0	2397
COLUMBUS	0	0	0	87	333	543	552	434	338	96	0	0	2383
MACON	0	0	0	71	297	502	505	403	295	63	0	0	2136
ROME	0	0	24	161	474	701	710	577	468	177	34	0	3326
SAVANNAH	0	0	0	47	246	437	437	353	254	45	0	0	1819
THOMASVILLE	0	0	0	25	198	366	394	305	208	33	0	0	1529
IDAHO BOISE	0	0	132	415	792	1017	1113	854	722	438	245	81	5809
IDAHO FALLS 46W	16	34	270	623	1056	1370	1538	1249	1085	651	391	192	8475
IDAHO FALLS 42NW	16	40	282	648	1107	1432	1600	1291	1107	657	388	192	8760
LEWISTON	0	0	123	403	756	933	1063	815	694	426	239	90	5542
POCATELLO	0	0	172	493	900	1166	1324	1058	905	555	319	141	7033
ILL. CAIRO	0	0	36	164	513	791	856	680	539	195	47	0	3821
CHICAGO	0	0	81	326	753	1113	1209	1044	890	480	211	48	6155
MOLINE	0	9	99	335	774	1181	1314	1100	918	450	189	39	6408
PEORIA	0	6	87	326	759	1113	1218	1025	849	426	183	33	6025
ROCKFORD	6	9	114	400	837	1221	1333	1137	961	516	236	60	6830
SPRINGFIELD	0	0	72	291	696	1023	1135	935	769	354	136	18	5429
IND. EVANSVILLE	0	0	66	220	606	896	955	767	620	237	68	0	4435
FORT WAYNE	0	9	105	378	783	1135	1178	1028	890	471	189	39	6205
INDIANAPOLIS	0	0	90	316	723	1051	1113	949	809	432	177	39	5699
SOUTH BEND	0	6	111	372	777	1125	1221	1070	933	525	239	60	6439
IOWA Burlington	0	0	93	322	768	1135	1259	1042	859	426	177	33	6114
DES MOINES	0	9	99	363	837	1231	1398	1165	967	489	211	39	6808
DUBUQUE	12	31	156	450	906	1287	1420	1204	1026	546	260	78	7376
SIOUX CITY	0	9	108	369	867	1240	1435	1198	989	483	214	39	6951
WATERLOO	12	19	138	428	909	1296	1460	1221	1023	531	229	54	7320

NORMAL TOTAL HEATING DEGREE DAYS
(Base 65°)

STATE AND STATION	JULY	AUG.	SEP.	OCT.	NOV.	DEC.	JAN.	FEB.	MAR.	APR.	MAY	JUNE	ANNUAL
KANS. CONCORDIA	0	0	57	276	705	1023	1163	935	781	372	149	18	5479
DODGE CITY	0	0	33	251	666	939	1051	840	719	354	124	9	4986
GOODLAND	0	6	81	381	810	1073	1166	955	884	507	236	42	6141
TOPEKA	0	0	57	270	672	980	1122	893	722	330	124	12	5182
WICHITA	0	0	33	229	618	905	1023	804	645	270	87	6	4620
KY. COVINGTON	0	0	75	291	669	983	1035	893	756	390	149	24	5265
LEXINGTON	0	0	54	239	609	902	946	818	685	325	105	0	4683
LOUISVILLE	0	0	54	248	609	890	930	818	682	315	105	9	4660
LA. ALEXANDRIA	0	0	0	56	273	431	471	361	260	69	0	0	1921
BATON ROUGE	0	0	0	31	216	369	409	294	208	33	0	0	1560
BURRWOOD	0	0	0	0	96	214	298	218	171	27	0	0	1024
LAKE CHARLES	0	0	0	19	210	341	381	274	195	39	0	0	1459
NEW ORLEANS	0	0	0	19	192	322	363	258	192	39	0	0	1385
SHREVEPORT	0	0	0	47	297	477	552	426	304	81	0	0	2184
MAINE CARIBOU	78	115	336	682	1044	1535	1690	1470	1308	858	468	183	9767
PORTLAND	12	53	195	508	807	1215	1339	1182	1042	675	372	111	7511
MD. BALTIMORE	0	0	48	264	585	905	936	820	679	327	90	0	4654
FREDERICK	0	0	66	307	624	955	995	876	741	384	127	12	5087
MASS. BLUE HILL OBSY	0	22	108	381	690	1085	1178	1053	936	579	267	69	6368
BOSTON	0	9	60	316	603	983	1088	972	846	513	208	36	5634
NANTUCKET	12	22	93	332	573	896	992	941	896	621	384	129	5891
PITTSFIELD	25	59	219	524	831	1231	1339	1196	1063	660	326	105	7578
WORCESTER	6	34	147	450	774	1172	1271	1123	998	612	304	78	6969
MICH. ALPENA	68	105	273	580	912	1268	1404	1299	1218	777	446	156	8506
DETROIT (CITY)	0	0	87	360	738	1088	1181	1058	936	522	220	42	6232
ESCANABA	59	87	243	539	924	1293	1445	1296	1203	777	456	159	8481
FLINT	16	40	159	465	843	1212	1330	1198	1066	639	319	90	7377
GRAND RAPIDS	9	28	135	434	804	1147	1259	1134	1011	579	279	75	6894
LANSING	6	22	138	431	813	1163	1262	1142	1011	579	273	69	6909
MARQUETTE	59	81	240	527	936	1268	1411	1268	1187	771	468	177	8393
MUSKEGON	12	28	120	400	762	1088	1209	1100	995	594	310	78	6696
SAULT STE. MARIE	96	105	279	580	951	1367	1525	1380	1277	810	477	201	9048
MINN. DULUTH	71	109	330	632	1131	1581	1745	1518	1355	840	490	198	10000
INTERNATIONAL FALLS	71	112	363	701	1236	1724	1919	1621	1414	828	443	174	10606
MINNEAPOLIS	22	31	189	505	1014	1454	1631	1380	1166	621	288	81	8382
ROCHESTER	25	34	186	474	1005	1438	1593	1366	1150	630	301	93	8295
SAINT CLOUD	28	47	225	549	1065	1500	1702	1445	1221	666	326	105	8879
MISS. JACKSON	0	0	0	65	315	502	546	414	310	87	0	0	2239
MERIDIAN	0	0	0	81	339	518	543	417	310	81	0	0	2289
VICKSBURG	0	0	0	53	279	462	512	384	282	69	0	0	2041
MO. COLUMBIA	0	0	54	251	651	967	1076	874	716	324	121	12	5046
KANSAS	0	0	39	220	612	905	1032	818	682	294	109	0	4711
ST. JOSEPH	0	6	60	285	708	1039	1172	949	769	348	133	15	5484
ST. LOUIS	0	0	60	251	627	936	1026	848	704	312	121	15	4900
SPRINGFIELD	0	0	45	223	600	877	973	781	660	291	105	6	4561
MONT. BILLINGS	6	15	186	487	897	1135	1296	1100	970	570	285	102	7049
GLASGOW	31	47	270	608	1104	1466	1711	1439	1187	648	335	150	8996
GREAT FALLS	28	53	258	543	921	1169	1349	1154	1063	642	384	186	7750
HAVRE	28	53	306	595	1065	1367	1584	1364	1181	657	338	162	8700
HELENA	31	59	294	601	1002	1265	1438	1170	1042	651	381	195	8129
KALISPELL	50	99	321	654	1020	1240	1401	1134	1029	639	397	207	8191
MILES CITY	6	6	174	502	972	1296	1504	1252	1057	579	276	99	7723
MISSOULA	34	74	303	651	1035	1287	1420	1120	970	621	391	219	8125
NEBR. GRAND ISLAND	0	6	108	381	834	1172	1314	1089	908	462	211	45	6530
LINCOLN	0	6	75	301	726	1066	1237	1016	834	402	171	30	5864
NORFOLK	9	0	111	397	873	1234	1414	1179	983	498	233	48	6979
NORTH PLATTE	0	6	123	440	885	1166	1271	1039	930	519	248	57	6684
OMAHA	0	12	105	357	828	1175	1355	1126	939	465	208	42	6612
SCOTTSBLUFF	0	0	138	459	876	1128	1231	1008	921	552	285	75	6673
VALENTINE	9	12	165	493	942	1237	1395	1176	1045	579	288	84	7425
NEV. ELKO	9	34	225	561	924	1197	1314	1036	911	621	409	192	7433
ELY	28	43	234	592	939	1184	1308	1075	977	672	456	225	7733
LAS VEGAS	0	0	0	78	387	617	688	487	335	111	6	0	2709
RENO	43	87	204	490	801	1026	1073	823	729	510	357	189	6332
WINNEMUCCA	0	34	210	536	876	1091	1172	916	837	573	363	153	6761
N. H. CONCORD	6	50	177	505	822	1240	1358	1184	1032	636	298	75	7383
MT. WASH. OBSY.	493	536	720	1057	1341	1742	1820	1663	1652	1260	930	603	13817
N. J. ATLANTIC CITY	0	0	39	251	549	880	936	848	741	420	133	15	4812
NEWARK	0	0	30	248	573	921	983	876	729	381	118	0	4859
TRENTON	0	0	57	264	576	924	989	885	753	399	121	12	4980
N. MEX. ALBUQUERQUE	0	0	12	229	642	868	930	703	595	288	81	0	4348
CLAYTON	0	6	66	310	699	899	986	812	747	429	183	21	5158
RATON	9	28	126	431	825	1048	1116	904	834	543	301	63	6228
ROSWELL	0	0	18	202	573	806	840	641	481	201	31	0	3793
SILVER CITY	0	0	6	183	525	729	791	605	518	261	87	0	3705
N. Y. ALBANY	0	19	138	440	777	1194	1311	1156	992	564	239	45	6875
BINGHAMTON (AP)	22	65	201	471	810	1184	1277	1154	1045	645	313	99	7286
BINGHAMTON (PO)	0	28	141	406	732	1107	1190	1081	949	543	229	45	6451
BUFFALO	19	37	141	440	777	1156	1256	1145	1039	645	329	78	7062
CENTRAL PARK	0	0	30	233	540	902	986	885	760	408	118	9	4871
J. F. KENNEDY INTL.	0	0	36	248	564	933	1029	935	815	480	167	12	5219
LAGUARDIA	0	0	27	223	528	887	973	879	750	414	124	6	4811
ROCHESTER	9	31	126	415	747	1125	1234	1123	1014	597	279	48	6748
SCHENECTADY	0	22	123	422	756	1159	1283	1131	970	543	211	30	6650
SYRACUSE	6	28	132	415	744	1153	1271	1140	1004	570	248	45	6756
N.C. ASHEVILLE	0	0	48	245	555	775	784	683	592	273	87	0	4042
CAPE HATTERAS	0	0	0	78	273	521	580	518	440	177	25	0	2612
CHARLOTTE	0	0	6	124	438	691	691	582	481	156	22	0	3191
GREENSBORO	0	0	33	192	513	778	784	672	552	234	47	0	3805
RALEIGH	0	0	21	164	450	716	725	616	487	180	34	0	3393
WILMINGTON	0	0	0	74	291	521	546	462	357	96	0	0	2347
WINSTON SALEM	0	0	21	171	483	747	753	652	524	207	37	0	3595
N. DAK. BISMARCK	34	28	222	577	1083	1463	1708	1442	1203	645	329	117	8851
DEVILS LAKE	40	53	273	642	1191	1634	1872	1579	1345	753	381	138	9901
FARGO	28	37	219	574	1107	1569	1789	1520	1262	690	332	99	9226
WILLISTON	31	43	261	601	1122	1513	1758	1473	1262	681	357	141	9243

NORMAL TOTAL HEATING DEGREE DAYS
(Base 65°)

STATE AND STATION	JULY	AUG.	SEP.	OCT.	NOV.	DEC.	JAN.	FEB.	MAR.	APR.	MAY	JUNE	ANNUAL
OHIO AKRON	0	9	96	381	726	1070	1138	1016	871	489	202	39	6037
CINCINNATI	0	0	54	248	612	921	970	837	701	336	118	9	4806
CLEVELAND	9	25	105	384	738	1088	1159	1047	918	552	260	66	6351
COLUMBUS	0	6	84	347	714	1039	1088	949	809	426	171	27	5660
DAYTON	0	6	78	310	696	1045	1097	955	809	429	167	30	5622
MANSFIELD	9	22	114	397	768	1110	1169	1042	924	543	245	60	6403
SANDUSKY	0	6	66	313	684	1032	1107	991	868	495	198	36	5796
TOLEDO	0	16	117	406	792	1138	1200	1056	924	543	242	60	6494
YOUNGSTOWN	6	19	120	412	771	1104	1169	1047	921	540	248	60	6417
OKLA. OKLAHOMA CITY	0	0	15	164	498	766	868	664	527	189	34	0	3725
TULSA	0	0	18	158	522	787	893	683	539	213	47	0	3860
OREG. ASTORIA	146	130	210	375	561	679	753	622	636	480	363	231	5186
BURNS	12	37	210	515	867	1113	1246	988	856	570	366	177	6957
EUGENE	34	34	129	366	585	719	803	627	589	426	279	135	4726
MEACHAM	84	124	288	580	918	1091	1209	1005	983	726	527	339	7874
MEDFORD	0	0	78	372	678	871	918	697	642	432	242	78	5008
PENDLETON	0	0	111	350	711	884	1017	773	617	396	205	63	5127
PORTLAND	25	28	114	335	597	735	825	644	586	396	245	105	4635
ROSEBURG	22	16	105	329	567	713	766	608	570	405	267	123	4491
SALEM	37	31	111	338	594	729	822	647	611	417	273	144	4754
SEXTON SUMMIT	81	81	171	443	666	874	958	809	818	609	465	279	6254
PA. ALLENTOWN	0	0	90	353	693	1045	1116	1002	849	471	167	24	5810
ERIE	0	25	102	391	714	1063	1169	1081	973	585	288	60	6451
HARRISBURG	0	0	63	298	648	992	1045	907	766	396	124	12	5251
PHILADELPHIA	0	0	60	291	621	964	1014	890	744	390	115	12	5101
PITTSBURGH	0	9	105	375	726	1063	1119	1002	874	480	195	39	5987
READING	0	0	54	257	597	939	1001	885	735	372	105	0	4945
SCRANTON	0	19	132	434	762	1104	1156	1028	893	498	195	33	6254
WILLIAMSPORT	0	9	111	375	717	1073	1122	1002	856	468	177	24	5934
R. I. BLOCK IS.	0	16	78	307	594	902	1020	955	877	612	344	99	5804
PROVIDENCE	0	16	96	372	660	1023	1110	988	868	534	236	51	5954
S. C. CHARLESTON	0	0	0	59	282	471	487	389	291	54	0	0	2033
COLUMBIA	0	0	0	84	345	577	570	470	357	81	0	0	2484
FLORENCE	0	0	0	78	315	552	552	459	347	84	0	0	2387
GREENVILLE	0	0	0	112	387	636	648	535	434	120	12	0	2884
SPARTANBURG	0	0	15	130	417	667	663	560	453	144	25	0	3074
S. DAK. HURON	9	12	165	508	1014	1432	1628	1355	1125	600	288	87	8223
RAPID CITY	22	12	165	481	897	1172	1333	1145	1051	615	326	126	7345
SIOUX FALLS	19	25	168	462	972	1361	1544	1285	1082	573	270	78	7839
TENN. BRISTOL	0	0	51	236	573	828	828	700	598	261	68	0	4143
CHATTANOOGA	0	0	18	143	468	698	722	577	453	150	25	0	3254
KNOXVILLE	0	0	30	171	489	725	732	613	493	198	43	0	3494
MEMPHIS	0	0	18	130	447	698	729	585	456	147	22	0	3232
NASHVILLE	0	0	30	158	495	732	778	644	512	189	40	0	3578
OAK RIDGE (CO)	0	0	39	192	531	772	778	669	552	228	56	0	3817
TEX. ABILENE	0	0	0	99	366	586	642	470	347	114	0	0	2624
AMARILLO	0	0	18	205	570	797	877	664	546	252	56	0	3985
AUSTIN	0	0	0	31	225	388	468	325	223	51	0	0	1711
BROWNSVILLE	0	0	0	0	66	149	205	106	74	0	0	0	600
CORPUS CHRISTI	0	0	0	0	120	220	291	174	109	0	0	0	914
DALLAS	0	0	0	62	321	524	601	440	319	90	6	0	2363
EL PASO	0	0	0	84	414	648	685	445	319	105	0	0	2700
FORT WORTH	0	0	0	65	324	536	614	448	319	99	0	0	2405
GALVESTON	0	0	0	0	138	270	350	258	189	30	0	0	1235
HOUSTON	0	0	0	6	183	307	384	288	192	36	0	0	1396
LAREDO	0	0	0	0	105	217	267	134	74	0	0	0	797
LUBBOCK	0	0	18	174	513	744	800	613	484	201	31	0	3578
MIDLAND	0	0	0	87	381	592	651	468	322	90	0	0	2591
PORT ARTHUR	0	0	0	22	207	329	384	274	192	39	0	0	1447
SAN ANGELO	0	0	0	68	318	536	567	412	288	66	0	0	2255
SAN ANTONIO	0	0	0	31	207	363	428	286	195	39	0	0	1549
VICTORIA	0	0	0	6	150	270	344	230	152	21	0	0	1173
WACO	0	0	0	43	270	456	536	389	270	66	0	0	2030
WICHITA FALLS	0	0	0	99	381	632	698	518	378	120	6	0	2832
UTAH MILFORD	0	0	99	443	867	1141	1252	988	822	519	279	87	6497
SALT LAKE CITY	0	0	81	419	849	1082	1172	910	763	459	233	84	6052
WENDOVER	0	0	48	372	822	1091	1178	902	729	408	177	51	5778
VT. BURLINGTON	28	65	207	539	891	1349	1513	1333	1187	714	353	90	8269
VA. CAPE HENRY	0	0	0	112	360	645	694	633	536	246	53	0	3279
LYNCHBURG	0	0	51	223	540	822	849	731	605	267	78	0	4166
NORFOLK	0	0	0	136	408	698	738	655	533	216	37	0	3421
RICHMOND	0	0	36	214	495	784	815	703	546	219	53	0	3865
ROANOKE	0	0	51	229	549	825	834	722	614	261	65	0	4150
WASH. NAT'L. AP.	0	0	33	217	519	834	871	762	626	288	74	0	4224
WASH. OLYMPIA	68	71	198	422	636	753	834	675	645	450	307	177	5236
SEATTLE	50	47	129	329	543	657	738	599	577	396	242	117	4424
SEATTLE BOEING	34	40	147	384	624	763	831	655	608	411	242	99	4838
SEATTLE TACOMA	56	62	162	391	633	750	828	678	657	474	295	159	5145
SPOKANE	9	25	168	493	879	1082	1231	980	834	531	288	135	6655
STAMPEDE PASS	273	291	393	701	1008	1178	1287	1075	1085	855	654	483	9283
TATOOSH IS.	295	279	306	406	534	639	713	613	645	525	431	333	5719
WALLA WALLA	0	0	87	310	681	843	986	745	589	342	177	45	4805
YAKIMA	0	12	144	450	828	1039	1163	868	713	435	220	69	5941
W. VA. CHARLESTON	0	0	63	254	591	865	880	770	648	300	96	9	4476
ELKINS	9	25	135	400	729	992	1008	896	791	444	198	48	5675
HUNTINGTON	0	0	63	257	585	856	880	764	636	294	99	12	4446
PARKERSBURG	0	0	60	264	606	905	942	826	691	339	115	6	4754
WIS. GREEN BAY	28	50	174	484	924	1333	1494	1313	1141	654	335	99	8029
LA CROSSE	12	19	153	437	924	1339	1504	1277	1070	540	245	69	7589
MADISON	25	40	174	474	930	1330	1473	1274	1113	618	310	102	7863
MILWAUKEE	43	47	174	471	876	1252	1376	1193	1054	642	372	135	7635
WYO. CASPER	6	16	192	524	942	1169	1290	1084	1020	657	381	129	7410
CHEYENNE	19	31	210	543	924	1101	1228	1056	1011	672	381	102	7278
LANDER	6	19	204	555	1020	1299	1417	1145	1017	654	381	153	7870
SHERIAN	25	31	219	539	948	1200	1355	1154	1054	642	366	150	7683

PRECIPITATION

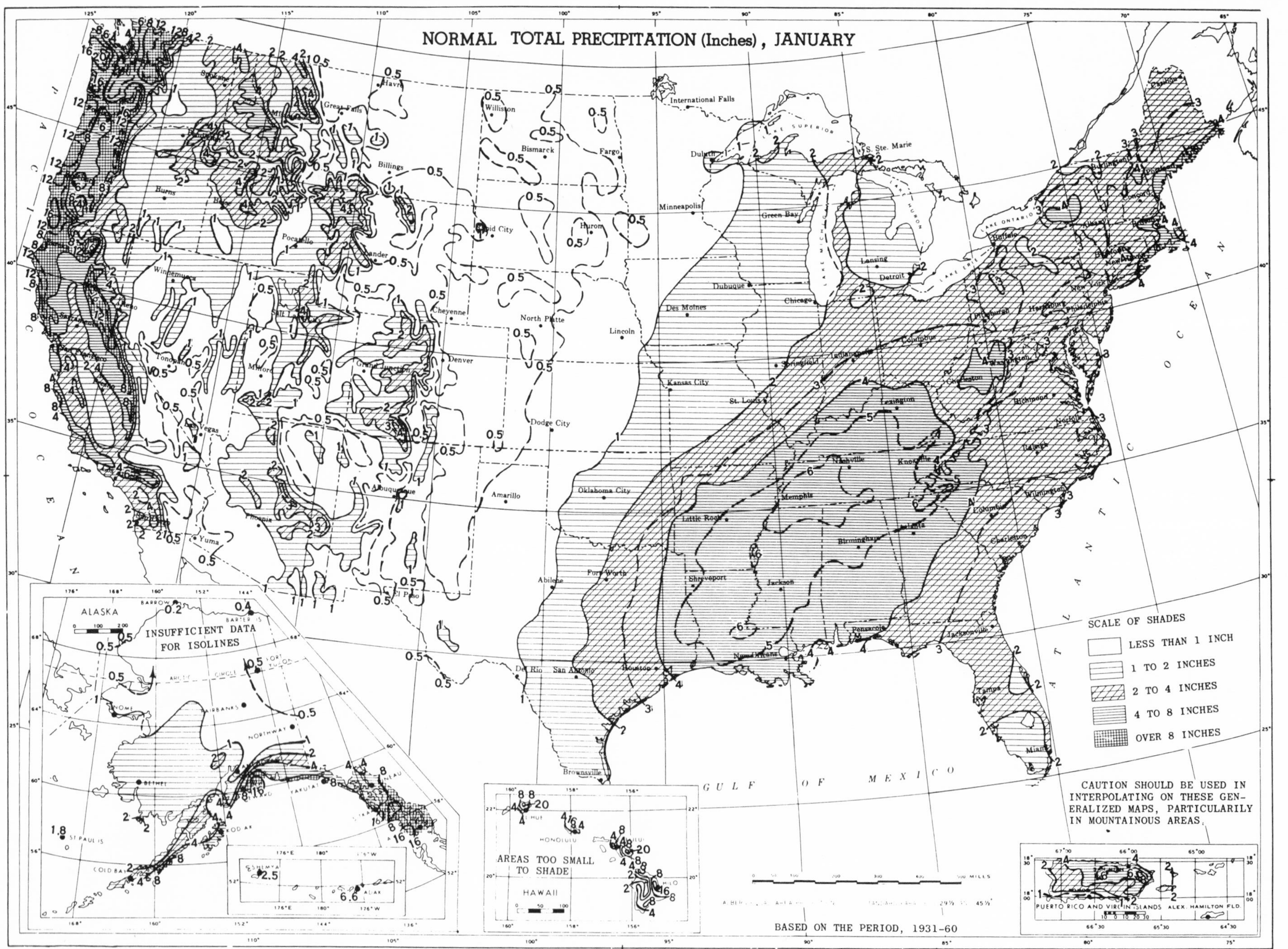
NORMAL TOTAL PRECIPITATION (Inches), JANUARY
SCALE OF SHADES
LESS THAN 1 INCH
1 TO 2 INCHES
2 TO 4 INCHES
4 TO 8 INCHES
OVER 8 INCHES
CAUTION SHOULD BE USED IN INTERPOLATING ON THESE GENERALIZED MAPS, PARTICULARILY IN MOUNTAINOUS AREAS.
BASED ON THE PERIOD, 1931-60
ATLANTIC OCEAN
GULF OF MEXICO
PACIFIC OCEAN
PUERTO RICO AND VIRGIN ISLANDS
ALEX. HAMILTON FLD.
AREAS TOO SMALL TO SHADE
HAWAII
HONOLULU
HILO
ALASKA
INSUFFICIENT DATA FOR ISOLINES
BARROW
BARTER IS
FORT YUKON
ARCTIC CIRCLE
NOME
FAIRBANKS
NORTHWAY
BETHEL
KODIAK
ST PAUL IS
COLD BAY
YAKUTAT
SHEMYA
ADAK
International Falls
Duluth
S. Ste. Marie
Minneapolis
Green Bay
Lansing
Detroit
Dubuque
Chicago
Des Moines
Lincoln
North Platte
Cheyenne
Denver
Kansas City
St. Louis
Springfield
Indianapolis
Columbus
Pittsburgh
Harrisburg
Philadelphia
New York
Buffalo
Albany
Concord
Burlington
Portland
Caribou
Washington
Charleston
Richmond
Norfolk
Raleigh
Wilmington
Columbia
Lexington
Nashville
Knoxville
Memphis
Birmingham
Atlanta
Little Rock
Shreveport
Jackson
Pensacola
New Orleans
Jacksonville
Tampa
Miami
Oklahoma City
Amarillo
Dodge City
Fort Worth
Abilene
Houston
San Antonio
Del Rio
Brownsville
El Paso
Albuquerque
Grand Junction
Phoenix
Yuma
Las Vegas
Tonopah
Milford
Salt Lake City
Winnemucca
Pocatello
Lander
Billings
Great Falls
Havre
Williston
Bismarck
Fargo
Rapid City
Huron
Burns
Boise
Spokane
Lake Superior
Lake Michigan
Lake Huron
Lake Ontario
Lake Erie

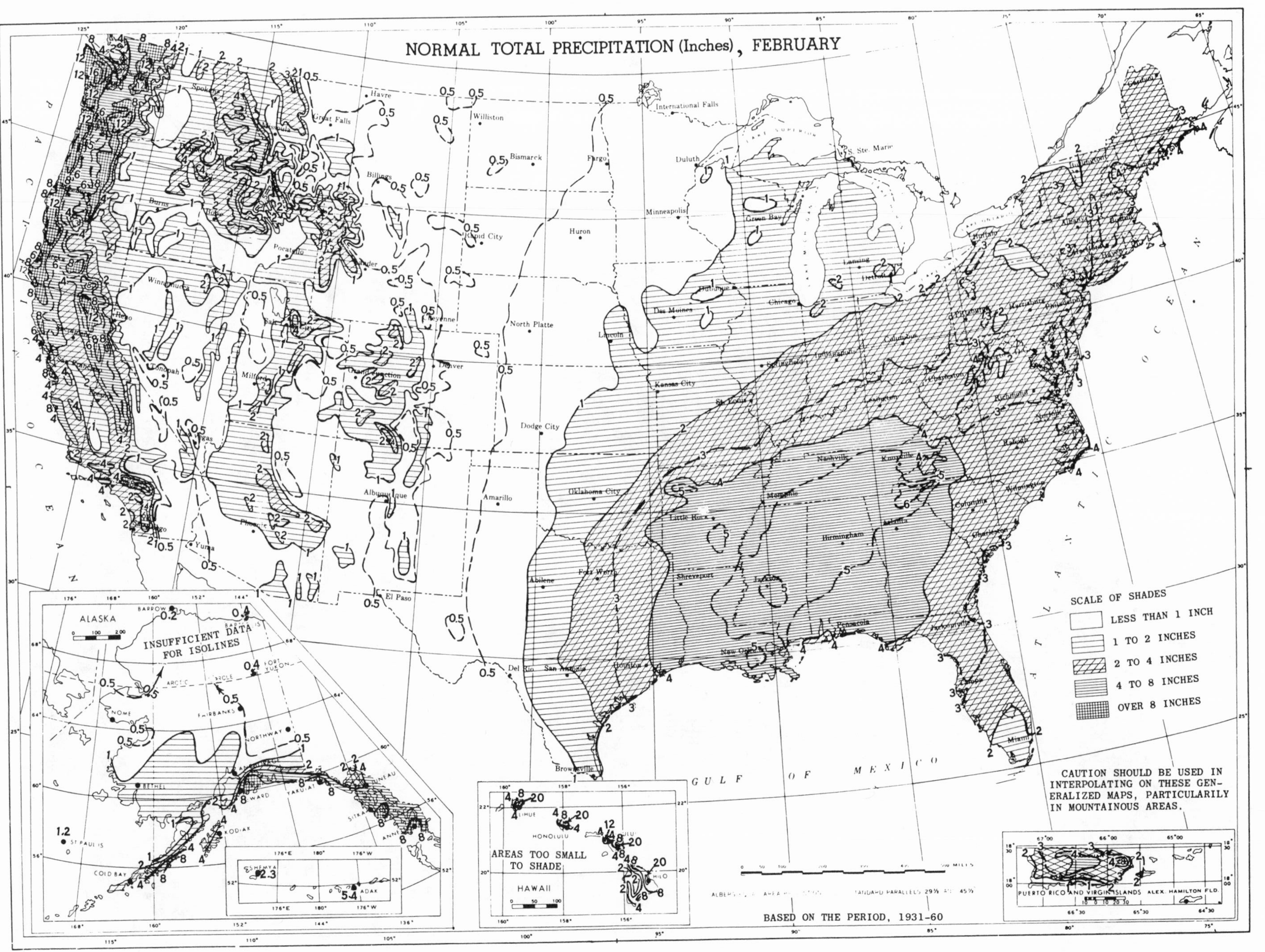
NORMAL TOTAL PRECIPITATION (Inches), FEBRUARY
SCALE OF SHADES
LESS THAN 1 INCH
1 TO 2 INCHES
2 TO 4 INCHES
4 TO 8 INCHES
OVER 8 INCHES
CAUTION SHOULD BE USED IN INTERPOLATING ON THESE GENERALIZED MAPS, PARTICULARLY IN MOUNTAINOUS AREAS.
BASED ON THE PERIOD, 1931-60
ALASKA
INSUFFICIENT DATA FOR ISOLINES
AREAS TOO SMALL TO SHADE
HAWAII
PUERTO RICO AND VIRGIN ISLANDS
GULF OF MEXICO
PACIFIC OCEAN
ATLANTIC OCEAN

NORMAL TOTAL PRECIPITATION (Inches), MARCH

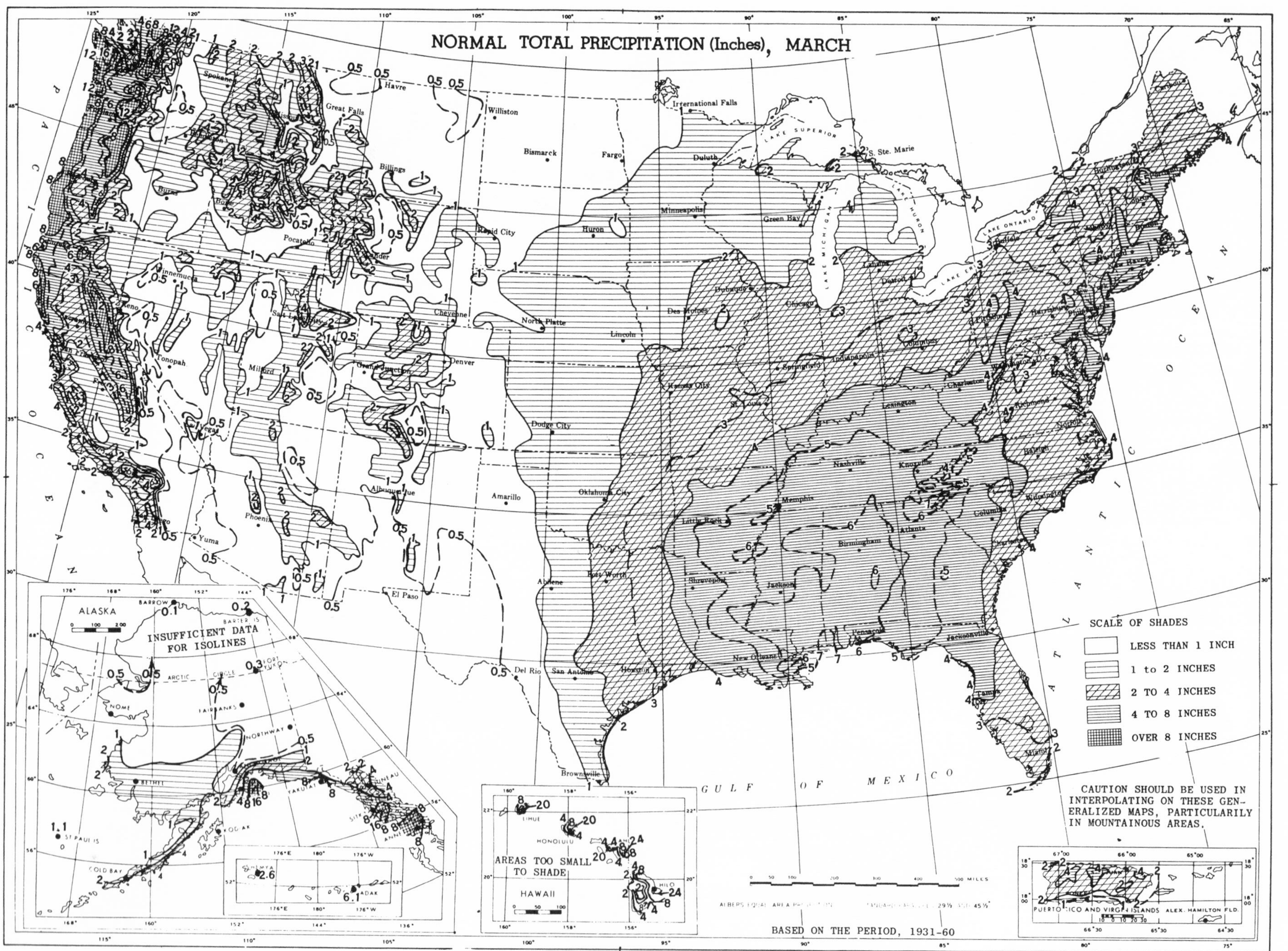

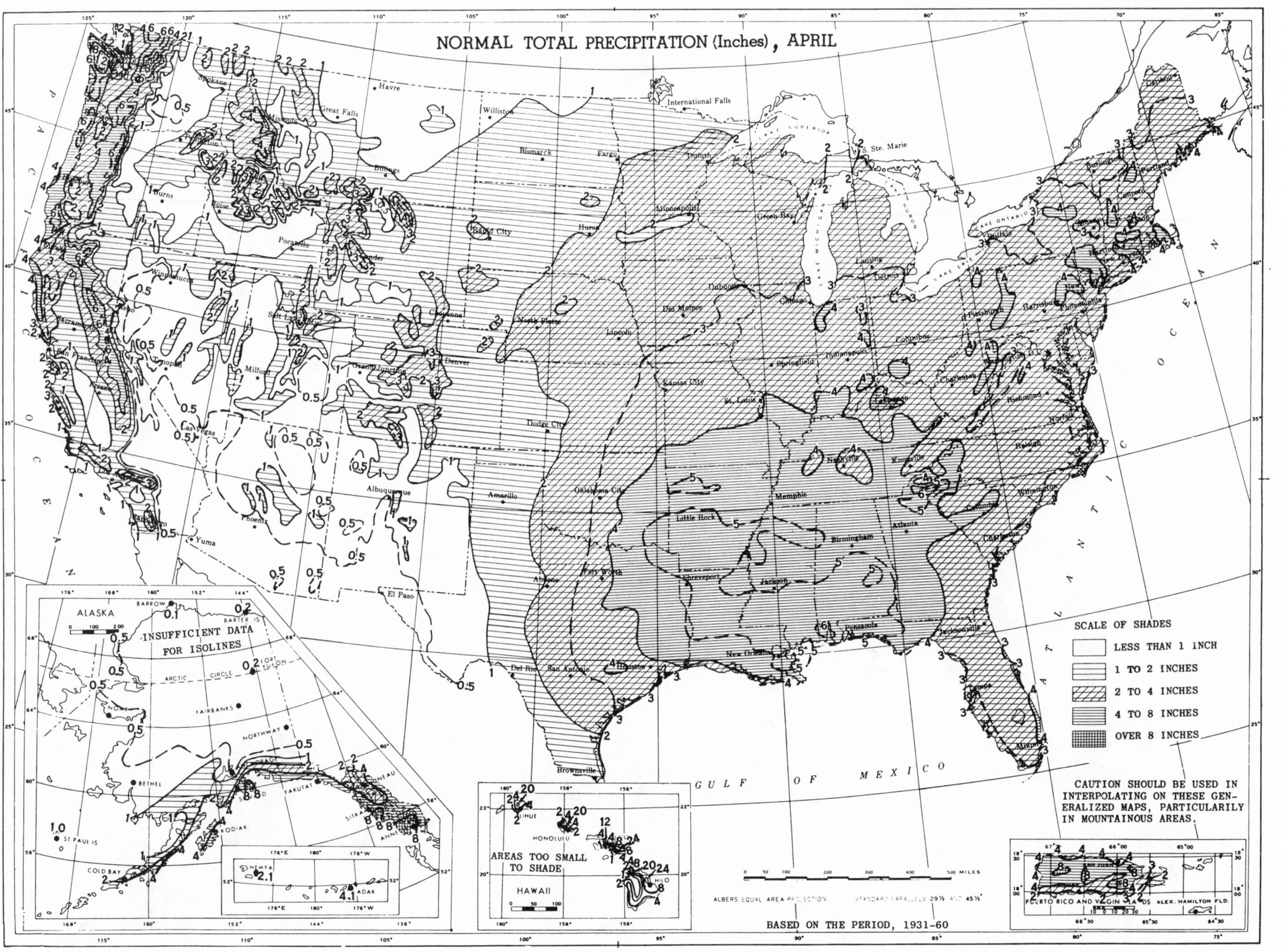
NORMAL TOTAL PRECIPITATION (Inches), APRIL
SCALE OF SHADES
LESS THAN 1 INCH
1 TO 2 INCHES
2 TO 4 INCHES
4 TO 8 INCHES
OVER 8 INCHES
CAUTION SHOULD BE USED IN INTERPOLATING ON THESE GENERALIZED MAPS, PARTICULARILY IN MOUNTAINOUS AREAS.
ALASKA
INSUFFICIENT DATA FOR ISOLINES
HAWAII
AREAS TOO SMALL TO SHADE
PUERTO RICO AND VIRGIN ISLANDS
ALBERS EQUAL AREA PROJECTION
BASED ON THE PERIOD, 1931-60
GULF OF MEXICO
ATLANTIC OCEAN
PACIFIC OCEAN

NORMAL TOTAL PRECIPITATION (Inches), MAY
PACIFIC OCEAN
ATLANTIC OCEAN
GULF OF MEXICO
SCALE OF SHADES
LESS THAN 1 INCH
1 TO 2 INCHES
2 TO 4 INCHES
4 TO 8 INCHES
OVER 8 INCHES
CAUTION SHOULD BE USED IN INTERPOLATING ON THESE GENERALIZED MAPS, PARTICULARILY IN MOUNTAINOUS AREAS.
ALASKA
INSUFFICIENT DATA FOR ISOLINES
AREAS TOO SMALL TO SHADE
HAWAII
PUERTO RICO AND VIRGIN ISLANDS
ALEX. HAMILTON FLD.
BASED ON THE PERIOD, 1931-60

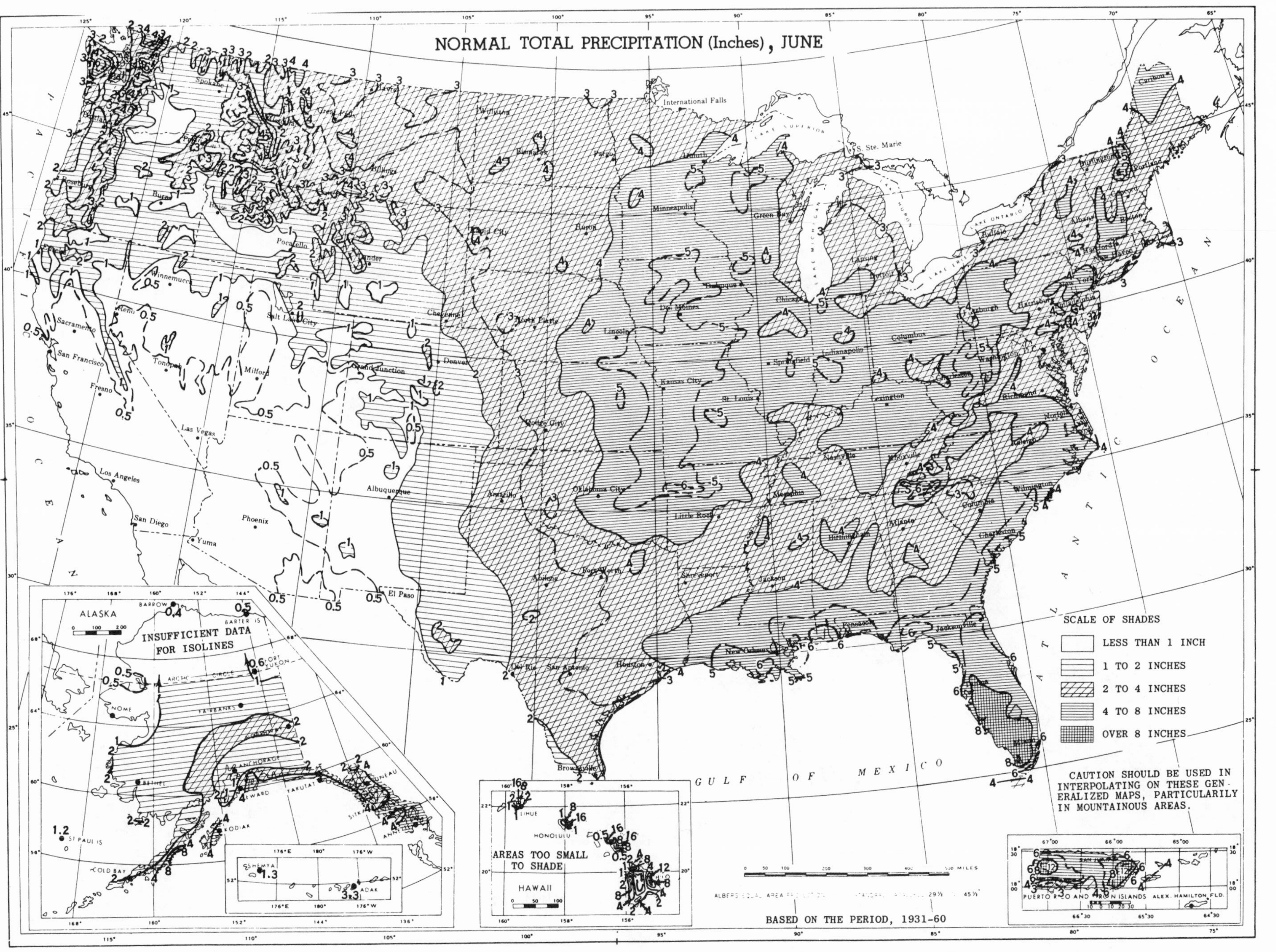
NORMAL TOTAL PRECIPITATION (Inches), JUNE
SCALE OF SHADES
LESS THAN 1 INCH
1 TO 2 INCHES
2 TO 4 INCHES
4 TO 8 INCHES
OVER 8 INCHES
CAUTION SHOULD BE USED IN INTERPOLATING ON THESE GENERALIZED MAPS, PARTICULARILY IN MOUNTAINOUS AREAS.
BASED ON THE PERIOD, 1931-60
ALASKA
INSUFFICIENT DATA FOR ISOLINES
HAWAII
AREAS TOO SMALL TO SHADE
PUERTO RICO AND VIRGIN ISLANDS
GULF OF MEXICO
PACIFIC OCEAN
ATLANTIC OCEAN
International Falls
Sault Ste. Marie
Minneapolis
Chicago
Denver
Salt Lake City
Albuquerque
Phoenix
Los Angeles
San Diego
San Francisco
Sacramento
Fresno
Reno
Las Vegas
El Paso
Kansas City
St. Louis
Oklahoma City
Little Rock
New Orleans
Houston
San Antonio
Brownsville
Jacksonville
Atlanta
Columbus
Indianapolis
Pittsburgh
Washington, D.C.
Richmond
Norfolk
Wilmington
Charleston
Miami
Albany
Boston
Portland
Caribou
Buffalo
New York
Harrisburg
Philadelphia
Duluth
Fargo
Bismarck
Huron
Des Moines
Lincoln
North Platte
Cheyenne
Dodge City
Amarillo
Abilene
Fort Worth
Shreveport
Jackson
Memphis
Nashville
Knoxville
Birmingham
Pensacola
Spokane
Great Falls
Havre
Williston
Billings
Boise
Pocatello
Burns
Winnemucca
Tonopah
Milford
Grand Junction
Yuma
Lansing
Detroit
Green Bay
Dubuque
Springfield
Lexington
Raleigh
Columbia
Lander
Rapid City
Burlington
Hartford
New Haven
Nome
Barrow
Fairbanks
Anchorage
Bethel
Kodiak
Cold Bay
St. Paul Is.
Juneau
Sitka
Yakutat
Seward
Barter Is.
Fort Yukon
Arctic Circle
Adak
Shemya
Lihue
Honolulu
San Juan
Alex. Hamilton Fld.
ALBERS EQUAL AREA PROJECTION
STANDARD PARALLELS 29½° AND 45½°

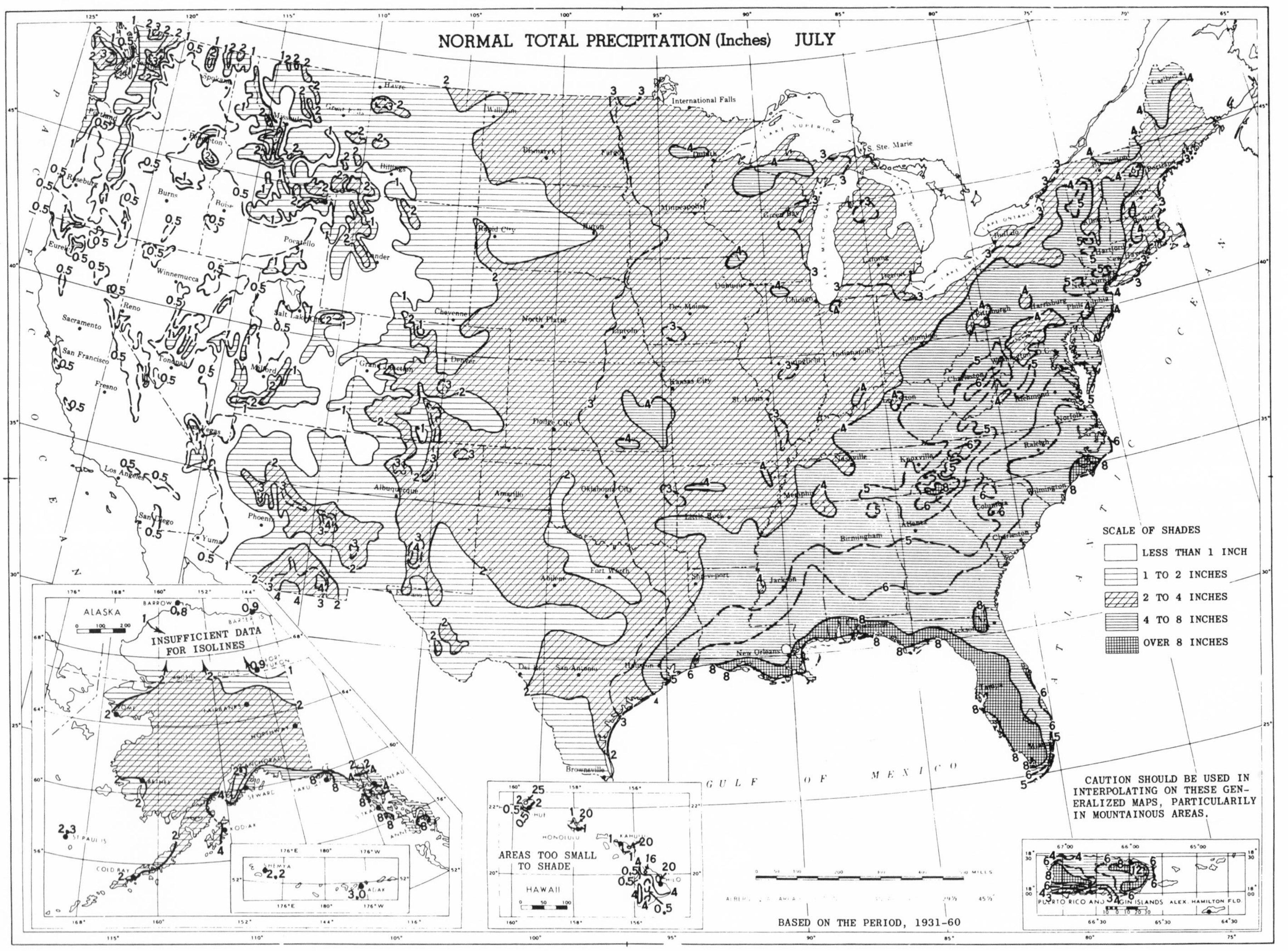
NORMAL TOTAL PRECIPITATION (Inches) JULY
SCALE OF SHADES
LESS THAN 1 INCH
1 TO 2 INCHES
2 TO 4 INCHES
4 TO 8 INCHES
OVER 8 INCHES
CAUTION SHOULD BE USED IN INTERPOLATING ON THESE GENERALIZED MAPS, PARTICULARILY IN MOUNTAINOUS AREAS.
BASED ON THE PERIOD, 1931-60
ALASKA
INSUFFICIENT DATA FOR ISOLINES
AREAS TOO SMALL TO SHADE
HAWAII
PUERTO RICO AND VIRGIN ISLANDS
ALEX. HAMILTON FLD
GULF OF MEXICO
PACIFIC OCEAN
ATLANTIC OCEAN

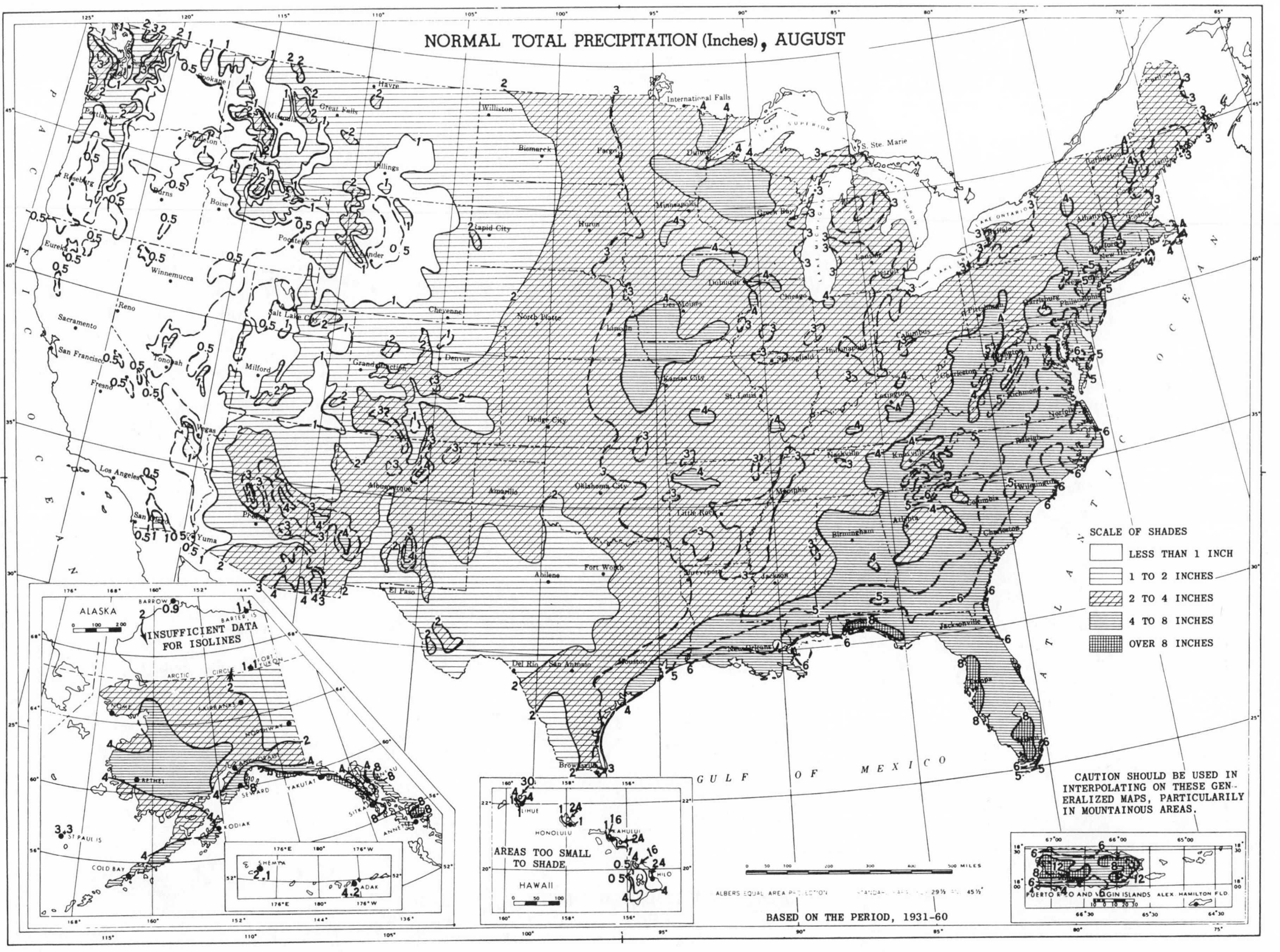
NORMAL TOTAL PRECIPITATION (Inches), AUGUST
SCALE OF SHADES
LESS THAN 1 INCH
1 TO 2 INCHES
2 TO 4 INCHES
4 TO 8 INCHES
OVER 8 INCHES
CAUTION SHOULD BE USED IN INTERPOLATING ON THESE GENERALIZED MAPS, PARTICULARILY IN MOUNTAINOUS AREAS.
BASED ON THE PERIOD, 1931-60
ALBERS EQUAL AREA PROJECTION
PACIFIC OCEAN
ATLANTIC OCEAN
GULF OF MEXICO
ALASKA
INSUFFICIENT DATA FOR ISOLINES
HAWAII
AREAS TOO SMALL TO SHADE
PUERTO RICO AND VIRGIN ISLANDS

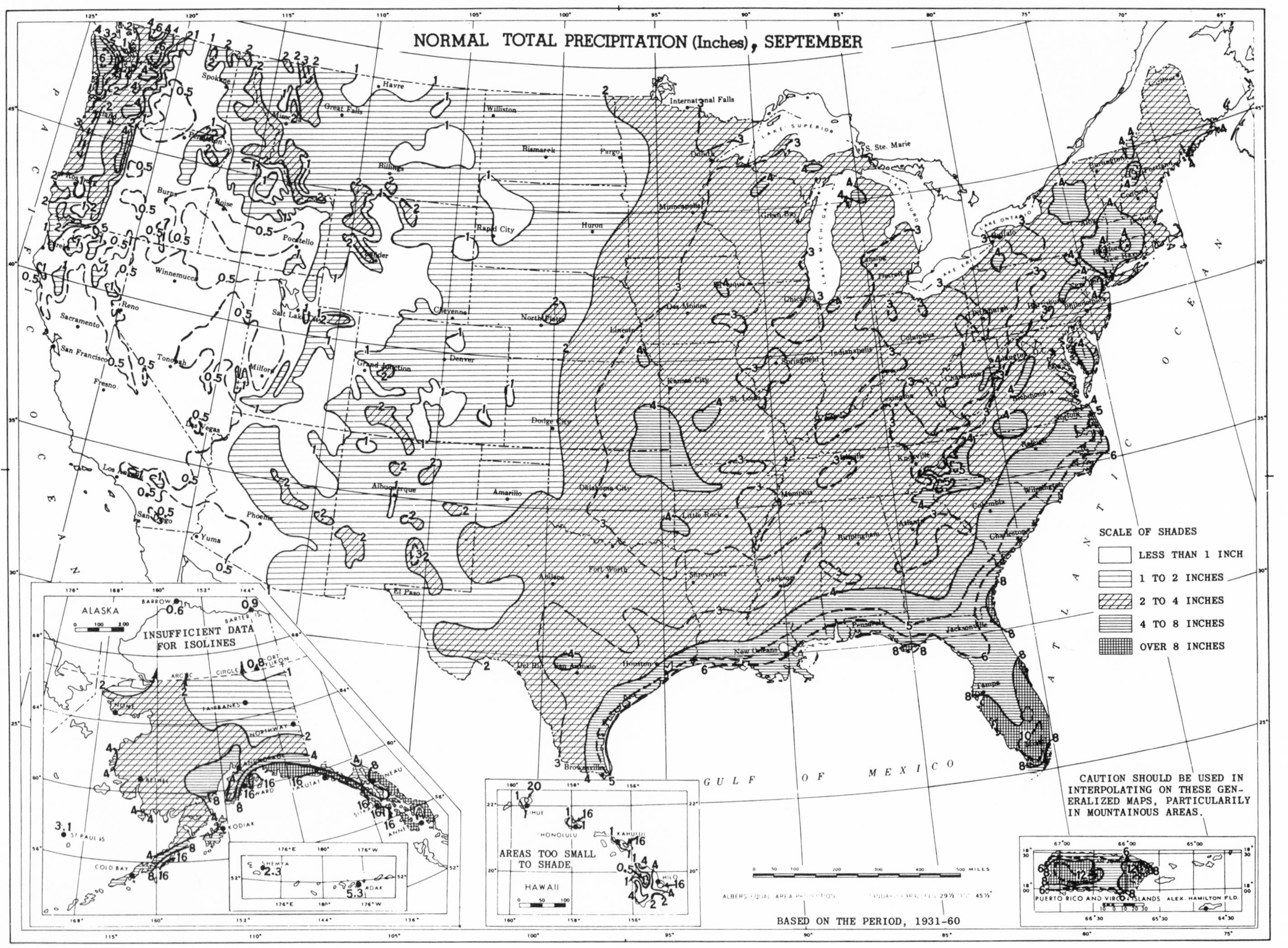
NORMAL TOTAL PRECIPITATION (Inches), SEPTEMBER
SCALE OF SHADES
LESS THAN 1 INCH
1 TO 2 INCHES
2 TO 4 INCHES
4 TO 8 INCHES
OVER 8 INCHES
CAUTION SHOULD BE USED IN INTERPOLATING ON THESE GENERALIZED MAPS, PARTICULARILY IN MOUNTAINOUS AREAS.
BASED ON THE PERIOD, 1931-60
ALASKA
INSUFFICIENT DATA FOR ISOLINES
HAWAII
AREAS TOO SMALL TO SHADE
PUERTO RICO AND VIRGIN ISLANDS
PACIFIC OCEAN
ATLANTIC OCEAN
GULF OF MEXICO

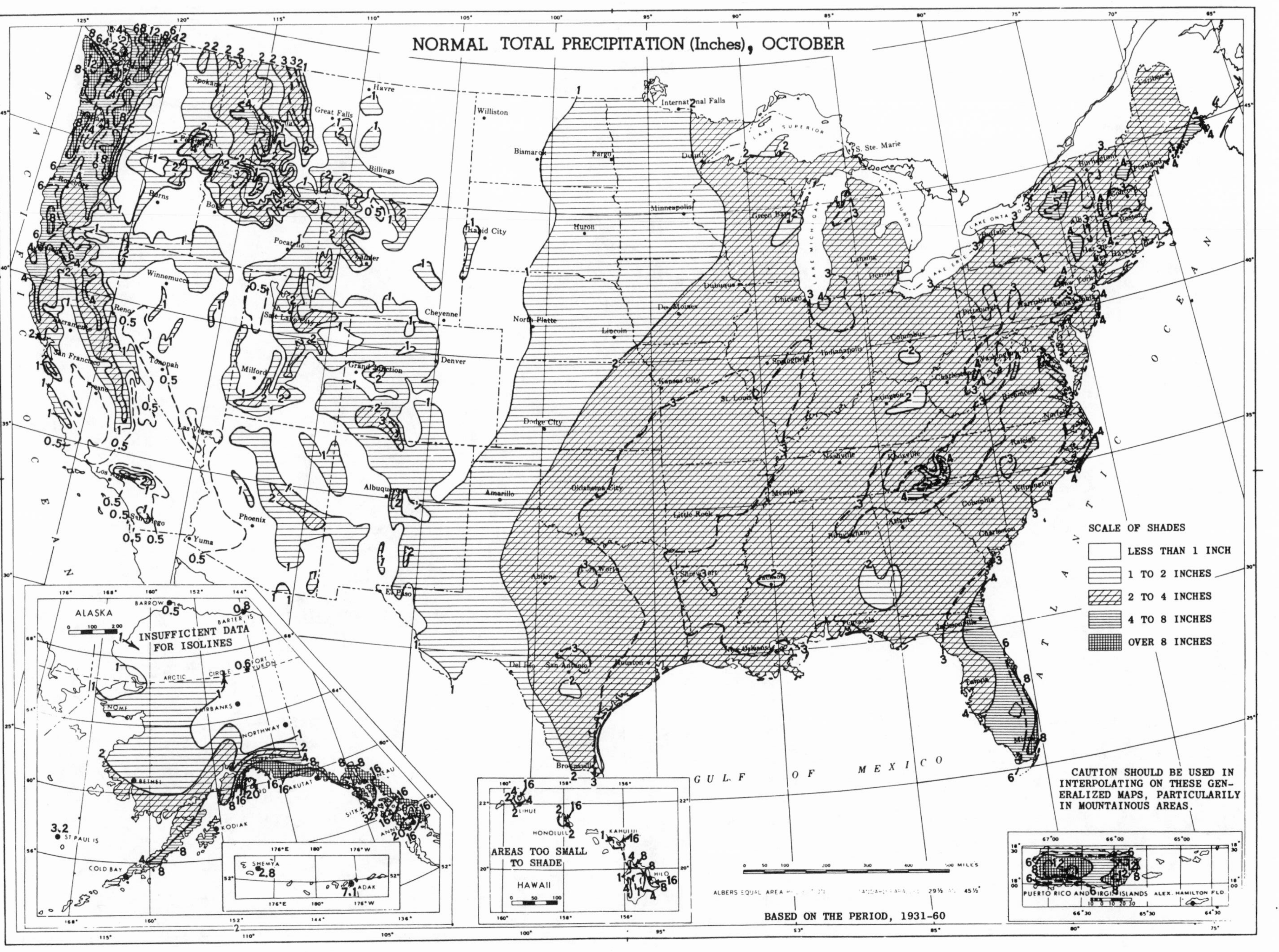

NORMAL TOTAL PRECIPITATION (Inches), OCTOBER
SCALE OF SHADES
LESS THAN 1 INCH
1 TO 2 INCHES
2 TO 4 INCHES
4 TO 8 INCHES
OVER 8 INCHES
CAUTION SHOULD BE USED IN INTERPOLATING ON THESE GENERALIZED MAPS, PARTICULARILY IN MOUNTAINOUS AREAS.
BASED ON THE PERIOD, 1931-60
GULF OF MEXICO
ATLANTIC OCEAN
PACIFIC OCEAN
ALASKA
INSUFFICIENT DATA FOR ISOLINES
AREAS TOO SMALL TO SHADE
HAWAII
PUERTO RICO AND VIRGIN ISLANDS

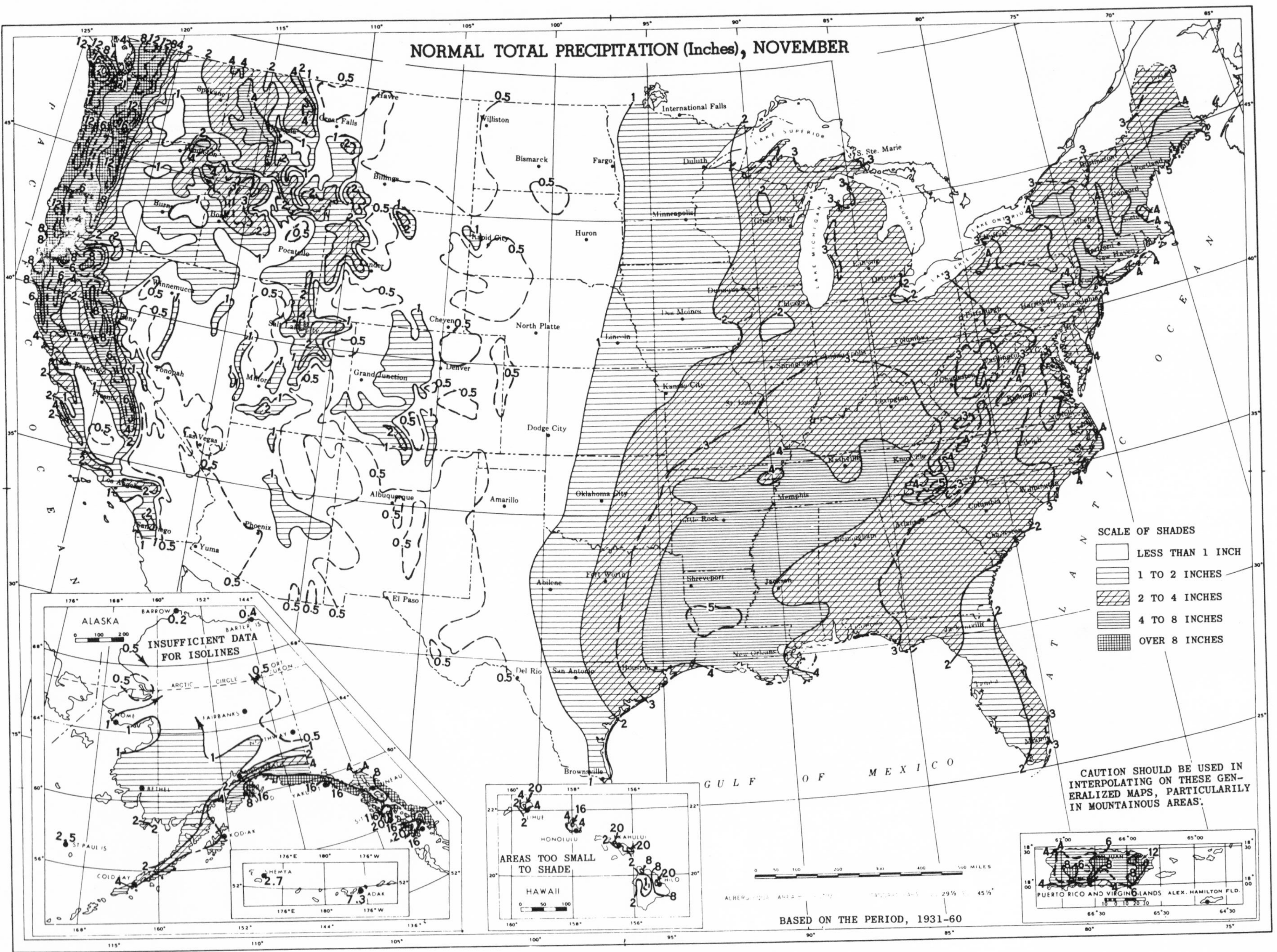
NORMAL TOTAL PRECIPITATION (Inches), NOVEMBER
SCALE OF SHADES
LESS THAN 1 INCH
1 TO 2 INCHES
2 TO 4 INCHES
4 TO 8 INCHES
OVER 8 INCHES
CAUTION SHOULD BE USED IN INTERPOLATING ON THESE GENERALIZED MAPS, PARTICULARILY IN MOUNTAINOUS AREAS.
BASED ON THE PERIOD, 1931-60
ALASKA
INSUFFICIENT DATA FOR ISOLINES
AREAS TOO SMALL TO SHADE
HAWAII
PUERTO RICO AND VIRGIN ISLANDS
GULF OF MEXICO
PACIFIC OCEAN
ATLANTIC OCEAN

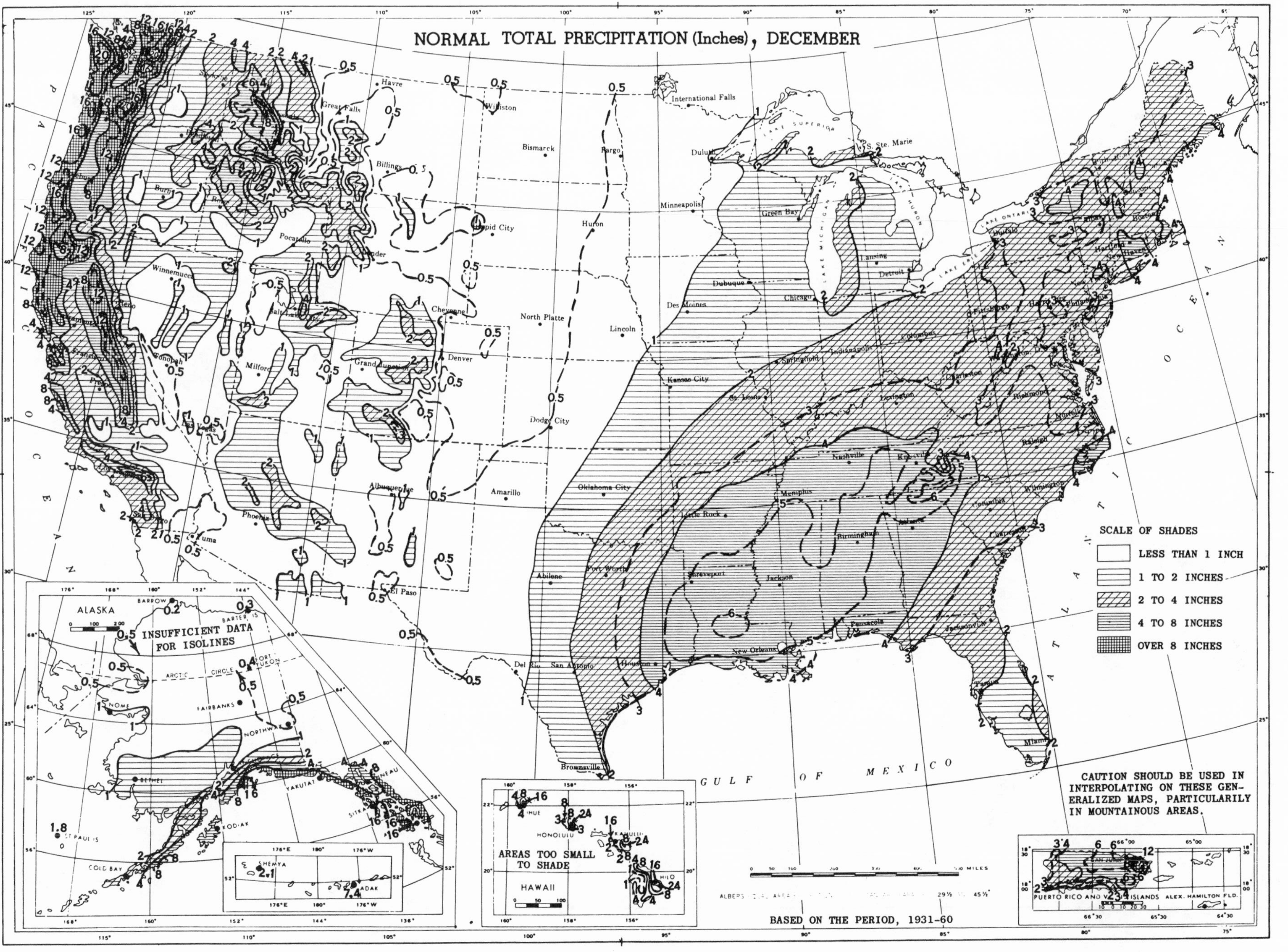
NORMAL TOTAL PRECIPITATION (Inches), DECEMBER
SCALE OF SHADES
LESS THAN 1 INCH
1 TO 2 INCHES
2 TO 4 INCHES
4 TO 8 INCHES
OVER 8 INCHES
CAUTION SHOULD BE USED IN INTERPOLATING ON THESE GENERALIZED MAPS, PARTICULARLY IN MOUNTAINOUS AREAS.
BASED ON THE PERIOD, 1931-60
GULF OF MEXICO
ATLANTIC OCEAN
PACIFIC OCEAN
AREAS TOO SMALL TO SHADE
HAWAII
HONOLULU
HILO
LIHUE
ALASKA
INSUFFICIENT DATA FOR ISOLINES
BARROW
BARTER IS
FORT YUKON
ARCTIC CIRCLE
FAIRBANKS
NOME
NORTHWAY
BETHEL
KODIAK
YAKUTAT
JUNEAU
SITKA
ST. PAUL IS
COLD BAY
SHEMYA
ADAK
PUERTO RICO AND V. ISLANDS
ALEX. HAMILTON FLD.
International Falls
Duluth
Minneapolis
Green Bay
Chicago
Detroit
Lansing
S. Ste. Marie
LAKE SUPERIOR
LAKE MICHIGAN
LAKE HURON
LAKE ERIE
LAKE ONTARIO
Buffalo
Boston
Hartford
New Haven
Dubuque
Des Moines
Kansas City
St. Louis
Springfield
Indianapolis
Columbus
Pittsburgh
Charleston
Lexington
Richmond
Norfolk
Raleigh
Wilmington
Columbia
Nashville
Knoxville
Memphis
Little Rock
Birmingham
Shreveport
Jackson
New Orleans
Pensacola
Jacksonville
Tampa
Miami
Fargo
Bismarck
Huron
Lincoln
North Platte
Dodge City
Oklahoma City
Amarillo
Fort Worth
Abilene
Del Rio
San Antonio
Houston
Brownsville
Havre
Great Falls
Williston
Billings
Rapid City
Lander
Pocatello
Cheyenne
Denver
Salt Lake City
Grand Junction
Milford
Winnemucca
Reno
Tonopah
Albuquerque
El Paso
Phoenix
Yuma
Burns
Spokane

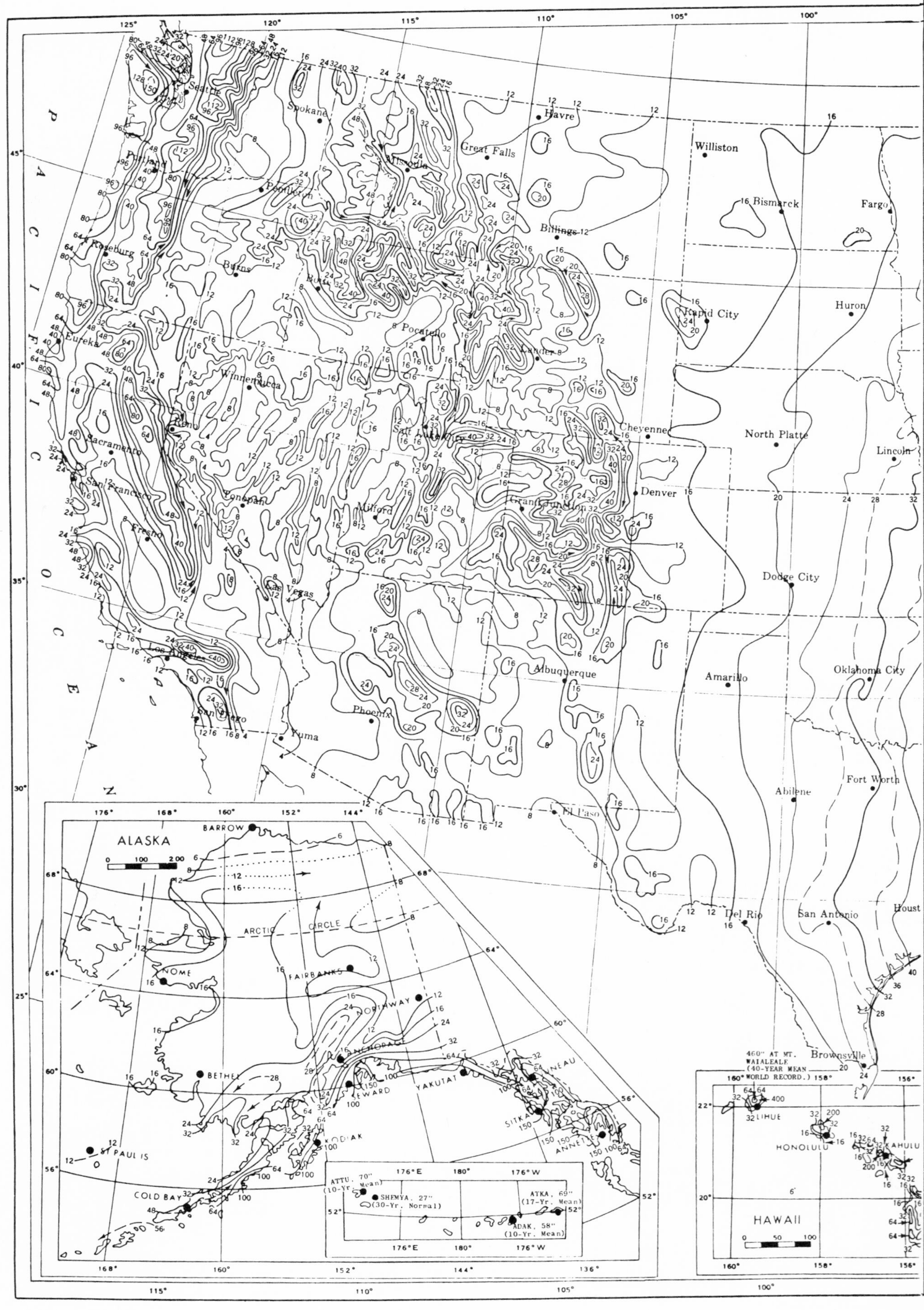
PACIFIC OCEAN
Seattle
Spokane
Havre
Great Falls
Williston
Bismarck
Fargo
Portland
Pendleton
Missoula
Billings
Roseburg
Burns
Boise
Rapid City
Huron
Eureka
Pocatello
Lander
Winnemucca
Reno
Salt Lake City
Cheyenne
North Platte
Lincoln
Sacramento
San Francisco
Tonopah
Milford
Grand Junction
Denver
Fresno
Las Vegas
Dodge City
Los Angeles
San Diego
Yuma
Phoenix
Albuquerque
Amarillo
Oklahoma City
Abilene
Fort Worth
El Paso
Del Rio
San Antonio
Houston
Brownsville
ALASKA
BARROW
ARCTIC CIRCLE
NOME
FAIRBANKS
NORTHWAY
ANCHORAGE
BETHEL
SEWARD
YAKUTAT
JUNEAU
SITKA
ANNETTE
KODIAK
ST PAUL IS
COLD BAY
ATTU, 70" (10-Yr. Mean)
SHEMYA, 27" (30-Yr. Normal)
ATKA, 69" (17-Yr. Mean)
ADAK, 58" (10-Yr. Mean)
460" AT MT. WAIALEALE (40-YEAR MEAN WORLD RECORD.)
LIHUE
HONOLULU
KAHULUI
HAWAII

PRECIPITATION (Inches)

NORMAL MONTHLY TOTAL

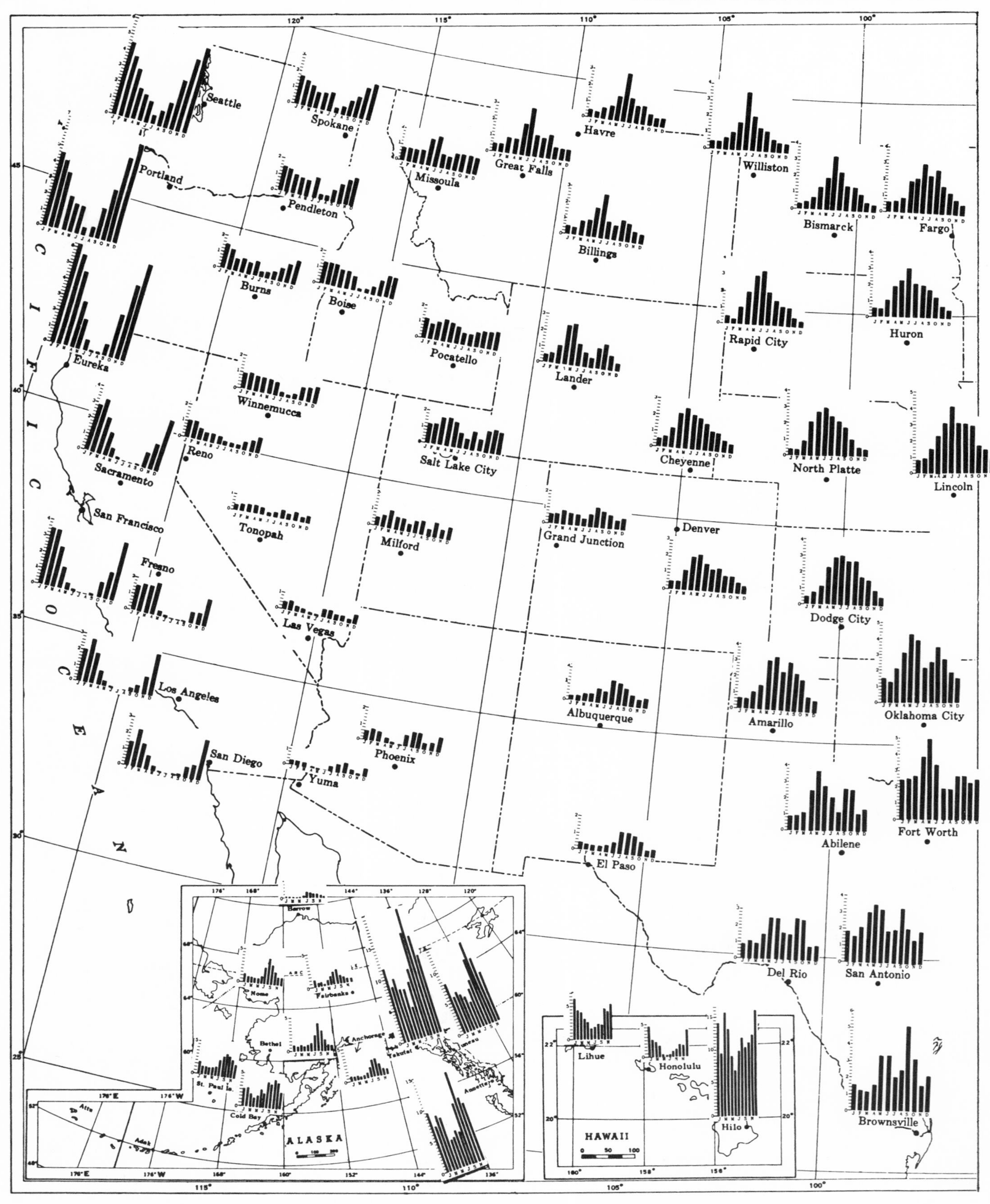

PRECIPITATION (Inches)

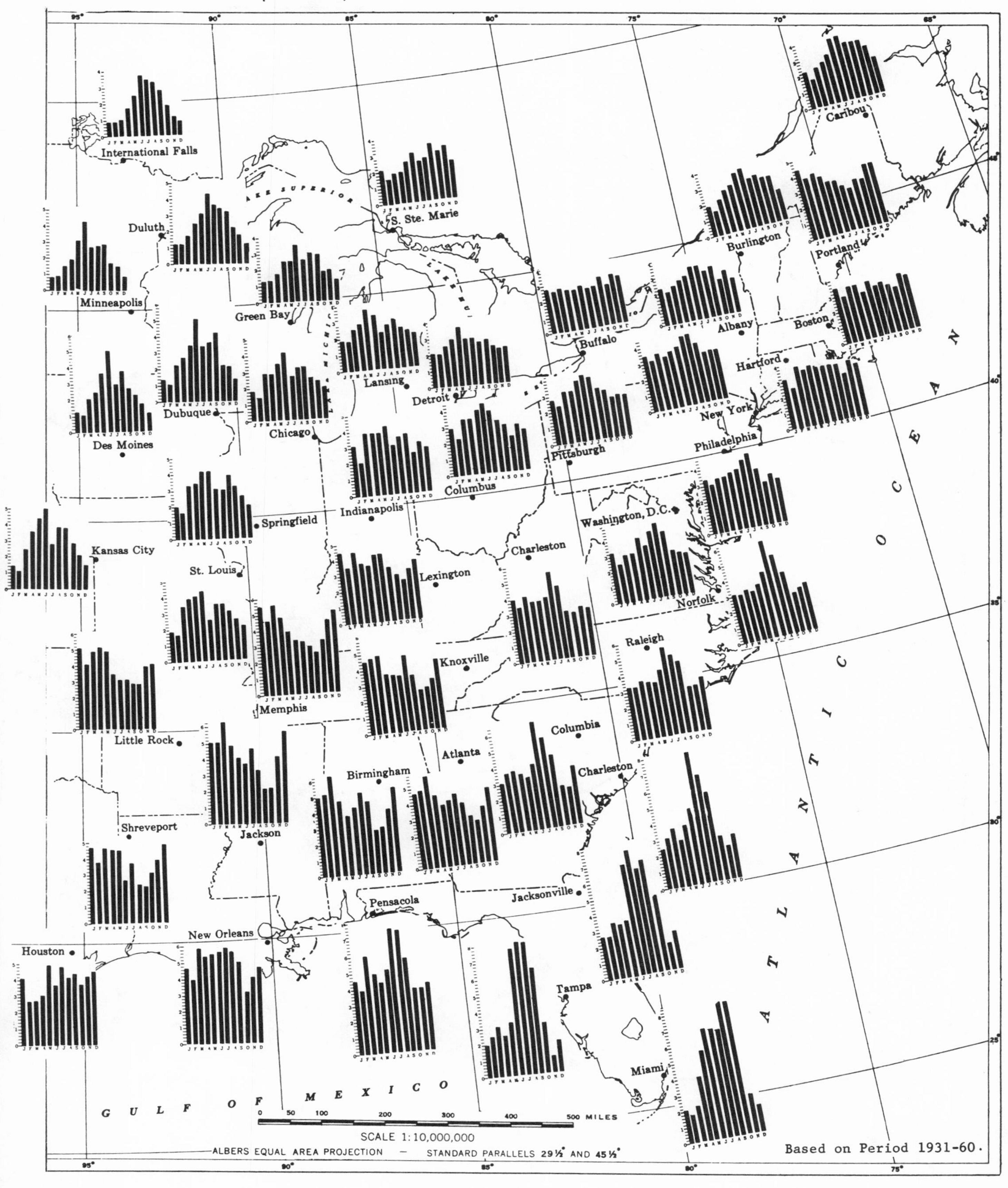

Based on Period 1931-60.

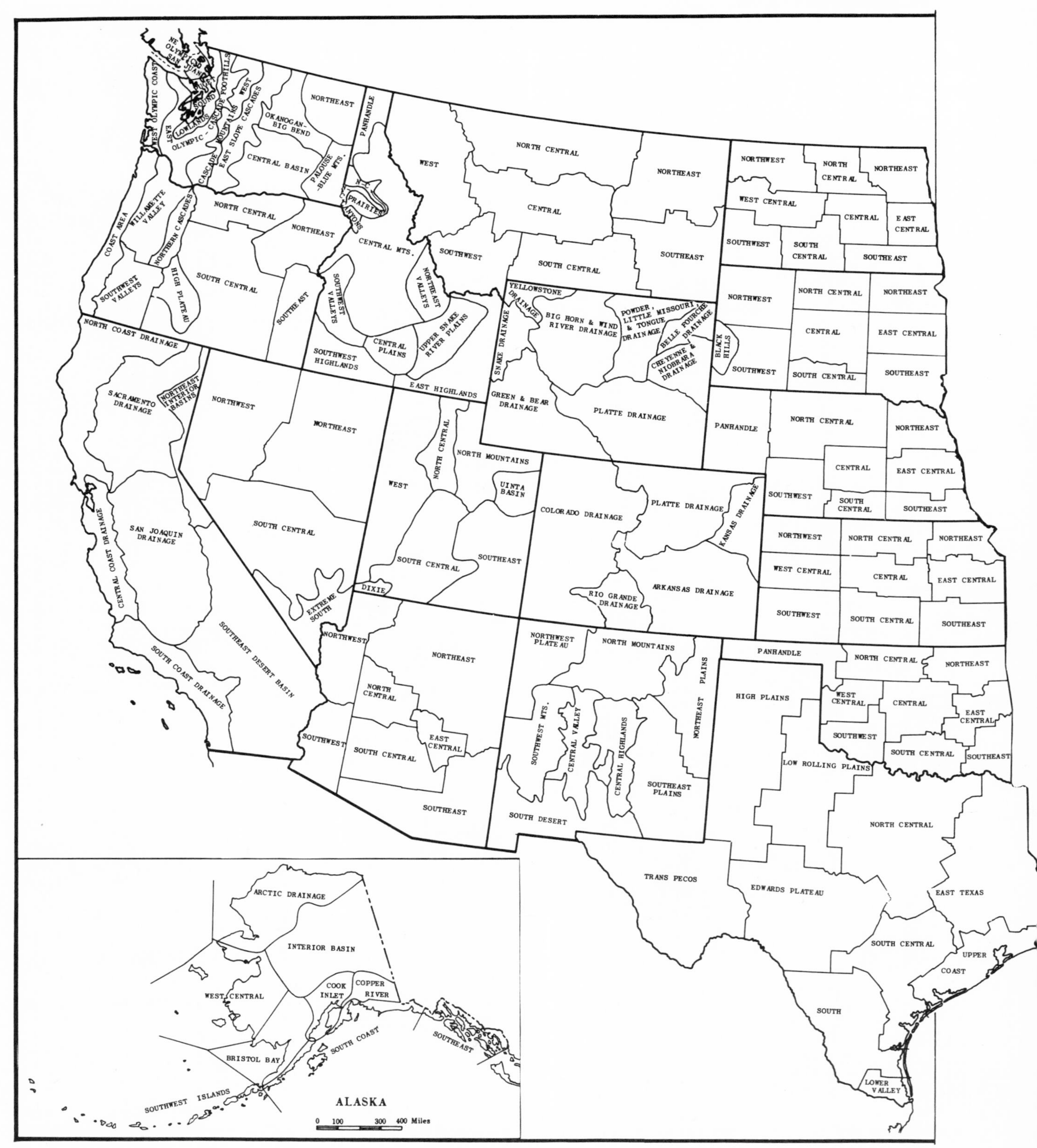
WEST OLYMPIC COAST
EAST OLYMPIC-CASCADE FOOTHILLS
CASCADE MOUNTAINS WEST
EAST SLOPE CASCADES
NORTHEAST
OKANOGAN-BIG BEND
CENTRAL BASIN
PALOUSE-BLUE MTS.
PANHANDLE
COAST AREA
WILLAMETTE VALLEY
NORTHERN CASCADES
NORTH CENTRAL
NORTHEAST
SOUTHWEST VALLEYS
HIGH PLATEAU
SOUTH CENTRAL
SOUTHEAST
CENTRAL MTS.
SOUTHWEST VALLEYS
NORTHEAST VALLEYS
CENTRAL PLAINS
UPPER SNAKE RIVER PLAINS
SOUTHWEST HIGHLANDS
EAST HIGHLANDS
WEST
NORTH CENTRAL
NORTHEAST
CENTRAL
SOUTHWEST
SOUTH CENTRAL
SOUTHEAST
NORTHWEST
NORTH CENTRAL
NORTHEAST
WEST CENTRAL
CENTRAL
EAST CENTRAL
SOUTHWEST
SOUTH CENTRAL
SOUTHEAST
YELLOWSTONE DRAINAGE
SNAKE DRAINAGE
BIG HORN & WIND RIVER DRAINAGE
POWDER, LITTLE MISSOURI & TONGUE DRAINAGE
BELLE FOURCHE DRAINAGE
CHEYENNE, NIOBRARA DRAINAGE
BLACK HILLS
GREEN & BEAR DRAINAGE
PLATTE DRAINAGE
NORTHWEST
NORTH CENTRAL
NORTHEAST
CENTRAL
EAST CENTRAL
SOUTHWEST
SOUTH CENTRAL
SOUTHEAST
NORTH COAST DRAINAGE
SACRAMENTO DRAINAGE
NORTHEAST INTERIOR BASINS
NORTHWEST
NORTHEAST
SOUTH CENTRAL
SAN JOAQUIN DRAINAGE
CENTRAL COAST DRAINAGE
SOUTH COAST DRAINAGE
SOUTHEAST DESERT BASIN
EXTREME SOUTH
WEST
NORTH CENTRAL
NORTH MOUNTAINS
UINTA BASIN
SOUTH CENTRAL
SOUTHEAST
DIXIE
PANHANDLE
NORTH CENTRAL
NORTHEAST
CENTRAL
EAST CENTRAL
SOUTHWEST
SOUTH CENTRAL
SOUTHEAST
COLORADO DRAINAGE
PLATTE DRAINAGE
KANSAS DRAINAGE
ARKANSAS DRAINAGE
RIO GRANDE DRAINAGE
NORTHWEST
NORTH CENTRAL
NORTHEAST
WEST CENTRAL
CENTRAL
EAST CENTRAL
SOUTHWEST
SOUTH CENTRAL
SOUTHEAST
NORTHWEST
NORTHEAST
NORTH CENTRAL
EAST CENTRAL
SOUTHWEST
SOUTH CENTRAL
SOUTHEAST
NORTHWEST PLATEAU
NORTH MOUNTAINS
PLAINS
NORTHEAST
SOUTHWEST MTS.
CENTRAL VALLEY
CENTRAL HIGHLANDS
SOUTHEAST PLAINS
SOUTH DESERT
PANHANDLE
NORTH CENTRAL
NORTHEAST
WEST CENTRAL
CENTRAL
EAST CENTRAL
SOUTHWEST
SOUTH CENTRAL
SOUTHEAST
HIGH PLAINS
LOW ROLLING PLAINS
NORTH CENTRAL
TRANS PECOS
EDWARDS PLATEAU
EAST TEXAS
SOUTH CENTRAL
UPPER COAST
SOUTH
LOWER VALLEY
ARCTIC DRAINAGE
INTERIOR BASIN
WEST CENTRAL
COOK INLET
COPPER RIVER
BRISTOL BAY
SOUTH COAST
SOUTHEAST
SOUTHWEST ISLANDS
ALASKA
0 100 300 400 Miles

DIVISIONS

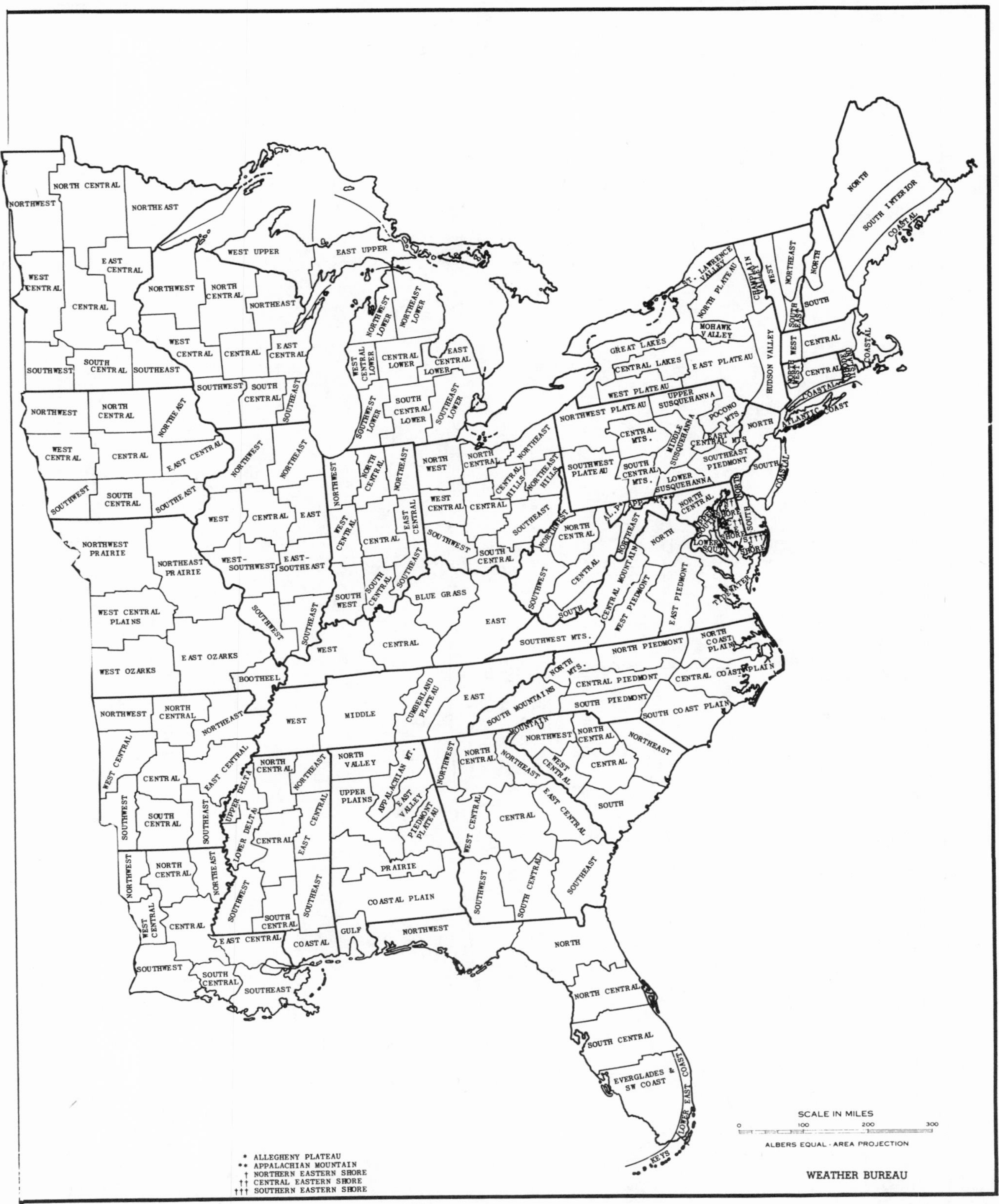
NORTH CENTRAL
NORTHWEST
NORTHEAST
WEST CENTRAL
EAST CENTRAL
CENTRAL
SOUTHWEST
SOUTH CENTRAL
SOUTHEAST
WEST UPPER
EAST UPPER
NORTHWEST LOWER
NORTHEAST LOWER
WEST CENTRAL LOWER
CENTRAL LOWER
EAST CENTRAL LOWER
SOUTHWEST LOWER
SOUTH CENTRAL LOWER
SOUTHEAST LOWER
WEST
EAST
WEST-SOUTHWEST
EAST-SOUTHEAST
NORTHWEST PRAIRIE
NORTHEAST PRAIRIE
WEST CENTRAL PLAINS
WEST OZARKS
EAST OZARKS
BOOTHEEL
NORTH
SOUTH
BLUE GRASS
SOUTHWEST MTS.
NORTHEAST HILLS
CENTRAL HILLS
CENTRAL MOUNTAIN
WEST PIEDMONT
EAST PIEDMONT
NORTH PIEDMONT
NORTH COAST PLAIN
CENTRAL PIEDMONT
CENTRAL COAST PLAIN
SOUTH PIEDMONT
SOUTH COAST PLAIN
NORTH MTS.
SOUTH MOUNTAINS
TIDEWATER
MIDDLE
CUMBERLAND PLATEAU
NORTH VALLEY
APPALACHIAN MT.
UPPER PLAINS
EAST VALLEY
PIEDMONT PLATEAU
PRAIRIE
COASTAL PLAIN
GULF
UPPER DELTA
LOWER DELTA
COASTAL
NORTHWEST PLATEAU
SOUTHWEST PLATEAU
CENTRAL MTS.
SOUTH CENTRAL MTS.
UPPER SUSQUEHANNA
MIDDLE SUSQUEHANNA
LOWER SUSQUEHANNA
POCONO MTS.
EAST CENTRAL MTS.
SOUTHEAST PIEDMONT
GREAT LAKES
CENTRAL LAKES
EAST PLATEAU
WEST PLATEAU
MOHAWK VALLEY
HUDSON VALLEY
ST. LAWRENCE VALLEY
NORTH PLATEAU
CHAMPLAIN VALLEY
ATLANTIC COAST
SOUTH INTERIOR
EVERGLADES & SW COAST
LOWER EAST COAST
KEYS
* ALLEGHENY PLATEAU
** APPALACHIAN MOUNTAIN
† NORTHERN EASTERN SHORE
†† CENTRAL EASTERN SHORE
††† SOUTHERN EASTERN SHORE
SCALE IN MILES
0 100 200 300
ALBERS EQUAL-AREA PROJECTION
WEATHER BUREAU

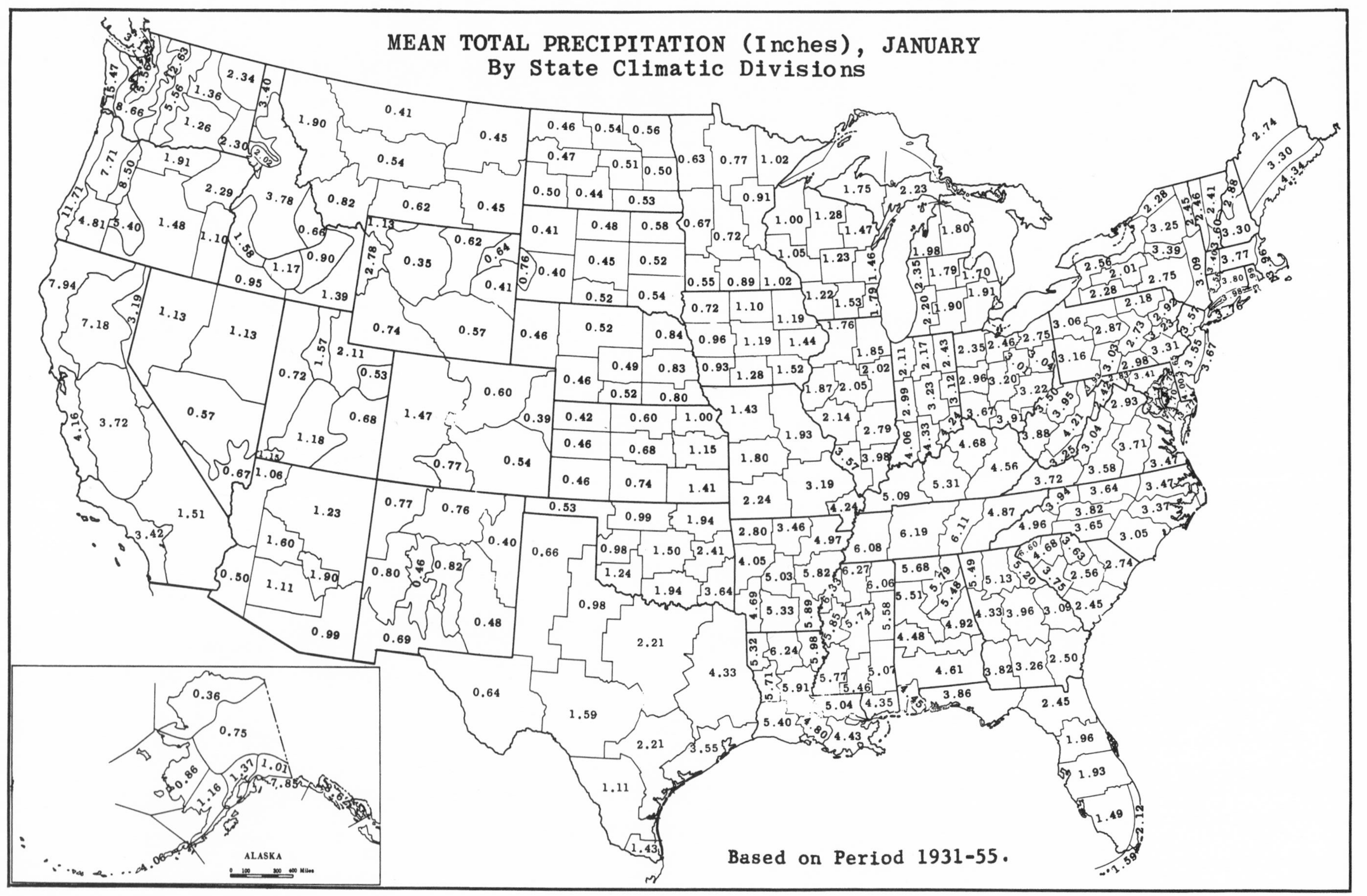
MEAN TOTAL PRECIPITATION (Inches), JANUARY
By State Climatic Divisions
Based on Period 1931-55.
ALASKA

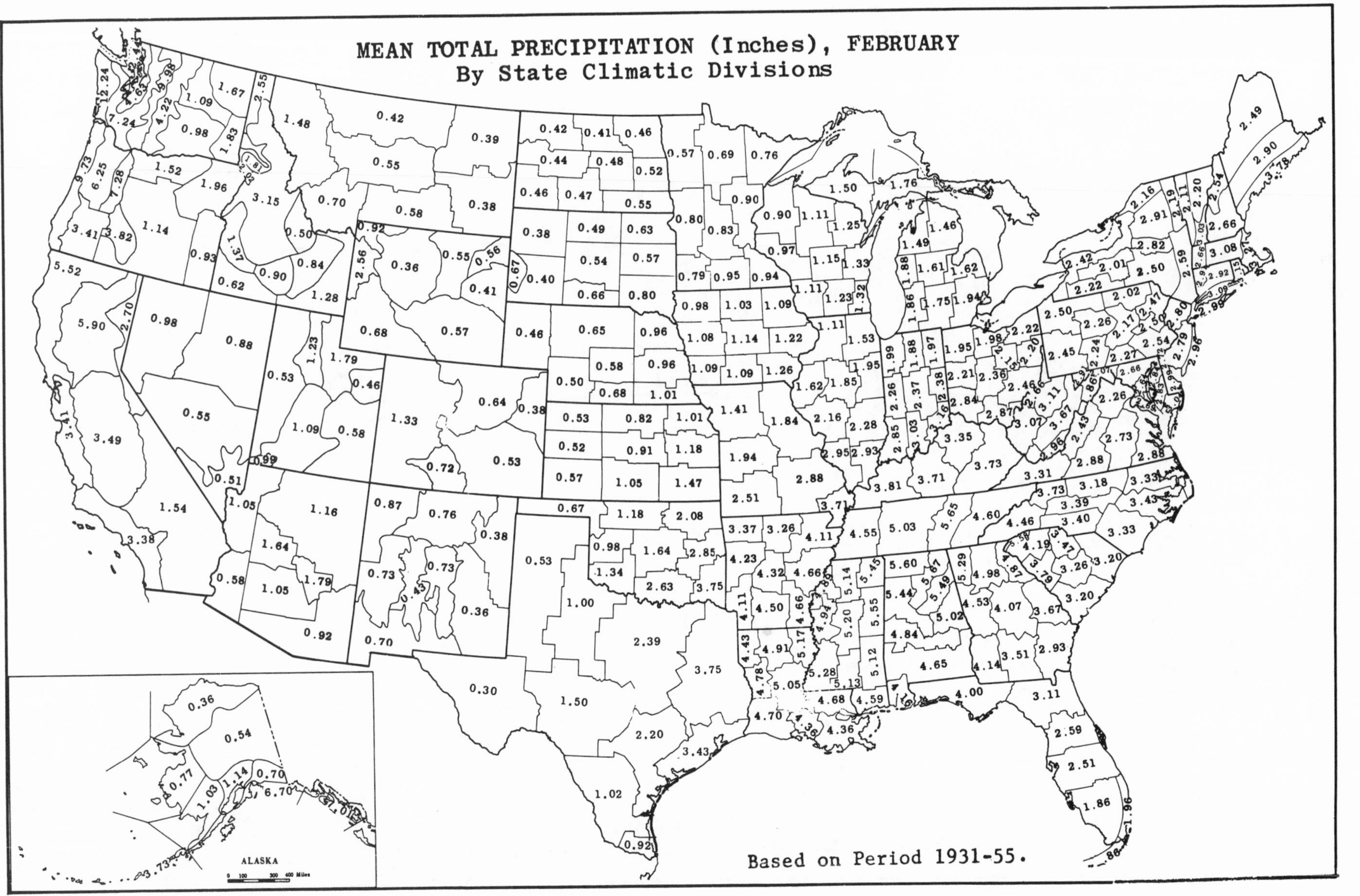
MEAN TOTAL PRECIPITATION (Inches), FEBRUARY
By State Climatic Divisions
Based on Period 1931-55.
ALASKA

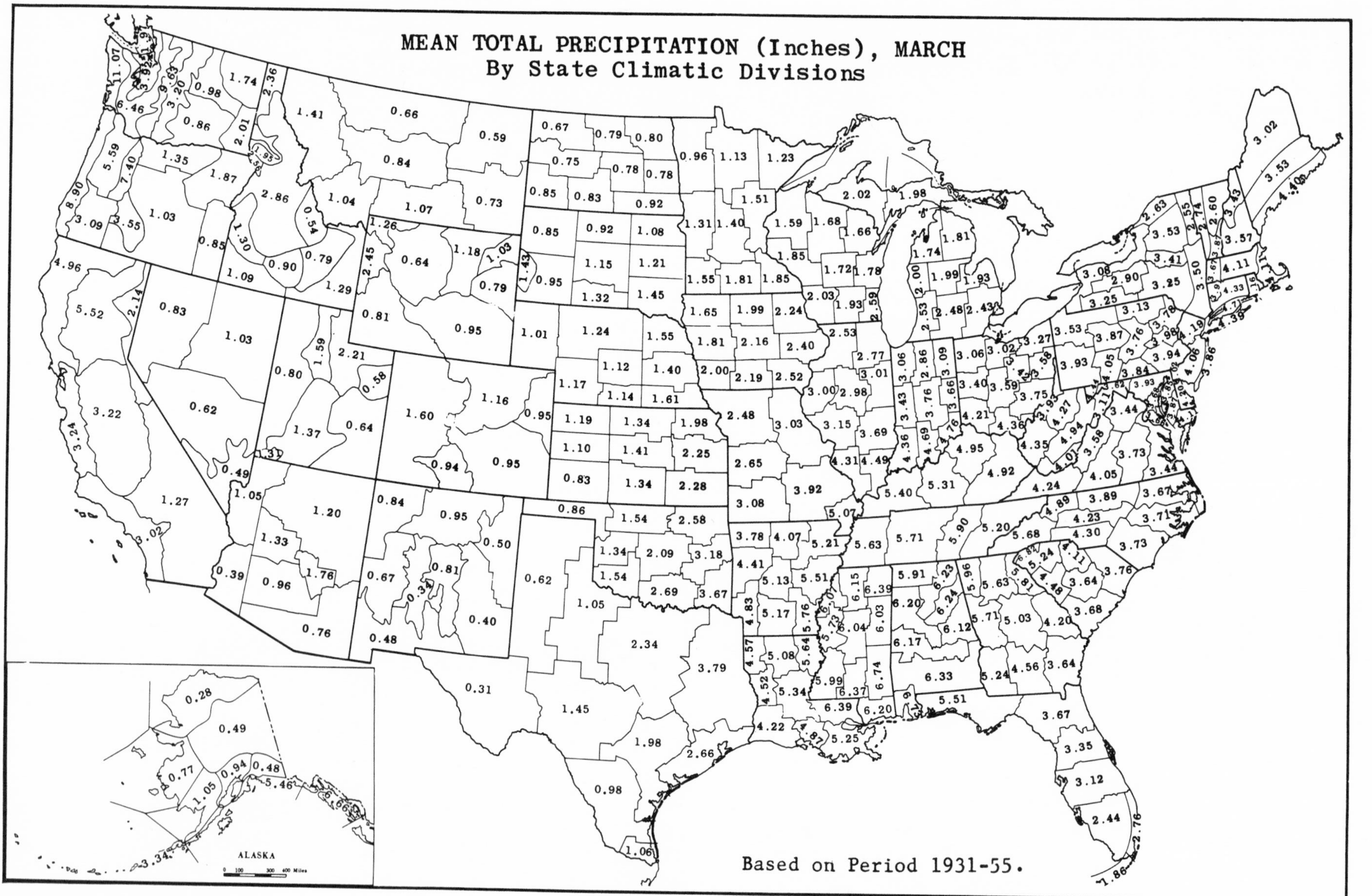
MEAN TOTAL PRECIPITATION (Inches), MARCH
By State Climatic Divisions
Based on Period 1931-55.
ALASKA

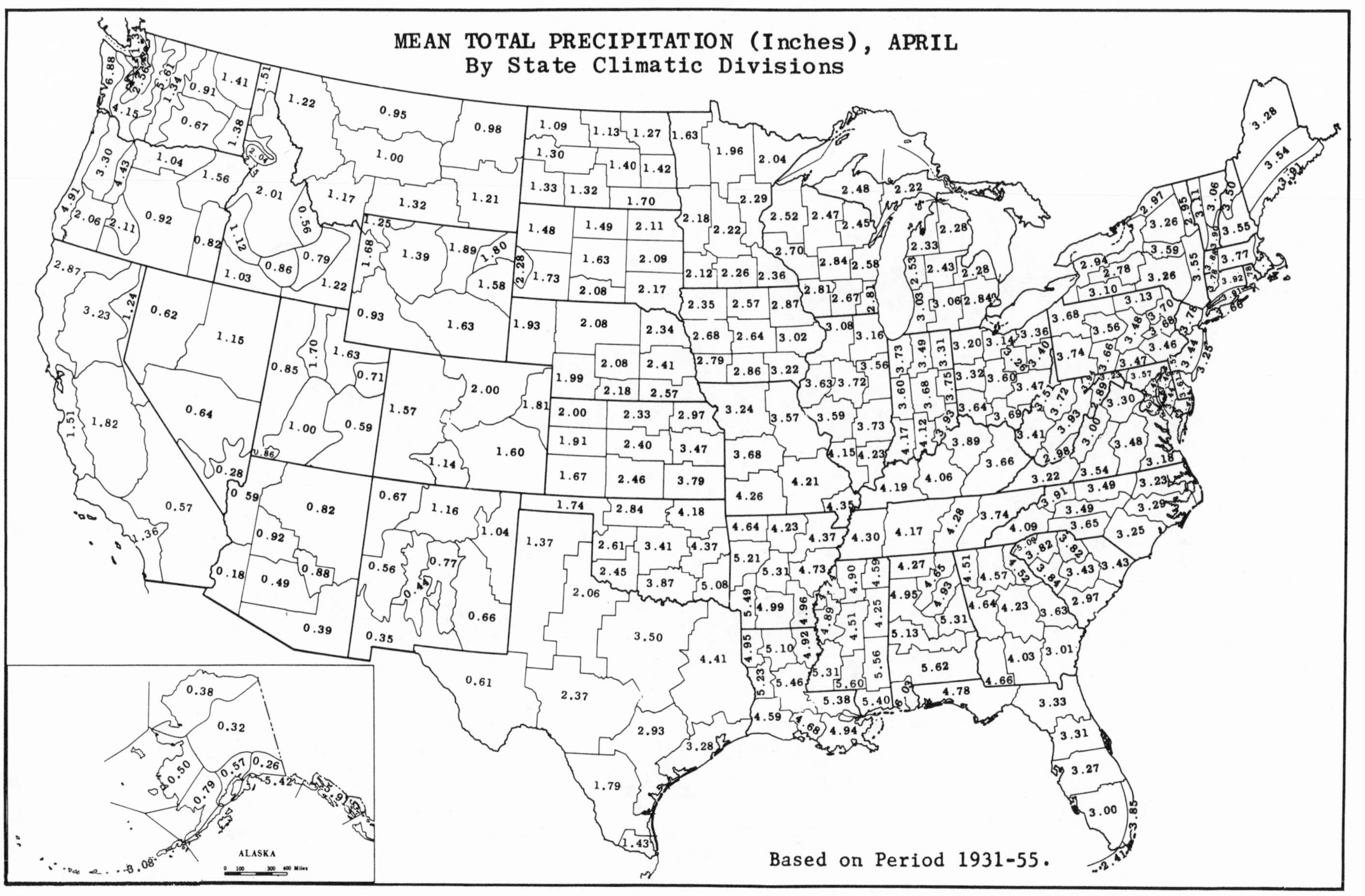
MEAN TOTAL PRECIPITATION (Inches), APRIL
By State Climatic Divisions
Based on Period 1931-55.
ALASKA

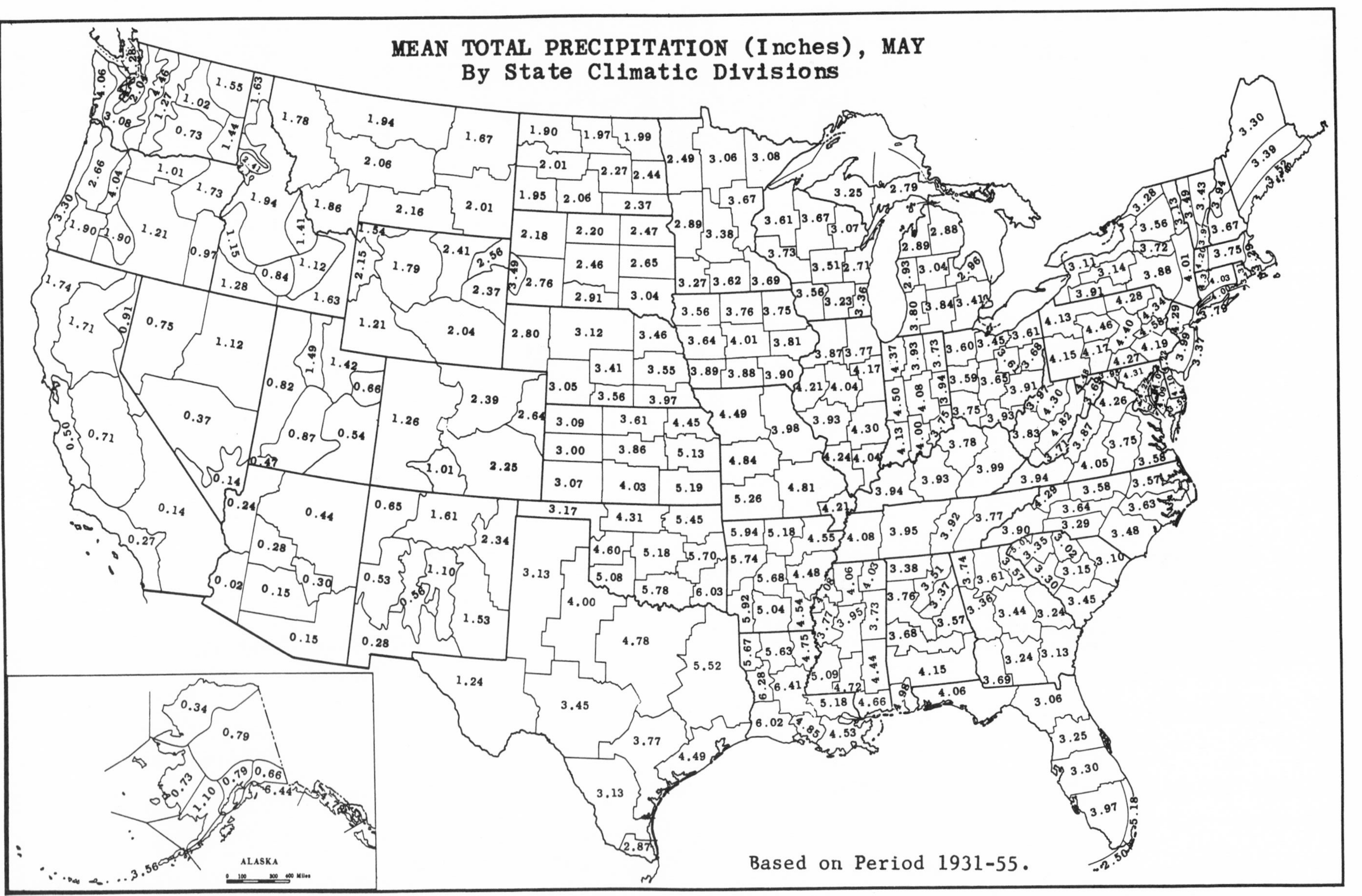
MEAN TOTAL PRECIPITATION (Inches), MAY
By State Climatic Divisions
ALASKA
Based on Period 1931-55.

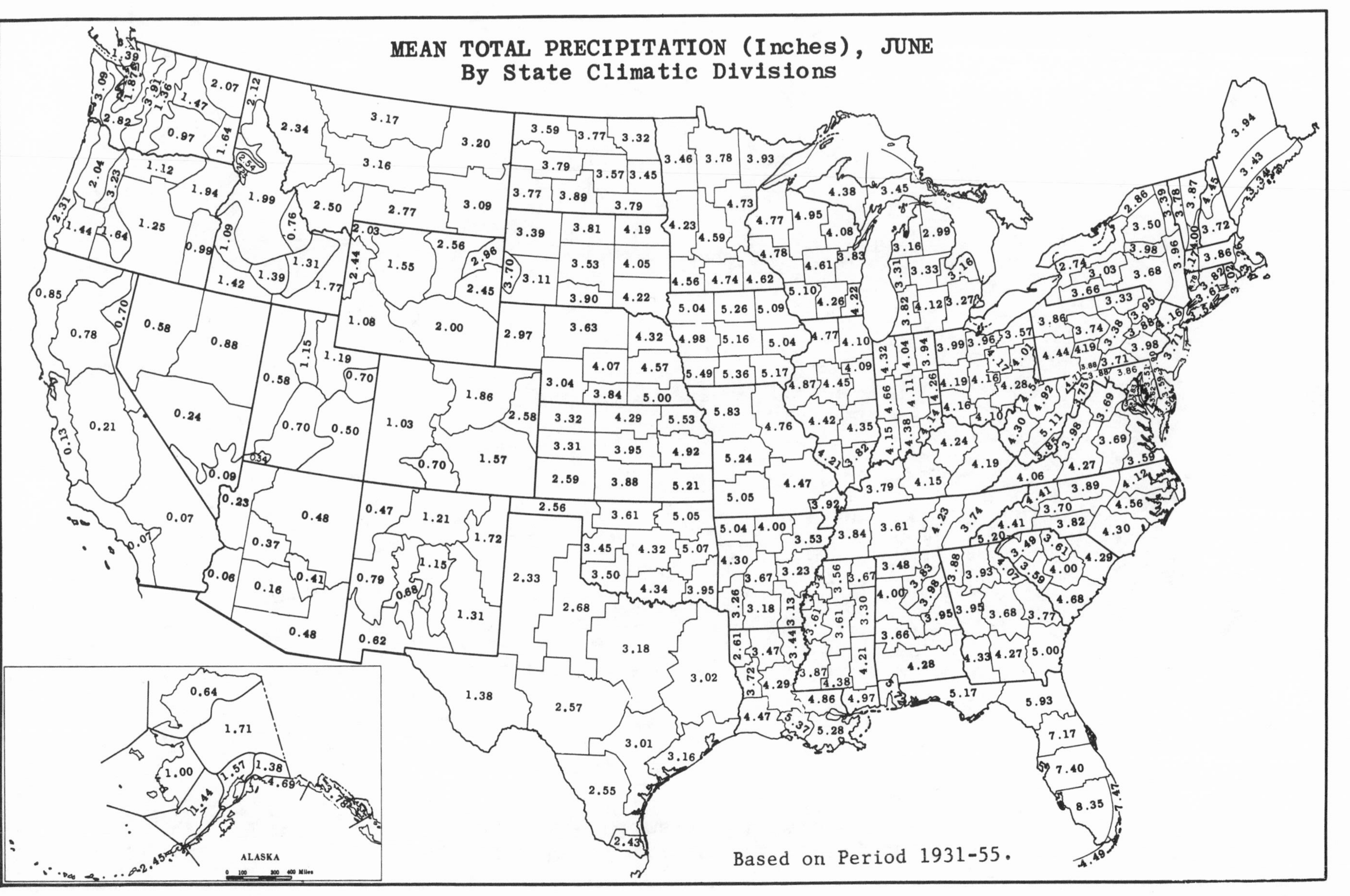
MEAN TOTAL PRECIPITATION (Inches), JUNE
By State Climatic Divisions
Based on Period 1931-55.
ALASKA

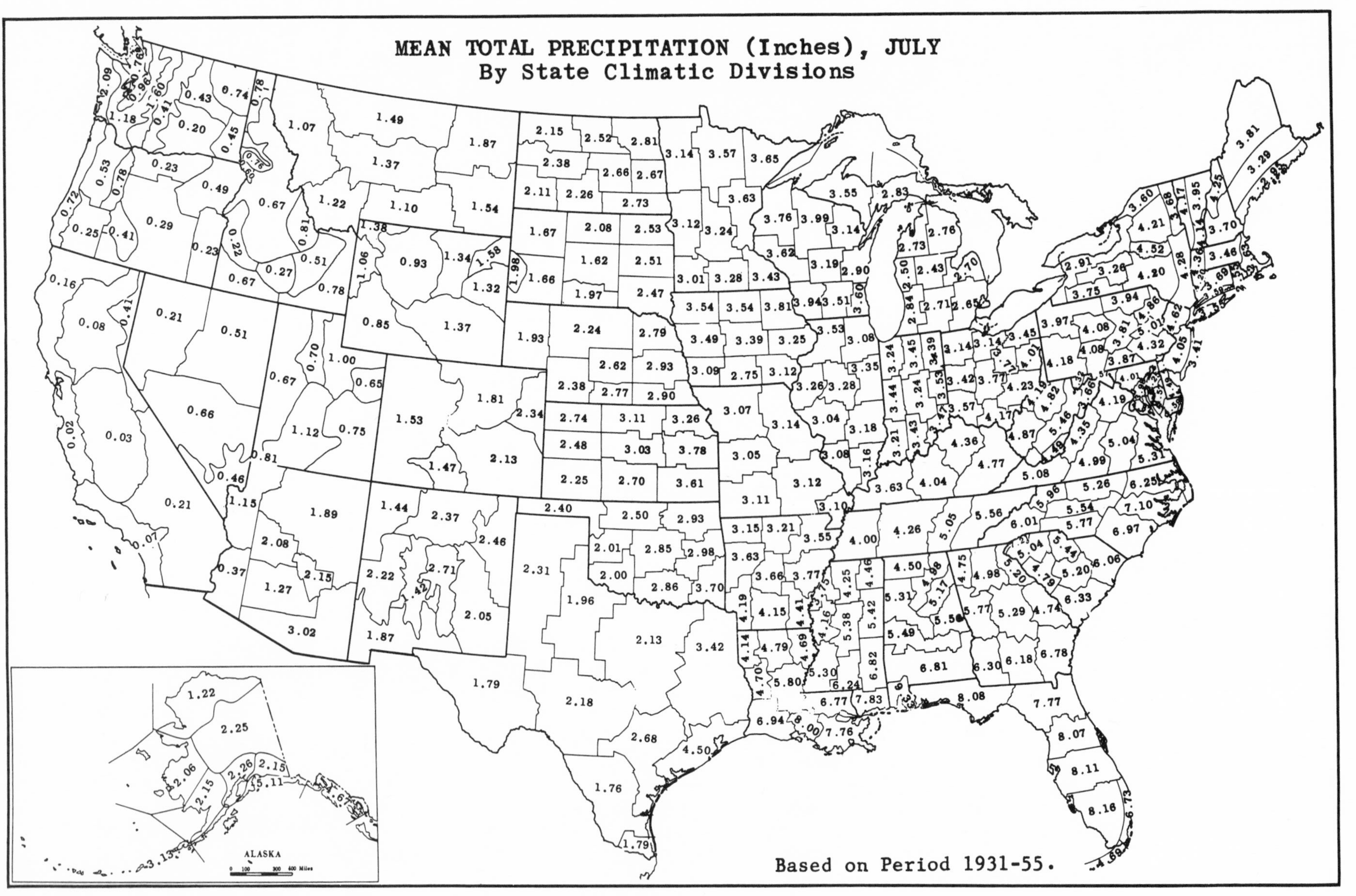
MEAN TOTAL PRECIPITATION (Inches), JULY
By State Climatic Divisions
Based on Period 1931-55.
ALASKA

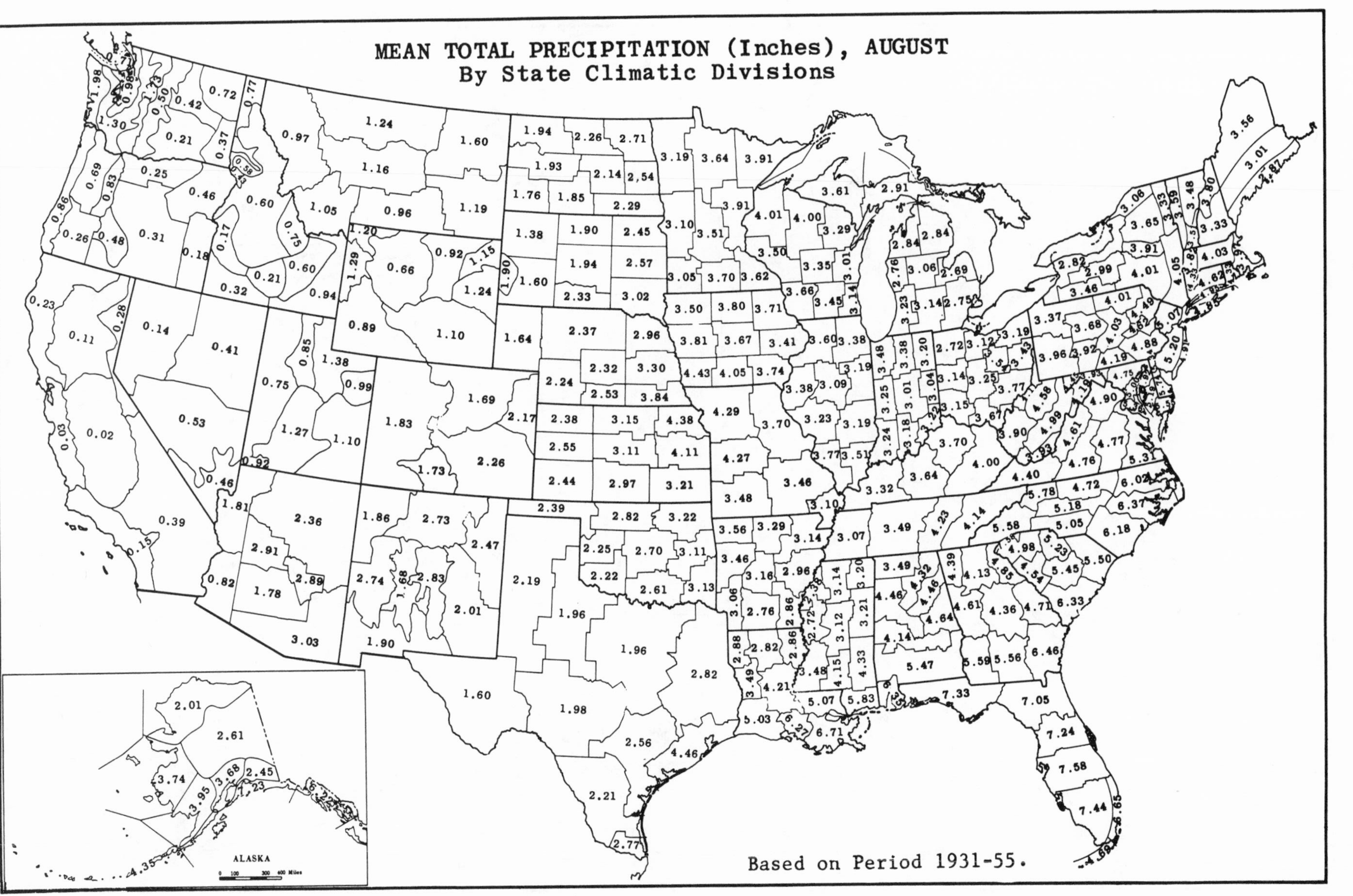
MEAN TOTAL PRECIPITATION (Inches), AUGUST
By State Climatic Divisions
Based on Period 1931-55.
ALASKA

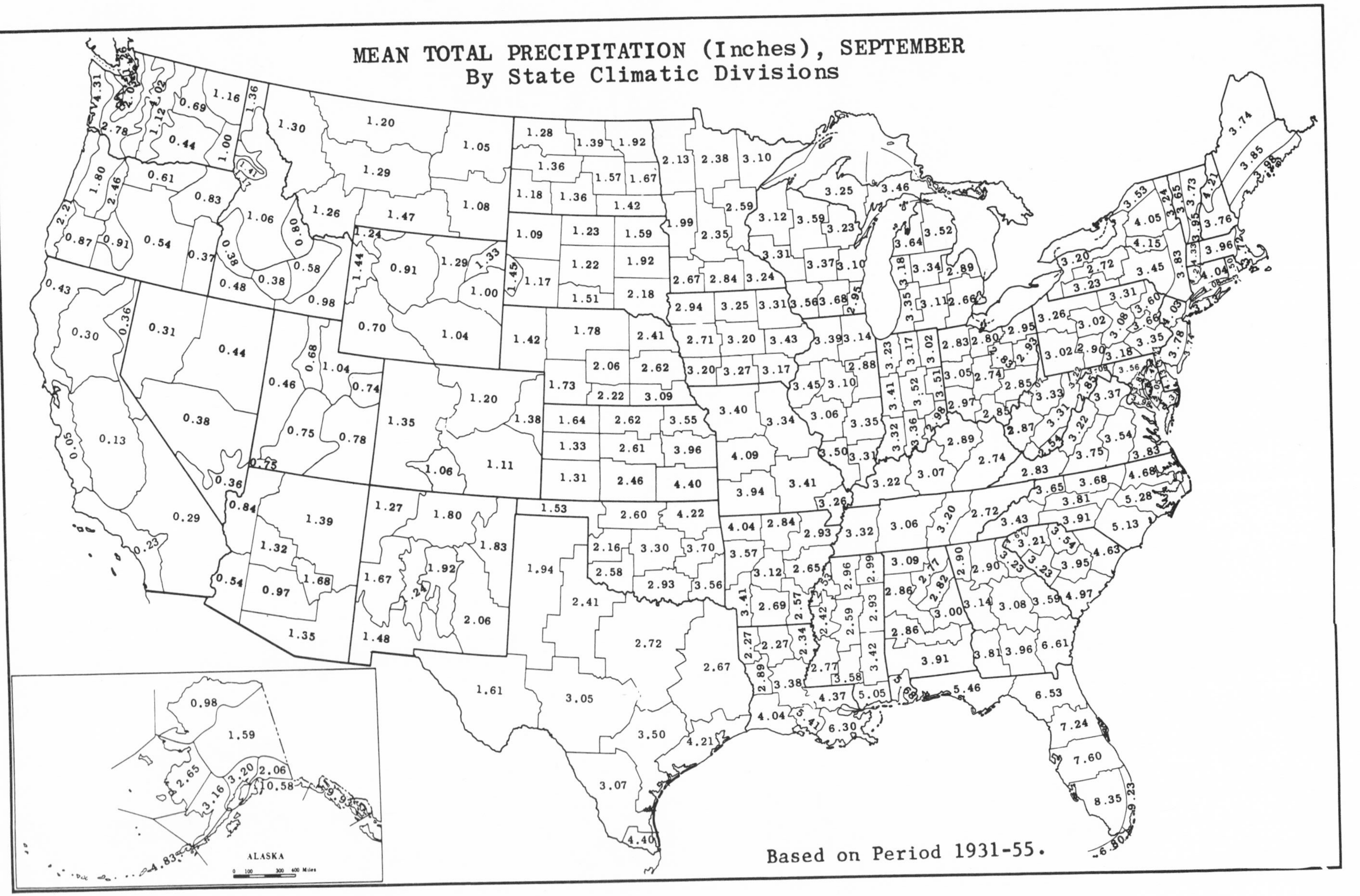
MEAN TOTAL PRECIPITATION (Inches), SEPTEMBER
By State Climatic Divisions
Based on Period 1931-55.
ALASKA

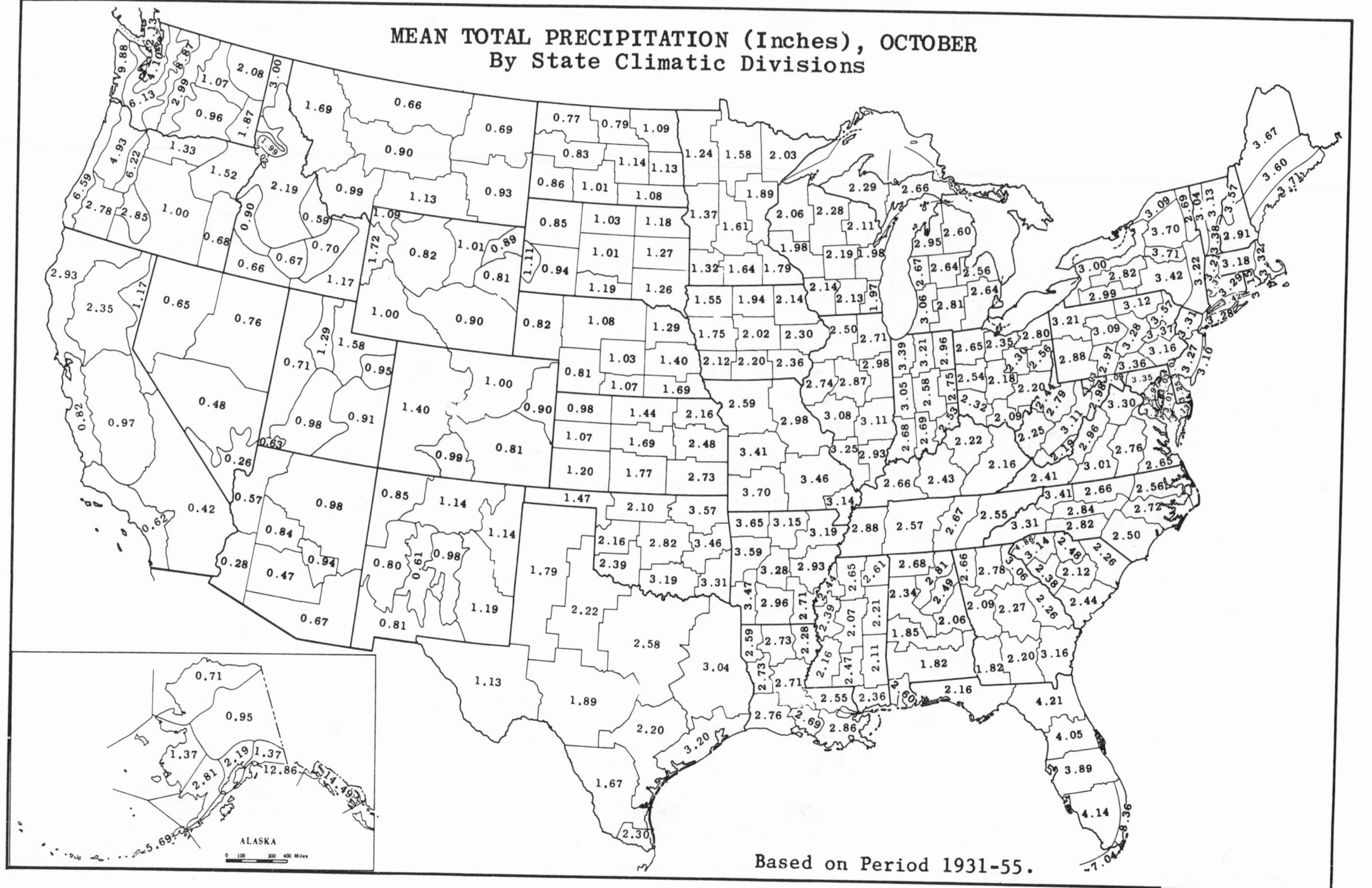
MEAN TOTAL PRECIPITATION (Inches), OCTOBER
By State Climatic Divisions
Based on Period 1931-55.
ALASKA

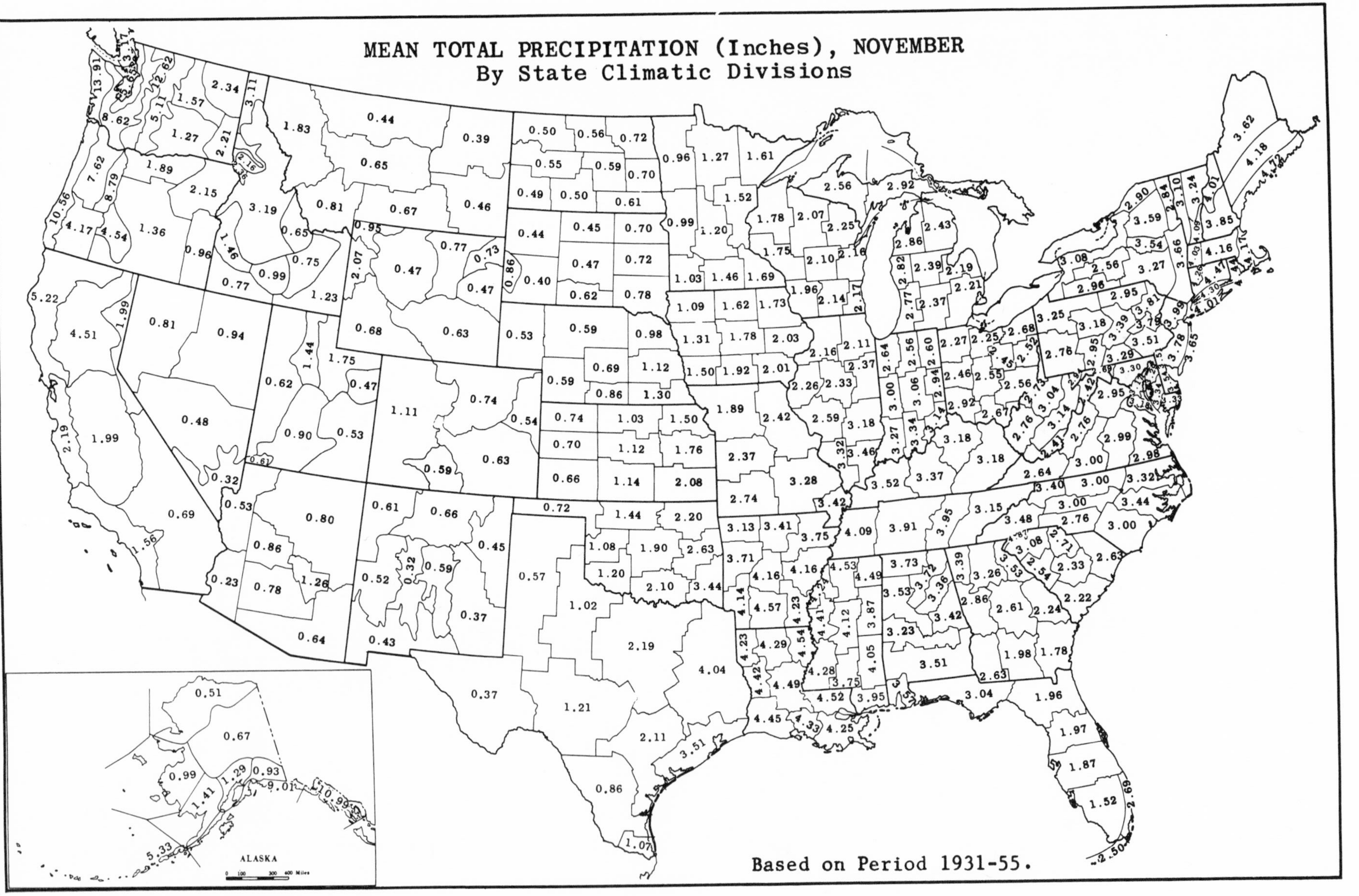
MEAN TOTAL PRECIPITATION (Inches), NOVEMBER
By State Climatic Divisions
Based on Period 1931-55.
ALASKA

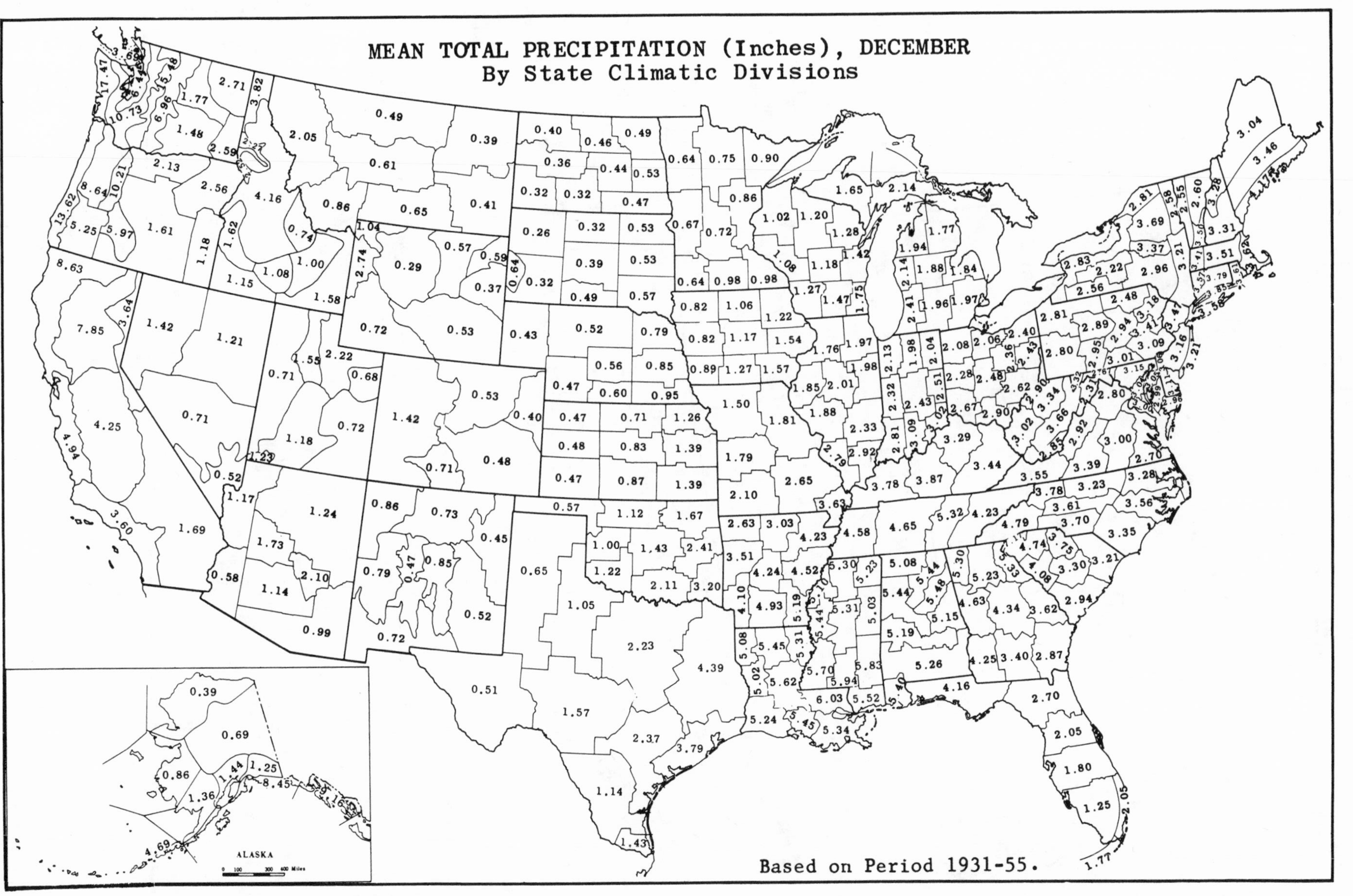
MEAN TOTAL PRECIPITATION (Inches), DECEMBER
By State Climatic Divisions
Based on Period 1931-55.
ALASKA

MEAN MONTHLY TOTAL PRECIPITATION (INCHES)

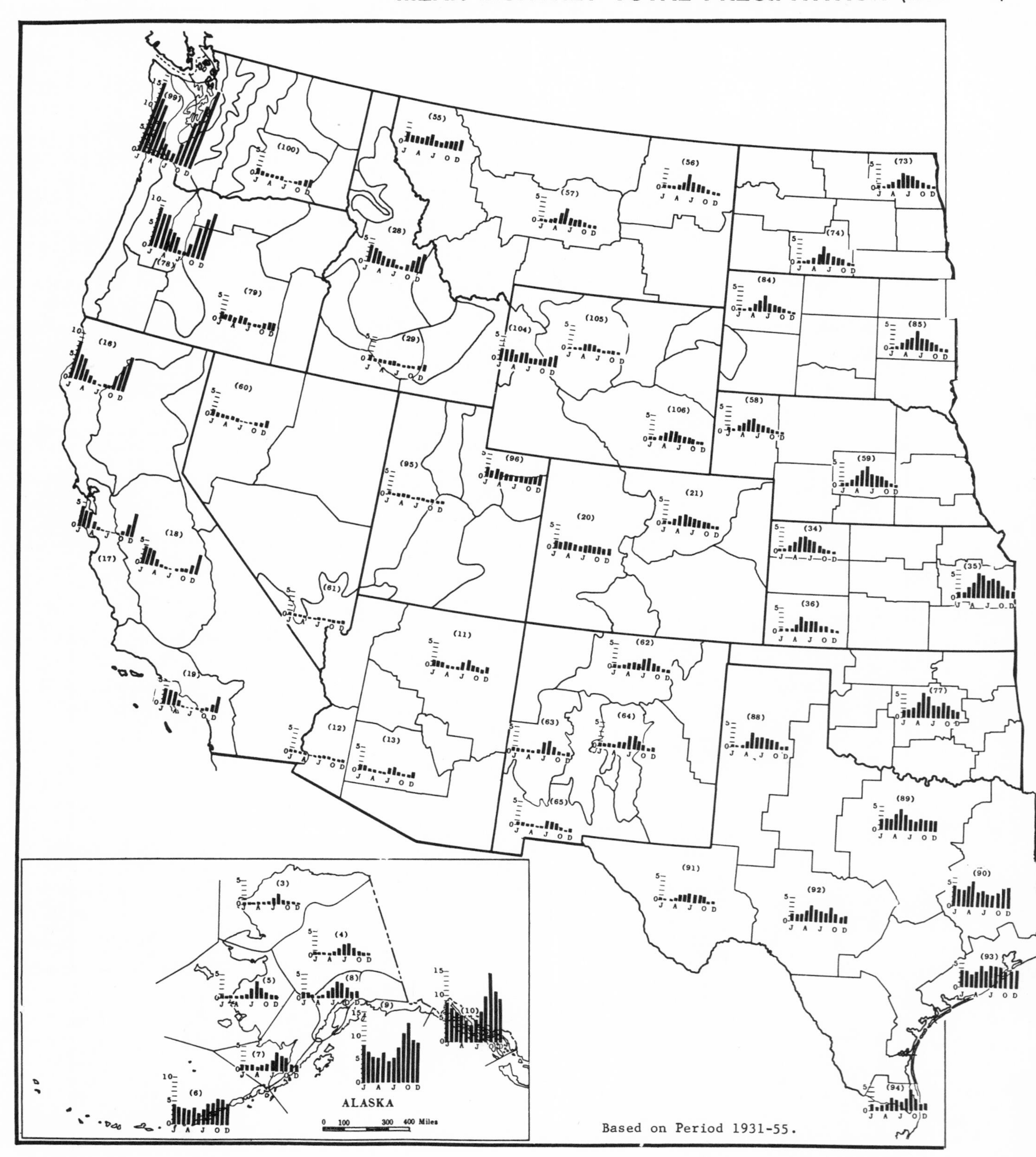

FOR SELECTED STATE CLIMATIC DIVISIONS

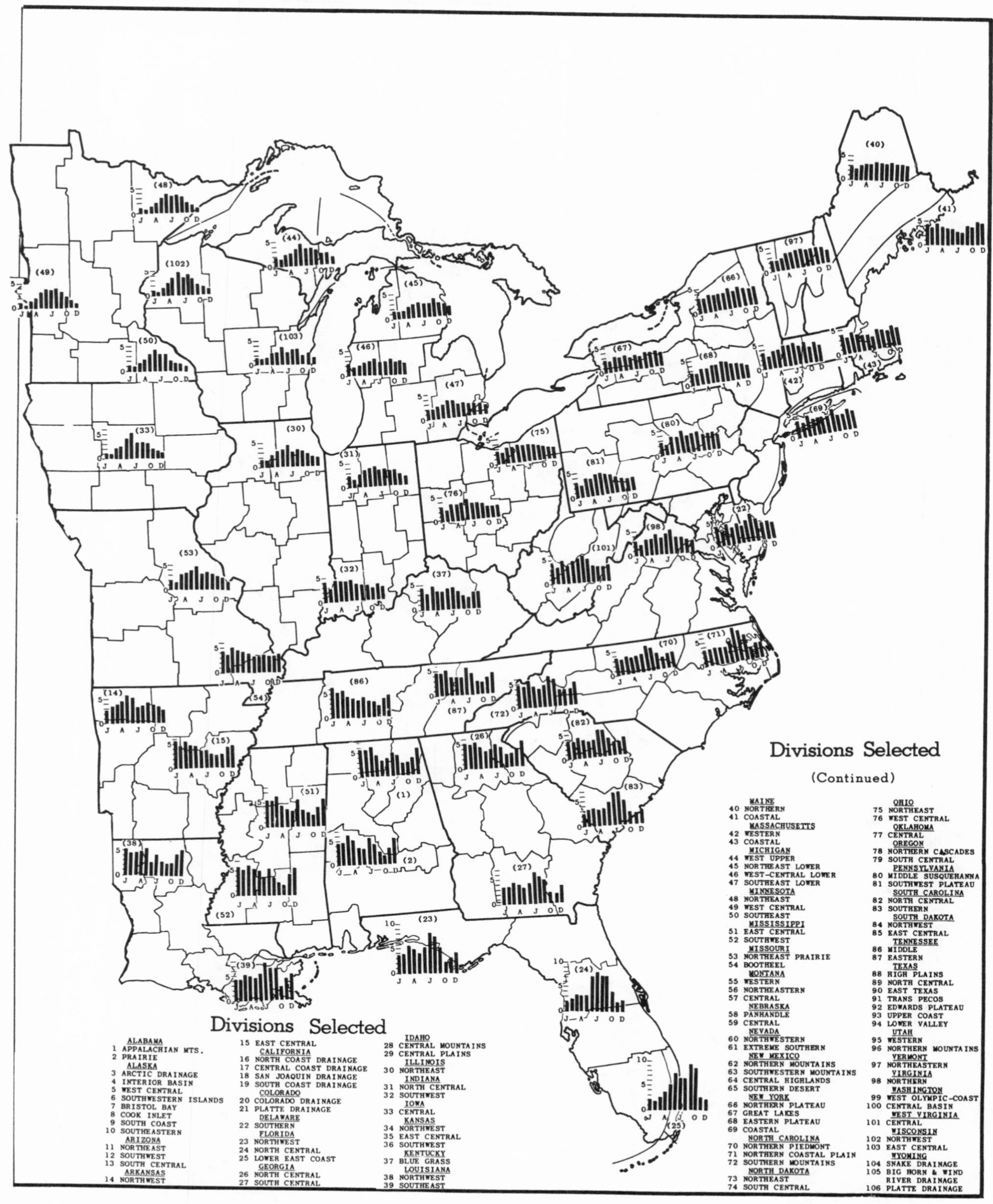

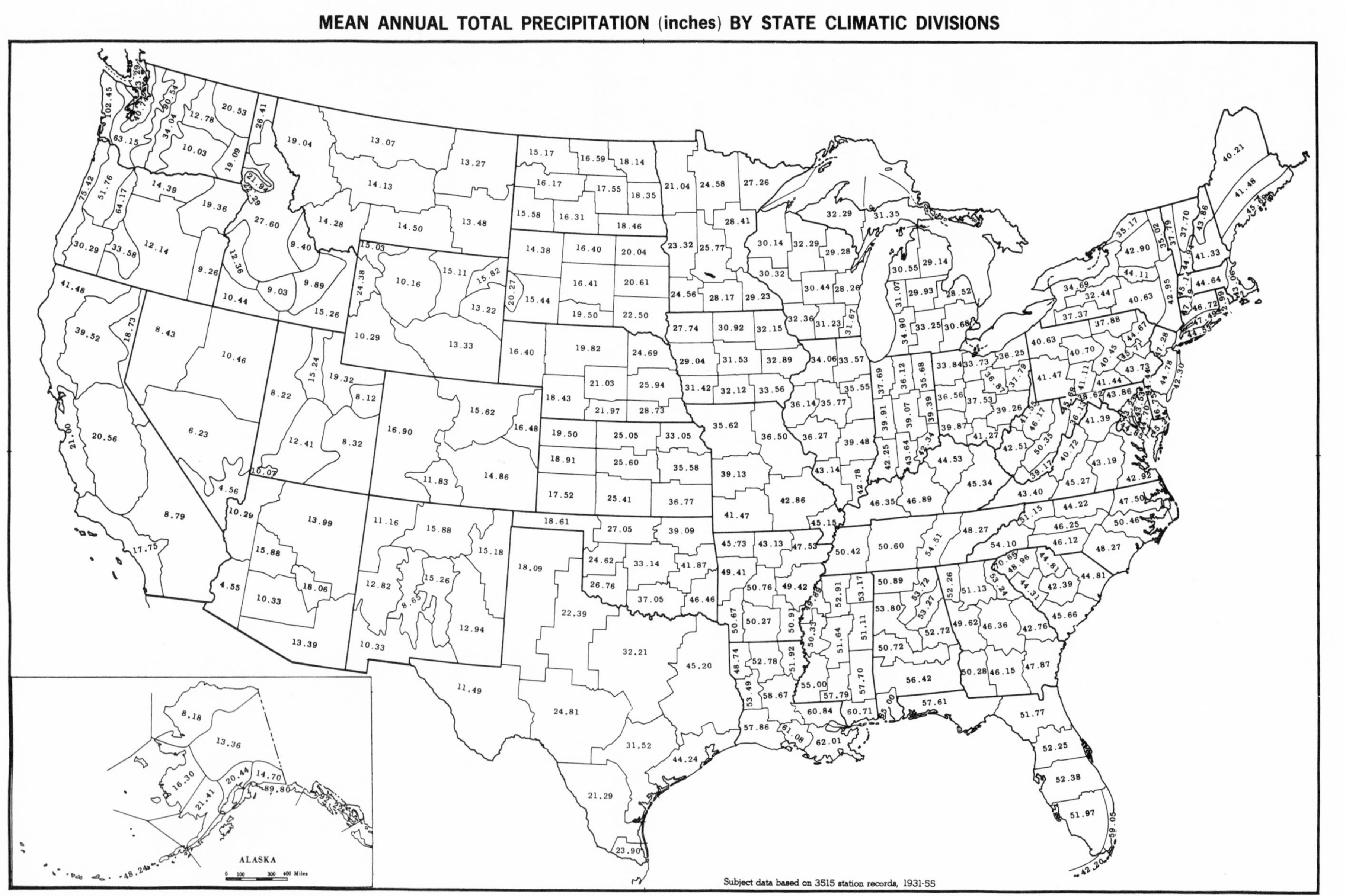
MEAN ANNUAL TOTAL PRECIPITATION (inches) BY STATE CLIMATIC DIVISIONS
ALASKA
0 100 300 400 Miles
Subject data based on 3515 station records, 1931-55

MEAN ANNUAL PRECIPITATION IN MILLIONS OF GALLONS OF WATER PER SQUARE MILE
By State Climatic Divisions

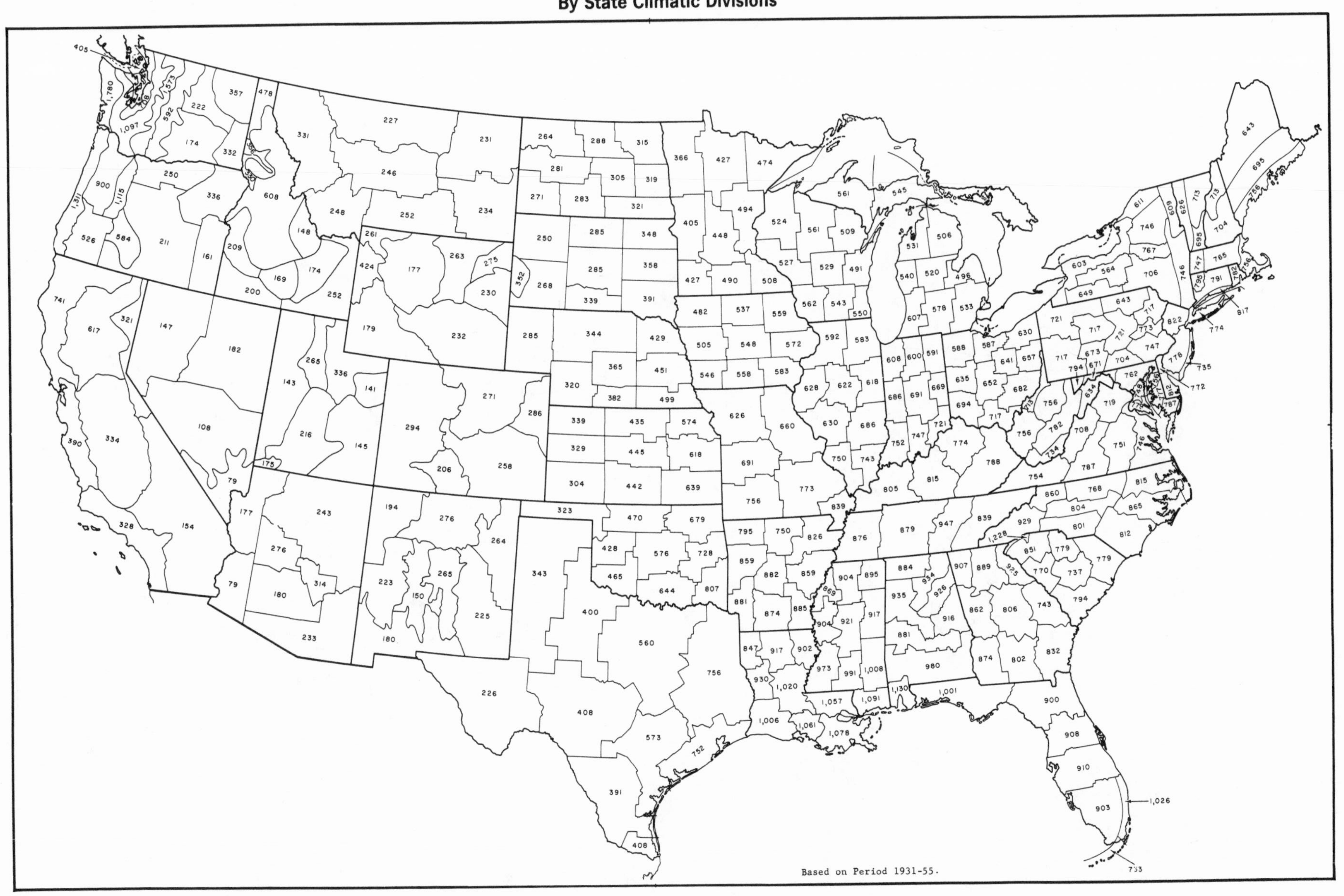

MEAN ANNUAL PRECIPITATION IN MILLIONS OF GALLONS OF WATER PER CAPITA
BY State Climatic Divisions
Based on Period 1931-55.

52

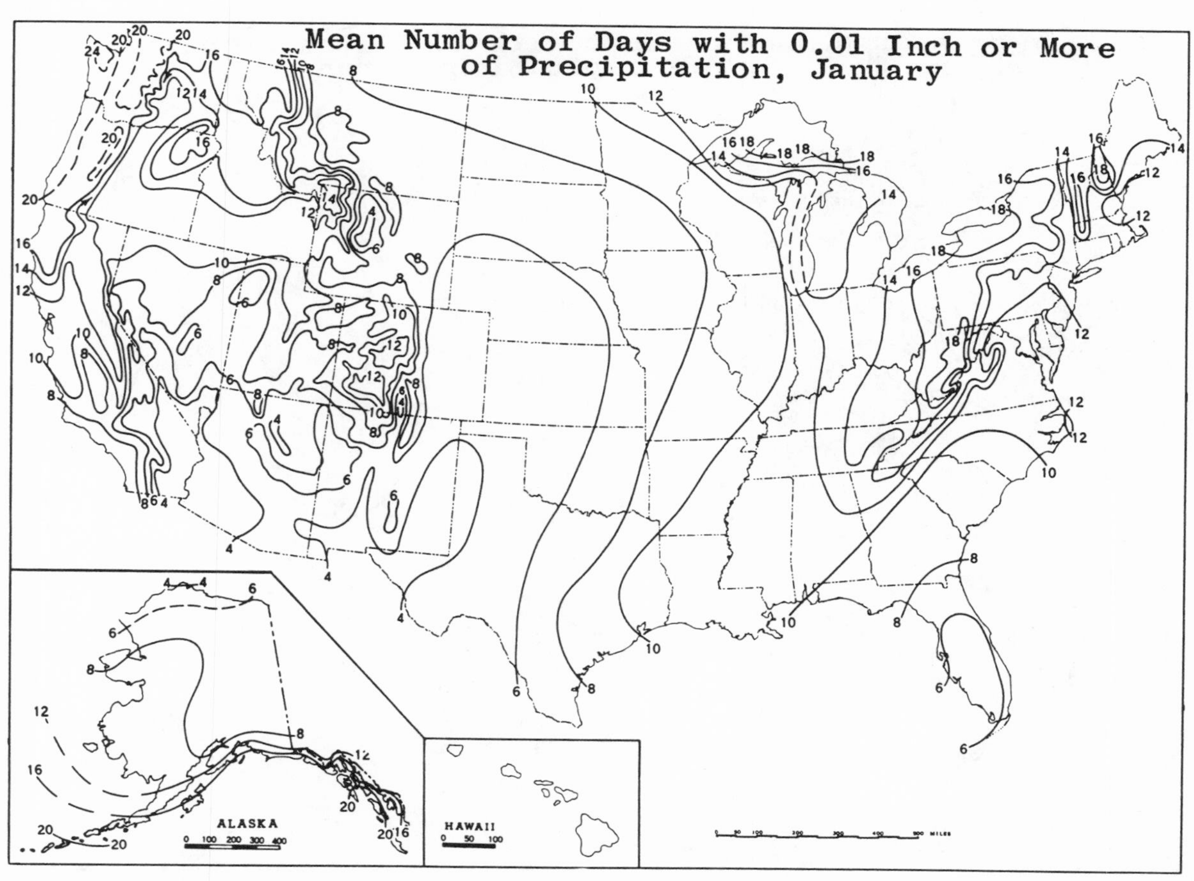
Mean Number of Days with 0.01 Inch or More of Precipitation, January
ALASKA
HAWAII
MILES

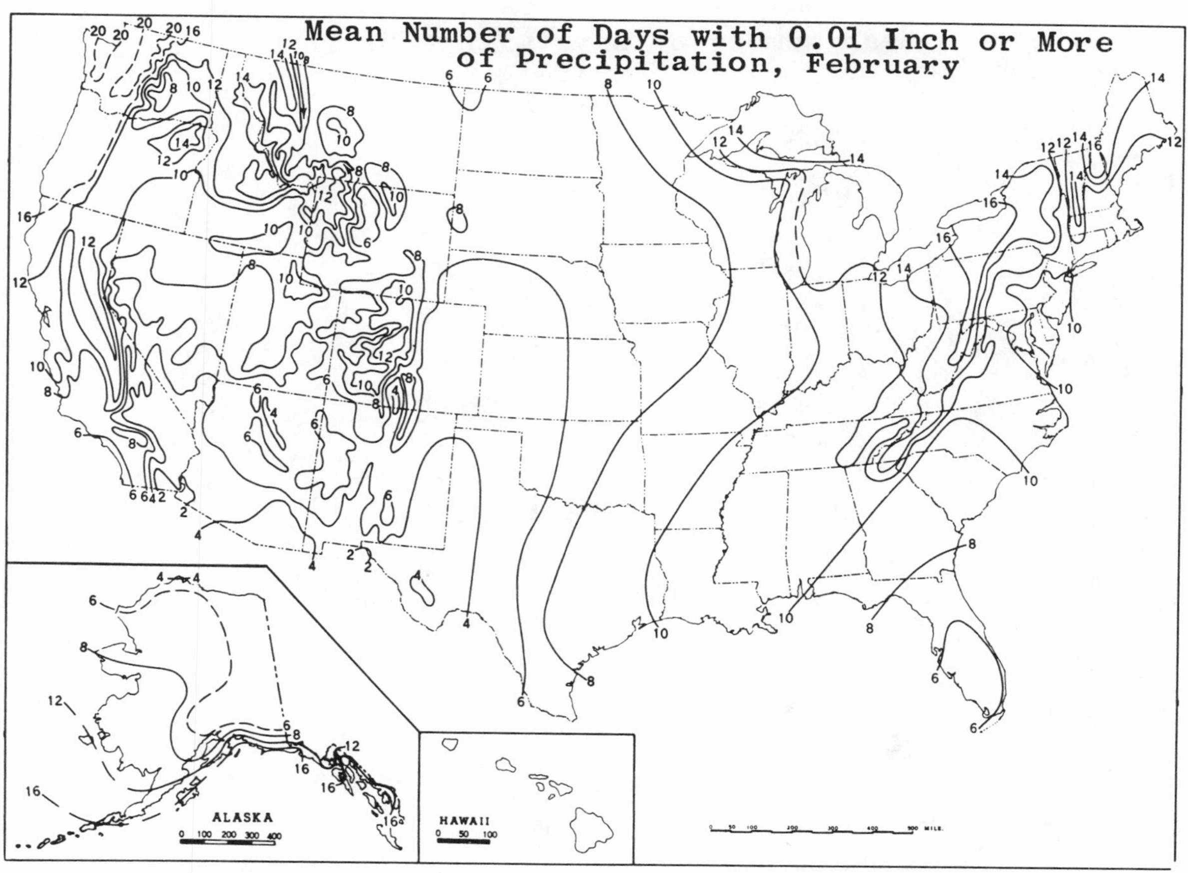
Mean Number of Days with 0.01 Inch or More of Precipitation, February
ALASKA
HAWAII

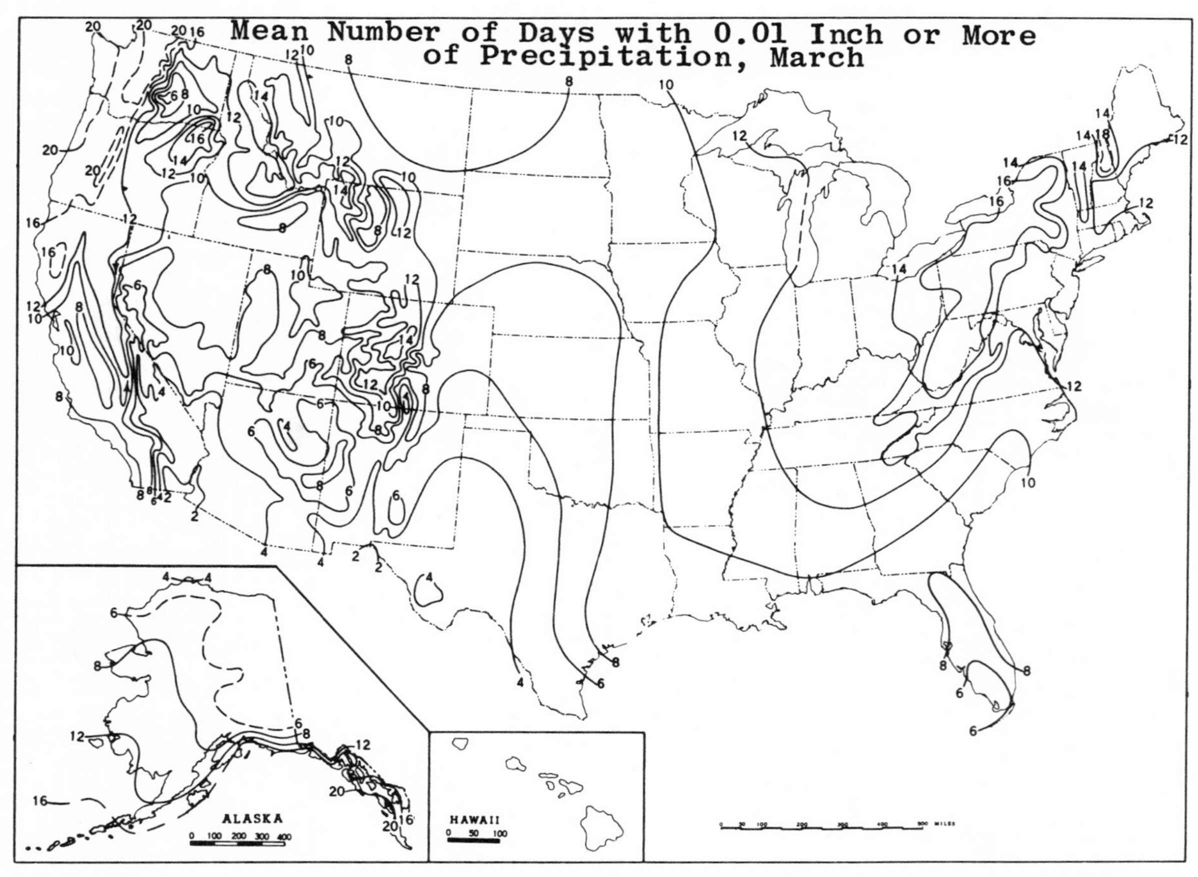
Mean Number of Days with 0.01 Inch or More of Precipitation, March
ALASKA
0 100 200 300 400
HAWAII
0 50 100

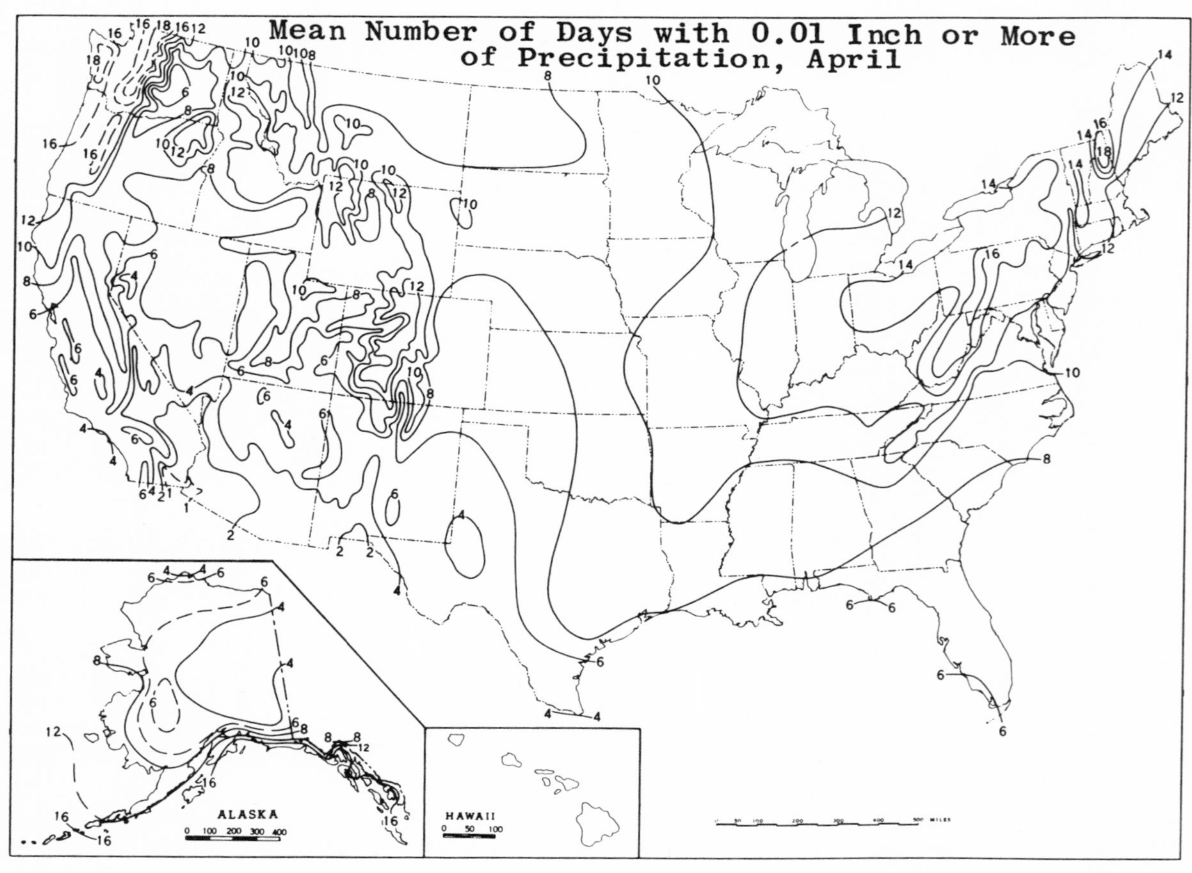
Mean Number of Days with 0.01 Inch or More of Precipitation, April
ALASKA
0 100 200 300 400
HAWAII
0 50 100

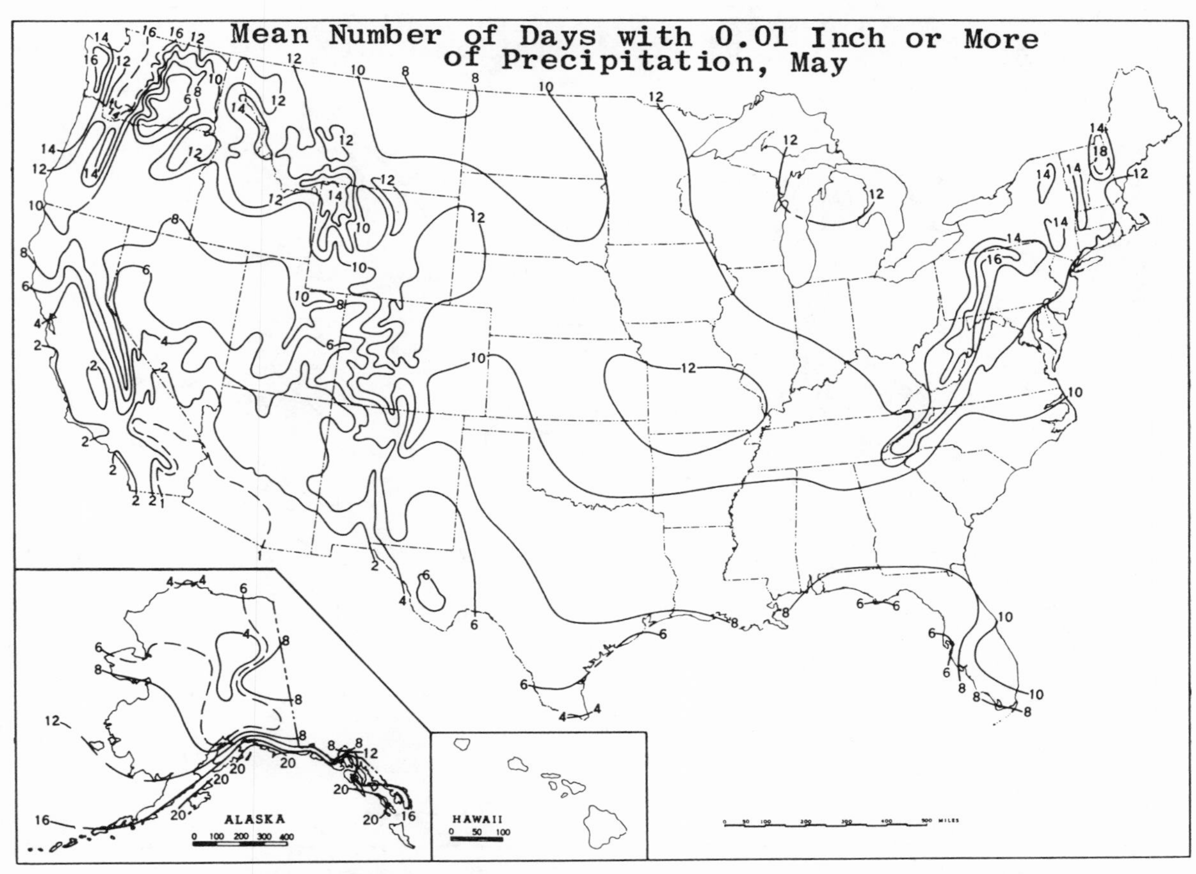
Mean Number of Days with 0.01 Inch or More of Precipitation, May
ALASKA
0 100 200 300 400
HAWAII
0 50 100

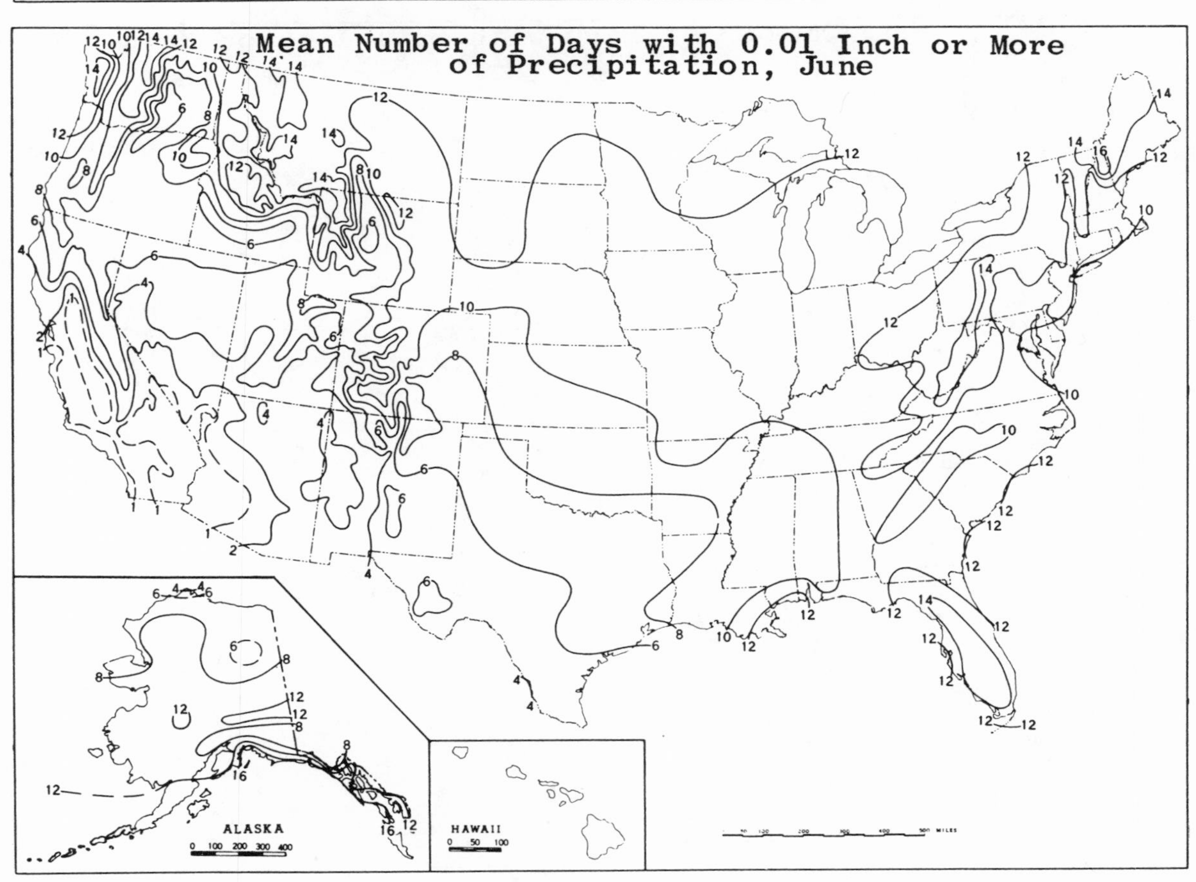
Mean Number of Days with 0.01 Inch or More of Precipitation, June
ALASKA
0 100 200 300 400
HAWAII
0 50 100

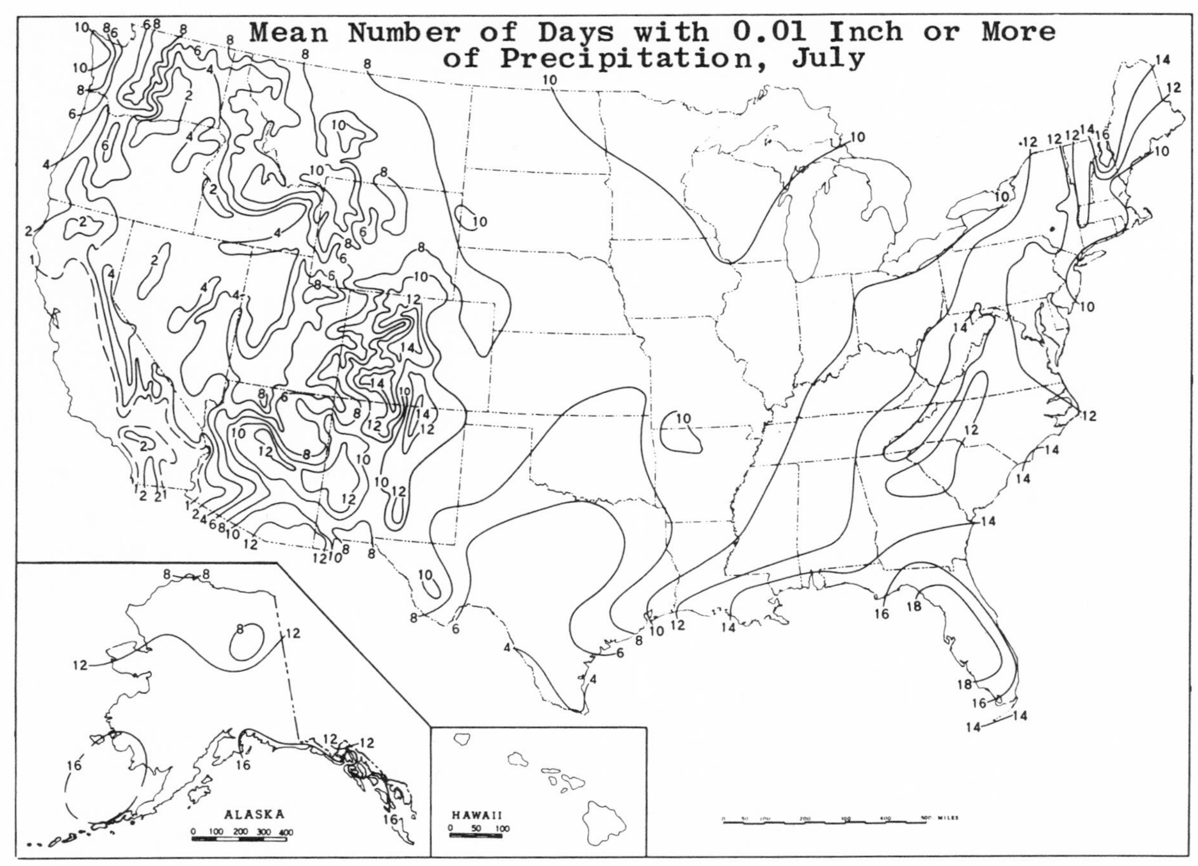
Mean Number of Days with 0.01 Inch or More of Precipitation, July
ALASKA
0 100 200 300 400
HAWAII
0 50 100

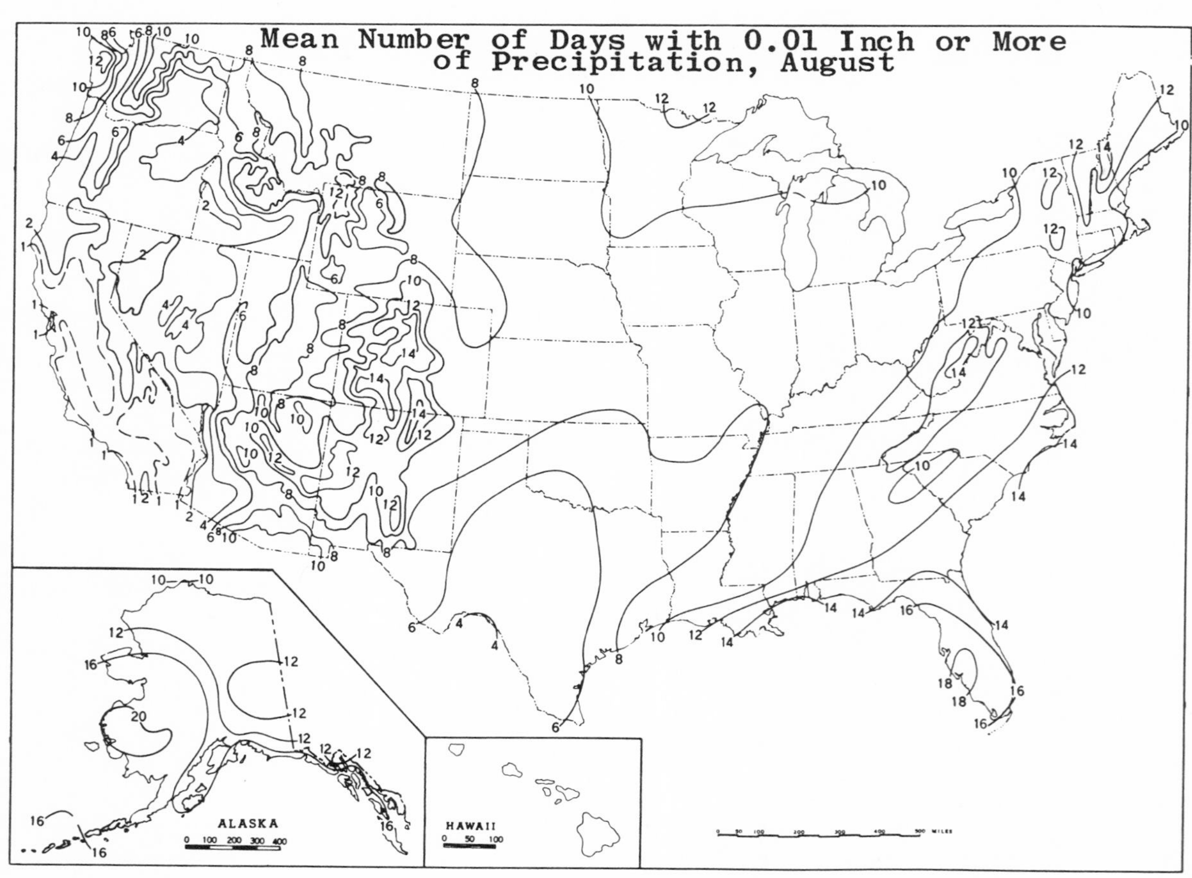
Mean Number of Days with 0.01 Inch or More of Precipitation, August
ALASKA
0 100 200 300 400
HAWAII
0 50 100

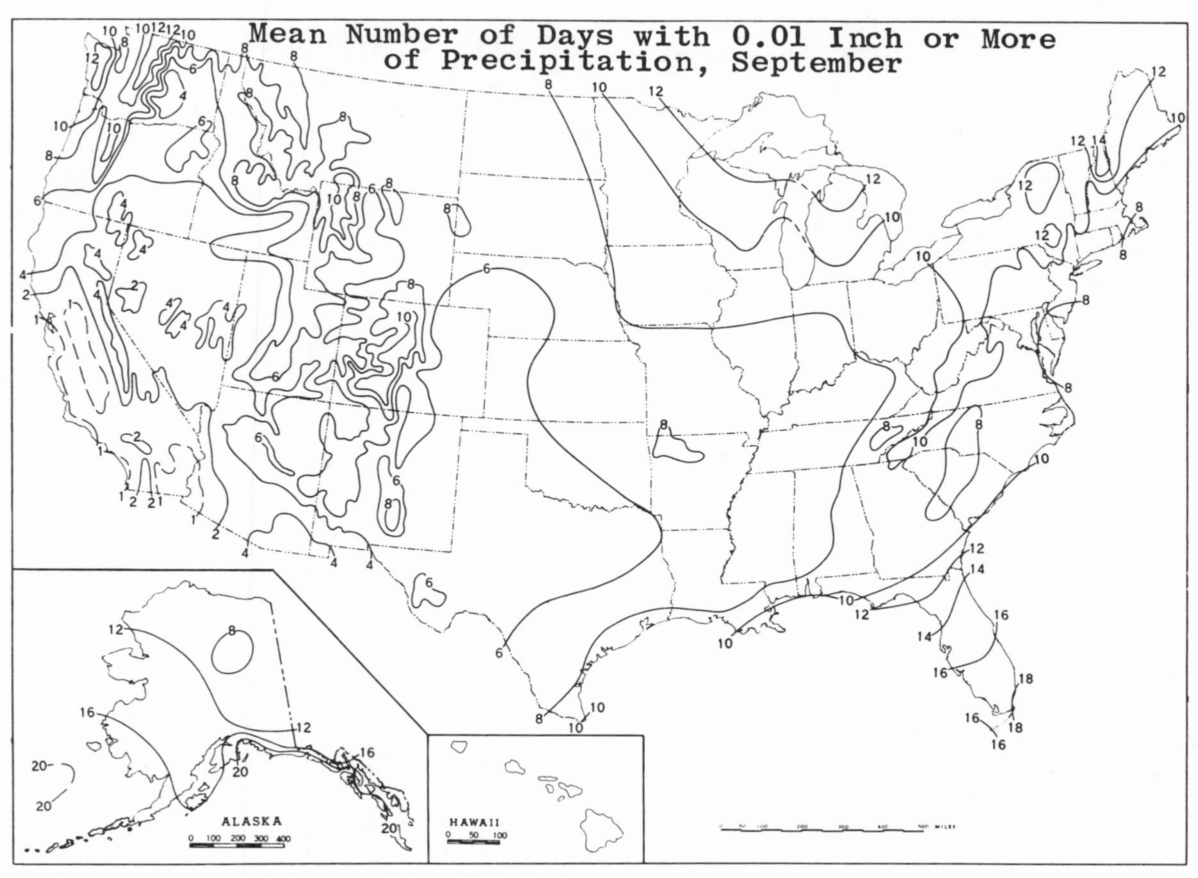
Mean Number of Days with 0.01 Inch or More of Precipitation, September
ALASKA
HAWAII

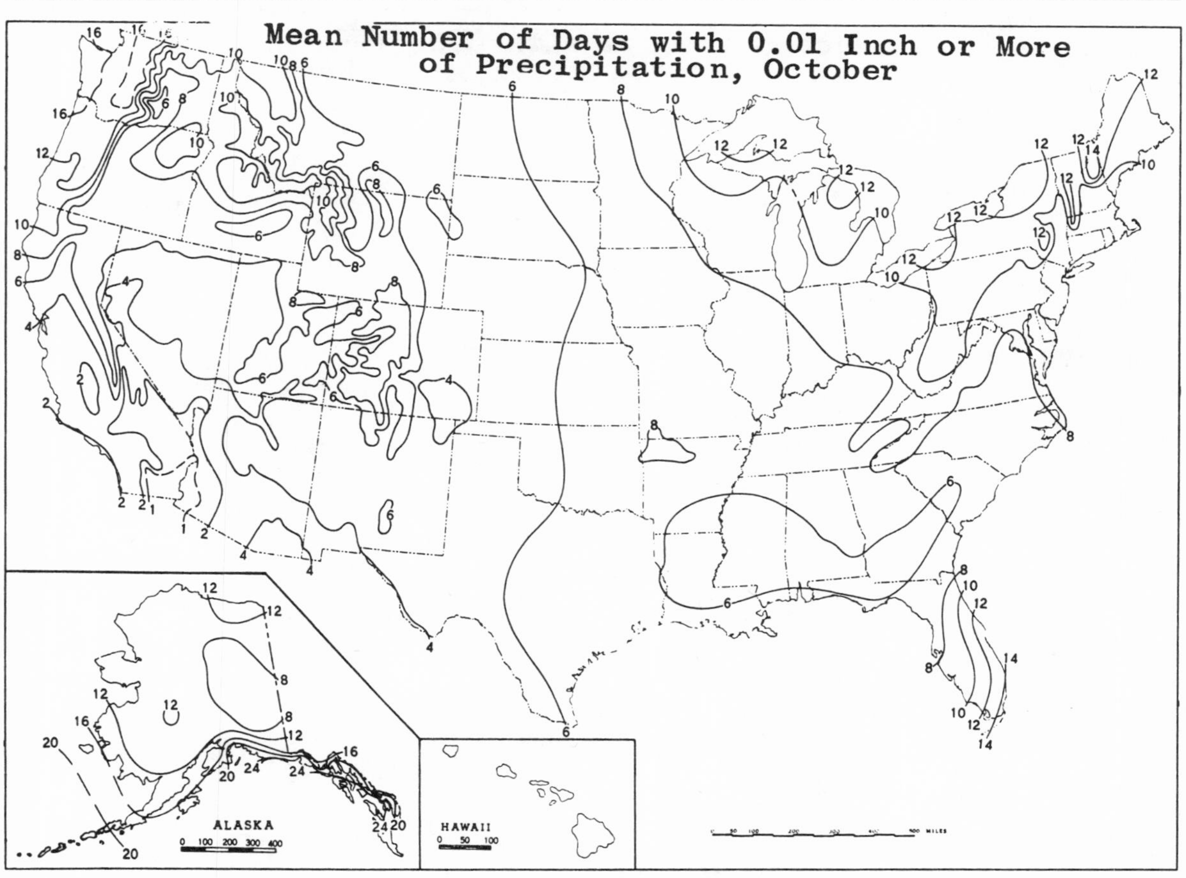
Mean Number of Days with 0.01 Inch or More of Precipitation, October
ALASKA
HAWAII

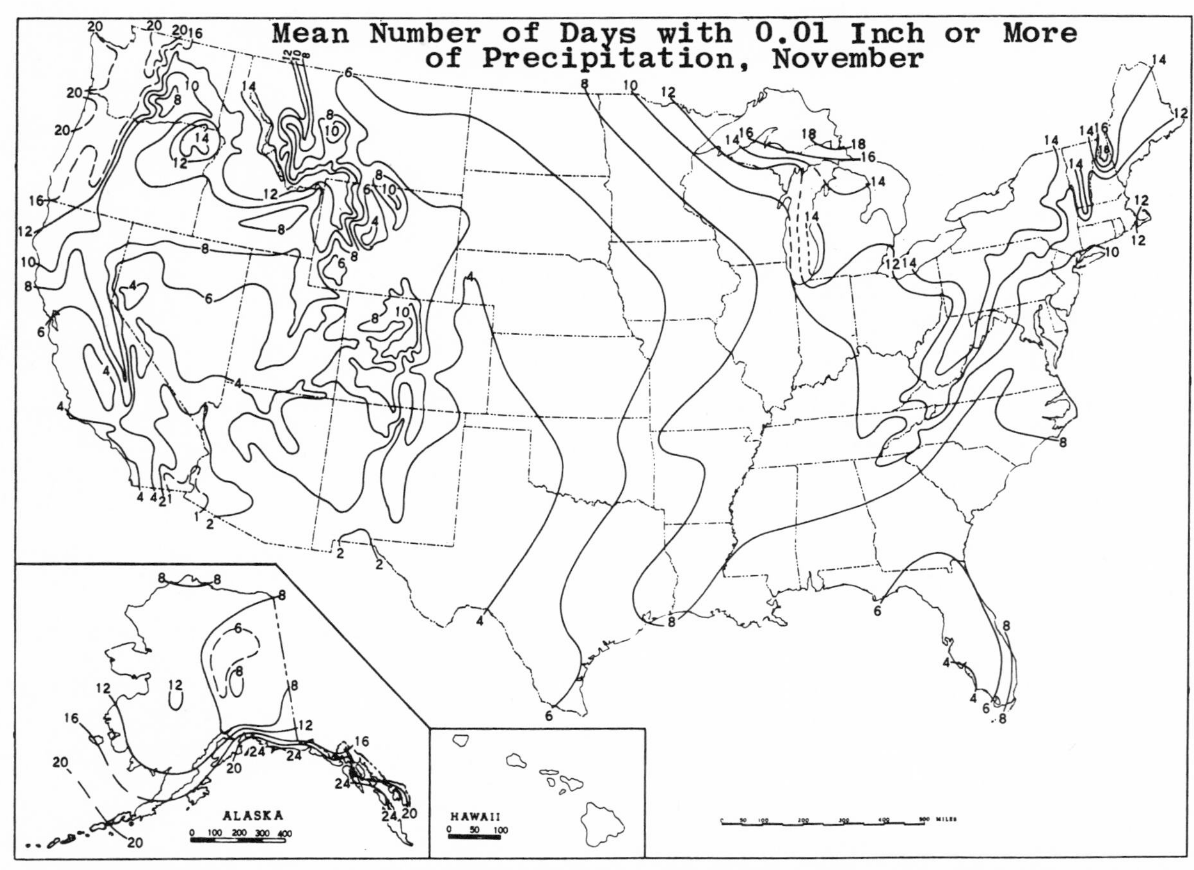
Mean Number of Days with 0.01 Inch or More
of Precipitation, November
ALASKA
0 100 200 300 400
HAWAII
0 50 100
MILES

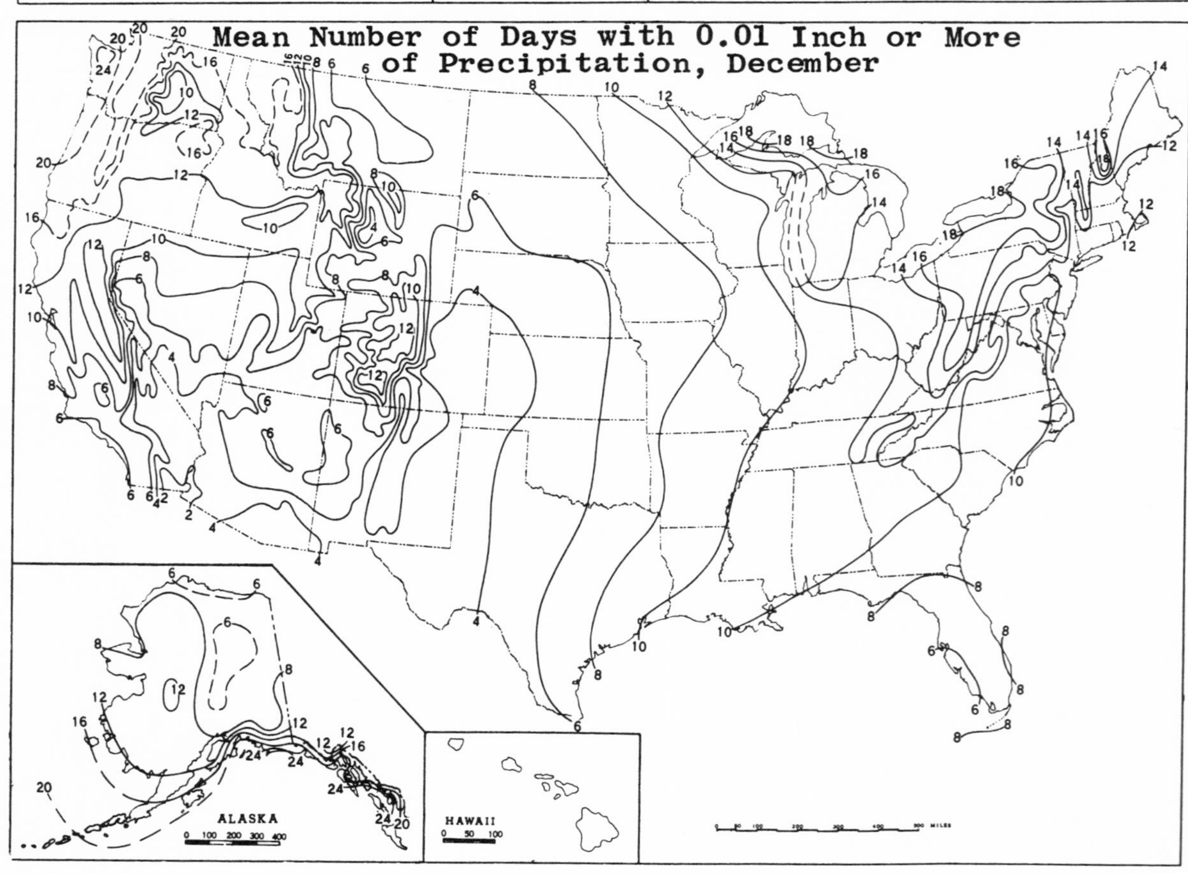
Mean Number of Days with 0.01 Inch or More
of Precipitation, December
ALASKA
0 100 200 300 400
HAWAII
0 50 100
MILES

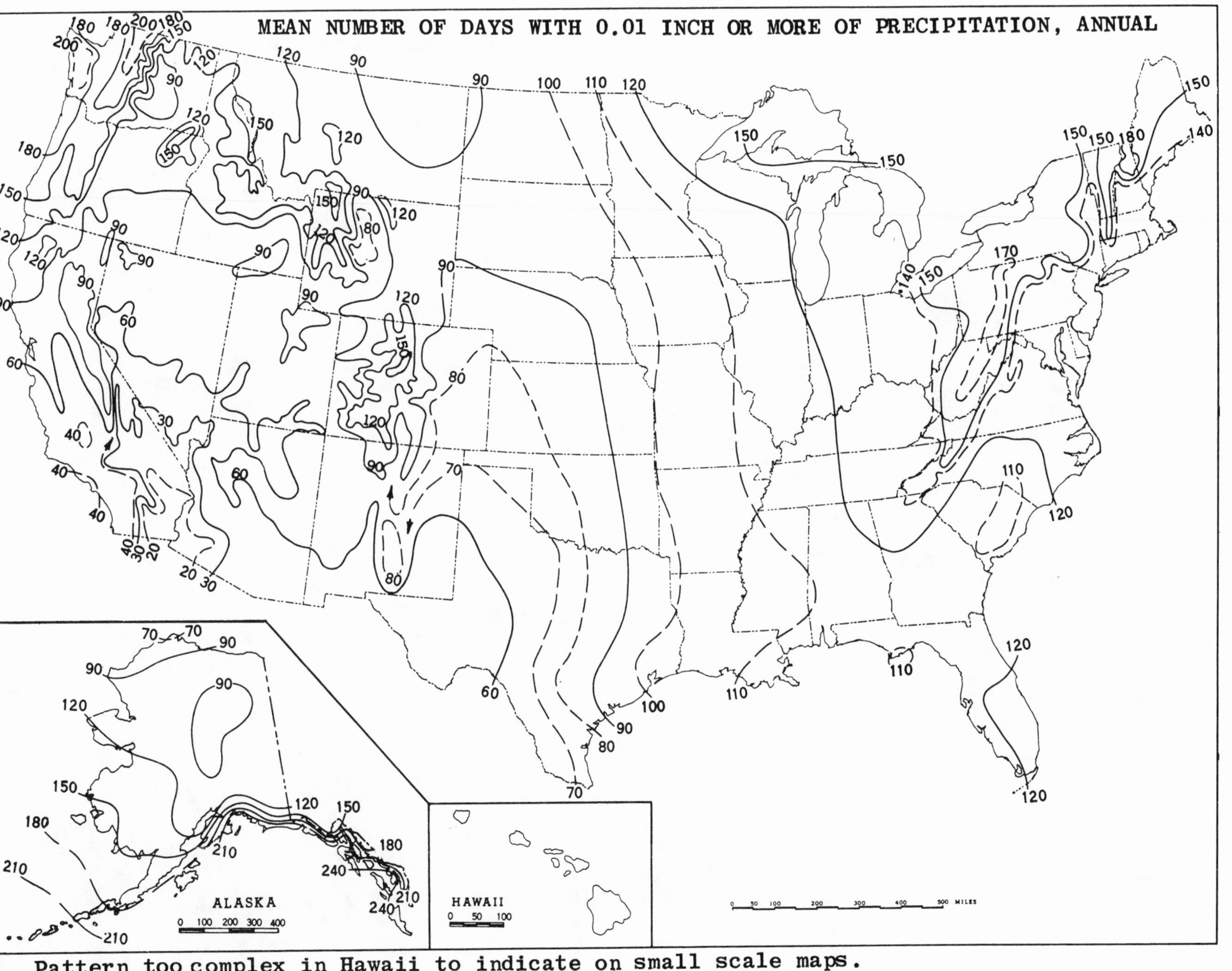

Pattern too complex in Hawaii to indicate on small scale maps.

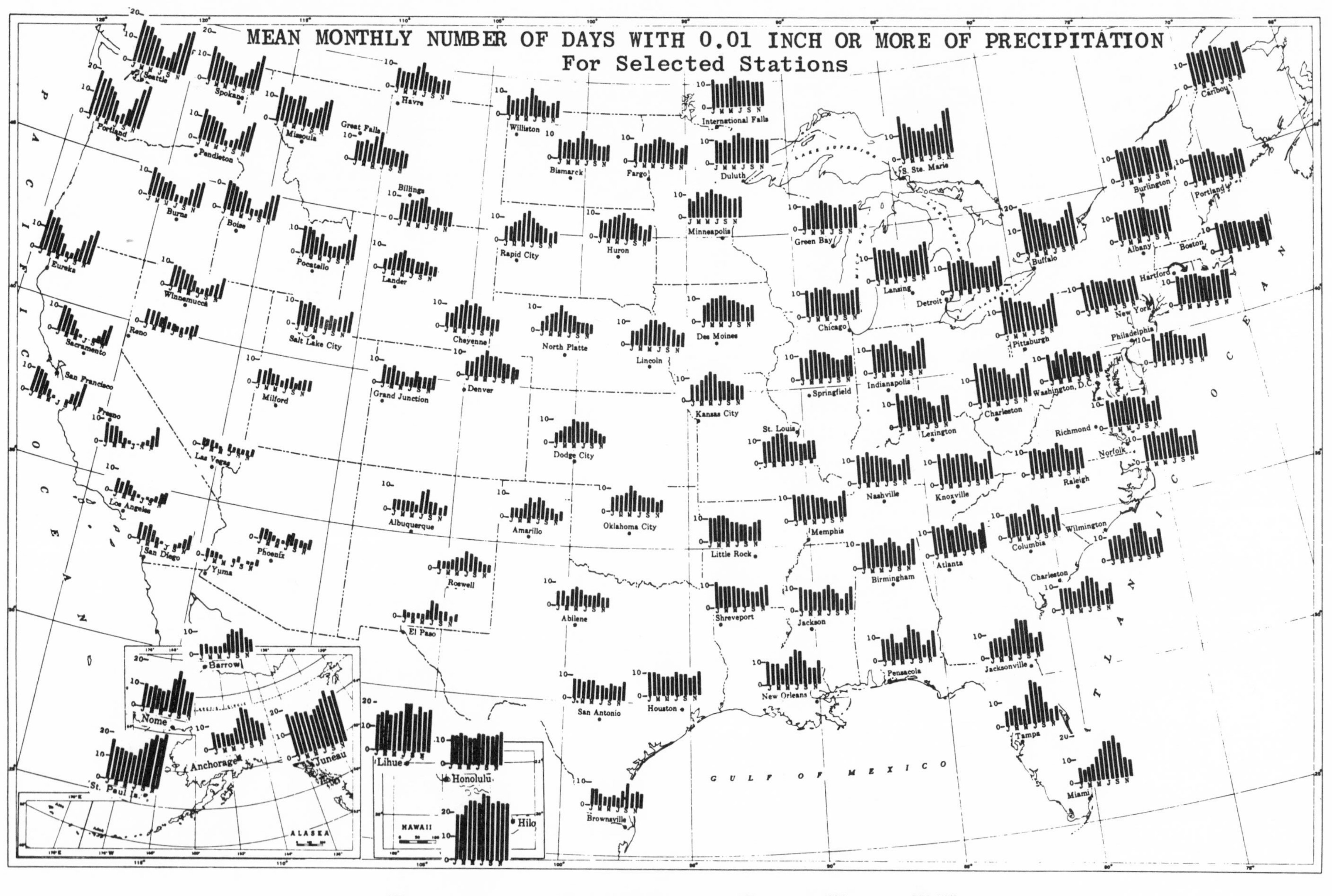
MEAN MONTHLY NUMBER OF DAYS WITH 0.01 INCH OR MORE OF PRECIPITATION
For Selected Stations
PACIFIC OCEAN
ATLANTIC OCEAN
GULF OF MEXICO
ALASKA
HAWAII
200 0 200 400 600 800 1000 Miles
SCALE 1:20,000,000
ALBERS EQUAL AREA PROJECTION — STANDARD PARALLELS 29½° AND 45½°

MEAN NUMBER OF DAYS WITH 0.01 INCH OR MORE OF PRECIPITATION

STATE AND STATION	YEARS	JAN.	FEB.	MAR.	APR.	MAY	JUNE	JULY	AUG.	SEPT.	OCT.	NOV.	DEC.	ANNUAL
ALA. BIRMINGHAM	18	11	11	12	9	9	10	12	10	8	6	9	11	118
MOBILE	20	10	11	11	8	8	12	18	13	10	6	8	11	125
MONTGOMERY U	83	11	10	10	9	9	11	12	11	8	6	7	10	114
ALASKA ANCHORAGE	35	7	6	6	5	5	7	11	15	14	10	8	7	104
BARROW	41	4	4	3	3	3	4	8	10	9	10	6	5	69
BETHEL	18	11	11	13	9	11	12	16	21	16	13	13	13	158
COLD BAY	15	17	15	17	12	16	15	17	19	19	22	20	19	207
FAIRBANKS	32	8	7	7	4	7	10	13	14	10	11	9	7	108
FT. YUKON	35	6	5	5	2	3	6	7	9	8	7	5	5	68
JUNEAU	18	18	17	18	17	17	15	17	18	20	23	20	22	222
NOME	39	10	9	10	9	8	9	14	18	15	11	10	10	133
ST. PAUL ISLAND	43	17	14	14	13	13	12	15	19	20	22	21	19	200
ARIZ. FLAGSTAFF	12	6	6	6	5	3	3	12	11	4	4	5	5	70
PHOENIX	22	4	4	3	2	1	1	4	5	3	3	2	4	34
TUCSON	20	4	4	4	2	1	1	11	9	3	4	2	4	49
WINSLOW	30	4	4	4	3	3	2	7	9	5	4	3	4	53
YUMA	11	3	2	2	1	*	*	1	2	*	1	1	2	14
ARK. FT. SMITH	16	8	9	9	10	11	8	8	7	7	6	6	7	96
LITTLE ROCK	19	10	10	11	11	10	8	8	7	7	6	8	9	105
CALIF. BAKERSFIELD	24	6	7	7	4	2	*	*	*	1	2	3	6	37
EUREKA	51	17	14	15	12	9	5	2	2	5	9	12	15	118
FRESNO	21	8	7	7	5	2	1	*	*	1	2	4	8	46
LOS ANGELES U	21	6	5	6	4	2	1	*	1	1	2	4	5	36
RED BLUFF	17	12	9	9	6	6	3	1	1	2	5	7	10	71
SACRAMENTO	28	11	10	8	6	3	1	*	*	1	3	6	9	58
SAN DIEGO	21	7	6	7	5	2	1	*	*	1	2	4	6	42
SAN FRANCISCO	34	11	10	9	6	3	1	*	*	1	4	6	10	62
COLO. ALAMOSA	16	4	4	4	5	7	4	9	10	5	5	4	3	64
DENVER	27	6	6	8	9	11	9	9	8	6	6	5	4	87
GRAND JUNCTION	15	8	7	8	6	6	4	4	7	5	5	5	6	71
PUEBLO	21	5	5	6	7	9	7	9	9	4	4	3	3	72
CONN. HARTFORD U	51	12	11	12	12	12	11	10	10	9	8	10	10	127
NEW HAVEN	18	12	11	13	13	13	11	11	10	9	8	11	11	132
DEL. WILMINGTON	14	12	10	13	12	13	10	9	9	8	9	9	10	123
D. C. WASHINGTON	20	11	8	12	10	12	9	10	10	8	7	8	9	115
FLA. APALACHICOLA	31	8	8	8	6	6	10	16	14	12	6	6	8	108
JACKSONVILLE	20	7	8	8	7	8	11	15	14	14	8	6	8	115
KEY WEST	13	6	5	6	5	7	11	14	15	15	13	7	8	113
MIAMI	19	6	5	6	7	10	13	16	16	18	15	8	7	127
ORLANDO	19	5	6	7	7	8	14	18	16	15	9	5	6	117
TAMPA	15	6	7	8	7	6	12	18	17	15	8	5	6	113
GA. ATLANTA U	77	12	11	12	10	9	11	12	11	8	7	8	11	122
AUGUSTA U	85	10	10	10	8	9	11	12	12	7	6	7	10	112
MACON	13	9	10	11	8	9	10	13	11	9	6	6	9	111
SAVANNAH U	85	9	9	9	7	9	12	14	13	10	7	6	9	114
THOMASVILLE	39	9	9	9	7	9	13	16	14	10	5	6	9	116
HAWAII HILO	19	19	19	23	25	25	24	28	27	24	25	24	24	287
HONOLULU	39	12	12	13	12	11	11	13	13	12	12	13	14	148
LIHUE	10	15	16	17	15	16	17	20	20	16	19	17	18	204
IDAHO BOISE	22	12	11	10	8	9	6	2	2	3	7	10	11	90
POCATELLO	23	11	11	10	7	9	6	4	4	5	6	8	11	92
ILL. CAIRO U	19	11	10	12	12	12	10	9	8	8	7	9	9	117
CHICAGO U	85	11	10	12	11	12	11	9	9	9	9	10	11	124
SPRINGFIELD	14	9	9	12	12	11	11	10	8	7	7	9	9	114
IND. EVANSVILLE	21	11	9	12	12	11	10	10	8	7	7	9	10	116
FT. WAYNE	15	13	11	13	14	12	11	10	9	8	8	10	12	131
INDIANAPOLIS	22	12	10	12	12	13	11	9	8	7	8	10	11	123
IOWA BURLINGTON U	64	7	7	10	10	12	11	8	8	9	7	7	7	103
DES MOINES	22	7	8	10	10	11	11	9	9	8	7	6	7	103
DUBUQUE	11	9	8	11	11	11	10	10	9	8	8	7	9	111
SIOUX CITY	20	6	7	9	9	11	12	9	9	7	6	5	7	97
KANS. CONCORDIA U	76	5	6	7	9	11	11	8	7	7	6	4	5	86
DODGE CITY	19	4	5	7	7	10	9	9	9	6	5	4	3	78
GOODLAND	41	4	5	6	8	10	10	8	8	5	5	4	4	77
WICHITA U	67	5	6	7	8	11	9	8	8	8	7	5	5	87
KY. LEXINGTON	17	14	11	13	13	12	11	11	9	7	7	11	11	130
LOUISVILLE	14	13	11	13	13	11	11	10	8	7	7	9	11	123
LA. LAKE CHARLES	23	10	10	8	8	8	9	12	10	8	5	8	10	106
NEW ORLEANS	46	10	9	9	7	9	12	15	14	10	7	7	10	119
SHREVEPORT U	84	11	9	9	9	9	8	8	7	6	6	8	9	99
MAINE CARIBOU	22	14	14	13	14	13	15	14	13	12	12	14	14	162
PORTLAND	21	12	11	11	12	13	11	9	9	8	9	11	10	126
MD. BALTIMORE U	91	11	10	12	11	11	11	11	11	8	8	9	10	123
MASS. BOSTON U	85	12	11	12	11	11	10	10	10	9	9	10	11	126
NANTUCKET	15	13	11	13	12	11	8	8	9	7	9	12	12	125
MICH. ALPENA U	45	15	12	12	11	11	11	9	10	12	12	14	15	144
DETROIT	28	13	12	13	13	12	11	9	9	9	9	11	13	134
ESCANABA U	52	11	10	10	10	11	11	11	11	12	10	11	11	129
GRAND RAPIDS U	52	14	12	12	12	12	10	8	9	10	10	12	14	135
MARQUETTE	24	17	15	12	11	12	12	10	11	13	11	16	15	155
S. STE. MARIE	20	18	15	13	11	11	12	10	10	13	11	18	19	161
MINN. DULUTH U	85	10	9	10	9	12	13	11	11	11	10	10	10	126
INTERNATIONAL FALLS	22	12	10	10	10	12	13	11	12	12	9	12	12	135
MINNEAPOLIS	23	8	7	11	9	11	12	10	10	9	8	8	9	112
MISS. MERIDIAN	16	10	10	10	8	9	8	11	9	7	5	7	9	105
VICKSBURG	24	10	10	11	9	8	9	10	8	7	5	8	10	105
MO. KANSAS CITY	28	7	7	9	11	12	11	8	8	8	7	6	6	100
ST. LOUIS U	23	9	9	11	12	12	11	8	8	7	7	8	8	110
SPRINGFIELD	16	8	9	10	10	12	10	10	8	7	7	8	8	107
MONT. BILLINGS	27	7	7	9	9	10	11	7	6	7	6	6	6	92
HAVRE U	57	8	7	7	7	10	12	8	7	7	5	6	6	90
HELENA	21	7	7	8	8	11	12	8	8	6	6	7	8	97
KALISPELL	12	16	12	12	9	11	12	6	8	8	11	11	15	132
MILES CITY	24	7	6	7	7	10	11	8	7	7	5	5	6	88
NEBR. NORTH PLATTE	10	4	6	7	8	11	9	8	8	5	5	5	4	80
OMAHA	26	6	7	8	9	11	11	9	10	8	6	5	6	96
SCOTTS BLUFF	18	4	5	8	8	13	11	8	7	6	5	4	4	83
VALENTINE U	67	6	6	8	9	11	12	9	9	7	5	5	6	93
NEV. LAS VEGAS	19	3	3	3	2	1	*	3	2	2	2	2	3	26
RENO	19	6	6	6	4	5	3	3	2	2	3	4	5	47
WINNEMUCCA U	77	9	9	8	7	7	5	2	2	3	5	6	9	72
N. H. CONCORD U	53	11	9	10	11	10	10	10	10	10	9	10	10	120
MT. WASHINGTON R	29	18	17	19	18	18	16	17	15	15	14	19	19	205
N. J. ATLANTIC CITY U	85	12	10	12	11	11	10	10	10	8	9	9	10	122
N. MEX. ALBUQUERQUE	22	4	4	4	4	4	3	9	10	5	5	3	4	58
ROSWELL U	51	3	3	3	4	5	5	8	7	6	4	3	3	54
N. Y. ALBANY U	82	12	11	12	12	12	12	12	11	10	10	11	11	136
BINGHAMTON U	65	15	13	14	14	13	12	12	11	10	11	13	14	152
BUFFALO U	85	19	17	16	13	13	11	10	9	11	12	15	18	164
NEW YORK U	90	12	10	12	11	11	10	11	10	9	9	9	10	124
SYRACUSE	12	19	16	18	17	13	11	12	10	10	11	16	19	171
N. C. ASHEVILLE	31	11	11	12	10	11	12	15	12	8	8	8	10	129
CAPE HATTERAS R	81	12	10	11	9	9	10	12	12	9	8	9	11	122
CHARLOTTE	22	9	10	12	9	9	9	12	10	7	7	7	10	111
RALEIGH	17	10	11	10	9	10	10	12	11	9	8	9	9	117
WILMINGTON U	85	9	10	10	8	9	12	14	14	10	7	7	10	120
N. DAK. BISMARCK	22	8	7	8	7	9	12	9	9	7	6	6	7	95
FARGO	20	8	7	8	8	10	11	10	9	9	6	7	8	101
WILLISTON U	45	8	6	7	7	8	12	9	7	7	5	6	7	89
OHIO AKRON-CANTON	13	17	14	16	16	12	11	11	9	8	9	13	14	152
CINCINNATI (ABBE)	46	13	11	13	13	13	13	10	9	9	9	10	11	134
CLEVELAND	20	16	14	16	15	14	11	10	9	9	10	15	15	154
COLUMBUS	22	14	12	14	14	13	12	11	9	8	8	11	12	137
OKLA. OKLAHOMA CITY	21	6	7	7	8	11	9	7	6	6	6	4	5	82
TULSA	23	6	8	8	9	11	9	7	7	6	7	6	7	91
OREG. ASTORIA	8	24	21	22	18	16	15	7	9	11	18	18	22	201
BURNS	19	12	10	9	7	8	7	3	3	3	7	10	11	90
PORTLAND U	59	19	16	17	14	12	9	3	4	7	12	17	19	152
ROSEBURG U	78	18	16	16	13	11	7	2	2	6	11	15	17	134
PA. HARRISBURG	23	12	9	12	12	13	11	10	10	8	8	9	10	125
PHILADELPHIA	21	11	9	12	11	12	10	10	10	8	8	9	10	119
PITTSBURGH U	84	16	14	15	13	13	12	12	10	9	10	12	14	150
WILLIAMSPORT	17	13	12	14	14	14	11	12	11	9	10	13	13	144
R. I. BLOCK ISLAND U	77	12	11	12	11	11	9	9	9	8	9	10	11	122
PROVIDENCE U	51	12	10	12	11	11	10	10	9	8	8	10	10	121
S. C. CHARLESTON U	85	9	9	9	7	8	11	13	12	10	6	7	9	110
COLUMBIA U	68	9	10	10	8	9	11	13	12	8	6	7	10	113
GREENVILLE U & AP	38	11	13	11	10	10	10	13	11	8	7	8	10	122
S. DAK. HURON U	74	7	6	8	9	10	11	9	9	7	6	5	7	94
RAPID CITY U	56	6	6	8	9	12	13	10	8	7	6	4	5	94
TENN. CHATTANOOGA U	77	13	12	13	11	11	12	13	11	8	7	9	12	132
KNOXVILLE U	84	13	11	13	11	12	12	12	11	8	7	9	12	131
MEMPHIS U	84	11	10	11	10	10	9	9	8	7	7	9	11	112
NASHVILLE	20	12	11	13	11	11	10	10	9	7	7	9	11	120
TEX. ABILENE	21	6	6	4	7	8	7	5	5	5	6	4	4	67
AMARILLO	20	4	4	4	6	9	7	9	8	5	5	3	4	68
BROWNSVILLE	19	7	7	5	4	4	5	4	7	10	6	7	6	72
CORPUS CHRISTI	22	8	8	6	5	6	5	4	5	8	7	6	7	75
DALLAS	21	8	8	8	9	8	6	5	6	5	6	6	6	81
EL PASO	22	3	2	2	2	2	4	8	7	4	4	2	3	43
HOUSTON	24	10	10	9	8	8	8	10	9	9	7	8	10	106
LAREDO	18	6	6	4	5	6	4	3	5	6	5	5	6	61
LUBBOCK	15	4	4	4	4	8	7	8	5	4	5	3	3	59
MIDLAND	26	4	3	2	3	6	4	5	4	4	6	3	4	48
SAN ANTONIO U	71	8	8	7	8	8	6	6	5	7	6	6	8	83
UTAH MILFORD	30	6	6	7	5	5	3	4	5	3	4	5	5	58
SALT LAKE CITY	33	10	9	10	9	8	5	4	6	5	6	7	9	86
VT. BURLINGTON U	49	13	12	13	13	13	12	12	11	11	11	13	13	147
VA. NORFOLK U	85	11	11	12	10	11	11	12	12	8	8	8	10	124
RICHMOND U	58	10	11	12	10	12	11	11	11	10	7	8	10	122
ROANOKE	14	11	11	12	10	12	11	11	12	9	8	9	9	124
WASH. SEATTLE U	64	18	16	16	13	12	9	5	5	8	13	18	19	152
SPOKANE U	74	14	12	11	9	9	8	4	4	6	8	12	15	112
STAMPEDE PASS	18	22	21	22	19	16	15	8	11	12	17	19	23	205
TATOOSH ISLAND R	59	22	18	20	17	14	12	10	10	11	17	21	23	197
WALLA WALLA U	47	13	11	12	9	9	7	3	3	5	9	11	14	106
YAKIMA U	46	9	7	5	4	5	5	2	2	3	5	8	9	64
W. VA. CHARLESTON	14	17	14	16	14	14	12	13	10	8	10	12	14	152
HUNTINGTON U	15	15	12	13	13	13	11	11	9	8	8	10	12	134
PARKERSBURG U	53	15	13	14	13	12	13	12	10	9	9	11	13	143
WIS. GREEN BAY U	69	10	9	10	11	12	11	10	9	10	9	9	10	120
LA CROSSE U	83	10	8	10	10	12	12	10	9	10	9	8	9	117
MADISON U	77	9	8	10	11	12	11	9	9	10	8	8	10	115
MILWAUKEE U	85	11	10	11	11	12	11	9	9	9	9	9	11	122
WYO. CASPER	22	7	7	10	10	11	8	7	5	6	6	7	6	92
CHEYENNE U	85	6	6	9	10	12	10	11	10	6	6	5	5	96
LANDER U	64	4	5	7	8	9	6	6	5	5	6	4	4	69
SHERIDAN	21	7	9	12	11	12	11	7	7	8	7	8	8	107
YELLOWSTONE	44	13	11	12	10	13	12	10	9	8	9	10	12	129
P. R. SAN JUAN	62	20	11	14	14	16	17	19	20	18	18	19	20	209

* LESS THAN ONCE IN 2 YEARS.

DATA FROM AIRPORT, EXCEPT THOSE MARKED WITH U FOR URBAN AND R FOR RURAL.

Charts and tabulation based on data generally for periods of record through 1961, from State and Local <u>Climatological Data</u>.

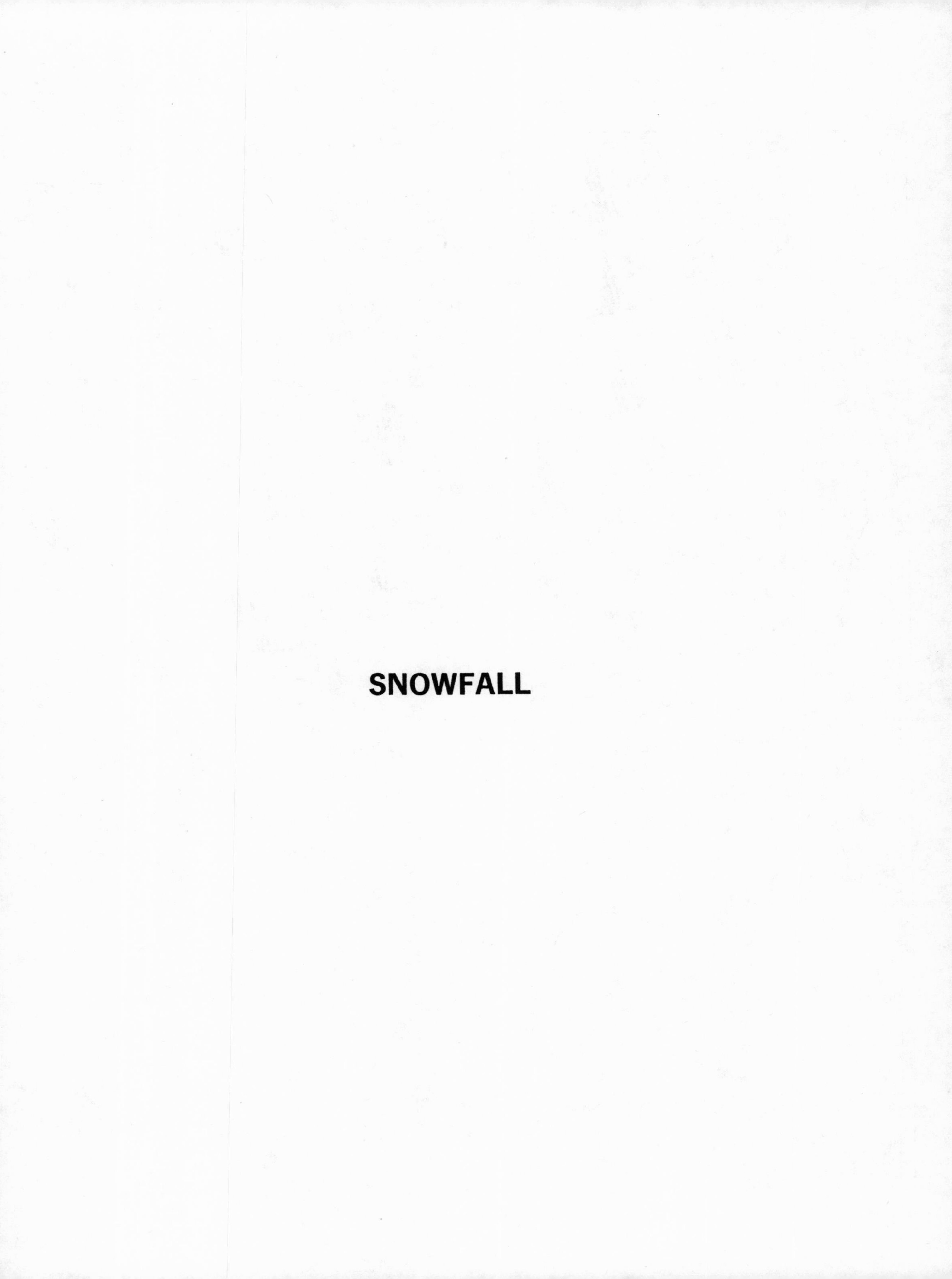

SNOWFALL

MEAN MONTHLY
For Selected

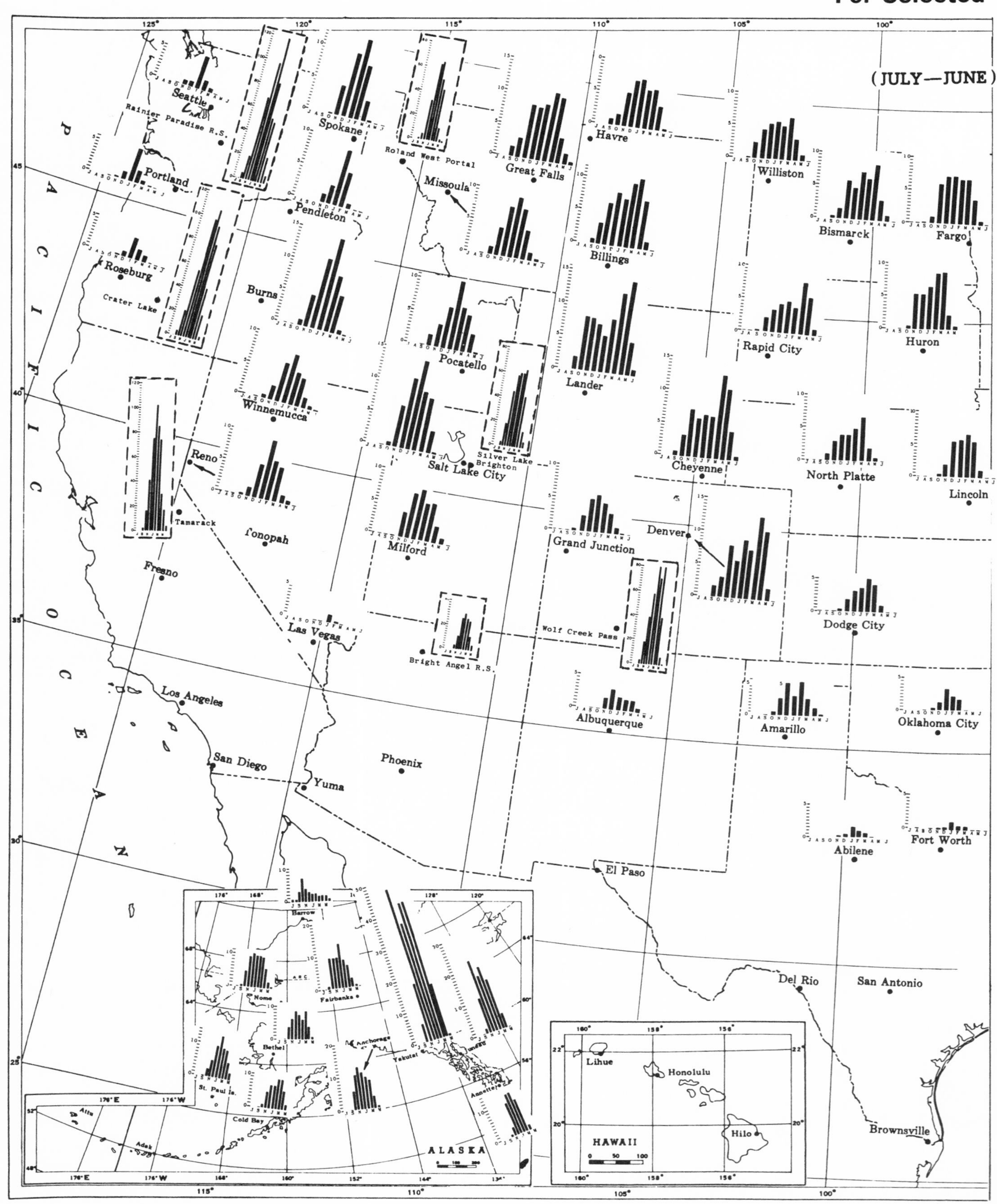

TOTAL SNOWFALL (Inches)
Stations*

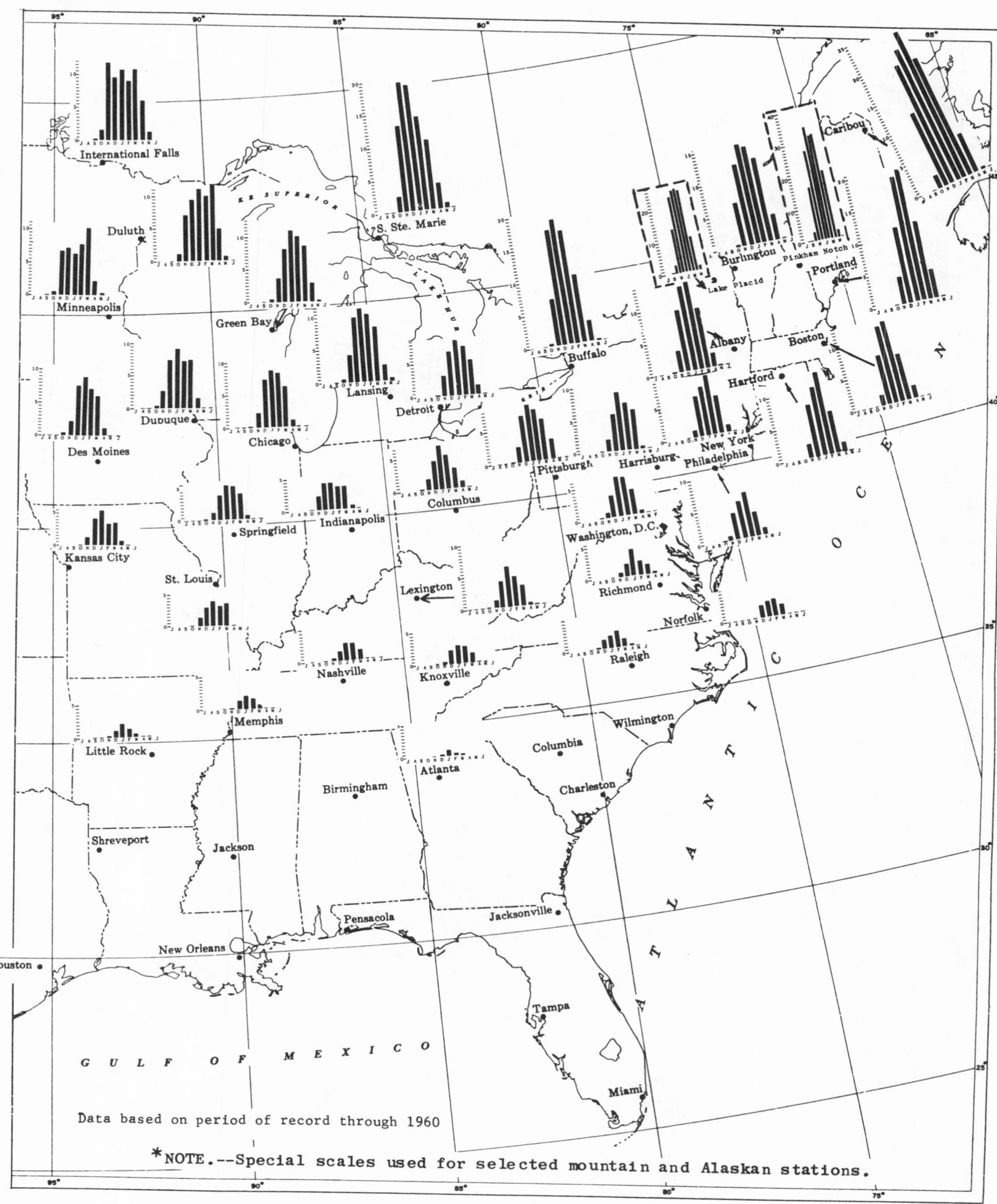

MEAN ANNUAL TOTAL SNOWFALL (Inches)

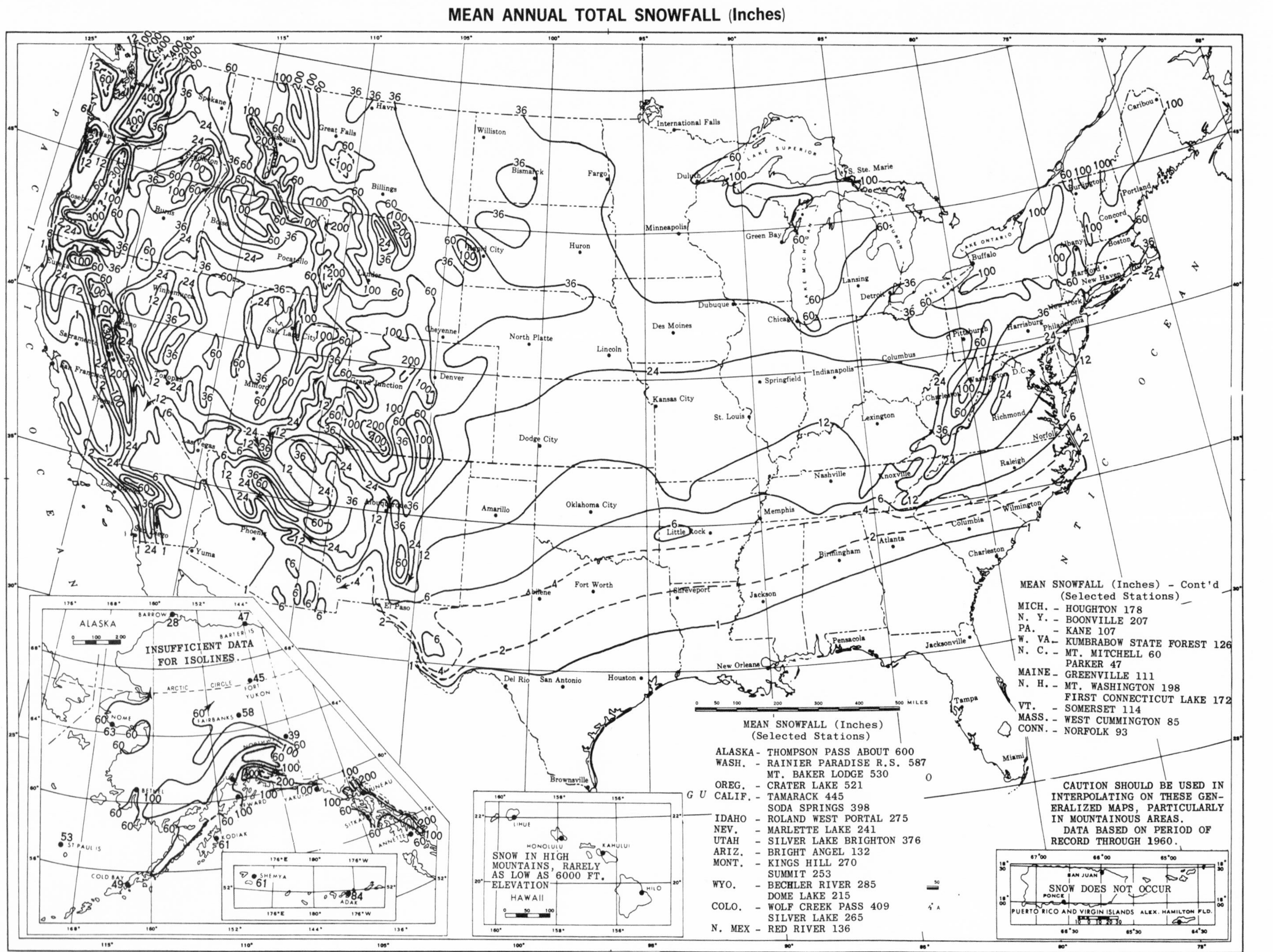

PACIFIC OCEAN
ATLANTIC OCEAN
GULF OF MEXICO
ALASKA
INSUFFICIENT DATA FOR ISOLINES.
ARCTIC CIRCLE
SNOW IN HIGH MOUNTAINS, RARELY AS LOW AS 6000 FT. ELEVATION
HAWAII
SNOW DOES NOT OCCUR
PUERTO RICO AND VIRGIN ISLANDS
MEAN SNOWFALL (Inches) (Selected Stations)
ALASKA - THOMPSON PASS ABOUT 600
WASH. - RAINIER PARADISE R.S. 587
MT. BAKER LODGE 530
OREG. - CRATER LAKE 521
CALIF. - TAMARACK 445
SODA SPRINGS 398
IDAHO - ROLAND WEST PORTAL 275
NEV. - MARLETTE LAKE 241
UTAH - SILVER LAKE BRIGHTON 376
ARIZ. - BRIGHT ANGEL 132
MONT. - KINGS HILL 270
SUMMIT 253
WYO. - BECHLER RIVER 285
DOME LAKE 215
COLO. - WOLF CREEK PASS 409
SILVER LAKE 265
N. MEX - RED RIVER 136
MEAN SNOWFALL (Inches) - Cont'd (Selected Stations)
MICH. - HOUGHTON 178
N. Y. - BOONVILLE 207
PA. - KANE 107
W. VA. - KUMBRABOW STATE FOREST 126
N. C. - MT. MITCHELL 60
PARKER 47
MAINE - GREENVILLE 111
N. H. - MT. WASHINGTON 198
FIRST CONNECTICUT LAKE 172
VT. - SOMERSET 114
MASS. - WEST CUMMINGTON 85
CONN. - NORFOLK 93
CAUTION SHOULD BE USED IN INTERPOLATING ON THESE GENERALIZED MAPS, PARTICULARLY IN MOUNTAINOUS AREAS.
DATA BASED ON PERIOD OF RECORD THROUGH 1960.

DEW POINT

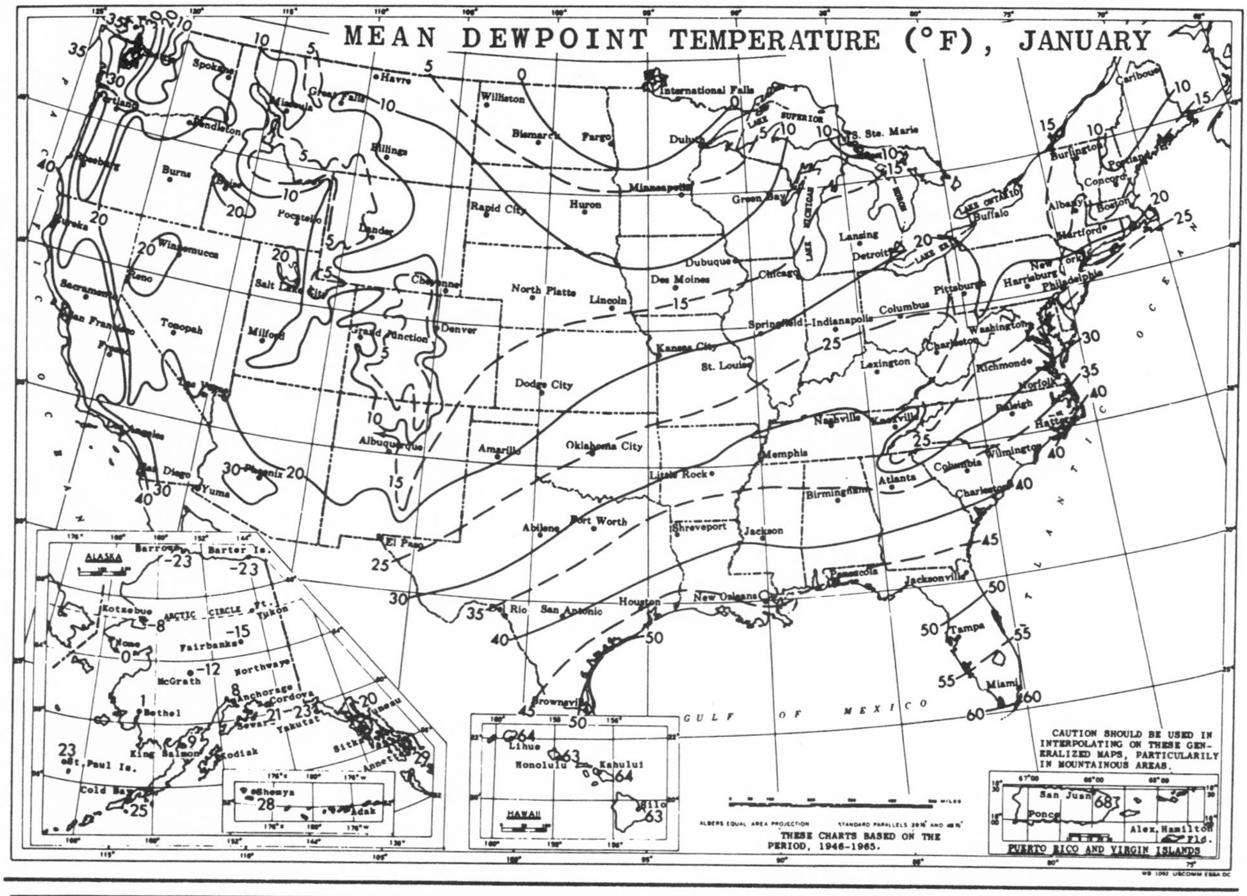

MEAN DEWPOINT TEMPERATURE (°F), JANUARY

CAUTION SHOULD BE USED IN INTERPOLATING ON THESE GENERALIZED MAPS, PARTICULARLY IN MOUNTAINOUS AREAS.

ALBERS EQUAL AREA PROJECTION — STANDARD PARALLELS 29½° AND 45½°

THESE CHARTS BASED ON THE PERIOD, 1946-1965.

ALASKA — HAWAII — PUERTO RICO AND VIRGIN ISLANDS

GULF OF MEXICO

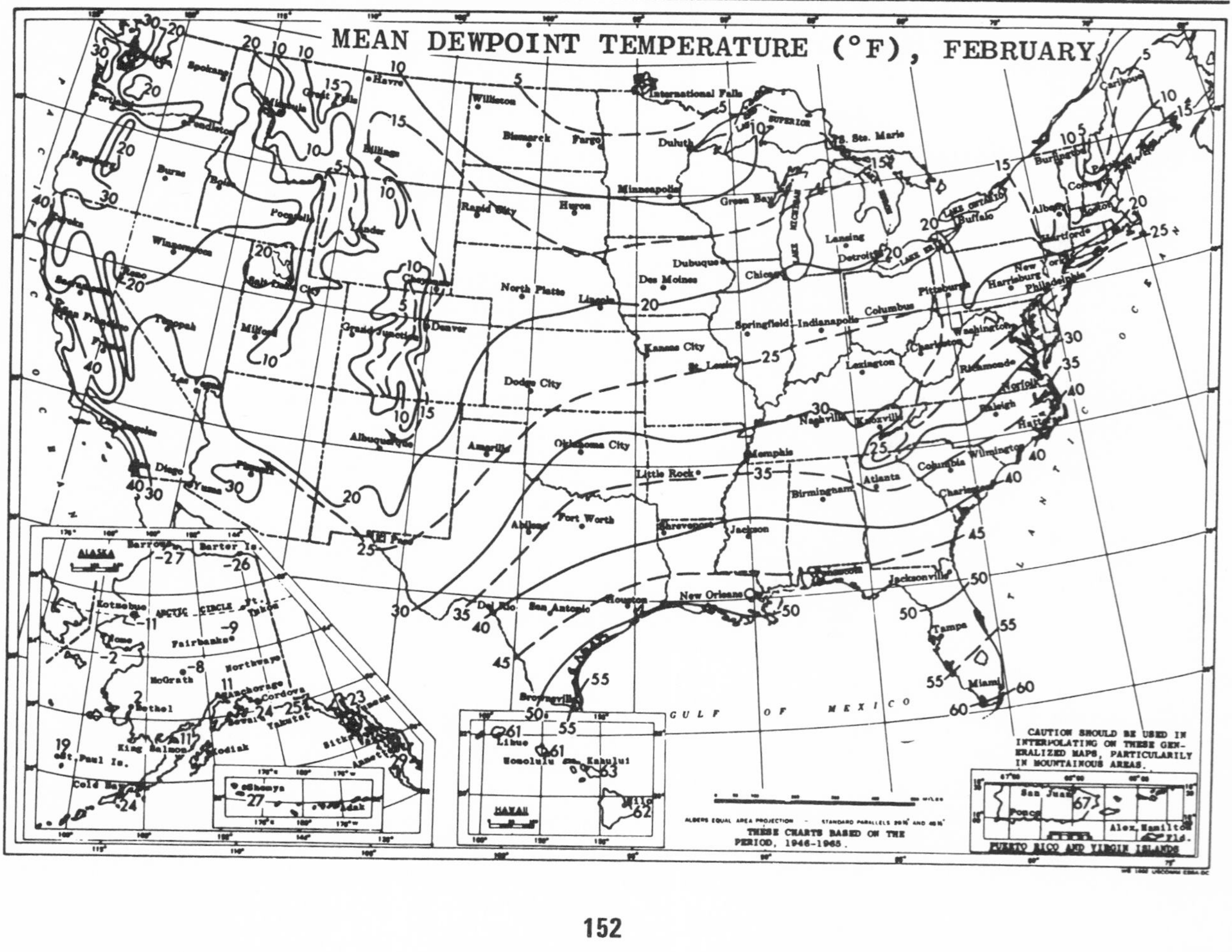

MEAN DEWPOINT TEMPERATURE (°F), FEBRUARY

CAUTION SHOULD BE USED IN INTERPOLATING ON THESE GENERALIZED MAPS, PARTICULARLY IN MOUNTAINOUS AREAS.

ALBERS EQUAL AREA PROJECTION — STANDARD PARALLELS 29½° AND 45½°

THESE CHARTS BASED ON THE PERIOD, 1946-1965.

ALASKA — HAWAII — PUERTO RICO AND VIRGIN ISLANDS

GULF OF MEXICO

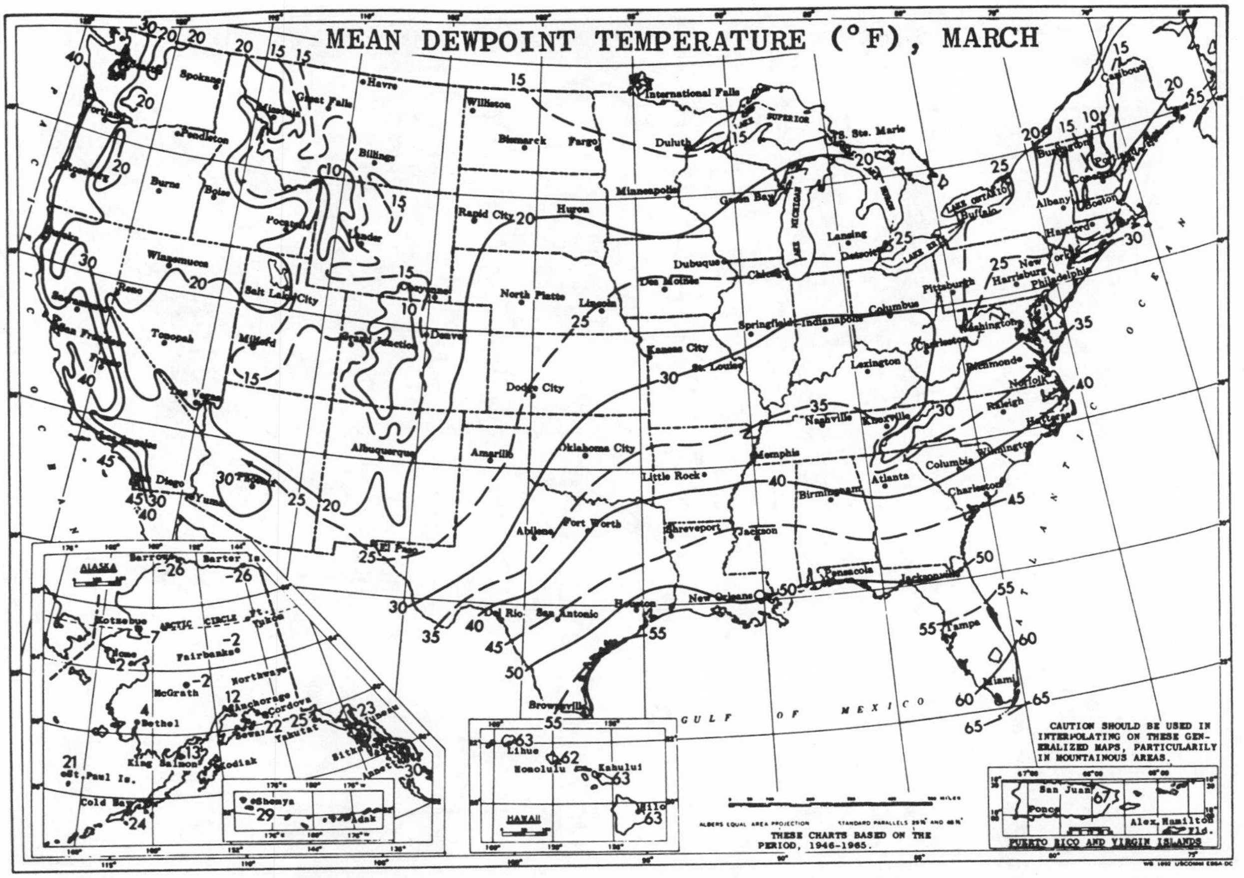

MEAN DEWPOINT TEMPERATURE (°F), MARCH

CAUTION SHOULD BE USED IN INTERPOLATING ON THESE GENERALIZED MAPS, PARTICULARLY IN MOUNTAINOUS AREAS.

ALBERS EQUAL AREA PROJECTION — STANDARD PARALLELS 29½° AND 45½°

THESE CHARTS BASED ON THE PERIOD, 1946-1965.

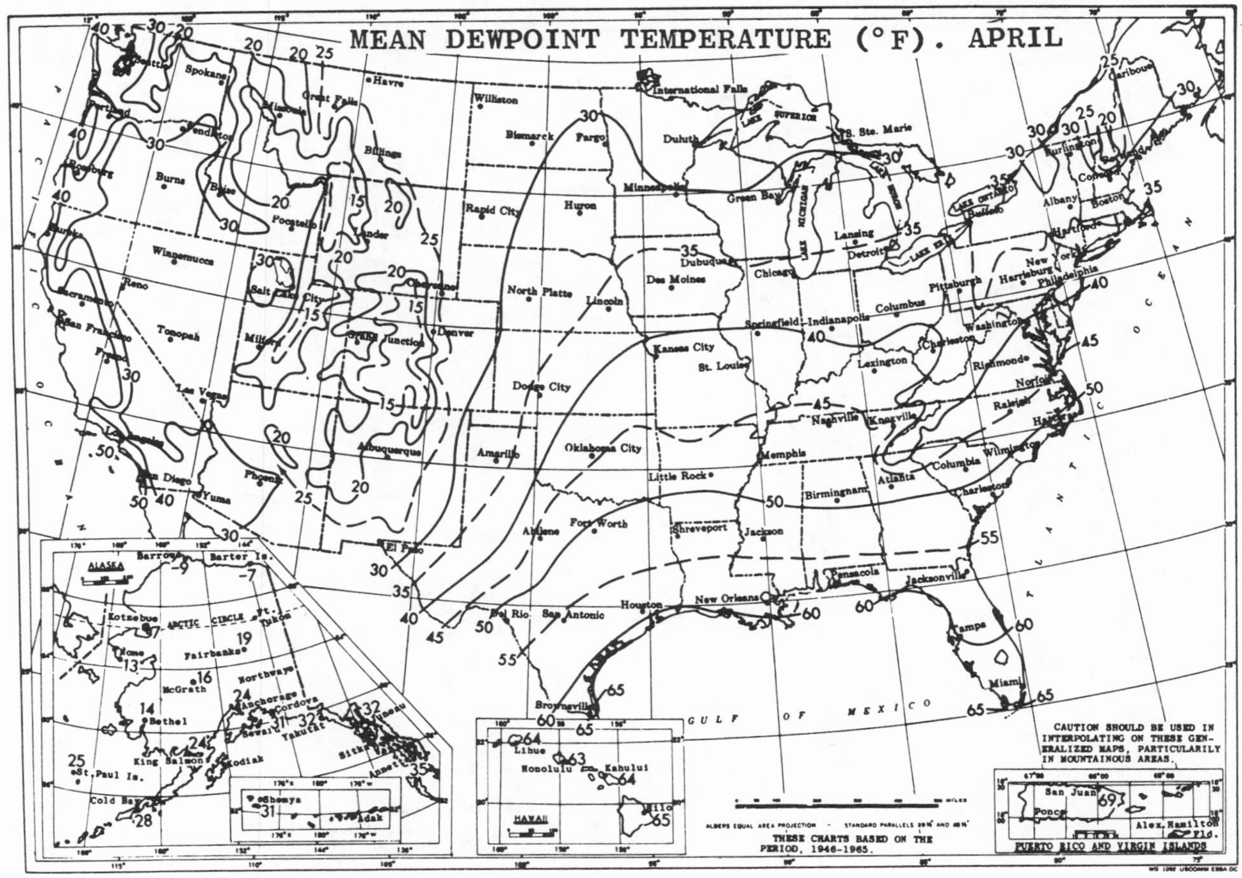

ALASKA — HAWAII — PUERTO RICO AND VIRGIN ISLANDS

WB 1092 USCOMM ESSA DC

MEAN DEWPOINT TEMPERATURE (°F). APRIL

CAUTION SHOULD BE USED IN INTERPOLATING ON THESE GENERALIZED MAPS, PARTICULARLY IN MOUNTAINOUS AREAS.

ALBERS EQUAL AREA PROJECTION — STANDARD PARALLELS 29½° AND 45½°

THESE CHARTS BASED ON THE PERIOD, 1946-1965.

ALASKA — HAWAII — PUERTO RICO AND VIRGIN ISLANDS

WB 1092 USCOMM ESSA DC

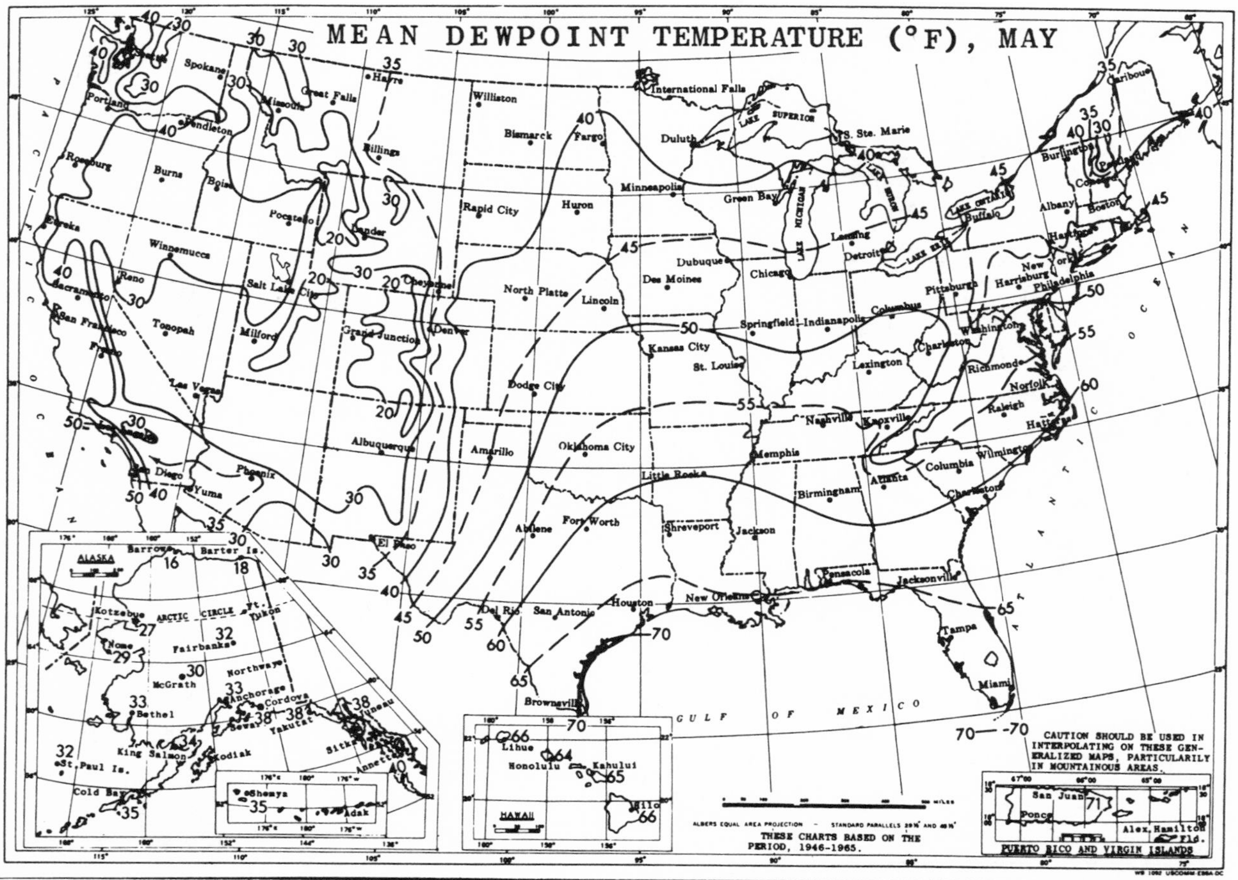

MEAN DEWPOINT TEMPERATURE (°F), MAY

CAUTION SHOULD BE USED IN INTERPOLATING ON THESE GENERALIZED MAPS, PARTICULARLY IN MOUNTAINOUS AREAS.

ALBERS EQUAL AREA PROJECTION – STANDARD PARALLELS 29½° AND 45½°

THESE CHARTS BASED ON THE PERIOD, 1946-1965.

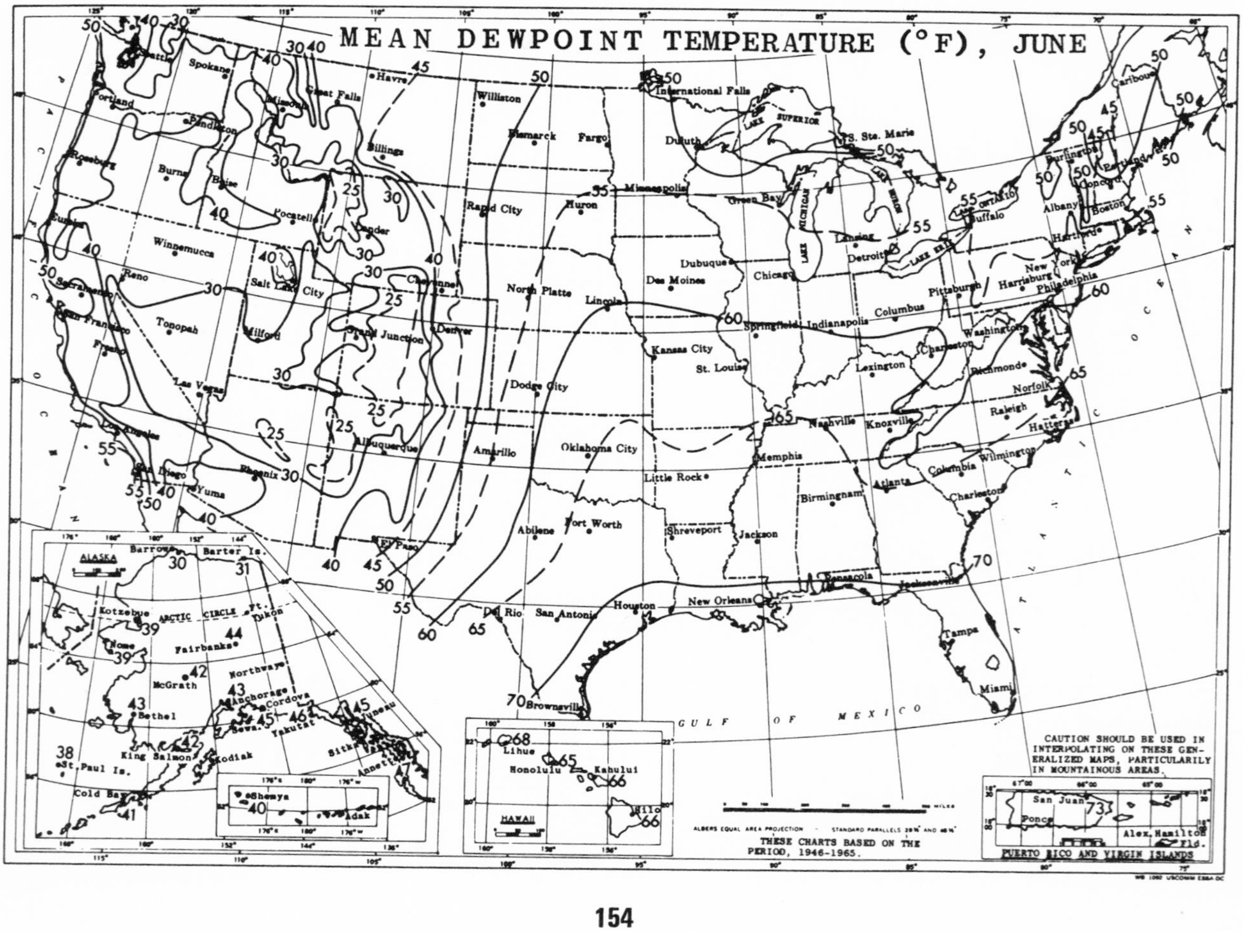

MEAN DEWPOINT TEMPERATURE (°F), JUNE

CAUTION SHOULD BE USED IN INTERPOLATING ON THESE GENERALIZED MAPS, PARTICULARLY IN MOUNTAINOUS AREAS.

ALBERS EQUAL AREA PROJECTION – STANDARD PARALLELS 29½° AND 45½°

THESE CHARTS BASED ON THE PERIOD, 1946-1965.

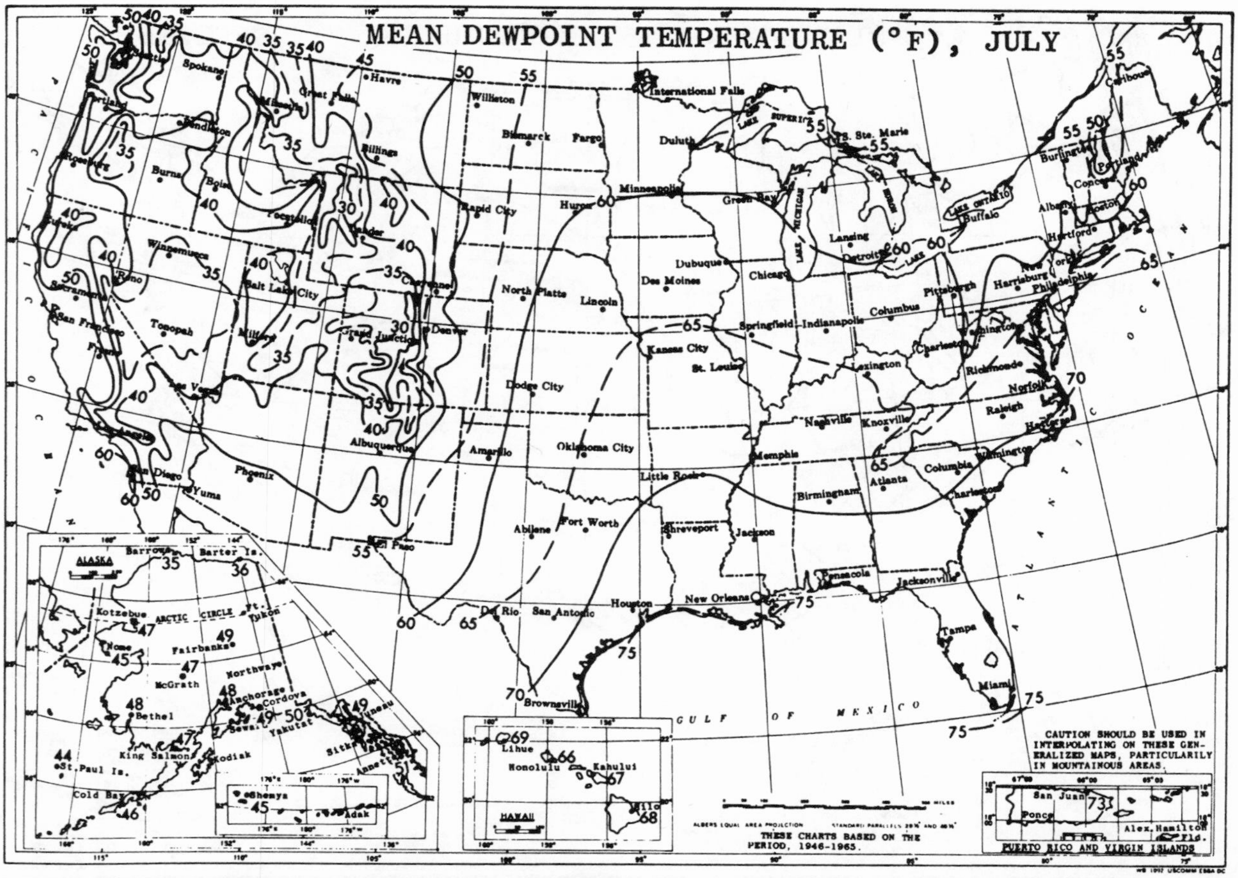

MEAN DEWPOINT TEMPERATURE (°F), JULY

CAUTION SHOULD BE USED IN INTERPOLATING ON THESE GENERALIZED MAPS, PARTICULARILY IN MOUNTAINOUS AREAS.

ALBERS EQUAL AREA PROJECTION — STANDARD PARALLELS 29½° AND 45½°

THESE CHARTS BASED ON THE PERIOD, 1946-1965.

ALASKA

HAWAII

PUERTO RICO AND VIRGIN ISLANDS

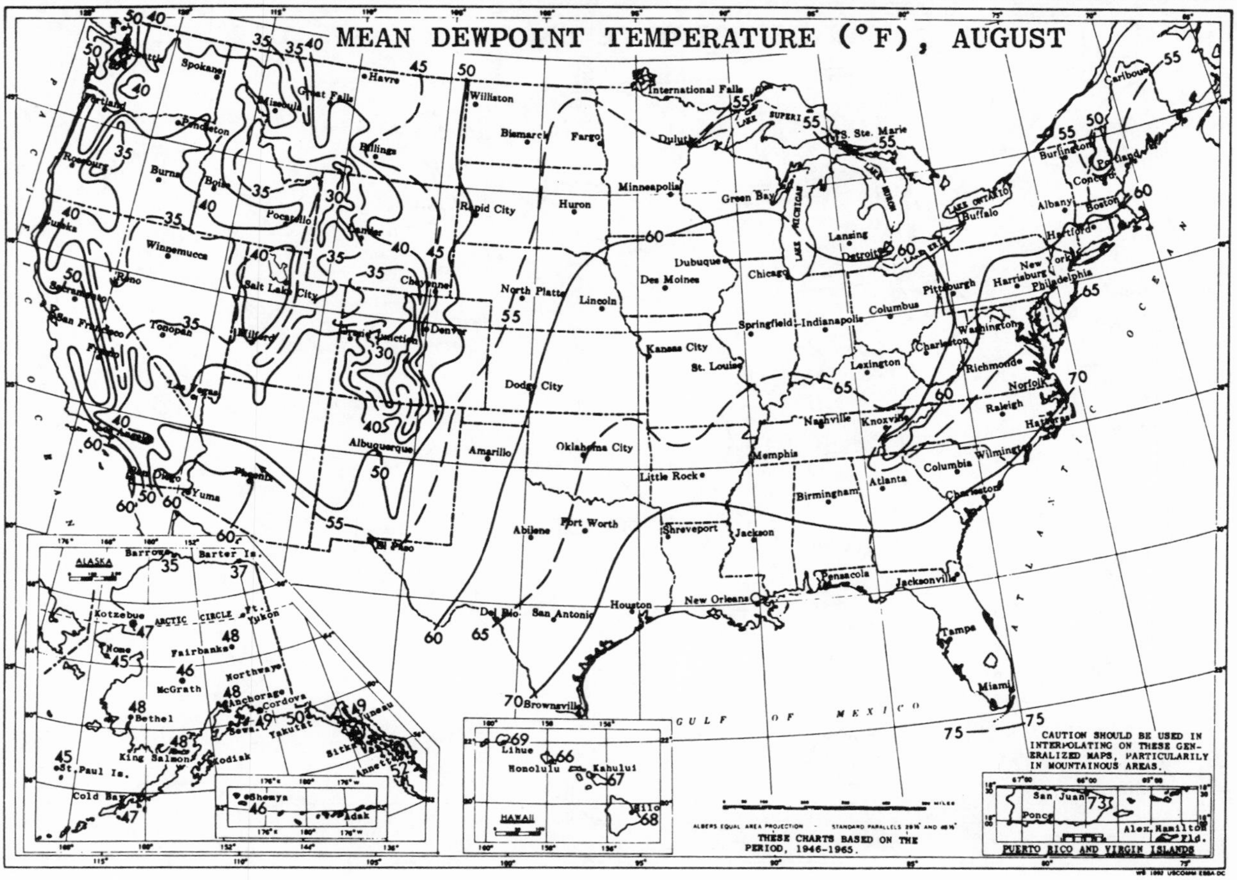

MEAN DEWPOINT TEMPERATURE (°F), AUGUST

CAUTION SHOULD BE USED IN INTERPOLATING ON THESE GENERALIZED MAPS, PARTICULARILY IN MOUNTAINOUS AREAS.

ALBERS EQUAL AREA PROJECTION — STANDARD PARALLELS 29½° AND 45½°

THESE CHARTS BASED ON THE PERIOD, 1946-1965.

ALASKA

HAWAII

PUERTO RICO AND VIRGIN ISLANDS

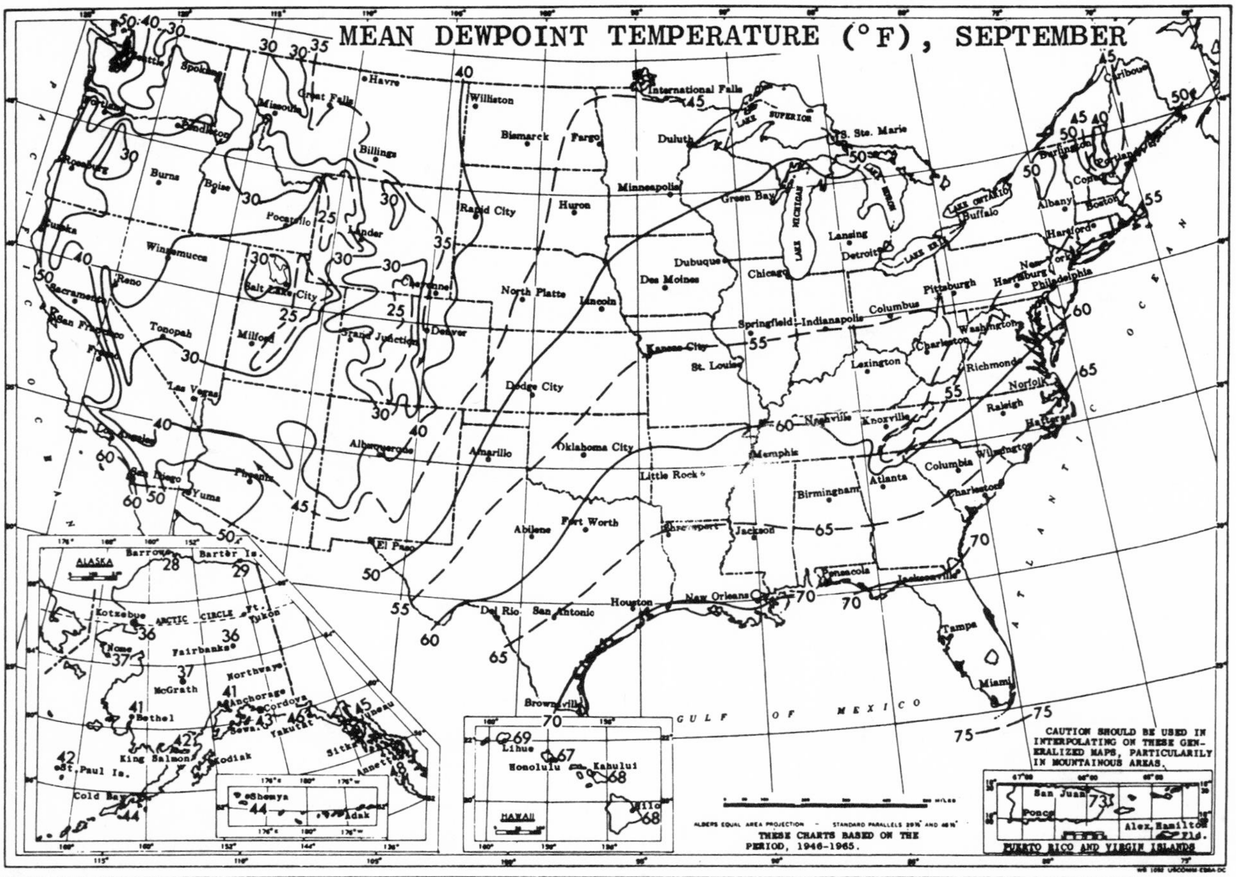

MEAN DEWPOINT TEMPERATURE (°F), SEPTEMBER

CAUTION SHOULD BE USED IN INTERPOLATING ON THESE GENERALIZED MAPS, PARTICULARILY IN MOUNTAINOUS AREAS.

ALBERS EQUAL AREA PROJECTION — STANDARD PARALLELS 29½° AND 45½°

THESE CHARTS BASED ON THE PERIOD, 1946-1965.

ALASKA

HAWAII

PUERTO RICO AND VIRGIN ISLANDS

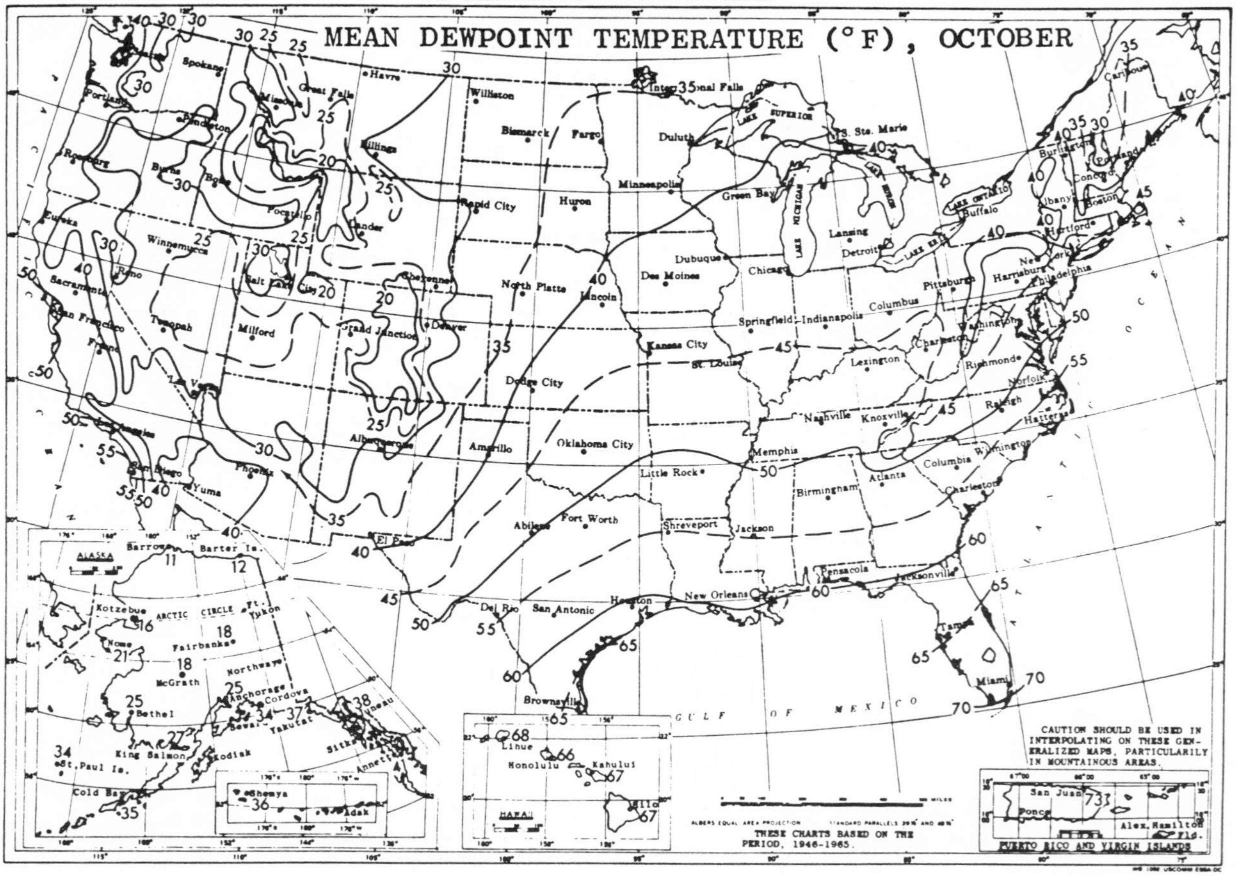

GULF OF MEXICO

MEAN DEWPOINT TEMPERATURE (°F), OCTOBER

CAUTION SHOULD BE USED IN INTERPOLATING ON THESE GENERALIZED MAPS, PARTICULARILY IN MOUNTAINOUS AREAS.

ALBERS EQUAL AREA PROJECTION — STANDARD PARALLELS 29½° AND 45½°

THESE CHARTS BASED ON THE PERIOD, 1946-1965.

ALASKA

HAWAII

PUERTO RICO AND VIRGIN ISLANDS

GULF OF MEXICO

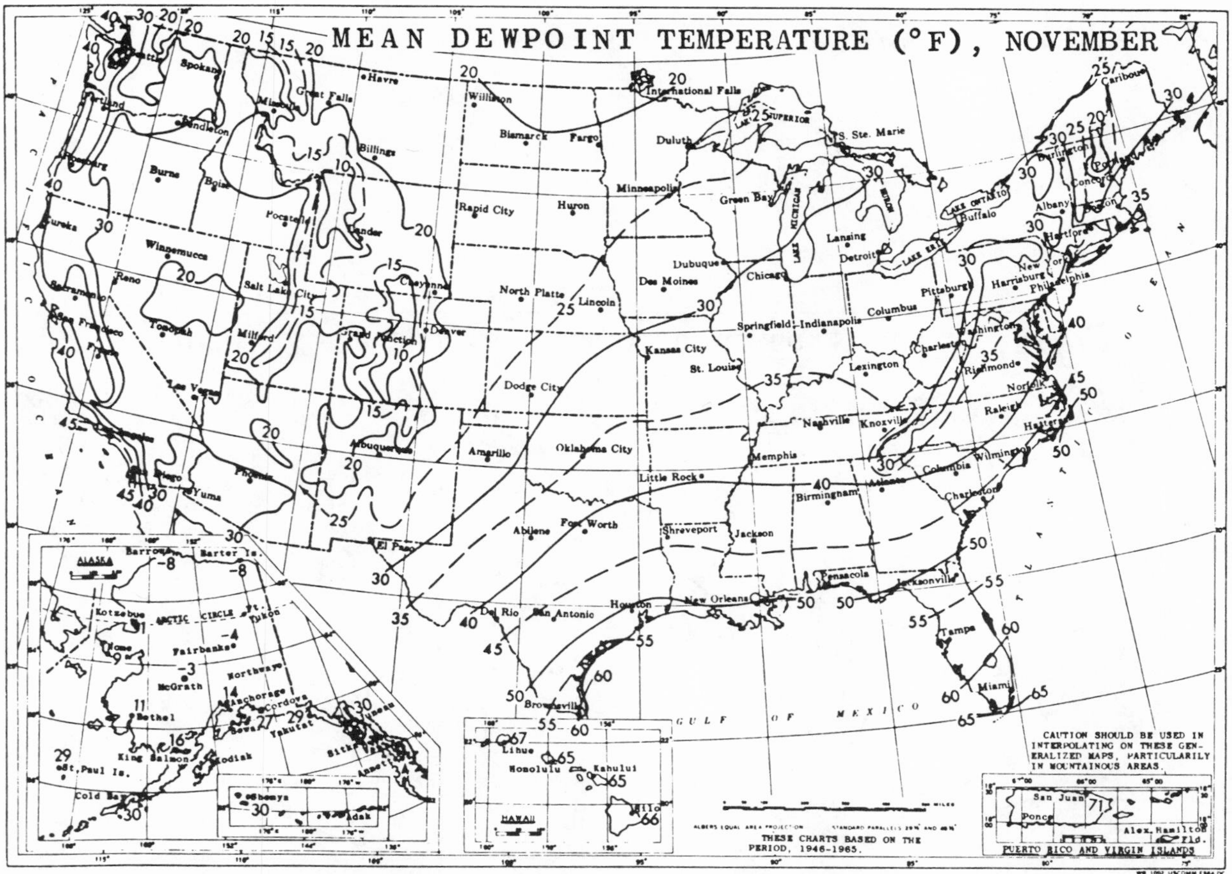

MEAN DEWPOINT TEMPERATURE (°F), NOVEMBER

CAUTION SHOULD BE USED IN INTERPOLATING ON THESE GENERALIZED MAPS, PARTICULARILY IN MOUNTAINOUS AREAS.

THESE CHARTS BASED ON THE PERIOD, 1946-1965.

ALASKA

HAWAII

PUERTO RICO AND VIRGIN ISLANDS

GULF OF MEXICO

MEAN DEWPOINT TEMPERATURE (°F), DECEMBER

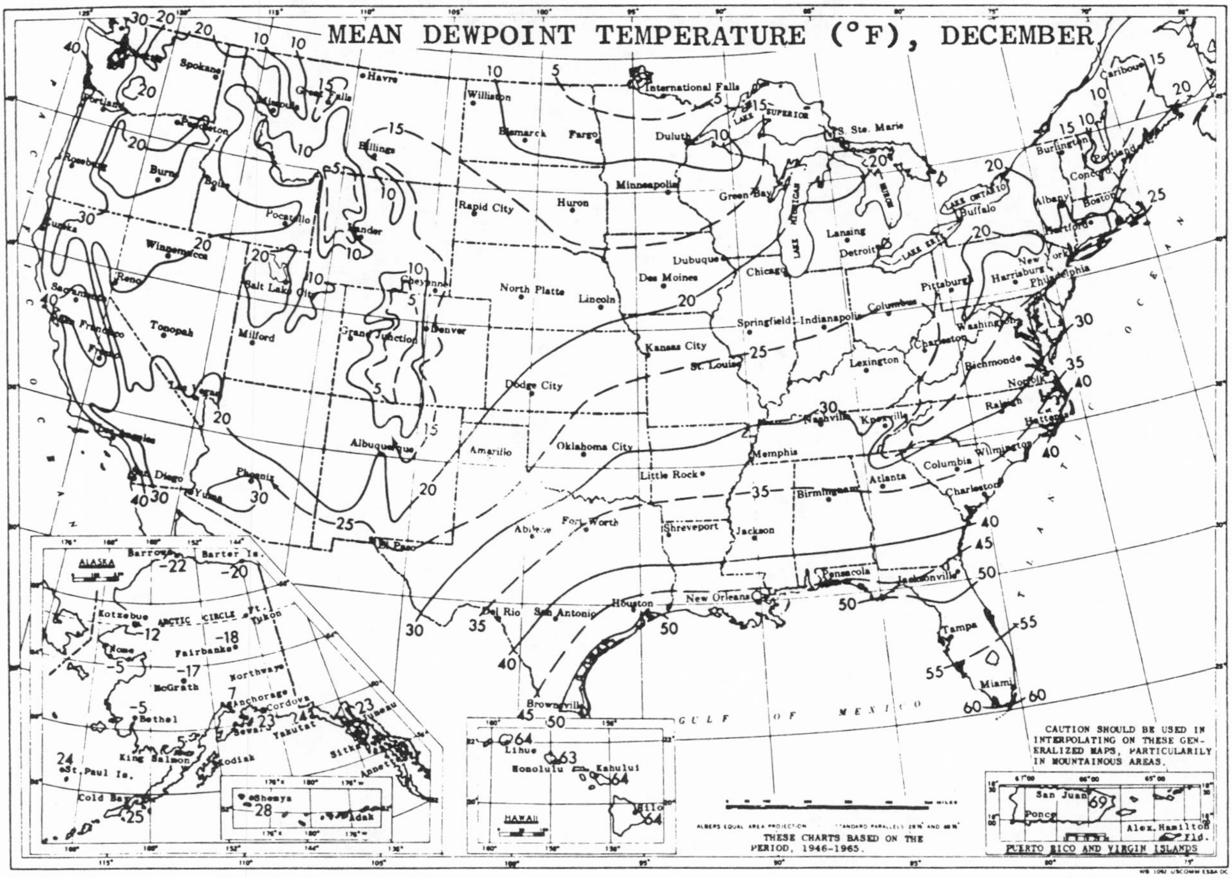

CAUTION SHOULD BE USED IN INTERPOLATING ON THESE GENERALIZED MAPS, PARTICULARILY IN MOUNTAINOUS AREAS.

THESE CHARTS BASED ON THE PERIOD, 1946-1965.

ALASKA

HAWAII

PUERTO RICO AND VIRGIN ISLANDS

GULF OF MEXICO

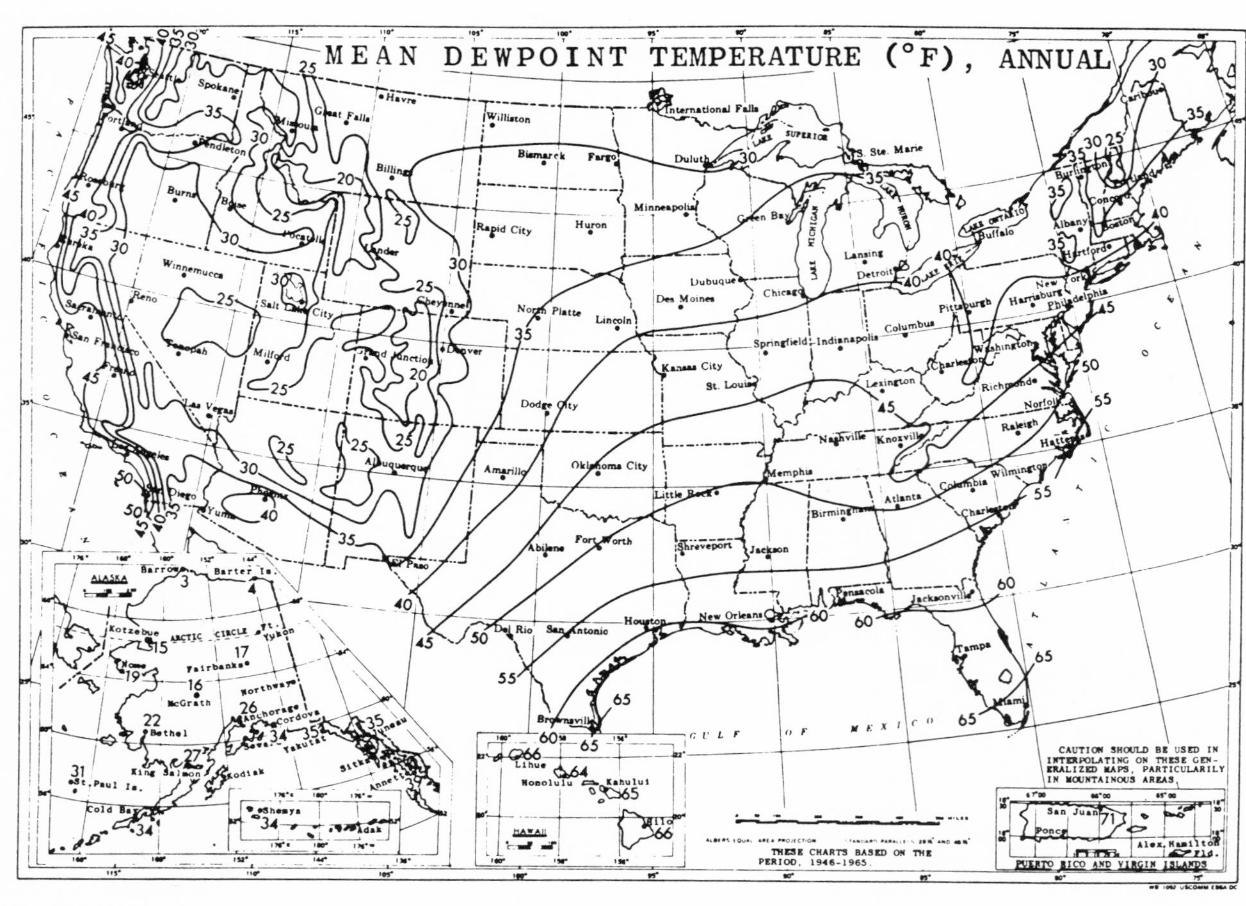
MEAN DEWPOINT TEMPERATURE (°F), ANNUAL
CAUTION SHOULD BE USED IN INTERPOLATING ON THESE GENERALIZED MAPS, PARTICULARILY IN MOUNTAINOUS AREAS.
THESE CHARTS BASED ON THE PERIOD, 1946-1965.
ALASKA
HAWAII
PUERTO RICO AND VIRGIN ISLANDS
GULF OF MEXICO

MEAN DEWPOINT TEMPERATURE (°F)

STATE AND STATION	YRS	JAN	FEB	MAR	APR	MAY	JUN	JUL	AUG	SEP	OCT	NOV	DEC	YEAR
ALA. BIRMINGHAM	20	36	37	41	49	58	66	69	68	62	52	41	36	51
MOBILE	20	44	45	48	57	64	70	72	72	68	57	48	44	57
MONTGOMERY	20	39	40	44	53	61	68	71	70	65	54	43	39	54
ALASKA ANCHORAGE	10	8	11	12	24	33	43	48	48	41	25	14	7	26
ANNETTE	10	29	29	30	35	40	47	51	52	48	41	34	31	39
BARROW	10	-23	-27	-26	-9	16	30	35	35	28	11	-8	-22	3
BARTER ISLAND	9	-23	-26	-26	-7	18	31	36	37	29	12	-8	-20	4
BETHEL	10	1	2	4	14	33	43	48	48	41	25	11	-5	22
COLD BAY	10	25	24	24	28	35	41	46	47	44	35	30	25	34
CORDOVA	9	21	24	22	31	38	45	49	49	43	34	27	23	34
FAIRBANKS	10	-15	-9	-2	19	32	44	49	48	36	18	-4	-18	17
JUNEAU	10	20	23	23	32	38	45	49	49	45	38	30	23	35
KING SALMON	10	9	11	13	24	34	42	47	48	42	27	16	5	27
KOTZEBUE	10	-8	-11	-7	7	27	39	47	47	36	16	1	-12	15
MCGRATH	10	-12	-8	-2	16	30	42	47	46	37	18	-3	-17	16
NOME	10	0	-2	2	13	29	39	45	45	37	21	9	-5	19
ST. PAUL ISLAND	10	23	19	21	25	32	38	44	45	42	34	29	24	31
SHEMYA	7	28	27	29	31	35	40	45	46	44	36	30	28	34
YAKUTAT	10	23	25	25	32	38	46	50	50	46	37	29	24	35
ARIZ. FLAGSTAFF	14	14	16	17	20	22	25	43	43	35	25	20	15	25
PHOENIX	20	33	33	33	35	36	42	58	60	53	44	36	33	41
PRESCOTT	15	19	21	21	23	25	29	48	50	41	31	24	21	29
TUCSON	20	28	26	27	26	27	35	56	59	48	39	30	28	36
WINSLOW	20	19	19	18	21	24	29	47	49	45	30	22	19	29
YUMA	20	27	28	29	33	36	42	57	61	54	43	32	29	39
ARK. FORT SMITH	20	30	33	37	48	59	67	69	67	61	51	38	32	49
LITTLE ROCK	20	32	34	39	49	60	67	70	69	62	52	40	34	51
TEXARKANA	11	35	38	43	52	63	69	70	70	64	53	43	36	53
CALIF. BAKERSFIELD	17	38	40	40	43	43	47	50	51	50	45	42	39	45
BLUE CANYON	7	23	24	25	30	34	36	36	35	35	31	30	27	31
BURBANK	17	36	38	40	45	49	53	57	57	54	49	41	36	46
EUREKA	10	41	40	40	44	48	50	52	53	53	49	48	43	47
FRESNO	20	38	41	41	44	45	48	51	52	51	46	42	40	45
LONG BEACH	7	38	42	44	48	50	55	58	60	58	53	46	42	50
LOS ANGELES	20	40	43	45	49	52	55	59	59	58	53	46	42	50
MT. SHASTA	7	25	28	26	30	35	40	41	38	38	34	34	30	33
OAKLAND	18	40	42	42	45	47	51	53	54	54	50	45	41	47
RED BLUFF	20	34	36	36	39	43	45	47	47	43	41	38	36	40
SACRAMENTO	20	39	41	41	45	47	50	53	53	50	47	42	40	46
SANDBERG	10	24	24	27	33	35	36	37	37	36	33	27	27	31
SAN DIEGO	20	42	44	46	50	52	56	60	61	60	55	46	43	51
SAN FRANCISCO	10	41	42	43	46	47	50	52	53	53	50	46	43	47
SANTA MARIA	16	40	42	42	46	48	51	54	54	53	49	43	40	47
COLO. COLORADO SPRINGS	15	10	14	15	22	35	40	47	47	37	26	17	12	27
DENVER	20	12	16	17	24	35	42	47	46	37	27	19	14	28
GRAND JUNCTION	20	17	20	21	25	29	31	39	43	35	29	24	19	28
PUEBLO	20	12	17	19	27	37	44	51	51	42	30	22	16	31
CONN. BRIDGEPORT	16	22	22	27	37	47	57	66	62	56	46	35	24	42
HARTFORD	20	18	19	25	35	46	57	62	61	54	44	37	22	40
NEW HAVEN	10	24	22	27	37	48	57	63	63	55	47	44	25	43
DEL. WILMINGTON	20	24	24	29	40	50	60	64	64	58	47	35	26	43
D.C. WASHINGTON	20	25	25	29	40	52	61	65	64	59	48	36	26	44
FLA. APALACHICOLA	10	47	48	54	60	66	72	73	74	71	62	50	50	61
DAYTONA BEACH	17	50	52	54	59	65	70	73	73	72	65	56	56	62
FT. MYERS	6	55	55	59	61	67	72	73	74	74	68	59	56	64
JACKSONVILLE	20	46	47	50	56	64	70	72	72	71	62	53	47	58
KEY WEST	16	61	63	65	66	69	73	74	75	74	71	67	64	69
MIAMI	16	57	59	61	63	68	72	73	74	74	69	63	58	66
ORLANDO	17	52	52	55	59	65	70	73	73	72	65	57	51	62
PENSACOLA	10	48	47	51	59	65	70	73	73	68	59	49	46	59
TALLAHASSEE	17	44	46	49	55	62	69	72	72	69	58	49	43	57
TAMPA	20	52	53	56	60	66	70	73	73	72	65	58	52	63
WEST PALM BEACH	17	57	58	61	63	68	72	74	74	74	68	62	57	66
GA. ATHENS	10	32	34	39	48	59	65	69	68	63	52	42	34	50
ATLANTA	20	34	34	39	48	57	65	68	67	62	51	40	34	50
AUGUSTA	20	37	37	42	50	60	66	70	70	65	54	43	36	53
MACON	20	37	38	42	51	59	66	70	69	64	53	43	37	52
SAVANNAH	20	42	43	46	54	63	69	72	72	68	58	48	41	56
HAWAII HILO	10	63	62	63	65	66	66	68	68	68	67	66	64	66
HONOLULU	10	63	61	62	63	64	65	66	66	67	66	65	63	64
KAHULUI	10	64	63	63	64	65	66	67	67	68	67	65	64	65
LIHUE	10	64	61	63	64	66	68	69	69	69	68	67	64	66
IDAHO BOISE	20	21	27	27	32	38	42	44	43	38	34	29	26	33
LEWISTON	10	21	28	28	32	40	45	44	44	42	39	33	28	35
POCATELLO	20	17	21	24	28	35	39	41	40	34	30	26	21	30

STATE AND STATION	YRS	JAN	FEB	MAR	APR	MAY	JUN	JUL	AUG	SEP	OCT	NOV	DEC	YEAR
ILL. CHICAGO	20	18	20	26	36	46	56	61	61	53	43	31	21	39
MOLINE	20	15	19	26	37	48	59	64	63	54	43	30	21	40
PEORIA	20	18	21	28	38	49	59	64	63	54	44	31	22	41
ROCKFORD	7	12	17	24	35	48	56	61	61	54	43	31	24	39
SPRINGFIELD	20	20	23	30	41	49	61	65	64	55	44	32	24	42
IND. EVANSVILLE	20	26	27	34	44	53	63	67	66	58	47	35	28	46
FT. WAYNE	20	20	22	28	36	48	58	62	61	54	44	33	23	41
INDIANAPOLIS	20	22	23	30	40	50	60	64	63	55	45	32	23	42
SOUTH BEND	20	20	22	27	37	48	57	62	61	53	44	33	24	41
TERRE HAUTE	10	26	25	32	42	49	61	65	64	55	45	33	26	44
IOWA BURLINGTON	20	17	21	27	39	49	60	65	64	54	43	30	21	41
DES MOINES	20	14	18	25	37	48	60	64	63	53	42	29	19	39
DUBUQUE	10	15	18	25	35	45	59	64	62	52	42	27	19	39
SIOUX CITY	20	10	16	24	35	47	58	63	62	51	40	26	17	37
KANS. CONCORDIA	15	17	22	27	39	50	61	64	62	54	44	30	21	41
DODGE CITY	20	18	23	25	36	49	57	61	59	51	41	29	22	39
GOODLAND	20	15	21	21	30	42	52	56	55	45	34	23	18	34
TOPEKA	20	19	23	29	41	53	63	66	65	56	45	31	23	43
WICHITA	20	21	25	30	41	53	62	65	63	55	45	33	25	43
KY. LEXINGTON	20	27	27	32	42	53	61	65	64	56	46	34	28	45
LOUISVILLE	20	27	28	33	43	54	63	66	65	57	47	35	27	45
LA. ALEXANDRIA	6	37	40	46	56	62	68	72	71	67	56	49	40	55
BATON ROUGE	20	44	46	49	57	64	70	73	72	68	57	48	44	58
LAKE CHARLES	20	46	48	51	58	66	71	73	73	69	58	50	46	59
NEW ORLEANS	20	46	48	52	59	66	72	73	73	70	60	52	47	60
SHREVEPORT	20	38	40	44	54	62	69	71	70	65	55	45	39	54
MAINE CARIBOU	18	7	5	15	28	38	50	56	54	47	36	27	12	31
EASTPORT	10	18	20	25	33	40	47	55	57	52	42	33	22	37
PORTLAND	20	16	16	23	33	43	53	59	58	49	39	32	19	37
MD. BALTIMORE	20	24	24	29	40	51	61	65	64	58	47	35	25	44
MASS. BLUE HILL	10	20	20	26	35	46	56	62	61	53	45	34	23	40
BOSTON	20	19	19	25	34	44	55	60	60	53	44	34	22	39
NANTUCKET	20	26	25	29	37	46	56	63	62	57	49	39	29	43
WORCESTER	6	14	13	20	29	42	52	56	56	51	39	30	18	35
MICH. ALPENA	10	16	16	21	32	41	53	58	58	50	42	30	20	36
DETROIT	20	19	19	25	35	45	56	60	59	53	43	32	23	39
ESCANABA	10	12	14	20	32	41	52	59	58	50	41	28	17	35
FLINT	10	15	16	23	34	45	54	58	58	52	42	32	21	38
GRAND RAPIDS	20	19	19	25	34	44	55	59	59	52	43	32	23	39
LANSING	17	18	19	24	34	46	56	59	59	52	41	31	22	38
MARQUETTE	10	12	13	19	29	38	50	56	56	49	40	27	17	34
MUSKEGON	20	20	20	24	34	44	55	60	60	52	43	32	24	39
SAULT STE. MARIE	20	9	10	18	29	39	50	55	55	49	40	29	17	33
MINN. DULUTH	20	2	5	15	27	37	49	56	55	46	36	22	9	30
INTERNAT'L FALLS	20	-5	2	11	26	37	50	56	54	45	35	20	4	28
MINN.-ST. PAUL	20	6	10	20	32	43	55	60	59	50	40	25	13	34
ROCHESTER	20	7	12	19	32	44	56	61	59	49	39	25	14	35
ST. CLOUD	16	2	8	17	30	40	53	58	58	48	37	22	12	32
ST. PAUL	5	6	10	19	31	41	56	60	58	50	38	23	12	34
MISS. JACKSON	20	40	41	44	53	58	68	71	70	65	53	42	38	54
MERIDIAN	16	39	40	45	54	61	68	71	71	66	55	45	39	55
VICKSBURG	10	40	41	46	55	63	70	73	71	65	54	43	36	55
MO. COLUMBIA	20	21	24	29	40	52	62	66	64	55	45	32	25	43
KANSAS CITY	20	20	24	29	40	53	62	65	64	55	45	33	24	43
ST. JOSEPH	10	19	24	29	40	51	63	65	65	53	44	30	23	42
ST. LOUIS	20	22	25	30	42	52	62	66	64	56	46	33	26	44
SPRINGFIELD	20	24	27	32	43	55	64	67	64	57	47	34	27	45
MONT. BILLINGS	20	11	16	20	28	38	46	48	46	38	31	22	15	30
BUTTE	10	6	12	16	23	32	38	39	39	33	27	18	13	25
GLASGOW	13	5	10	17	25	37	45	49	44	38	29	22	10	28
GREAT FALLS	20	10	15	18	26	34	42	44	42	36	29	21	15	28
HAVRE	12	6	12	16	26	33	42	45	43	37	29	21	13	27
HELENA	20	10	16	19	26	35	42	45	43	37	30	21	15	28
KALISPELL	9	14	20	23	29	36	43	47	46	39	34	27	20	32
MILES CITY	19	10	15	20	29	39	47	51	48	39	32	23	15	31
MISSOULA	20	15	21	23	29	37	43	45	43	39	34	26	20	31
NEBR. GRAND ISLAND	16	13	19	23	34	45	57	61	60	49	38	25	18	37
LINCOLN	8	16	21	28	37	46	60	63	62	51	41	28	20	39
NORFOLK	8	11	18	23	34	44	58	62	61	49	38	24	16	37
NORTH PLATTE	20	13	18	22	32	44	55	59	58	48	36	24	17	36
OMAHA	20	14	19	26	37	48	60	64	63	54	43	29	20	40
SCOTTSBLUFF	16	13	17	19	28	40	49	54	53	43	32	21	17	32
VALENTINE	10	13	15	22	31	39	51	56	55	43	34	22	16	33

MEAN DEWPOINT TEMPERATURE (°F) - Continued

STATE AND STATION	YRS	JAN	FEB	MAR	APR	MAY	JUN	JUL	AUG	SEP	OCT	NOV	DEC	YEAR
NEV. ELKO	17	15	20	22	26	32	35	36	34	29	25	23	18	26
ELY	20	12	18	19	24	26	29	33	34	27	24	19	16	23
LAS VEGAS	20	21	22	20	24	26	28	39	41	33	29	25	23	28
RENO	20	20	24	23	26	32	37	40	38	35	31	25	22	29
WINNEMUCCA	18	20	23	23	26	31	34	34	33	30	28	24	22	27
N. H. CONCORD	20	14	14	22	32	43	53	59	58	51	39	30	18	36
MT. WASHINGTON	10	4	5	7	20	29	41	45	45	36	28	17	7	24
N. J. ATLANTIC CITY	18	27	26	30	40	50	60	66	65	59	49	39	28	45
NEWARK	20	23	23	27	37	47	57	62	62	56	46	34	25	42
TRENTON	10	25	23	29	38	50	59	64	64	56	48	35	25	43
N. MEX. ALBUQUERQUE	20	19	19	19	23	29	35	49	50	42	33	23	20	30
CLAYTON	10	18	20	22	31	42	50	54	57	47	35	24	19	35
ROSWELL	17	20	22	22	28	37	48	56	55	48	39	26	22	35
N. Y. ALBANY	20	16	16	24	34	44	55	60	59	53	42	32	20	38
BINGHAMTON	15	17	18	23	34	44	54	58	58	51	41	30	20	37
BUFFALO	20	19	19	25	35	45	55	59	59	52	43	33	23	39
CANTON	5	16	14	21	30	43	53	57	57	50	42	30	17	36
NEW YORK	15	22	23	27	38	47	57	62	62	56	46	35	26	42
OSWEGO	7	20	18	25	33	42	54	60	60	52	44	32	21	38
ROCHESTER	20	19	19	25	35	45	55	59	59	53	43	33	23	39
SYRACUSE	20	18	18	25	35	45	55	59	59	53	42	32	22	39
N. C. ASHEVILLE	10	30	29	33	43	53	63	65	64	56	47	32	29	45
CAPE HATTERAS	20	40	40	44	52	61	68	72	72	68	59	50	41	56
CHARLOTTE	20	32	32	36	46	56	64	67	67	61	50	39	32	49
GREENSBORO	20	29	29	34	44	55	63	67	66	60	48	37	29	47
RALEIGH	20	32	31	35	45	56	64	68	67	61	50	38	30	48
WILMINGTON	16	39	39	44	52	60	68	71	70	67	56	46	37	54
WINSTON SALEM	19	29	29	32	42	54	59	66	68	59	51	36	28	46
N. DAK. BISMARCK	20	1	7	17	29	38	51	56	53	42	33	20	9	30
DEVILS LAKE	10	-4	3	14	30	40	51	57	56	45	35	19	5	29
FARGO	20	-1	6	17	31	40	54	59	57	46	36	22	8	31
WILLISTON	16	3	7	17	27	37	48	53	51	42	32	21	11	29
OHIO AKRON-CANTON	17	22	22	27	37	47	56	60	60	53	42	32	24	40
CINCINNATI	12	26	26	29	39	51	60	63	62	55	43	31	26	43
CLEVELAND	20	23	22	28	37	47	57	61	61	54	44	33	24	41
COLUMBUS	20	23	24	30	40	50	59	63	62	55	44	33	25	42
DAYTON	20	23	24	29	39	50	58	62	61	54	44	33	24	42
TOLEDO	20	21	21	27	37	47	57	61	61	54	43	32	23	40
YOUNGSTOWN	17	22	22	26	37	46	56	60	59	53	42	32	22	40
OKLA. OKLAHOMA CITY	17	26	30	33	45	58	65	67	65	58	48	35	28	47
TULSA	20	26	30	34	46	58	66	68	66	59	49	36	29	47
ORE. ASTORIA	10	37	40	39	42	46	51	54	55	52	49	42	40	46
BAKER	7	15	23	27	30	37	42	44	42	37	32	26	23	32
BURNS	15	19	23	24	27	33	37	38	37	32	30	26	19	29
EUGENE	7	33	37	35	40	44	48	50	50	48	45	42	38	43
MEDFORD	20	32	34	35	38	42	46	49	49	45	43	37	34	40
PENDLETON	17	24	30	30	34	40	42	43	43	41	39	33	29	36
PORTLAND	20	33	36	37	41	46	50	53	54	51	47	40	36	44
ROSEBURG	10	35	37	36	38	41	46	48	48	47	45	41	38	42
SALEM	17	34	38	38	41	45	50	52	52	50	46	41	38	44
PA. ALLENTOWN	18	21	22	27	38	48	58	62	62	55	44	34	23	41
ERIE	13	22	21	27	36	46	56	60	60	54	44	34	24	40
HARRISBURG	20	21	21	26	37	48	58	62	62	55	44	33	23	41
PHILADELPHIA	16	24	24	28	39	49	59	64	63	57	46	35	25	43
PITTSBURGH	20	22	22	27	37	47	57	60	60	53	42	31	24	40
READING	10	24	23	28	37	49	58	63	63	55	45	34	24	42
SCRANTON	20	19	19	25	36	46	55	60	59	52	43	32	22	39
WILLIAMSPORT	13	20	19	25	36	47	56	61	60	54	43	32	21	40
R.I. BLOCK ISLAND	10	28	26	31	39	48	57	65	64	57	50	39	30	45
PROVIDENCE	20	20	19	25	34	45	55	62	61	54	44	34	23	40
S.C. CHARLESTON	20	40	41	44	53	62	69	72	71	68	57	47	39	55
COLUMBIA	20	36	36	41	49	58	66	69	69	64	53	43	35	52
FLORENCE	13	37	37	40	49	58	67	71	70	65	56	42	34	52
GREENVILLE	17	30	30	36	44	55	63	67	66	60	50	38	32	48
SPARTANBURG	10	37	36	38	45	55	63	67	67	61	50	37	31	49

STATE AND STATION	YRS	JAN	FEB	MAR	APR	MAY	JUN	JUL	AUG	SEP	OCT	NOV	DEC	YEAR
S. DAK. HURON	20	6	12	21	32	43	56	60	58	47	37	23	13	34
RAPID CITY	20	12	15	20	28	39	50	53	50	39	31	22	16	31
SIOUX FALLS	17	5	12	20	31	43	56	60	59	47	37	23	12	34
TENN. BRISTOL	13	30	30	34	42	54	62	65	65	58	47	35	28	46
CHATTANOOGA	20	33	33	37	47	57	65	68	68	61	50	39	32	49
KNOXVILLE	20	32	32	36	45	55	63	66	66	59	49	38	31	48
MEMPHIS	16	33	35	38	48	59	66	69	68	62	49	38	33	50
NASHVILLE	20	32	33	37	47	57	65	68	67	60	49	38	32	49
TEXAS ABILENE	20	29	32	33	45	56	62	63	61	58	50	37	31	46
AMARILLO	20	19	23	23	32	45	55	58	57	51	40	27	22	38
AUSTIN	20	38	42	45	55	64	68	69	68	65	56	46	41	55
BROWNSVILLE	20	53	55	59	65	70	73	74	74	72	66	60	54	65
CORPUS CHRISTI	20	48	52	55	63	70	73	74	74	71	64	55	50	62
DALLAS	20	34	37	41	52	62	67	68	67	63	53	42	36	52
DEL RIO	14	38	40	42	51	59	66	66	65	63	55	43	38	52
EL PASO	20	24	23	23	26	31	42	55	55	49	39	28	25	35
FORT WORTH	20	33	36	39	51	61	66	67	66	61	53	41	35	51
GALVESTON	18	47	50	54	62	69	73	75	74	71	64	54	50	62
HOUSTON	20	45	48	51	60	66	72	73	73	69	60	51	47	60
LAREDO	19	43	46	51	57	65	69	68	68	67	60	50	44	57
LUBBOCK	20	25	26	27	37	49	57	61	60	55	45	31	24	41
MIDLAND	10	25	29	29	37	49	58	60	58	56	47	36	30	43
PALESTINE	10	40	42	45	54	64	71	72	71	66	57	46	43	56
PORT ARTHUR	20	47	49	52	60	67	73	75	74	70	61	51	47	61
SAN ANGELO	20	30	34	35	45	56	62	62	61	59	50	38	32	47
SAN ANTONIO	20	39	42	45	55	64	68	68	67	65	56	46	41	55
VICTORIA	17	46	49	52	59	67	72	72	71	69	61	51	47	60
WACO	20	37	40	43	54	63	68	69	67	63	55	44	38	53
WICHITA FALLS	20	28	35	34	46	58	64	64	63	58	49	36	30	47
UTAH MILFORD	7	16	19	15	23	27	30	40	40	29	25	22	19	25
SALT LAKE CITY	20	20	23	26	31	36	40	44	45	38	34	28	24	32
VT. BURLINGTON	20	12	12	20	32	43	54	59	58	51	40	30	17	36
VA. LYNCHBURG	10	29	27	32	42	53	61	66	65	58	47	34	27	45
NORFOLK	20	32	32	36	45	56	64	68	68	63	53	42	33	49
RICHMOND	20	28	29	33	43	55	63	67	67	60	49	38	29	47
ROANOKE	17	26	26	29	39	52	60	64	63	57	45	33	25	43
CAPE HENRY	8	38	35	39	45	58	65	70	69	64	57	44	36	52
WASH. ELLENBURG	5	14	24	28	31	38	44	46	47	43	38	31	25	34
NORTH HEAD	7	34	39	40	42	47	51	54	55	53	48	43	40	46
OLYMPIA	17	33	36	35	38	43	48	51	52	49	45	40	37	42
SEATTLE AP	6	33	36	35	38	43	48	52	53	51	47	40	37	43
SEATTLE-TACOMA	16	34	35	35	39	43	48	52	53	50	46	39	36	43
SPOKANE	20	20	25	27	32	38	43	44	44	40	37	30	26	34
STAMPEDE PASS	7	18	24	24	28	34	39	42	44	41	35	29	25	32
TACOMA	7	32	36	37	40	46	50	52	53	52	46	40	37	43
TATOOSH ISLAND	20	36	38	38	41	46	50	52	54	51	47	42	39	45
WALLA WALLA	7	23	30	31	34	40	44	46	47	43	41	35	30	37
YAKIMA	19	21	26	27	31	37	43	44	46	42	37	30	26	34
W. VA. CHARLESTON	16	27	27	30	40	52	61	65	64	57	47	34	27	44
ELKINS	17	24	24	30	38	49	58	61	60	54	42	32	25	41
PARKERSBURG	10	28	26	31	40	51	60	64	63	56	46	34	27	44
WISC. GREEN BAY	20	10	12	21	33	43	55	60	59	51	41	27	15	36
LA CROSSE	20	9	13	21	33	45	56	61	61	52	41	27	15	36
MADISON	20	11	15	22	34	45	56	61	60	51	41	28	17	37
MILWAUKEE	20	14	17	24	34	44	55	61	61	52	43	29	19	38
WYO. CASPER	17	11	15	18	25	34	39	43	40	33	26	20	15	27
CHEYENNE	20	10	13	17	25	35	42	47	45	36	27	18	14	27
LANDER	20	8	13	17	24	33	39	42	40	35	28	18	12	26
SHERIDAN	20	11	16	20	28	38	46	48	45	38	30	22	16	30
ROCK SPRINGS	8	11	15	18	24	31	35	39	39	31	27	19	15	25
PUERTO RICO. SAN JUAN	20	68	67	67	69	71	73	73	73	73	73	71	69	71

Based on years of data indicated in table during 1946-65

MAXIMUM PERSISTING 12-HOUR 1000-MB DEW POINTS

Hydrometeorological Section
U. S. Weather Bureau, Washington, D. C.

The Hydrometeorological Section for a number of years has used maps of maximum persisting 12-hour 1000-mb. dew points as indices of the maximum precipitable water that can be expected in various regions of the United States in various months. Since there are other applications of these data and there have been requests for copies of these maps from time to time, they are being incorporated in the National Atlas to make them available.

A maximum persisting 12-hour surface dew point is the highest value that has been equaled or exceeded for 12 consecutive hours. During years when observations were made only twice a day at a 12-hour interval, the persisting dew point is based on two consecutive observations, for observations at 6-hour intervals on three consecutive observations. For the more recent years when observations are made every hour, only the dew points at the primary observation times six hours apart were used. During much of the record scanned, minimum air temperatures were also referred to as a supplement to the dew point observations and the persisting dew point reduced to the intervening minimum temperature when this was lower than the successive dew point observations.

A 12-hour persisting 1000-mb. dew point is the surface value reduced to 1000-mb. (approximately sea level) along a moist adiabat. This is approximately at the rate of 2.4°F., per 1,000 feet. Values on the maps may be readjusted to ground level by subtracting 2.4°F. for each 1,000 feet of elevation.

The monthly charts of maximum 12-hour persisting 1000-mb. dew points are a blending together of the results of studies made at different times for different regions of the Nation. The period of record and density of stations analyzed therefore differ somewhat. Most of the area of the maps is based on the records at selected Weather Bureau first-order stations from the beginning of observations through 1946. New York and New England are updated through 1952, with some cognizance taken of maximum dew points at sea in shaping the lines. California is updated through 1958 for the months of October through April. Very few stations were surveyed for the full period of record in the Southeastern United States, use being made in that region by some more restricted surveys by the Tennessee Valley Authority.

Several smoothing steps were applied in constructing the charts. The higher 12-hour persisting values abstracted from the climatological record were plotted on an annual graph for each station with date of the year as abscissa. A smooth enveloping curve was drawn through the highest values on each graph. An occasional value that appeared to be out of line was "undercut," that is, was allowed to fall slightly above the curve. In the most recent California revision, a few values were also undercut because the weather charts for those dates showed that the particular dew points were not representative of storm conditions (most of the extreme values do occur in weather similar to storm conditions, that is, with moderate or strong windflows from the ocean). Next, values were read from the curves at the fifteenth day of the month, and plotted on a map at the respective station. Smooth lines were then drawn on the maps, placing most of the values on or south of the corresponding isoline. Again, however, for final smoothing a few values were undercut. A dew point of 78°F., was adopted as the highest value representative of moisture in depth (the objective of the study) in the Southern United States in summer; the few higher values were neglected in constructing the isolines.

The maximum persisting dew points at stations for the period of record through 1946 are published in Weather Bureau Technical Paper No. 5, "Maximum Persisting Dew Points in the Western United States," for the region indicated by the title. That publication contains persisting maximum values not only for 12 hours, but for other durations through 96 hours.

The original purpose of the dew point charts was to adjust past major rainstorms to maximum moisture, in making estimates of probable maximum precipitation in connection with the design of spillways of dams. This adjustment is made by comparing the persisting 1000-mb., 12-hour representative storm dew point in the warm air, generally to the south of the rainfall center, with the maximum dew point from the maps at the same season and geographical location. The ratio for adjustment is the ratio of the precipitable water in saturated pseudo-adiabatic columns of air between 1000 mbs., and 300 mbs., for the respective dew points. Geographical adjustments to transpose a storm from one location to another are also made in a similar fashion, based on differences between maximum dew points near the respective locations.

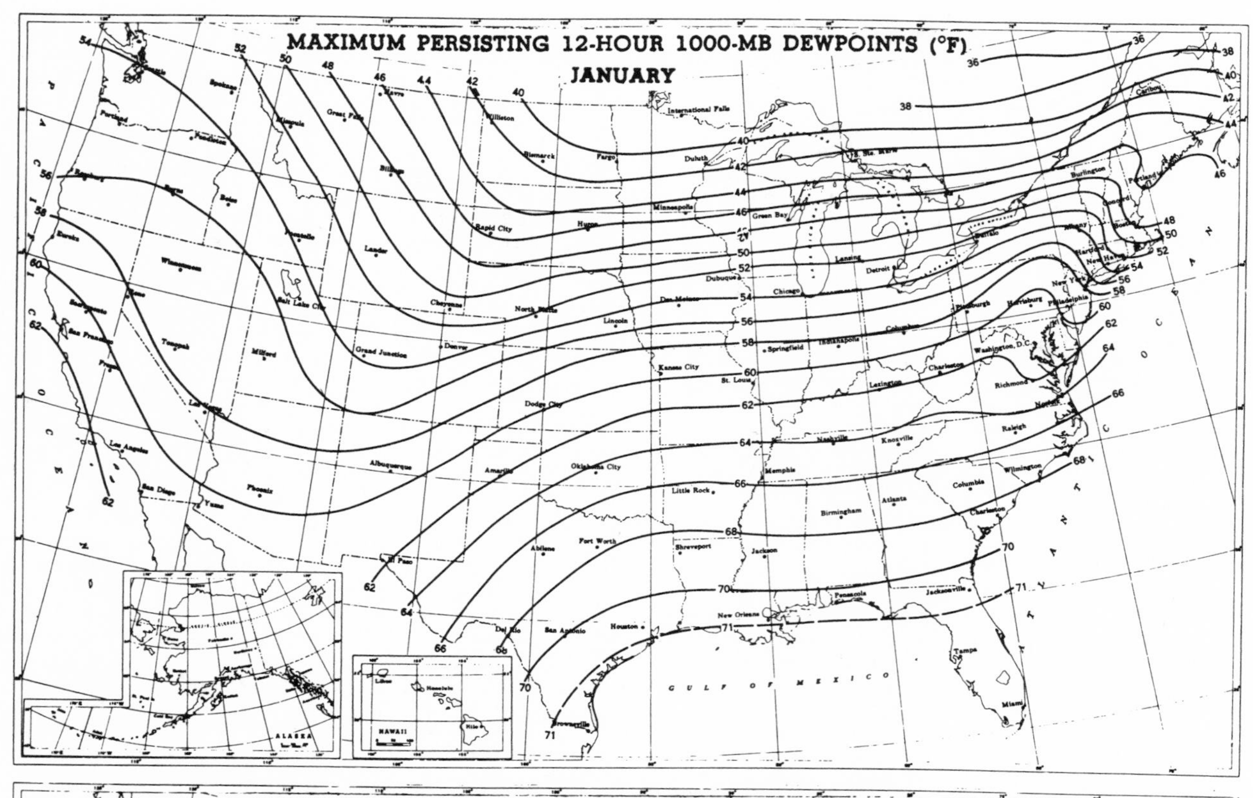
MAXIMUM PERSISTING 12-HOUR 1000-MB DEWPOINTS (°F)
JANUARY
GULF OF MEXICO
ALASKA
HAWAII

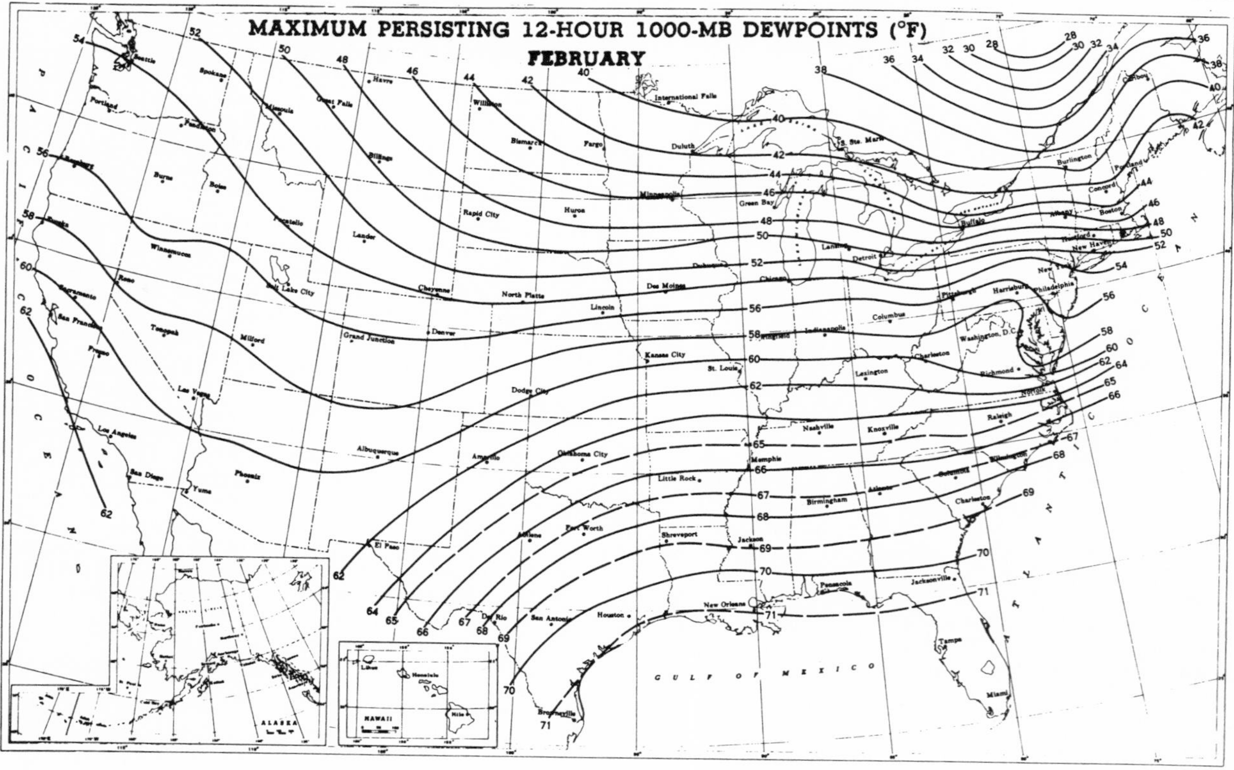
MAXIMUM PERSISTING 12-HOUR 1000-MB DEWPOINTS (°F)
FEBRUARY
GULF OF MEXICO
ALASKA
HAWAII

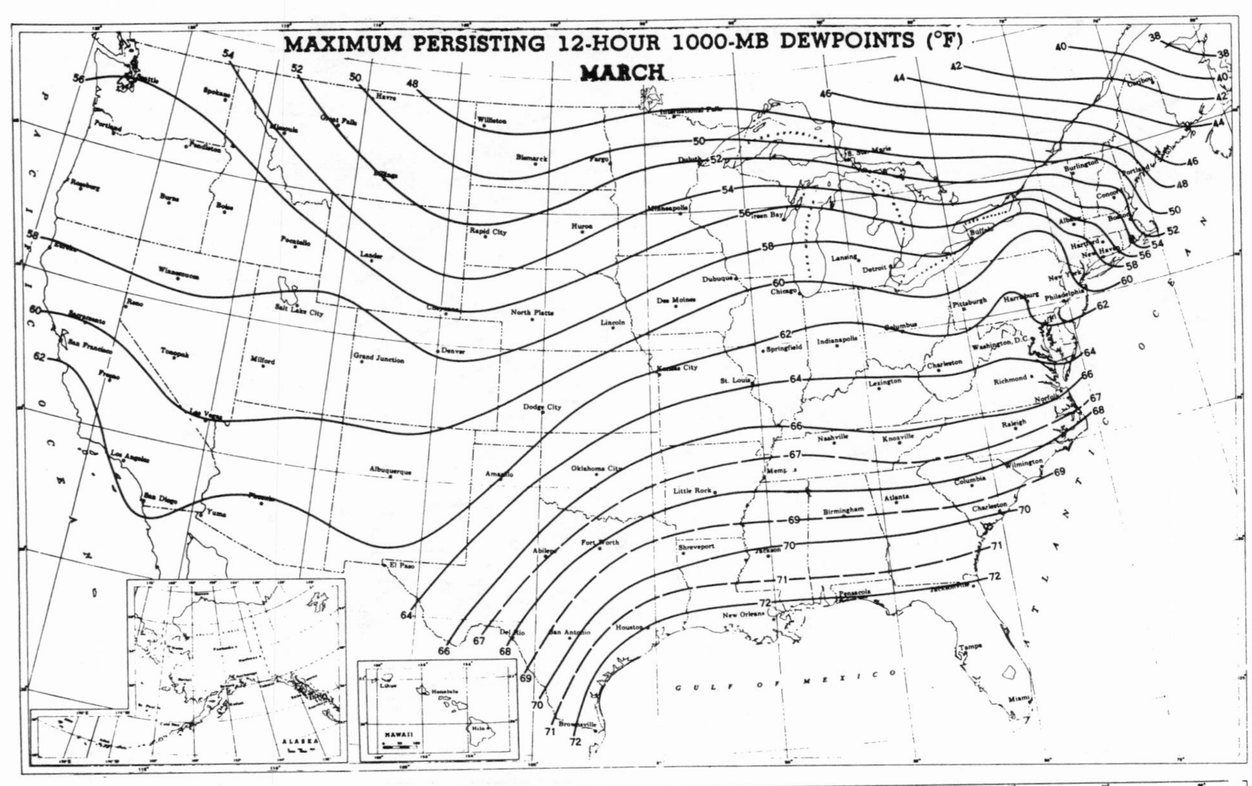
MAXIMUM PERSISTING 12-HOUR 1000-MB DEWPOINTS (°F)
MARCH
GULF OF MEXICO

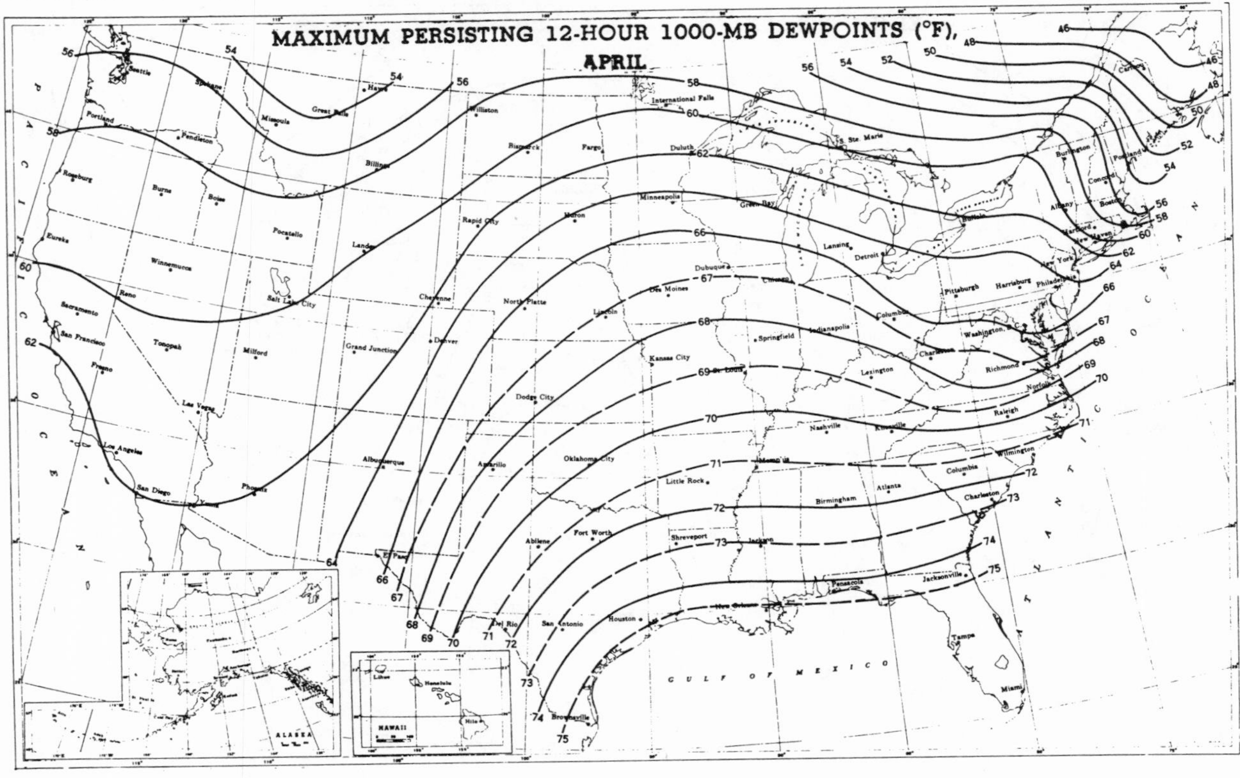
MAXIMUM PERSISTING 12-HOUR 1000-MB DEWPOINTS (°F),
APRIL
GULF OF MEXICO

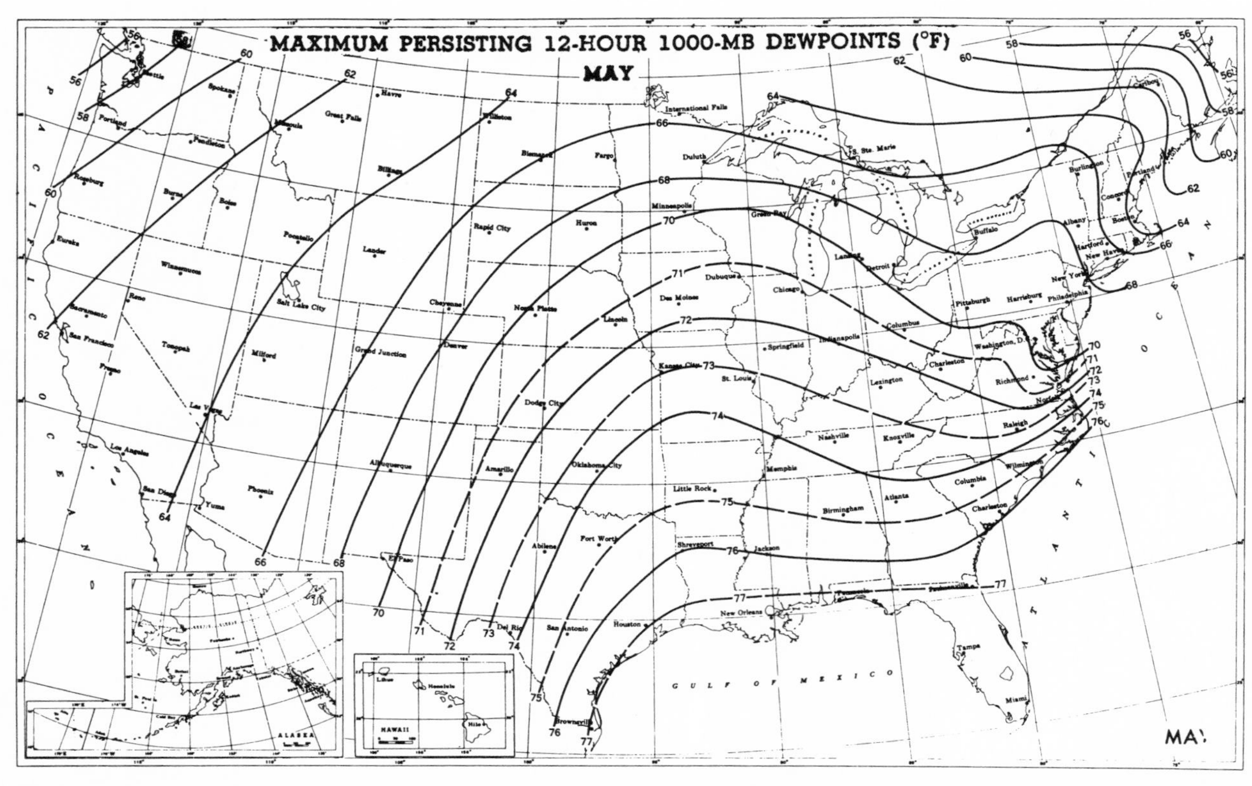
MAXIMUM PERSISTING 12-HOUR 1000-MB DEWPOINTS (°F)
MAY
GULF OF MEXICO

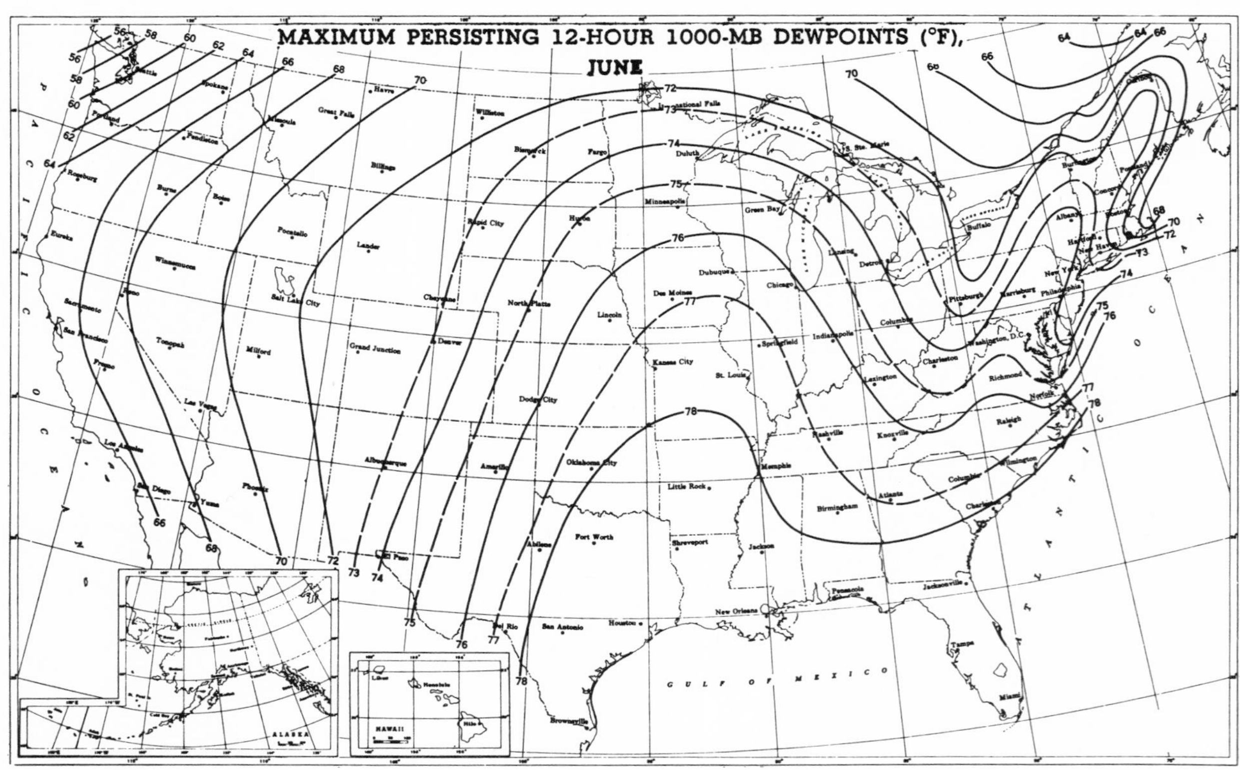
MAXIMUM PERSISTING 12-HOUR 1000-MB DEWPOINTS (°F),
JUNE
GULF OF MEXICO

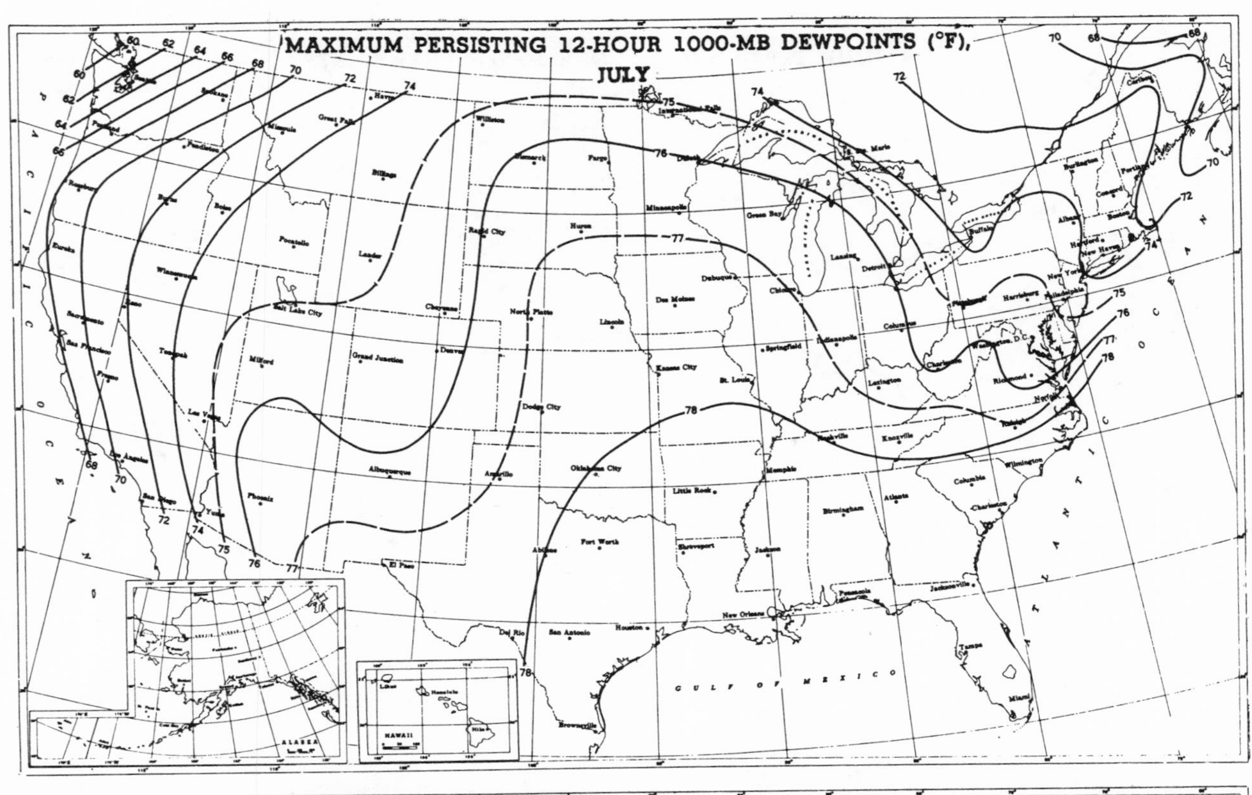
MAXIMUM PERSISTING 12-HOUR 1000-MB DEWPOINTS (°F),
JULY
GULF OF MEXICO
ALASKA
HAWAII

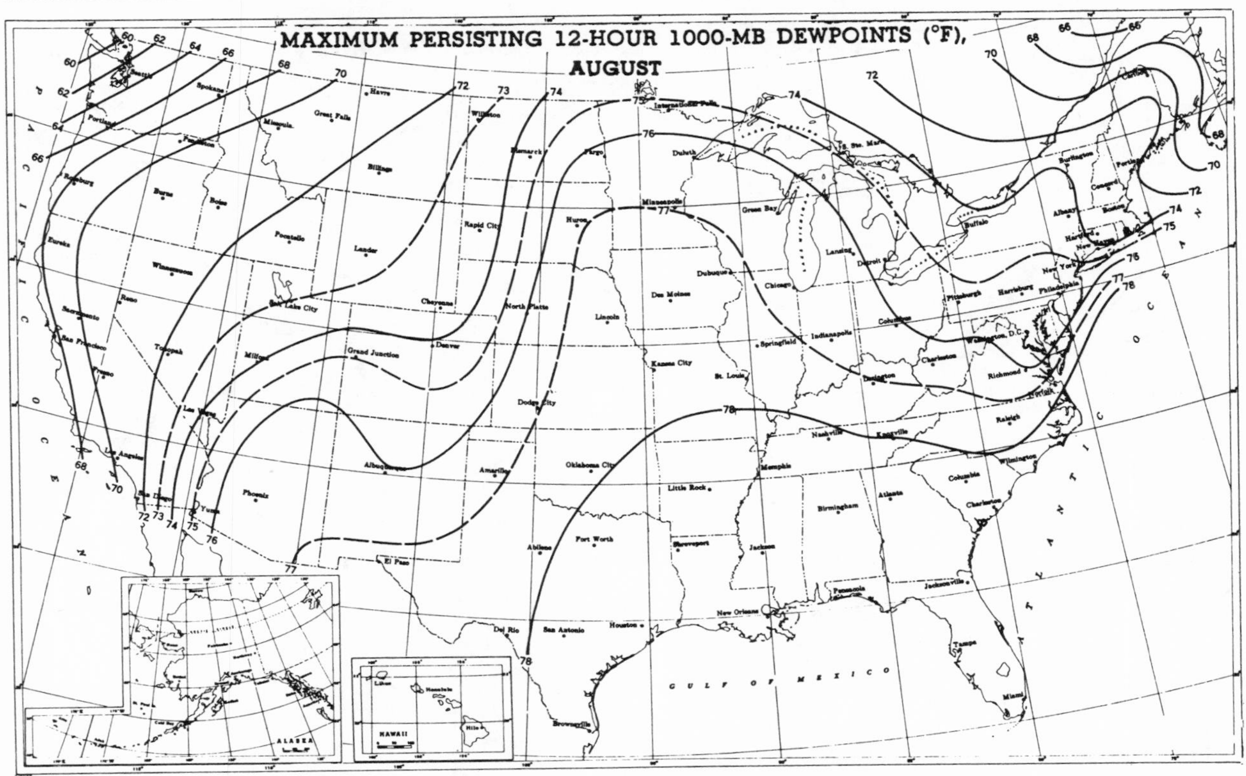
MAXIMUM PERSISTING 12-HOUR 1000-MB DEWPOINTS (°F),
AUGUST
GULF OF MEXICO
ALASKA
HAWAII

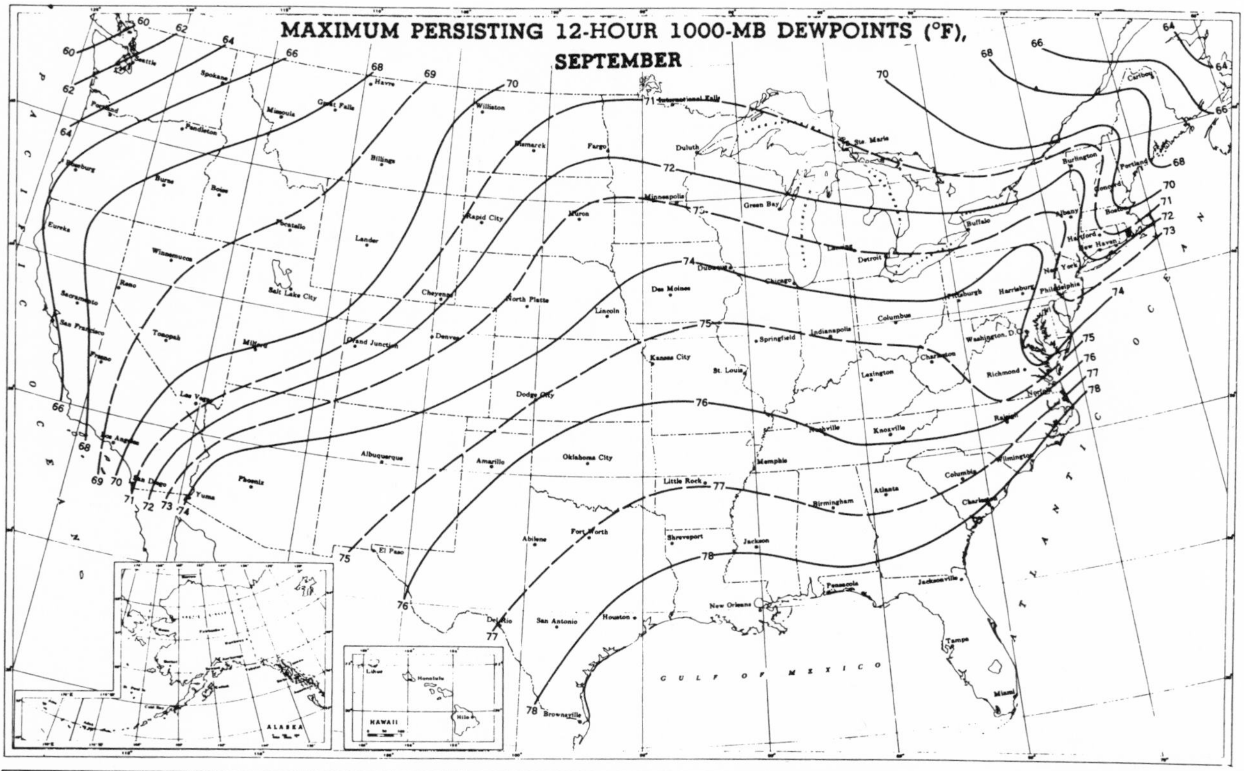

MAXIMUM PERSISTING 12-HOUR 1000-MB DEWPOINTS (°F), SEPTEMBER

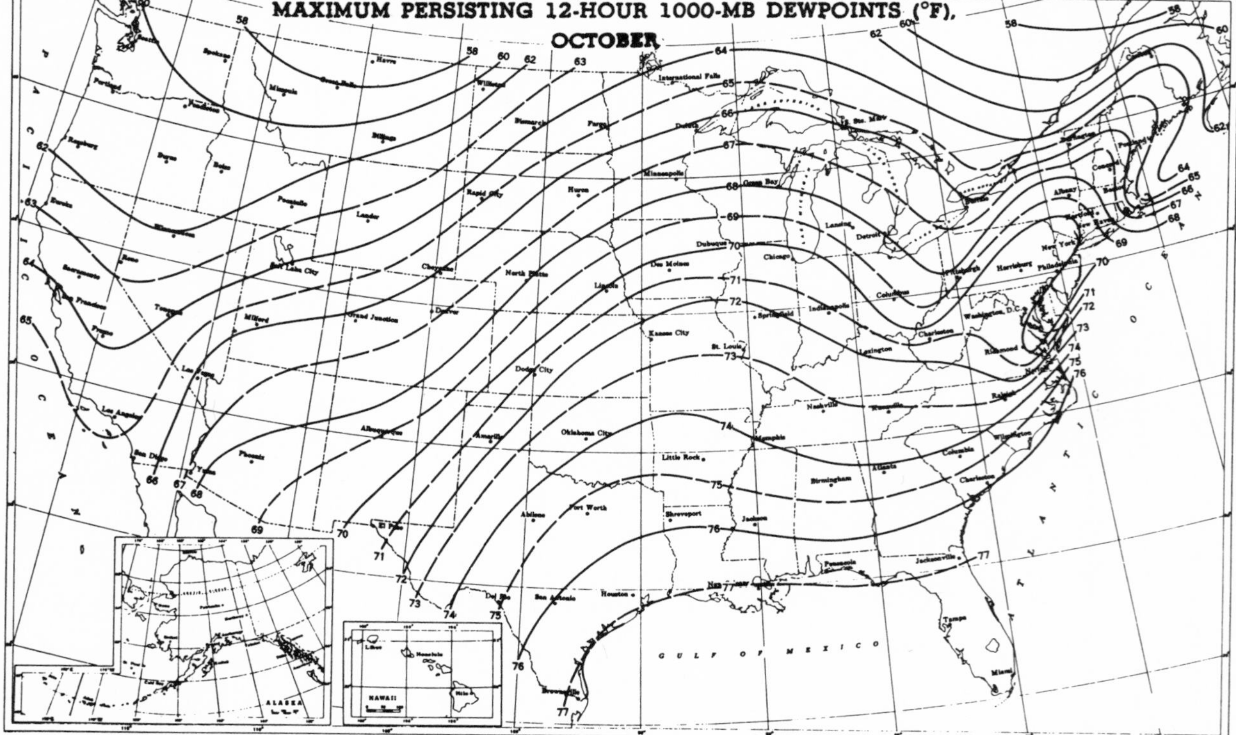

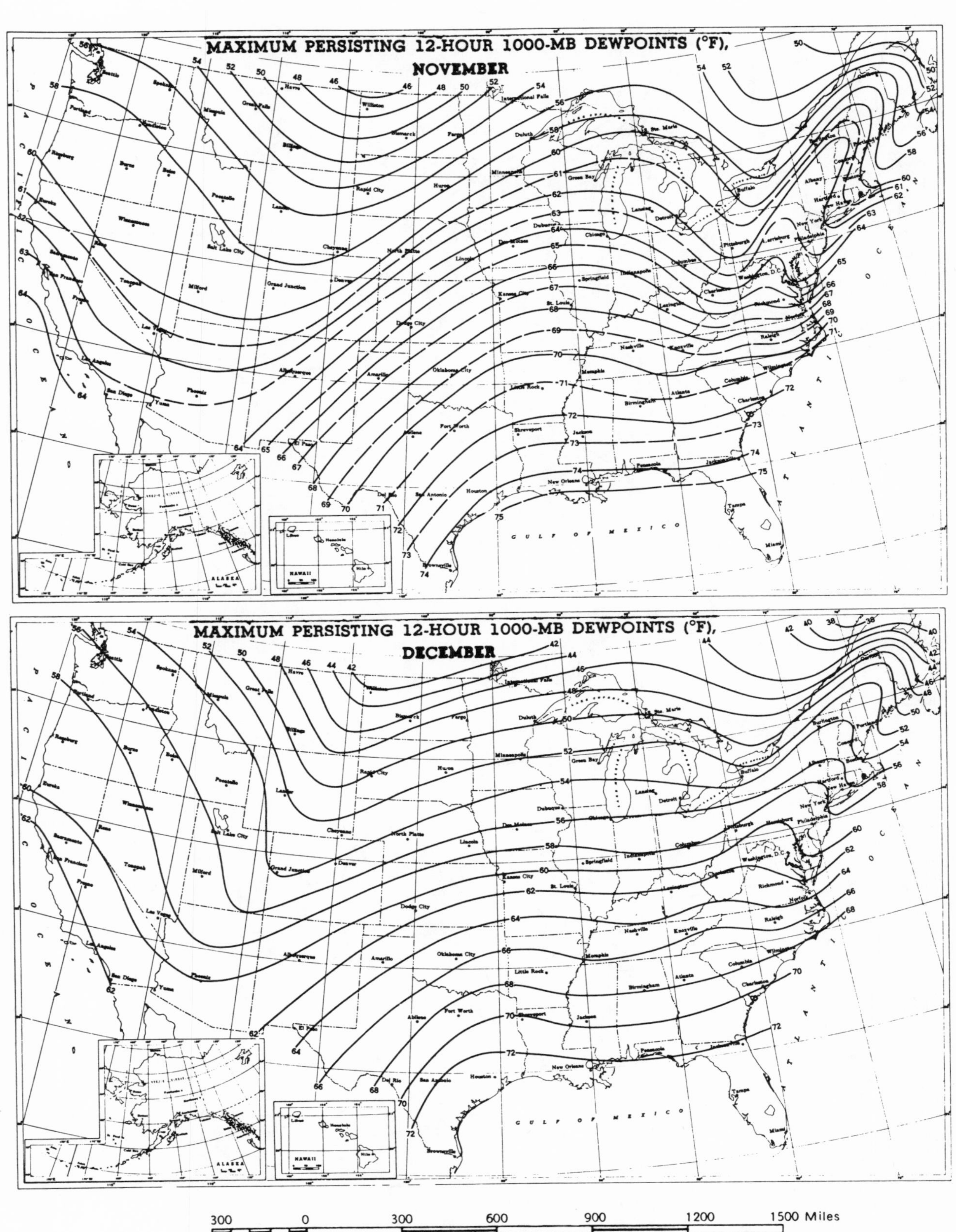

300 0 300 600 900 1200 1500 Miles

SCALE 1:30,000,000

ALBERS EQUAL AREA PROJECTION — STANDARD PARALLELS 29½° AND 45½°

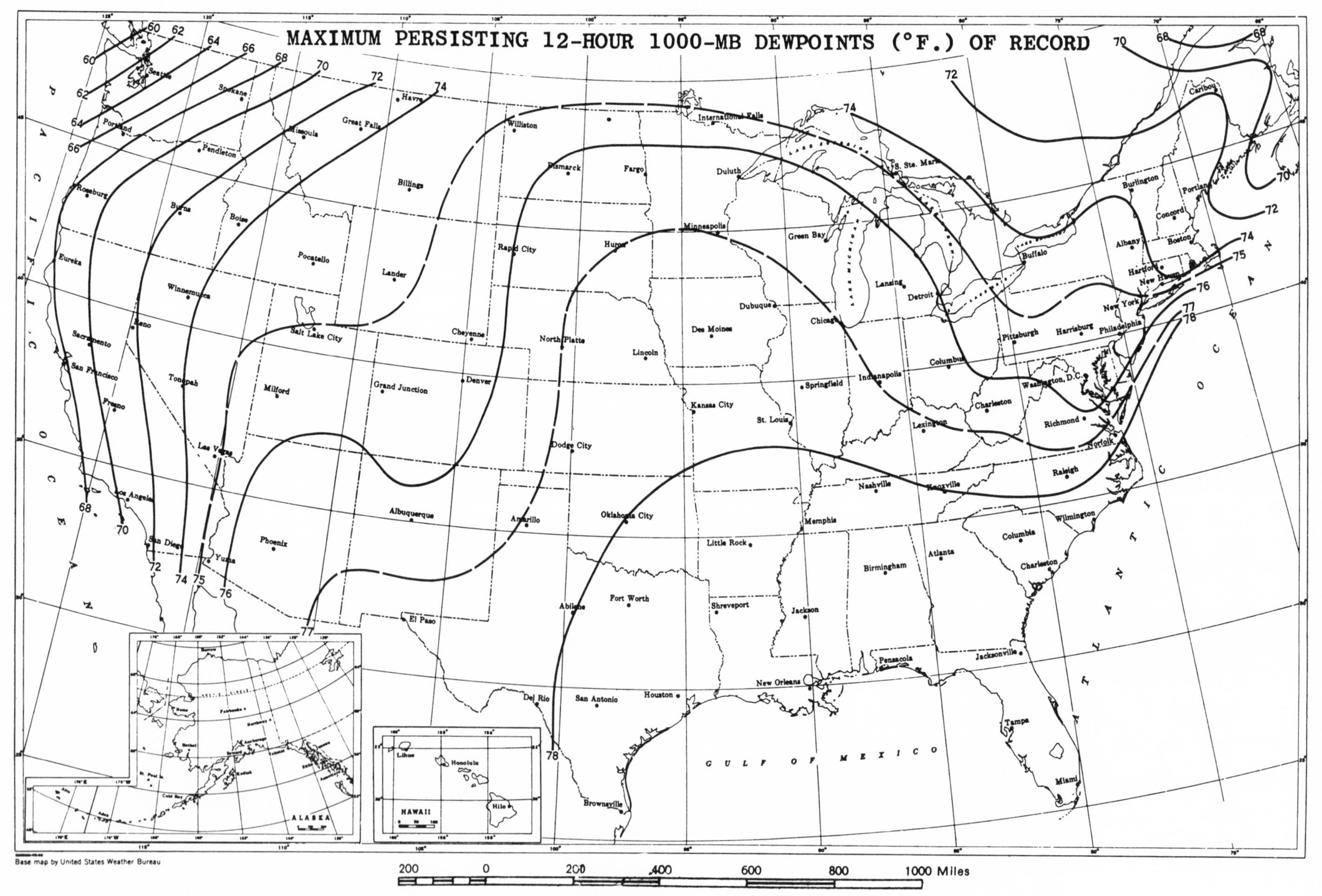
MAXIMUM PERSISTING 12-HOUR 1000-MB DEWPOINTS (°F.) OF RECORD
Base map by United States Weather Bureau
200 0 200 400 600 800 1000 Miles
SCALE 1:20,000,000
ALBERS EQUAL AREA PROJECTION-STANDARD PARALLELS 29½° AND 45½°
PACIFIC OCEAN
ATLANTIC OCEAN
GULF OF MEXICO
HAWAII
ALASKA

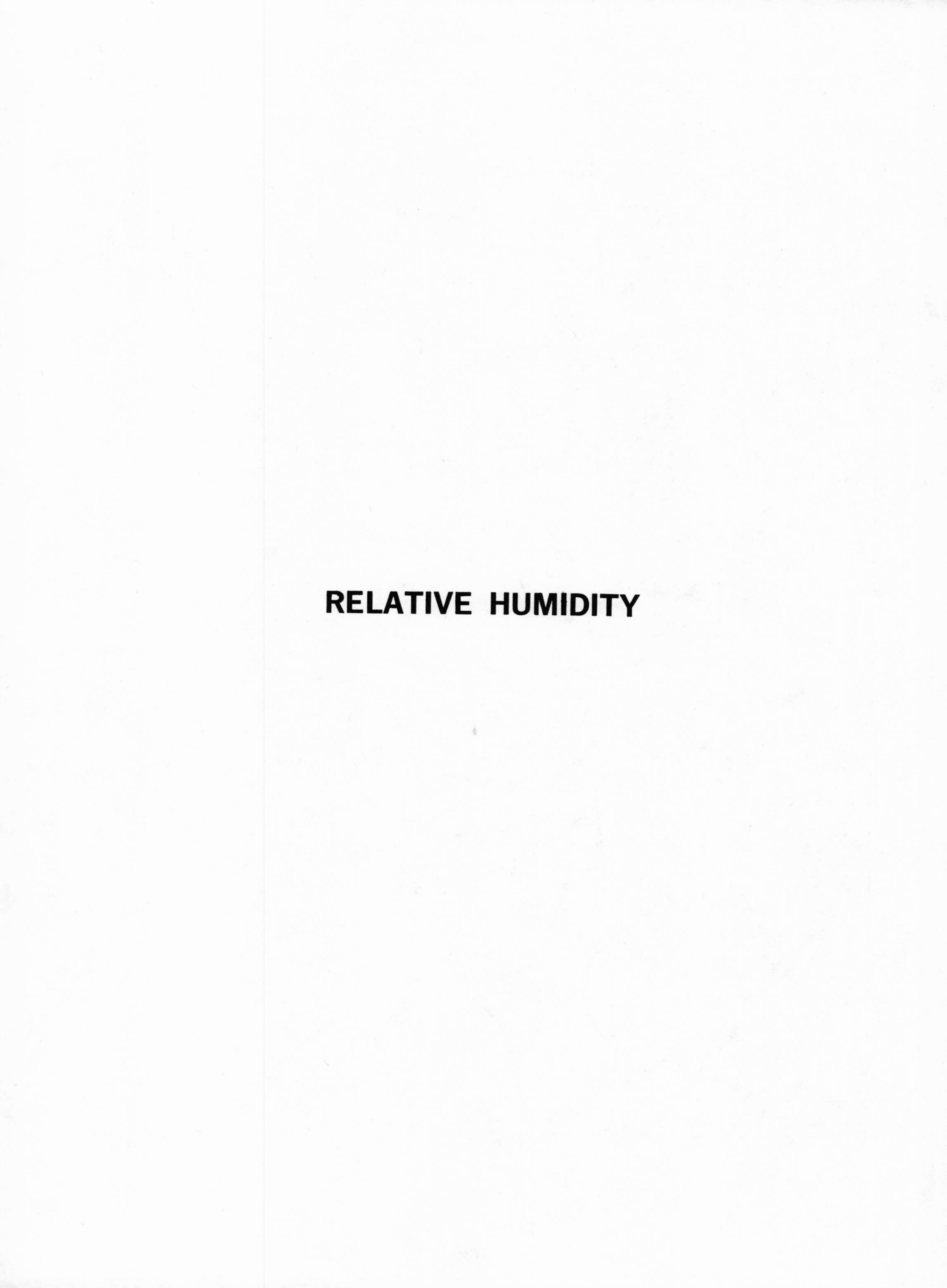

RELATIVE HUMIDITY

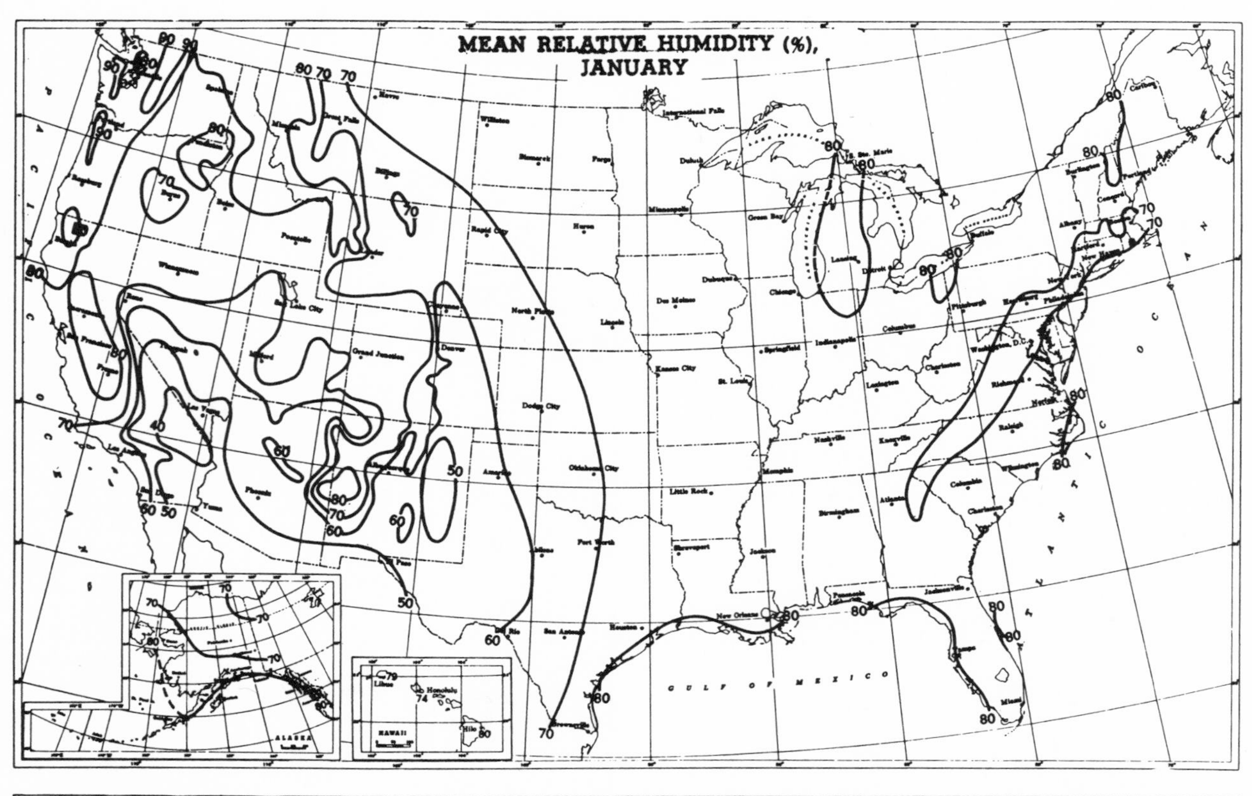
MEAN RELATIVE HUMIDITY (%),
JANUARY
GULF OF MEXICO
ALASKA
HAWAII

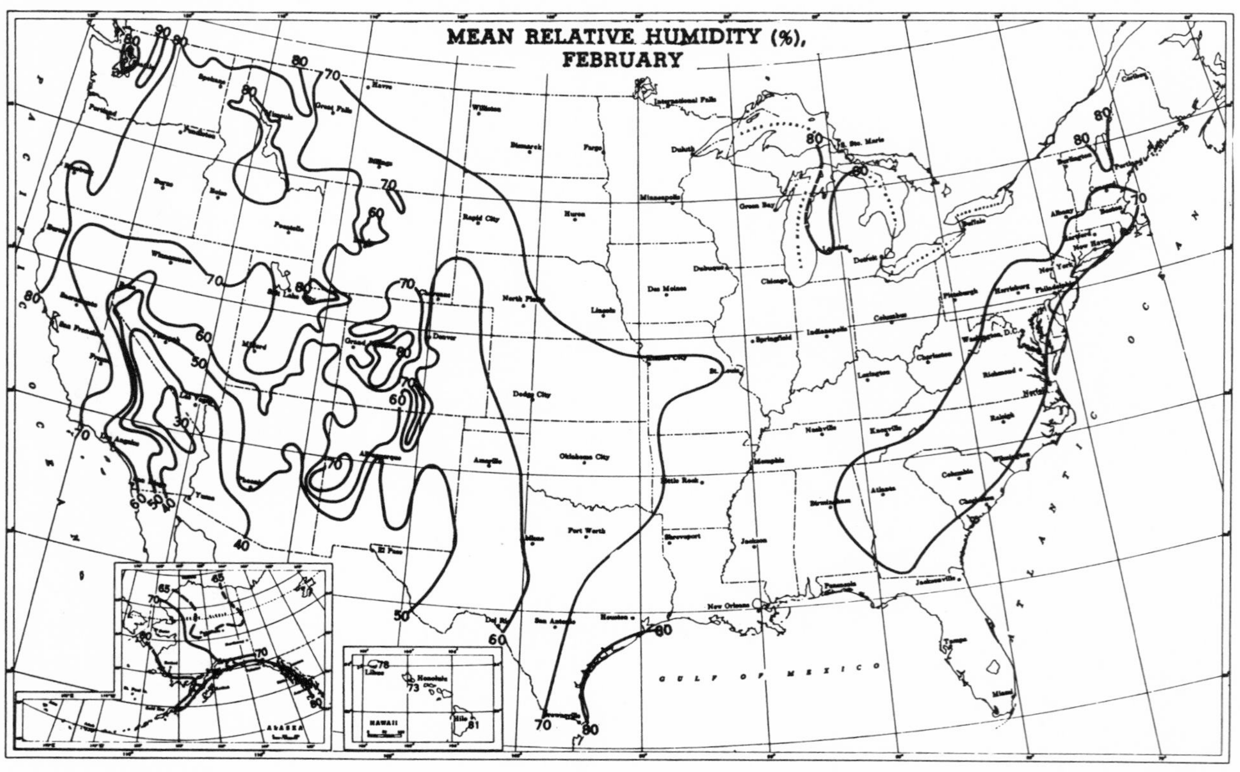
MEAN RELATIVE HUMIDITY (%),
FEBRUARY
GULF OF MEXICO
ALASKA
HAWAII

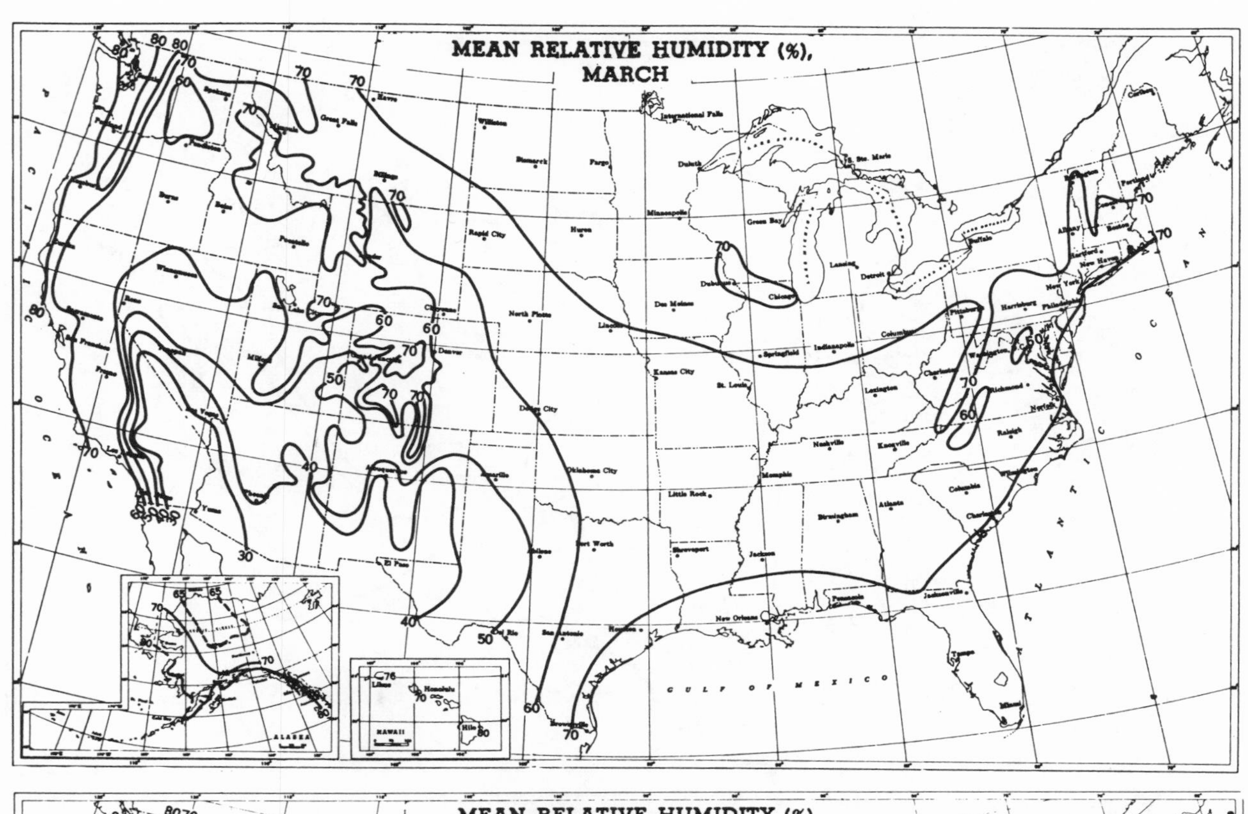
MEAN RELATIVE HUMIDITY (%),
MARCH
GULF OF MEXICO
ALASKA
HAWAII

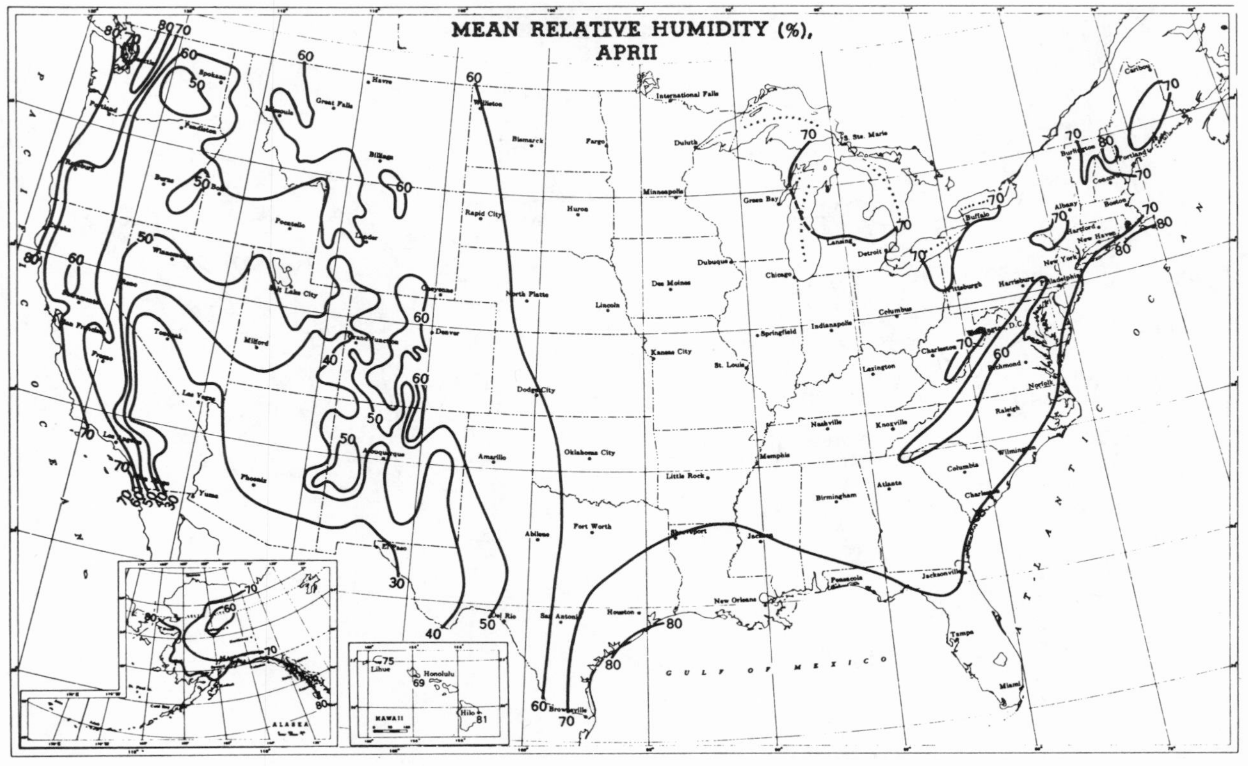
MEAN RELATIVE HUMIDITY (%),
APRII
GULF OF MEXICO
ALASKA
HAWAII

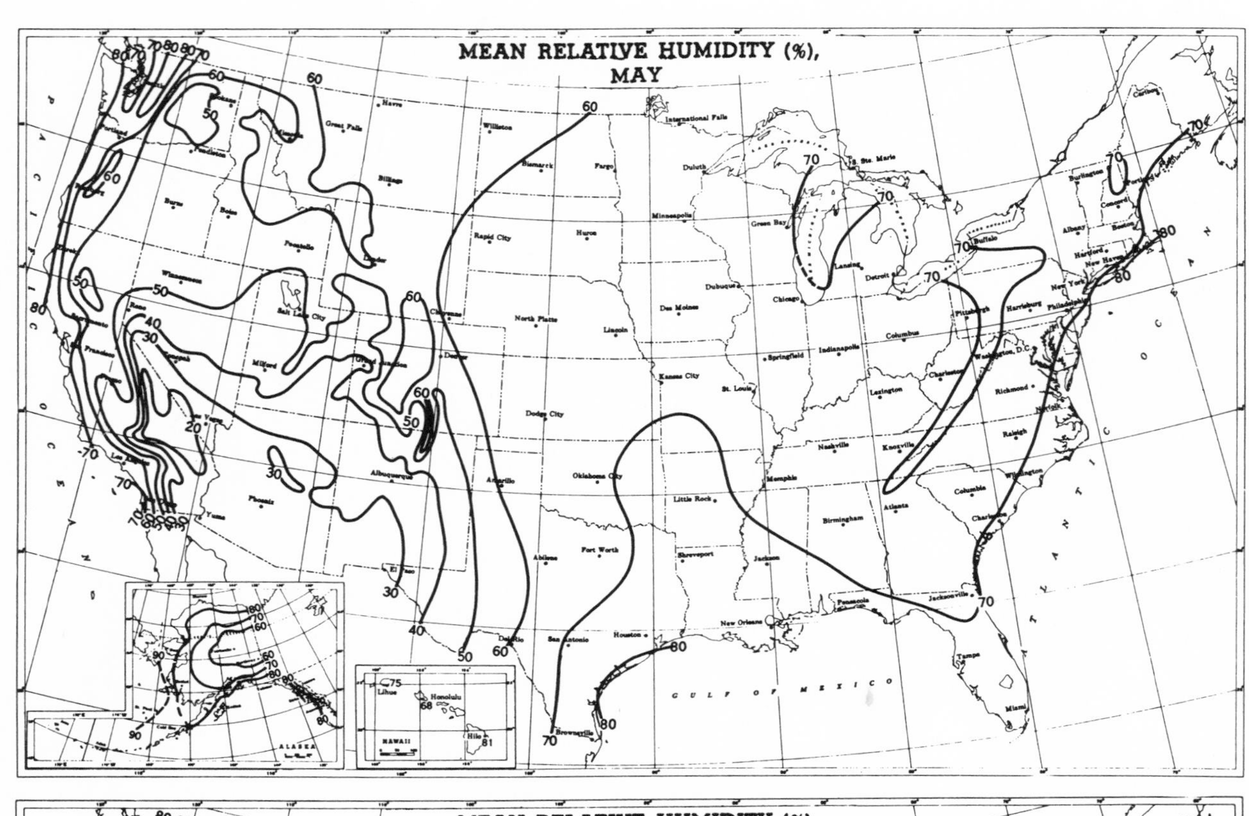
MEAN RELATIVE HUMIDITY (%),
MAY
GULF OF MEXICO
ALASKA
HAWAII

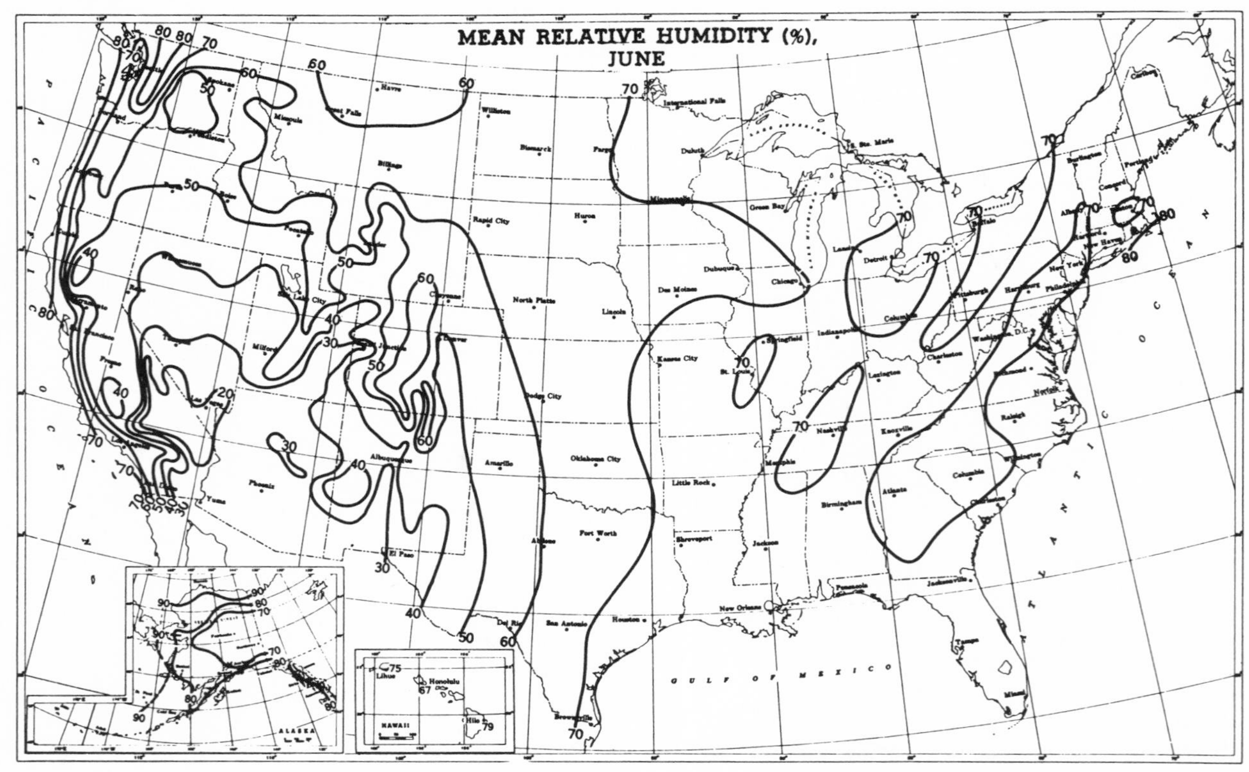
MEAN RELATIVE HUMIDITY (%),
JUNE
GULF OF MEXICO
ALASKA
HAWAII

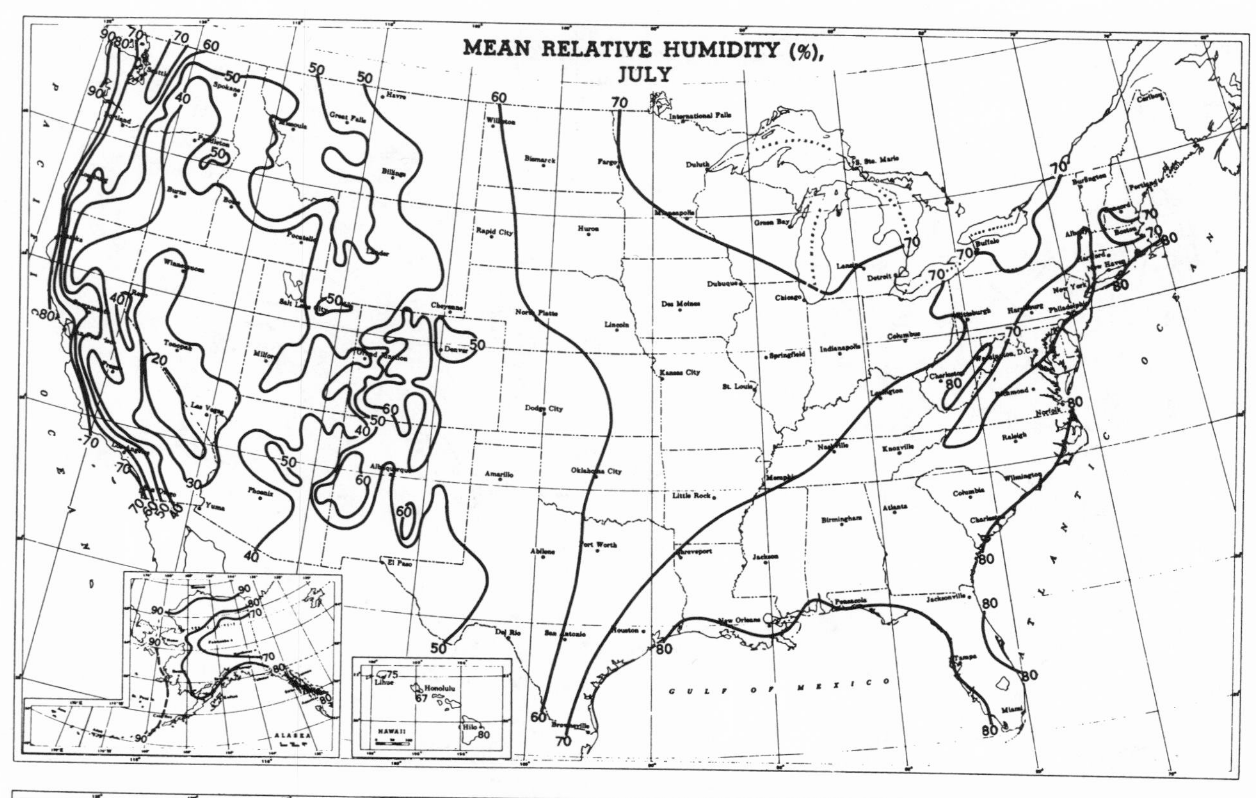

MEAN RELATIVE HUMIDITY (%),
JULY
International Falls
Bismarck
Rapid City
North Platte
Lincoln
Des Moines
Kansas City
Dodge City
Amarillo
Oklahoma City
Little Rock
Fort Worth
Abilene
El Paso
Shreveport
Del Rio
San Antonio
Houston
New Orleans
Memphis
Nashville
Knoxville
Atlanta
Birmingham
Jackson
Columbia
Charleston
Wilmington
Raleigh
Norfolk
Richmond
Jacksonville
Tampa
Miami
Phoenix
Yuma
Las Vegas
Denver
Cheyenne
Salt Lake City
Chicago
Indianapolis
Columbus
Pittsburgh
Harrisburg
Philadelphia
New York
Buffalo
Albany
Burlington
Portland
Duluth
Huron
Lansing
Detroit
Green Bay
Dubuque
Springfield
St. Louis
Lexington
GULF OF MEXICO
ALASKA
HAWAII
Lihue
Honolulu
Hilo

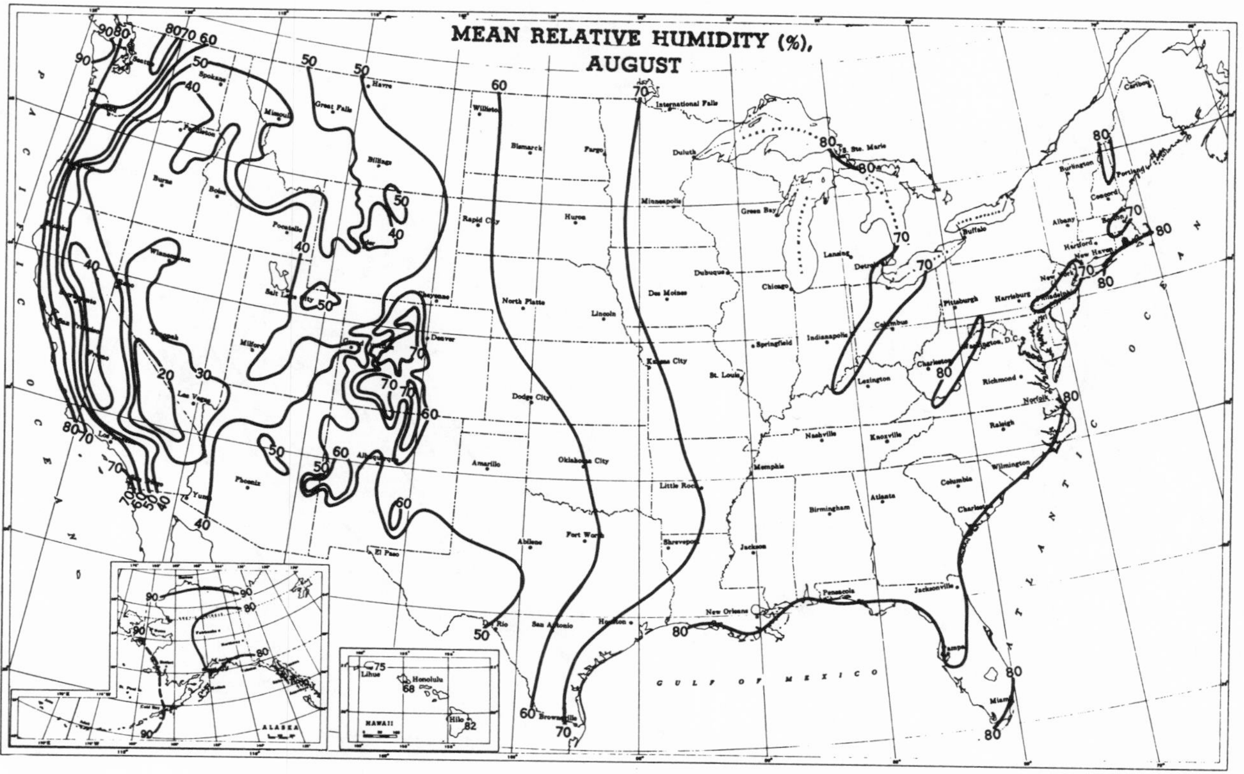

MEAN RELATIVE HUMIDITY (%),
AUGUST
International Falls
Bismarck
Rapid City
North Platte
Lincoln
Des Moines
Kansas City
Dodge City
Amarillo
Oklahoma City
Little Rock
Fort Worth
Abilene
El Paso
Shreveport
Del Rio
San Antonio
Houston
New Orleans
Memphis
Nashville
Knoxville
Atlanta
Birmingham
Jackson
Columbia
Charleston
Wilmington
Raleigh
Norfolk
Richmond
Jacksonville
Miami
Phoenix
Yuma
Las Vegas
Denver
Cheyenne
Salt Lake City
Chicago
Indianapolis
Columbus
Pittsburgh
Harrisburg
Buffalo
Albany
Burlington
Portland
Concord
Hartford
New Haven
Duluth
Huron
Lansing
Detroit
Green Bay
Dubuque
Springfield
St. Louis
Lexington
GULF OF MEXICO
ALASKA
HAWAII
Lihue
Honolulu
Hilo

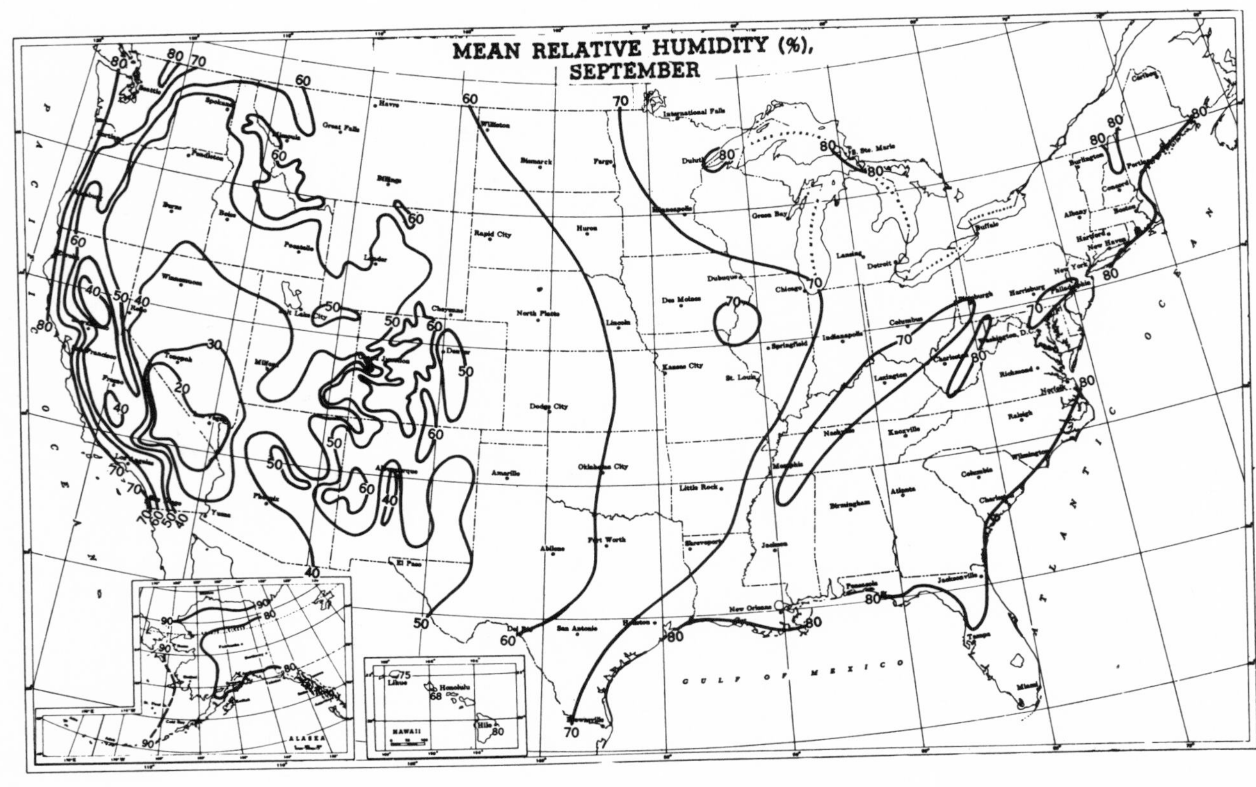
MEAN RELATIVE HUMIDITY (%),
SEPTEMBER
GULF OF MEXICO

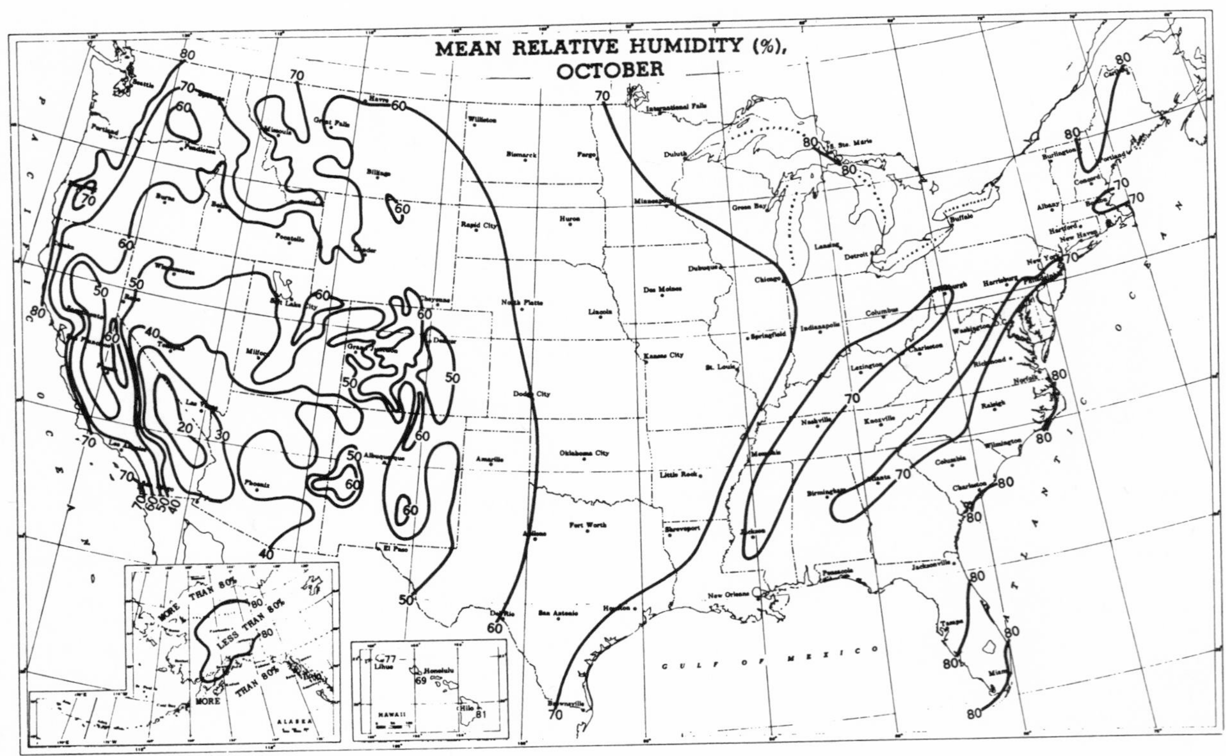
MEAN RELATIVE HUMIDITY (%),
OCTOBER
GULF OF MEXICO
MORE THAN 80%
LESS THAN 80%

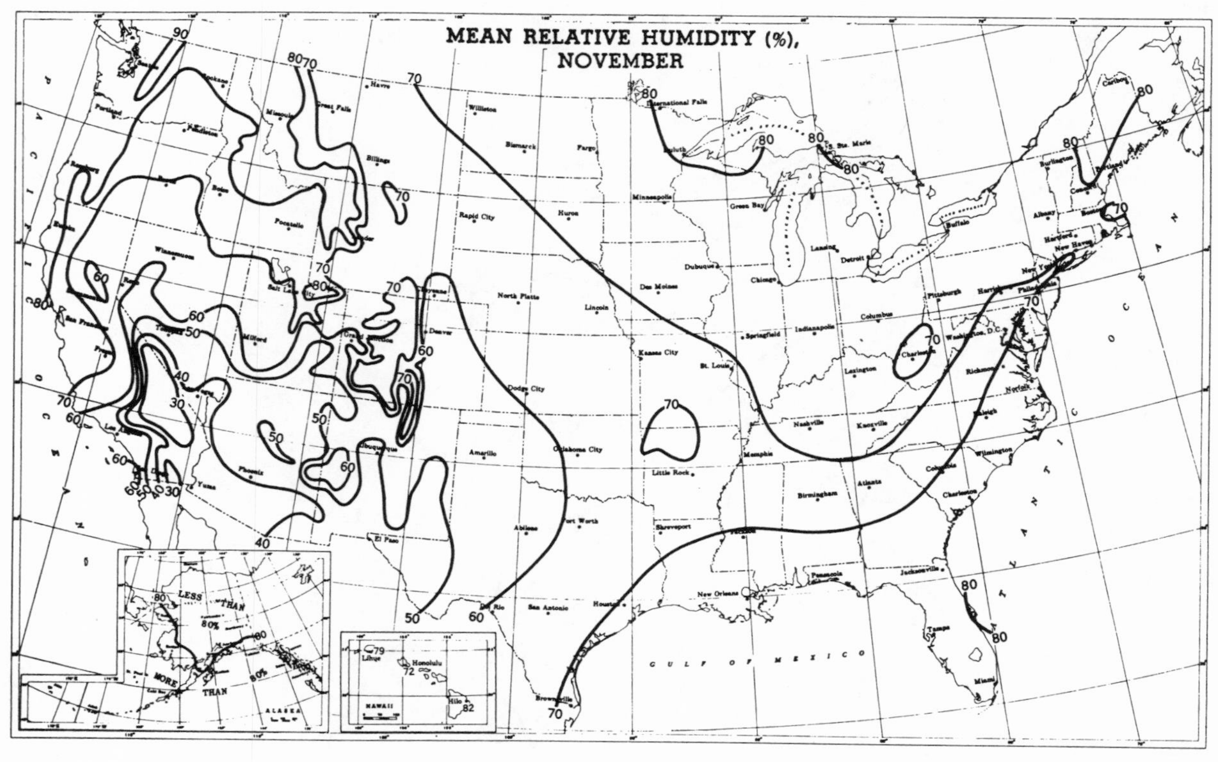

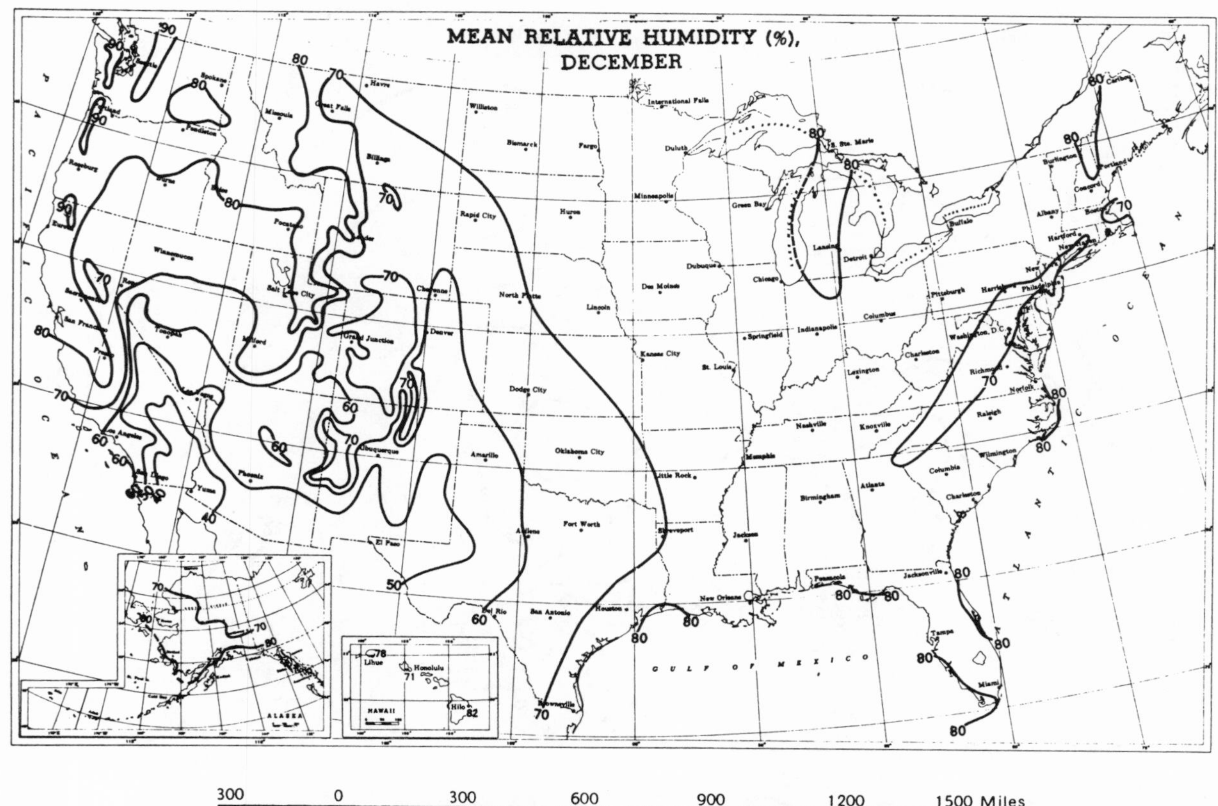

300 0 300 600 900 1200 1500 Miles

SCALE 1 : 30,000,000

ALBERS EQUAL AREA PROJECTION-STANDARD PARALLELS 29½° AND 45½°

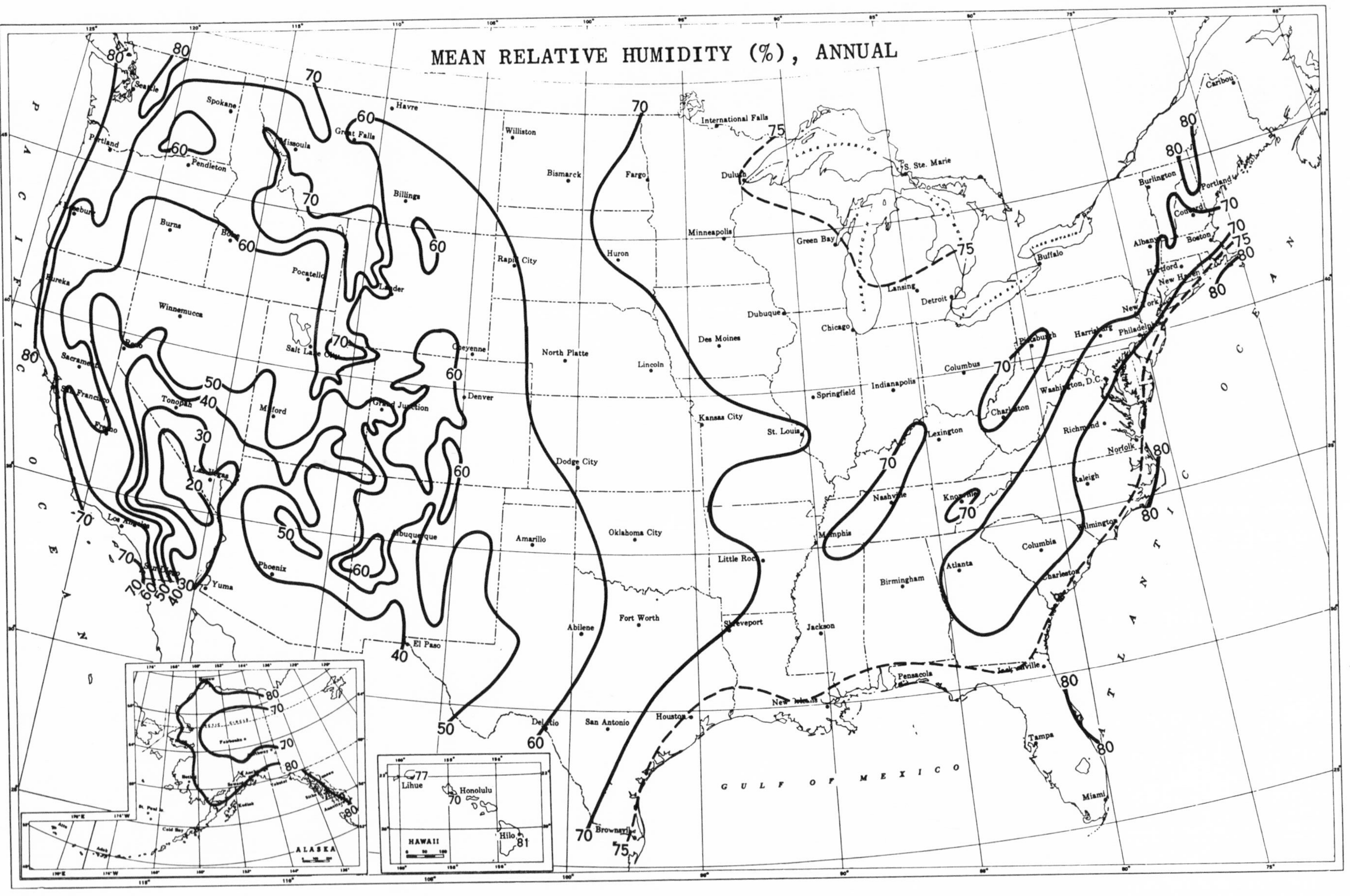

MEAN RELATIVE HUMIDITY (%), ANNUAL
P A C I F I C O C E A N
A T L A N T I C O C E A N
G U L F O F M E X I C O
HAWAII
ALASKA
Seattle
Portland
Spokane
Pendleton
Boise
Burns
Eureka
Winnemucca
Reno
Sacramento
San Francisco
Fresno
Los Angeles
San Diego
Tonopah
Las Vegas
Yuma
Phoenix
Milford
Salt Lake City
Pocatello
Missoula
Great Falls
Havre
Billings
Lander
Grand Junction
Cheyenne
Denver
Albuquerque
El Paso
Del Rio
Amarillo
Abilene
San Antonio
Brownsville
Houston
Fort Worth
Oklahoma City
Dodge City
North Platte
Rapid City
Williston
Bismarck
Fargo
Huron
Lincoln
Kansas City
Des Moines
Minneapolis
Duluth
International Falls
Dubuque
St. Louis
Springfield
Chicago
Green Bay
Lansing
Detroit
S. Ste. Marie
Indianapolis
Columbus
Lexington
Nashville
Memphis
Little Rock
Shreveport
Jackson
New Orleans
Birmingham
Pensacola
Atlanta
Knoxville
Charleston
Pittsburgh
Buffalo
Harrisburg
Washington, D.C.
Richmond
Raleigh
Norfolk
Wilmington
Columbia
Jacksonville
Tampa
Miami
Philadelphia
New York
Albany
Hartford
New Haven
Boston
Concord
Portland
Burlington
Caribou
Lihue
Honolulu
Hilo

PERCENT FREQUENCY OF SPECIFIED RELATIVE HUMIDITY VALUES AT SELECTED HOURS FOR MIDSEASONAL MONTHS

STATE and STATION	JANUARY 6 A.M. <30	<40	<50	≥50	≥70	≥90	JANUARY 3 P.M. <30	<40	<50	≥50	≥70	≥90	APRIL 5 A.M. <30	<40	<50	≥50	≥70	≥90	APRIL 3 P.M. <30	<40	<50	≥50	≥70	≥90	JULY 4 A.M. <30	<40	<50	≥50	≥70	≥90	JULY 3 P.M. <30	<40	<50	≥50	≥70	≥90	OCTOBER 5 A.M. <30	<40	<50	≥50	≥70	≥90	OCTOBER 3 P.M. <30	<40	<50	≥50	≥70	≥90
ALA. BIRMINGHAM	0	0	2	98	89	42	8	24	38	62	39	14	0	0	3	97	87	45	29	52	69	31	13	2	0	0	0	100	99	59	4	18	48	52	20	5	0	2	2	98	91	59	26	44	64	36	17	5
ARIZ. PHOENIX	2	4	9	91	71	29	42	68	82	18	8	3	3	12	42	58	19	3	81	90	95	5	2	0	3	14	37	63	20	3	66	89	95	5	2	0	1	15	50	50	9	3	89	97	98	2	1	0
CALIF. FRESNO	0	0	0	100	99	85	1	4	14	86	46	6	0	0	0	100	97	51	35	68	85	15	6	1	1	5	13	87	36	1	90	98	99	1	0	0	0	0	1	99	82	25	46	80	94	6	2	1
LOS ANGELES	1	6	11	89	69	28	7	17	29	71	31	9	1	3	5	95	90	33	1	4	13	87	33	2	0	0	0	100	99	26	0	1	3	97	22	0	1	3	6	94	87	44	7	9	12	88	46	0
SAN DIEGO	0	3	8	92	74	21	6	15	25	75	28	5	0	0	1	99	93	21	2	3	8	92	19	1	0	0	0	100	100	26	0	0	5	95	21	0	1	4	7	93	80	28	7	13	19	81	25	0
SAN FRANCISCO	0	0	1	99	96	68	1	4	9	91	58	6	0	0	1	99	97	45	1	3	7	93	46	1	0	0	0	100	100	70	1	2	10	90	20	0	1	2	3	97	95	59	6	14	24	76	33	0
COLO. DENVER	8	23	40	60	25	2	44	60	75	25	8	1	9	22	33	67	37	10	61	72	80	20	10	2	2	8	22	78	45	6	63	78	88	12	5	1	13	28	42	58	28	6	77	85	91	9	5	1
CONN. HARTFORD	0	1	2	98	65	26	1	9	32	68	32	10	0	0	3	97	82	46	17	33	54	46	26	11	0	0	0	100	99	66	4	22	50	50	15	3	0	0	1	99	93	60	6	27	46	54	19	6
D. C. WASHINGTON	0	1	9	91	52	23	6	31	50	50	18	6	0	1	11	89	58	24	25	53	67	33	17	6	0	0	1	99	88	21	5	39	67	33	8	2	0	0	1	99	84	36	13	35	57	43	16	3
FLA. JACKSONVILLE	0	0	1	99	91	57	13	21	44	56	13	3	0	1	2	98	89	51	20	39	58	42	10	3	0	0	0	100	100	53	0	6	22	78	28	8	0	0	1	99	98	65	1	11	26	74	30	5
MIAMI	0	0	0	100	87	40	4	9	32	68	10	3	0	0	0	100	87	29	1	7	32	68	16	3	0	0	0	100	99	48	0	1	3	97	30	4	0	0	0	100	98	59	0	1	11	89	36	5
TAMPA	0	0	1	99	92	67	7	9	41	59	10	1	0	0	1	99	90	53	10	31	48	52	7	2	0	0	0	100	100	77	0	2	16	84	26	4	0	1	1	99	98	80	1	7	33	67	22	6
GA. ATLANTA	0	0	0	100	82	46	9	23	33	67	33	14	0	0	3	97	72	30	31	53	72	28	13	3	0	0	1	99	94	50	5	21	47	53	20	3	0	1	2	98	85	46	11	32	56	44	19	6
HAWAII. HONOLULU	0	0	0	100	81	20	0	0	6	94	13	2	0	0	0	100	85	8	0	0	4	96	9	0	0	0	0	100	87	5	0	0	3	97	6	0	0	0	0	100	92	7	0	0	1	99	14	2
IDAHO. BOISE	0	0	3	97	83	28	1	3	12	88	48	14	0	3	8	92	50	10	45	76	87	13	5	1	1	18	44	56	9	0	88	99	100	0	0	0	2	10	22	78	34	8	41	73	85	15	4	1
ILL. CHICAGO	1	2	3	97	80	17	1	6	13	87	49	12	0	1	6	94	71	15	10	27	43	57	23	6	0	0	1	99	91	23	3	27	57	43	9	1	0	1	4	96	77	17	13	30	60	40	14	3
IOWA. DES MOINES	0	0	2	98	82	21	0	5	12	88	47	6	0	1	2	98	77	29	18	37	54	46	25	7	0	0	0	100	92	38	3	21	50	50	12	3	0	1	3	97	80	17	21	49	65	35	12	6
KANS. TOPEKA	1	1	4	96	71	21	4	18	43	57	21	4	0	1	3	97	74	25	26	42	55	45	19	7	0	0	2	98	87	45	15	34	55	45	16	5	1	1	4	96	77	32	30	52	71	29	11	3
LA. NEW ORLEANS	0	1	2	98	93	61	3	15	23	77	37	7	0	1	2	98	83	53	8	26	43	57	19	7	0	0	0	100	98	77	1	2	17	83	41	6	0	0	3	97	90	52	11	33	47	53	13	4
MASS. BOSTON	1	5	15	85	55	16	3	18	37	63	34	12	0	2	9	91	58	18	19	37	52	48	29	6	0	0	0	100	95	27	5	33	48	52	21	3	0	0	1	99	73	26	9	23	38	62	28	10
MICH. DETROIT	0	0	1	99	84	23	0	0	5	95	62	19	0	0	1	99	77	25	12	23	43	57	20	5	0	0	0	100	89	16	5	31	66	34	10	3	0	1	3	97	83	34	9	26	55	45	13	4
MINN. DULUTH	0	2	6	94	60	14	0	1	11	89	51	10	0	0	2	98	79	27	9	24	41	59	34	14	0	0	0	100	100	63	0	10	41	59	30	7	0	0	0	100	88	52	3	17	33	67	28	12
MINNEAPOLIS	0	1	1	99	81	14	0	1	10	90	45	5	0	0	3	97	71	24	20	37	50	50	21	9	0	0	0	100	94	35	0	28	61	39	11	1	0	1	2	98	75	32	12	37	60	40	13	4
MISS. JACKSON	0	0	1	99	92	58	8	21	34	66	35	12	0	0	0	100	90	50	17	40	62	38	17	3	0	0	0	100	98	68	1	14	49	51	19	5	1	1	2	98	93	48	31	53	68	32	14	3
MO. ST. LOUIS	0	1	2	98	76	23	2	12	30	70	35	14	0	0	1	99	72	20	16	42	62	38	17	5	0	0	1	99	75	22	14	40	64	36	9	2	0	3	8	92	76	22	20	52	69	31	13	6
MONT. BILLINGS	1	4	20	80	44	7	8	18	37	63	36	3	1	4	15	85	48	15	38	59	68	32	9	4	0	5	28	72	25	8	76	87	92	8	5	0	5	14	24	76	40	15	39	59	72	28	13	8
MISSOULA	0	0	1	99	87	37	0	1	6	94	54	11	0	1	6	94	70	21	49	69	80	20	8	3	0	0	2	98	61	15	66	82	90	10	3	0	0	0	1	99	84	39	18	46	65	35	11	1
N. J. NEWARK	0	1	4	96	61	28	1	9	30	70	34	10	0	1	5	95	64	15	19	36	53	47	20	7	0	0	0	100	84	26	5	43	65	35	13	2	0	0	1	99	82	37	7	30	55	45	19	2
N. MEX. ALBUQUERQUE	1	7	22	78	32	3	43	69	84	16	6	1	16	40	56	44	23	7	84	92	95	5	2	0	6	17	33	67	26	4	73	88	94	6	1	0	11	32	52	48	19	4	85	93	97	3	0	0
N. Y. ALBANY	0	0	1	99	75	21	0	6	17	83	33	8	0	0	1	99	80	29	17	41	58	42	18	7	0	0	0	100	95	58	3	32	62	38	11	5	0	0	0	100	90	54	5	20	48	52	19	5
BUFFALO	0	0	0	100	83	26	0	0	6	94	58	18	0	1	3	97	80	27	7	21	34	66	35	11	0	0	0	100	96	35	3	25	53	47	13	3	0	1	4	96	81	22	3	20	46	54	21	2
NEW YORK	0	0	8	92	44	19	1	15	38	62	29	8	0	1	8	92	62	18	15	31	45	55	20	8	0	0	1	99	75	15	5	37	58	42	14	1	0	0	1	99	67	21	6	25	49	51	21	1
N. C. RALEIGH	0	0	3	97	79	37	14	30	47	53	22	14	0	0	0	100	82	46	29	52	79	21	12	6	0	0	0	100	100	75	2	22	45	55	19	5	0	0	0	100	94	65	10	29	53	47	20	6
N. DAK. BISMARCK	0	0	0	100	83	6	0	0	2	98	55	1	0	0	0	100	89	32	20	34	49	51	21	6	0	0	0	100	90	42	9	37	67	33	6	3	0	0	4	96	82	34	20	46	62	38	15	3
OHIO. CINCINNATI	0	0	3	97	83	29	1	4	12	88	50	18	0	1	4	96	74	17	14	33	51	49	23	7	0	0	0	100	90	36	11	39	65	35	11	0	0	1	5	95	80	29	24	42	61	39	14	6
CLEVELAND	0	0	0	100	86	30	1	2	5	95	61	20	0	0	1	99	81	32	9	23	38	62	30	15	0	0	0	100	89	34	7	37	64	36	11	5	0	1	2	98	78	32	5	28	53	47	16	3
OKLA. OKLAHOMA CITY	0	0	8	92	71	28	15	38	52	48	27	13	0	1	9	91	64	25	30	53	64	36	18	4	0	0	0	100	96	44	1	15	46	54	19	5	1	6	11	89	69	26	27	49	68	32	11	6
OREG. MEDFORD	0	0	1	99	96	81	1	2	6	94	66	19	0	0	1	99	95	33	29	56	75	25	9	1	0	0	1	99	77	3	72	95	98	2	2	0	0	0	0	100	98	65	22	36	57	43	12	3
PA. PHILADELPHIA	0	3	8	92	64	30	3	13	32	68	28	9	1	1	4	96	70	22	18	43	63	37	18	6	0	0	0	100	91	37	9	41	64	36	11	2	0	0	1	99	90	36	9	31	57	43	13	3
S. C. CHARLESTON	0	0	1	99	89	56	9	21	46	54	21	6	0	1	2	98	86	57	20	37	60	40	16	3	0	0	0	100	99	91	0	1	9	91	33	8	0	0	1	99	95	80	2	13	28	72	24	6
COLUMBIA	0	0	1	99	88	50	14	35	56	44	19	7	0	1	1	99	87	49	34	58	72	28	13	6	0	0	0	100	99	61	4	23	50	50	12	1	0	0	1	99	96	65	14	38	59	41	12	6
TENN. KNOXVILLE	0	1	2	98	86	37	3	10	27	73	38	16	0	0	3	97	74	23	13	49	67	33	15	3	0	0	0	100	97	59	3	20	49	51	18	3	0	0	1	99	95	57	19	42	59	41	9	5
MEMPHIS	0	0	2	98	82	34	2	12	28	72	47	20	0	0	1	99	83	24	14	38	62	38	18	3	0	0	0	100	98	46	6	22	48	52	11	2	0	1	4	96	88	41	22	45	66	34	15	8
TEX. AMARILLO	1	6	21	79	39	6	51	68	76	24	9	2	9	20	33	67	41	15	61	76	85	15	5	1	0	1	11	89	55	22	46	66	79	21	3	0	1	6	18	82	47	15	58	72	83	17	8	5
BROWNSVILLE	0	3	7	93	88	65	9	15	23	77	29	4	0	0	1	99	95	48	4	12	22	78	13	1	0	0	0	100	100	52	0	1	20	80	3	1	0	0	0	100	94	43	5	11	34	66	15	3
DALLAS	0	1	5	95	73	30	18	39	48	52	25	10	0	1	4	96	76	25	27	40	55	45	19	3	0	0	0	100	85	14	1	34	66	34	10	1	1	1	6	94	75	29	26	53	72	28	11	5
EL PASO	7	20	41	59	30	8	58	71	81	19	9	0	32	56	70	30	11	2	90	93	96	4	1	1	1	9	24	76	33	3	60	83	91	9	1	0	8	34	59	41	17	3	79	90	95	5	2	1
GALVESTON	0	0	1	99	93	64	1	7	15	85	71	24	0	0	1	99	86	56	2	4	13	87	60	15	0	0	0	100	98	19	0	0	1	99	53	2	0	0	1	99	85	38	0	10	24	76	43	4
HOUSTON	0	0	3	97	87	65	9	21	28	72	39	15	0	0	1	99	88	63	9	22	35	65	24	7	0	0	0	100	100	78	0	2	19	81	21	3	0	1	2	98	90	54	17	31	47	53	18	5
SAN ANTONIO	0	3	9	91	71	46	28	40	53	47	29	12	1	3	8	92	81	41	28	43	60	40	18	6	0	0	0	100	95	32	7	64	89	11	4	2	0	3	6	94	78	40	34	60	80	20	8	2
UTAH. SALT LAKE CITY	0	1	4	96	79	26	1	6	21	79	35	6	3	6	17	83	46	11	45	64	79	21	9	3	2	8	29	71	24	3	77	89	93	7	3	0	4	9	17	83	42	6	51	79	89	11	5	1
VA. NORFOLK	0	0	4	96	73	35	6	18	39	61	28	12	0	1	2	98	76	36	11	33	52	48	17	4	0	0	0	100	97	52	0	7	33	67	25	7	0	0	1	99	86	49	4	15	30	70	34	7
WASH. SEATTLE	0	1	1	99	94	59	0	1	3	97	87	34	0	1	2	98	95	63	12	27	45	55	21	3	0	0	0	100	99	68	3	20	50	50	11	1	0	0	0	100	99	74	1	5	11	89	45	15
YAKIMA	0	1	2	98	91	44	0	4	11	89	58	16	0	1	6	94	71	6	66	88	92	8	2	0	0	0	3	97	65	3	77	94	97	3	1	0	0	0	0	100	92	46	22	50	74	26	7	1
WIS. LA CROSSE	0	0	2	98	79	14	0	3	10	90	44	7	0	2	5	95	77	26	19	34	48	52	29	12	0	0	0	100	98	67	0	16	51	49	23	5	0	1	4	96	78	34	13	38	65	35	12	2
P. R. SAN JUAN	0	0	0	100	85	28	0	0	2	98	35	0	0	0	0	100	95	27	0	0	5	95	42	3	0	0	0	100	100	41	0	0	0	100	62	3	0	0	0	100	99	68	0	0	0	100	59	4

THE HOURS SELECTED FOR THIS STUDY ARE THOSE CONFORMING MOST NEARLY TO THE AVERAGE NATIONAL DAILY MAXIMUM (EARLY MORNING) AND MINIMUM (3 P.M.) RELATIVE HUMIDITY.

IN THE COLUMN HEADINGS, THE SIGN < IS READ "LESS THAN," ≥ IS READ "EQUAL TO OR GREATER THAN." EXAMPLE: <50 \| ≥50 over 2 \| 98 IS READ: RELATIVE HUMIDITY OF LESS THAN 50% OCCURS 2% OF THE TIME, AND 50% OR HIGHER OCCURS 98% OF THE TIME. TIME IS LOCAL STANDARD.

THIS TABULATION IS BASED ON CLIMATOGRAPHY OF THE UNITED STATES NO. 30 SUMMARY OF HOURLY OBSERVATIONS; 5-YEAR PERIOD, MOSTLY 1949-54; TABLE E.

MEAN RELATIVE HUMIDITY (%)

STATE AND STATION	YEAR	JANUARY 1 A.M.	JANUARY 7 A.M.	JANUARY 1 P.M.	JANUARY 7 P.M.	APRIL 1 A.M.	APRIL 7 A.M.	APRIL 1 P.M.	APRIL 7 P.M.	JULY 1 A.M.	JULY 7 A.M.	JULY 1 P.M.	JULY 7 P.M.	OCTOBER 1 A.M.	OCTOBER 7 A.M.	OCTOBER 1 P.M.	OCTOBER 7 P.M.	ANNUAL 1 A.M.	ANNUAL 7 A.M.	ANNUAL 1 P.M.	ANNUAL 7 P.M.
ALA.BIRMINGHAM	54	80	81	61	67	77	77	50	54	86	84	56	68	84	83	49	64	81	81	54	63
MONTGOMERY	68	84	83	61	67	81	80	52	56	87	85	58	68	85	84	49	60	83	83	55	73
ALASKA.ANCHORAGE	20	74	74	72	74	75	67	53	67	84	74	62	72	81	81	68	79	79	74	63	73
FAIRBANKS	30	68	69	68	70	74	63	47	61	88	72	52	66	83	83	69	81	78	71	58	69
JUNEAU	16	78	79	76	79	85	72	64	80	88	78	68	79	88	85	79	87	85	79	72	82
NOME	25	77	80	76	79	82	82	77	81	91	87	83	85	83	83	76	81	83	83	78	81
ARIZ.PHOENIX	21	67	74	47	39	42	53	28	21	42	52	31	23	52	60	32	27	50	59	33	27
YUMA	68	51	57	32	32	38	55	21	19	41	60	28	25	45	58	27	28	43	58	27	26
ARK.LITTLE ROCK	18	76	80	66	68	73	80	55	56	81	86	56	59	79	85	51	60	77	83	57	61
CALIF.FRESNO	65	88	90	73	67	67	80	44	35	44	52	28	16	65	74	42	35	66	74	47	39
LOS ANGELES	20	71	65	46	52	74	78	50	55	79	84	49	53	74	76	47	57	73	75	47	53
RED BLUFF	15	77	81	71	59	58	68	44	34	34	49	29	16	52	62	42	31	56	66	47	36
SACRAMENTO	20	87	90	82	70	75	85	57	45	60	77	47	29	67	78	54	39	73	83	60	46
SAN DIEGO	19	77	76	58	60	78	81	62	61	83	86	68	64	80	82	62	65	79	80	61	62
SAN FRANCISCO	20	83	87	77	69	83	87	69	68	87	90	70	67	82	87	69	65	83	87	71	67
COLO.DENVER	20	60	60	44	49	60	67	40	39	57	67	32	34	56	62	34	35	59	64	38	40
GRAND JUNCTION	18	74	78	62	60	48	57	35	30	37	48	28	23	54	57	37	35	53	61	41	37
CONN.HARTFORD	43	75	73	61	67	79	71	52	61	88	78	55	70	86	82	54	70	82	76	56	68
D.C.WASHINGTON	65	70	73	56	64	69	68	45	55	84	79	53	68	82	80	50	70	76	75	52	65
FLA.APALACHICOLA	25	85	87	69	80	85	85	66	74	86	85	72	76	84	86	64	75	85	86	68	76
JACKSONVILLE	23	85	88	56	74	83	84	47	66	88	86	57	75	88	90	58	78	86	87	55	73
KEY WEST	19	82	84	69	79	79	79	66	75	79	77	68	74	83	84	70	78	81	81	68	76
MIAMI	17	83	86	55	74	81	81	56	70	87	84	65	76	87	88	64	78	85	85	60	75
TAMPA	64	85	87	60	74	82	80	53	68	88	83	62	79	87	66	58	74	86	84	58	74
GA.ATLANTA	71	76	80	64	69	72	75	51	56	85	83	57	68	78	79	53	63	77	79	56	64
HAWAII.HILO	10	85	81	68	85	87	82	69	84	89	82	68	82	87	81	68	86	87	82	69	84
HONOLULU	13	80	80	62	74	75	71	57	71	74	69	56	70	76	72	58	69	76	73	58	71
LIHUE	10	84	84	68	81	81	77	66	76	81	77	65	78	82	79	67	79	82	79	67	78
IDAHO.BOISE	20	81	82	75	74	63	73	48	38	39	54	34	23	61	68	49	43	63	71	53	46
POCATELLO	21	80	82	76	74	59	72	46	39	38	57	30	21	59	72	48	41	61	72	52	46
ILL.CAIRO	68	--	80	68	--	--	75	55	--	--	82	56	--	--	84	54	--	--	81	59	--
CHICAGO	36	79	80	71	75	73	75	58	64	77	76	56	61	74	79	56	65	77	78	61	67
SPRINGFIELD	17	84	85	73	79	78	82	57	61	81	85	53	57	79	85	52	63	81	85	60	66
IND.EVANSVILLE	56	81	82	69	73	75	75	53	57	84	78	52	57	80	82	51	61	80	80	58	63
FT. WAYNE	44	83	85	74	80	79	78	57	63	82	78	51	57	83	85	56	67	82	82	61	68
INDIANAPOLIS	20	81	83	72	78	76	80	57	62	83	84	55	61	80	87	53	65	81	84	60	68
IOWA.DES MOINES	19	78	79	71	73	71	79	54	54	78	85	56	56	71	80	52	55	77	82	61	62
DUBUQUE	79	--	81	70	76	--	75	53	58	--	78	53	60	--	82	55	66	--	80	59	67
SIOUX CITY	66	76	77	68	71	74	77	52	52	79	80	52	51	74	79	52	56	77	79	58	60
KANS.CONCORDIA	72	73	78	63	68	68	77	50	50	68	77	47	48	67	78	48	56	70	78	54	57
DODGE CITY	67	74	79	56	62	69	76	47	47	67	76	41	44	69	78	46	52	71	78	48	52
TOPEKA	18	75	78	64	67	72	79	53	53	76	81	51	50	73	80	50	54	75	81	56	58
WICHITA	66	76	78	63	66	71	75	52	51	72	78	48	49	60	76	51	54	73	78	54	56
KY.LOUISVILLE	71	78	78	68	69	74	72	54	56	83	77	52	57	80	81	52	59	79	77	57	61
LA.NEW ORLEANS	37	83	85	67	73	82	84	59	66	84	84	63	71	79	82	59	68	82	84	62	69
SHREVEPORT	66	82	83	67	67	79	82	56	57	83	86	56	62	82	85	54	59	81	84	58	61
MAINE.CARIBOU	15	75	73	69	73	79	77	59	69	88	82	58	71	85	86	61	76	82	79	62	73
EASTPORT	63	--	74	69	73	--	77	69	76	--	85	75	84	--	81	70	79	--	80	71	78
MASS.BOSTON	62	66	72	59	67	68	67	53	63	79	72	55	70	75	75	53	70	73	71	56	69
NANTUCKET	71	79	81	73	80	86	79	69	84	93	85	74	90	84	81	69	82	85	81	71	84
MICH.ALPENA	45	79	80	70	76	77	77	61	68	82	77	61	68	82	85	65	76	80	80	65	73
DETROIT	26	77	81	70	76	73	75	53	60	77	75	51	56	79	83	55	67	77	79	59	66
GRAND RAPIDS	51	86	85	78	80	77	76	53	60	83	76	50	54	84	84	56	72	83	81	61	68
MARQUETTE	55	74	76	72	74	70	72	65	68	77	74	65	69	76	78	65	73	76	76	68	72
SAULT STE. MARIE	64	78	78	75	78	80	78	61	69	91	84	61	70	88	87	68	79	85	82	67	75
MINN.DULUTH	18	77	78	73	73	75	80	59	60	87	88	61	65	82	88	62	72	81	84	65	69
INTERNAT'L FALLS	17	73	74	68	71	73	78	53	55	85	88	57	59	81	86	60	69	80	82	62	65
MPLS.-ST. PAUL	52	75	76	67	72	70	75	52	55	78	79	51	54	73	79	55	61	76	78	58	62
MISS.VICKSBURG	63	84	82	65	67	87	82	57	60	93	87	61	70	92	84	55	66	89	84	59	66
MO.KANSAS CITY	66	74	66	37	66	68	74	53	54	72	76	49	53	67	76	51	55	71	77	55	59
ST. LOUIS	65	77	77	65	68	72	73	54	58	74	73	50	55	74	76	52	58	75	76	57	61
SPRINGFIELD	68	80	82	68	72	74	77	56	57	84	82	57	60	78	82	54	61	80	81	60	63
MONT.HAVRE	50	76	82	71	78	65	79	47	44	57	74	38	35	67	79	50	54	69	79	54	55
HELENA	66	70	68	64	64	64	69	48	43	56	64	37	33	68	71	53	51	67	69	52	49
KALISPELL	50	80	79	79	74	68	78	54	54	59	74	46	34	77	84	67	61	71	79	63	55
NEBR.LINCOLN	14	78	79	67	72	75	81	54	54	76	82	50	50	72	78	47	56	77	81	57	61
NORTH PLATTE	15	80	83	61	62	73	82	50	47	74	84	50	47	75	84	47	51	77	84	54	54
NEV.ELY	7	72	75	60	57	54	66	36	30	37	50	22	21	53	62	34	29	55	64	38	34
LAS VEGAS	11	52	59	41	34	26	35	18	14	23	31	19	14	28	35	21	16	31	38	23	18
RENO	49	76	77	67	56	58	69	37	32	43	62	25	20	66	71	42	32	62	70	44	35
WINNEMUCCA	65	80	84	66	65	56	70	37	31	32	48	20	17	58	66	36	43	58	68	40	38
N.H.CONCORD	18	77	79	61	71	78	73	48	61	89	79	50	68	87	86	52	73	83	80	53	69
N.J.ATLANTIC CITY	70	--	79	68	74	--	76	65	75	--	81	72	82	--	80	64	75	--	79	68	77
N.MEX.ALBUQUERQUE	24	63	70	49	46	40	52	27	23	49	61	33	30	50	62	35	33	50	60	36	32
ROSWELL	51	61	70	45	41	42	58	28	24	57	73	38	35	61	74	41	40	54	68	38	35
N.Y.ALBANY	16	73	75	63	71	74	73	50	60	84	79	52	64	83	86	54	72	79	78	56	68
BINGHAMTON	11	81	82	68	77	79	79	54	64	88	86	56	71	85	88	55	75	84	84	59	73
BUFFALO	20	78	79	72	77	77	77	58	67	82	79	53	63	79	82	58	73	80	79	61	71
CANTON	43	82	80	71	80	76	75	60	66	82	75	56	65	81	82	62	74	81	78	63	72
NEW YORK	71	67	72	61	66	69	69	54	62	72	76	57	67	72	75	58	66	70	73	58	65
SYRACUSE	55	77	76	70	76	77	71	56	65	82	74	54	65	81	79	59	74	80	76	61	70
N.C.ASHEVILLE	45	--	82	59	69	--	77	47	57	--	88	56	73	--	87	50	67	--	84	54	67
RALEIGH	18	78	82	57	69	75	78	46	57	89	86	55	72	86	89	53	74	82	84	53	68
N.DAK.BISMARCK	55	76	74	67	70	76	79	50	51	79	81	48	48	74	81	51	57	77	79	56	59
DEVILS LAKE	22	--	76	72	75	--	82	56	54	--	86	53	53	--	83	57	60	--	82	62	63
FARGO	68	75	75	70	75	77	82	56	60	81	84	51	57	75	83	56	65	78	81	60	66
WILLISTON	43	74	74	67	69	68	76	49	47	71	77	45	41	70	79	53	53	73	77	55	55
OHIO.CINCINNATI	75	81	82	69	74	75	79	53	59	84	84	51	59	80	87	51	65	80	83	57	65
CLEVELAND	66	82	80	73	76	76	74	54	65	81	74	52	62	78	77	54	68	80	77	59	68
COLUMBUS	68	82	83	72	76	77	75	55	62	84	77	51	59	80	82	52	64	81	79	58	66
OKLA.OKLAHOMA CITY	63	75	79	62	65	71	76	50	52	76	80	49	51	71	79	58	57	74	79	54	57
OREG.BAKER	12	71	83	81	78	52	76	71	46	42	69	68	36	56	78	74	51	59	78	75	55
PORTLAND	19	83	87	82	78	76	87	66	54	71	85	63	47	86	91	79	67	79	87	73	61
ROSEBURG	71	90	91	87	79	77	89	63	53	62	85	53	38	86	93	77	61	78	90	70	58
PA.HARRISBURG	67	71	73	60	68	69	67	50	67	81	75	50	63	80	79	53	67	75	74	54	64
PHILADELPHIA	19	74	76	60	68	74	73	49	59	83	79	52	65	82	83	53	70	78	78	54	66
PITTSBURGH	65	76	77	67	73	69	72	50	59	80	77	53	60	75	80	52	63	75	77	56	65
R.I.PROVIDENCE	19	71	73	60	67	75	71	51	65	87	79	57	72	83	83	55	74	79	77	56	70
S.C.CHARLESTON	17	84	87	56	76	85	84	49	70	93	88	64	81	90	90	56	84	88	87	56	77
COLUMBIA	57	79	81	57	65	78	77	47	55	86	83	55	68	86	84	50	65	83	81	53	63
S.DAK.HURON	67	78	76	72	72	76	80	54	51	79	81	52	49	75	81	52	56	79	80	60	59
RAPID CITY	9	71	71	60	66	66	71	48	47	64	70	42	40	60	64	42	46	67	70	50	52
TENN.KNOXVILLE	67	80	83	65	69	71	75	49	54	84	83	55	66	82	86	51	63	79	82	56	63
MEMPHIS	64	80	79	68	70	74	74	56	58	82	81	58	63	81	80	54	62	79	79	59	64
NASHVILLE	69	80	84	66	72	74	76	51	55	82	81	53	61	81	84	51	58	79	81	56	62
TEX.ABILENE	68	71	73	50	55	65	72	40	48	61	74	39	40	69	77	47	52	66	75	45	48
AMARILLO	64	70	73	51	54	62	70	40	38	66	76	41	42	68	76	45	50	66	74	45	46
AUSTIN	29	78	83	63	64	80	84	55	54	79	88	50	51	78	85	50	54	78	85	55	56
BROWNSVILLE	32	87	88	66	76	87	79	59	82	86	91	57	68	87	90	50	73	87	89	61	72
DEL RIO	49	70	81	59	53	59	77	48	40	58	80	50	41	68	83	55	54	64	80	53	47
EL PASO	67	36	61	40	35	31	40	22	16	48	60	34	30	48	59	35	36	43	54	32	28
FT. WORTH	57	73	78	61	61	70	78	53	52	66	78	48	45	67	77	51	53	69	77	53	53
GALVESTON	69	86	87	77	83	86	86	73	80	83	82	69	74	79	82	66	72	83	84	71	77
HOUSTON	46	84	85	66	73	86	87	59	67	90	90	58	66	85	86	54	68	86	87	59	68
SAN ANTONIO	17	76	81	61	59	76	82	54	51	75	88	48	43	76	83	51	53	76	83	54	52
UTAH.SALT LAKE CITY	26	80	80	70	72	63	69	42	39	45	55	27	23	65	69	41	42	64	69	46	44
VT.BURLINGTON	18	78	80	69	77	77	74	54	64	84	76	54	66	82	82	59	74	81	78	60	71
VA.NORFOLK	66	77	79	61	71	77	74	52	67	88	81	61	76	85	81	60	74	81	79	58	72
RICHMOND	25	79	82	58	71	75	76	46	59	89	84	55	73	87	89	53	77	83	82	53	70
ROANOKE	12	69	72	54	62	67	70	44	53	83	82	52	65	79	82	49	63	74	77	51	61
WASH.NORTH HEAD	62	82	87	79	84	85	88	79	81	91	92	87	84	89	90	85	85	86	89	83	84
SEATTLE	20	82	85	80	73	74	85	63	52	70	85	62	46	86	91	79	68	78	87	72	60
SPOKANE	66	84	85	80	76	67	75	53	35	49	64	39	24	76	81	65	50	70	77	61	49
TATOOSH ISLAND	59	82	85	83	83	83	87	80	80	93	94	89	87	88	91	86	87	86	90	84	85
WALLA WALLA	52	--	82	73	80	--	69	44	43	--	54	29	24	--	72	50	54	--	70	51	51
YAKIMA	11	84	85	75	74	60	74	38	31	55	72	35	26	79	85	55	49	70	79	52	46
W.VA.ELKINS	14	83	82	65	77	80	81	53	62	96	94	61	76	90	93	53	75	88	88	59	73
PARKERSBURG	60	--	82	66	74	--	74	49	58	--	80	52	67	--	84	52	70	--	80	56	68
WIS.GREEN BAY	30	74	76	67	73	73	76	56	61	80	78	55	59	78	82	57	69	77	79	61	66
MADISON	23	77	80	71	76	75	78	54	59	81	79	52	56	79	83	56	66	79	81	61	67
MILWAUKEE	68	76	76	68	74	75	76	63	68	82	75	59	65	77	80	61	70	78	77	64	70
WYO.CHEYENNE	67	61	58	48	52	68	72	47	49	65	70	36	41	64	65	39	47	65	66	44	48
SHERIDAN	18	70	73	61	65	70	77	48	46	63	73	37	35	66	72	44	47	69	75	50	51
P.R.SAN JUAN	56	83	81	71	77	84	76	72	78	86	78	76	82	88	80	74	84	86	79	73	80
V.I.ST. CROIX	6	81	76	66	80	85	74	66	81	86	77	70	84	90	80	72	87	86	77	69	83

TIME IS EASTERN STANDARD (75th MERIDIAN); SUBTRACT 1 HOUR FOR CENTRAL (90th M.), 2 HOURS FOR MOUNTAIN (105th M.), AND 3 HOURS FOR PACIFIC (120th M.) TIME.

BASED ON RECORDS THROUGH 1959, EXCEPT IN A FEW INSTANCES. TAKEN FROM "NORMALS, MEANS, AND EXTREMES" TABLE IN U.S. WEATHER BUREAU PUBLICATION, LOCAL CLIMATOLOGICAL DATA.

MEAN DIURNAL RELATIVE HUMIDITY (%) FOR MIDSEASONAL MONTHS

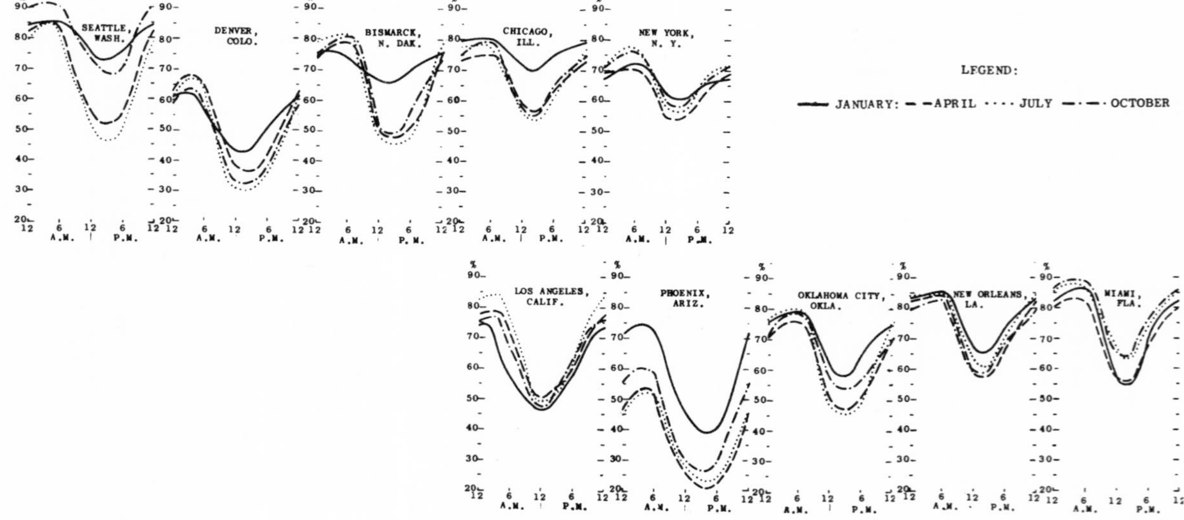

MEAN PAN AND LAKE EVAPORATION

EVAPORATION MAPS FOR THE UNITED STATES

Hydrologic Investigations Section, Hydrologic Services Division,
U. S. Weather Bureau, Washington, D. C.

INTRODUCTION.--Since evaporation inevitably extracts a portion of the gross water supply to a reservoir, the estimation of this loss is an important factor in reservoir design. In arid regions, the evaporation loss actually imposes a ceiling on the water supply obtainable through regulation. Speaking of storage on the main stem of the Colorado River, Langbein (1) states that "The gain in regulation to be achieved by increasing the present 29 million acre-feet to nearly 50 million acre-feet of capacity appears to be largely offset by a corresponding increase in evaporation."

In the final stages, the design of major storage projects requires detailed study of all data available, including observations made at the proposed reservoir sites. However, generalized estimates of free-water evaporation are invaluable in preliminary design studies of major projects, and are often fully adequate for the design of lesser projects. The maps presented herein have been prepared to serve these purposes, primarily, but they should be of value in other studies. For example, free-water evaporation (plate 2) is a good index to potential evapotranspiration, or consumptive use, and the pan coefficient (plate 3) is indicative of an aspect of climate.*

The following series of maps is presented for the United States (except Alaska and Hawaii):

- Plate 1 - Mean Annual Class A Pan Evaporation,
- Plate 2 - Mean Annual Lake Evaporation,
- Plate 3 - Mean Annual Class A Pan Coefficient,
- Plate 4 - Mean May-October Evaporation in Percent of Annual,
- Plate 5 - Standard Deviation of Annual Class A Pan Evaporation.

In 1942, A. F. Meyer (2) published a map comparable to that in plate 2, and in the following year R. E. Horton (3) published a map of Class A pan evaporation similar to plate 1. Subsequent to 1942, there has been a substantial increase in the Class A pan station network and significant progress in the development of techniques for estimating lake evaporation. However, the maps prepared by Horton and Meyer were carefully studied in the preparation of the new series -- any pronounced differences are considered to be reasonably substantiated by data now available.

Plate 3 shows the ratio of annual lake evaporation to that from the Class A pan. It can be used to estimate free-water evaporation for any site for which representative pan data are available. Plate 4 has been included to assist in the extrapolation of seasonal pan evaporation data to annual values, as well as to provide an indication of the seasonal distribution of evaporation from a shallow free-water body. Plate 5 shows the variability of pan evaporation, year-to-year, and can be used to estimate the frequency distribution of annual lake evaporation. The correct interpretation and use of these plates are discussed later.

METHODS FOR COMPUTING EVAPORATION.--The various methods for computing pan and lake evaporation are described in the Lake Hefner (4) and Lake Mead (5) Water-Loss Investigations Reports, and in Weather Bureau Research Paper No. 38 (6). There are four generally accepted methods of computing lake evaporation: (a) Water budget, (b) energy budget, (c) mass transfer, and (d) lake-to-pan relations. Very few reliable water-budget estimates are available because small errors in volume of inflow and outflow usually result in large errors in the residual evaporation value. The energy-budget approach requires such elaborate instrumentation that it is only feasible for special investigations. The mass-transfer method requires observations of lake surface-water temperature, dew point, and wind movement which are available for only a very few reservoirs. Methods (a), (b), and (c) are only applicable for existing lakes and reservoirs, and cannot be used in the design phase.

The few lake-evaporation determinations that have been made using water-budget, energy-budget, and mass-transfer methods were used in preparing plates 2 and 3. However, from a practical point of view, the lake-evaporation map is based essentially on pan evaporation and related meteorological data collected at Class A evaporation and first-order synoptic stations.

DEVELOPMENT OF MAPS.--The description of the development of the maps is given in Weather Bureau Technical Paper No. 37 "Evaporation Maps for the United States."

INTERPRETATION, USE, AND LIMITATIONS OF MAPS.--Although the utility of the derived maps hinges largely on their reliability, it is virtually impossible to make any meaningful generalizations in this respect. In deriving plates 1, 2, and 3, all available pertinent data were utilized to the greatest extent feasible with present-day knowledge of the relationships involved. It can be reasonably assumed, therefore, that the maps provide the most accurate generalized estimates yet available. The reliability of the maps is obviously poorer in the areas of high relief than in the plains region, and the density of the observation network is an important factor throughout.

It is known that some of the data collected over the years are from sheltered sites which are not representative. Through subjective evaluation of the station descriptions and wind data, an attempt was made to derive pan evaporation and coefficient maps indicative of a representative exposure, reasonably free of obstructions to wind and sunshine. Variations in the data were smoothed to a considerable extent, and it is entirely possible that the true areal variation in evaporation exceeds that shown on the maps. For example, a pan or small reservoir located in a canyon of northerly orientation and partially shielded from the sun would experience considerably less evaporation than indicated by the maps.

The effect of topography has been taken into account only in a general way, except where the data provided definite indications. Thus it will be noted that the isopleths tend to follow closely the topographic features in some portions of the maps while the resemblance is more casual in other areas. Both Class A pan and lake evaporation were assumed to decrease with elevation (7), (8), but the decrease assumed for lake evaporation is less. With an increase in elevation, dew point and air temperatures tend to decrease, while wind movement usually increases. Solar radiation, on the other hand, increases up-slope during cloudless days and may otherwise increase or decrease depending on the variation of cloudiness with elevation. There are but few reliable observations of the variation of all these factors up mountain slopes, but it is probable that the effect of these changes is less for lake evaporation than for pan evaporation.

There is good reason to expect that plate 4, showing seasonal distribution of pan evaporation, is more reliable than any other map in the series. Plate 5, on the other hand, is based on a sparse network, and time trends resulting from changes in site, exposure, etc., may have caused some bias in the derived values of standard deviation. Data which were obviously inconsistent were eliminated from the analysis, but any undetected inconsistencies result in values which tend to be too high. Even so, any bias in the final, smoothed isopleths should be small.

The use of plates 1-5 is self-evident in most respects and need not be considered further here. Certain limitations and less obvious features are discussed in the following paragraphs.

Plates 1, 2, and 3. Unless the user has at hand pan-evaporation data not considered in the development of this series of maps, average annual lake evaporation can be taken directly from plate 2. The value so determined will also suffice if pan-evaporation data collected at the site substantiate that given by plate 1. If the pan evaporation at the site exceeds that given by plate 1, application of the pan coefficient (plate 3) will probably provide a better estimate of lake evaporation than that given by plate 2. If, on the other hand, observed pan evaporation is less than that given by plate 1, a value of lake evaporation less than given by plate 2 should be accepted only after it has been determined that the pan site is reasonably free of obstructions to wind and sunshine. This is to say that pan evaporation and the pan coefficient are both dependent upon exposure.

It should be emphasized that values of free-water evaporation given by plate 2 (or plates 1 and 3) assume that there is no net advection (heat content of inflow less outflow) over a long period of time. The mean annual advection is usually small and can be neglected, but this is not always the case. It was found at Lake Mead, for example, that advection results in a 5-inch increase in mean annual evaporation. If the advection term is appreciable, adjustment should be made as discussed in references 5 and 6.

Plate 4. The Class A pans are not in operation during the winter months over much of the country because of freezing weather. Plate 4 provides means of estimating average annual evaporation from that observed during the open season, May through October. When used in conjunction with plate 1, it also provides a means of estimating average growing-season evaporation (Class A pan) which is so important in some studies.

Although the seasonal ratios of plate 4 are based on Class A pan data, it is believed that they are equally applicable to free-water evaporation for shallow lakes. The ratios based on monthly computed lake evaporation for the first-order stations showed no significant deviation from those based on the pan values. It should be emphasized that the seasonal ratios can be applied to annual lake evaporation only in case of shallow lakes where energy storage can be ignored. In deep lakes, the energy storage becomes an important factor in determining seasonal or monthly evaporation. For example, at Lake Mead the maximum lake evaporation occurs in August, but maximum Class A pan evaporation is observed in June; for Lake Ontario, the maximum lake evaporation is in September, and maximum pan evaporation in July (9). Corrections can be made for changes in energy storage and heat advection into or out of the lake in the manner described in references 5 and 6.

Plate 5. The standard deviation of annual Class A pan evaporation can be obtained for any selected site directly from plate 5. If the annual pan coefficient were constant, year-to-year, then the standard deviation of lake evaporation would be the product of that for pan evaporation and the pan coefficient. Because of variation in the annual pan coefficient, the standard deviation computed in this manner may be a few percent too low. Since the values given by plate 5 are probably biased on the high side (discussed previously), the two possible errors tend to compensate.

Having obtained the mean and standard deviation, the frequency distribution of annual lake (or pan) evaporation can be derived, assuming the data are normally distributed. If it is further assumed that the annual evaporation totals occurring in successive years are independent, the frequency distribution of n-year evaporation can also be derived (10).

REFERENCES

1. W. B. Langbein, "Water Yield and Reservoir Storage in the United States," Geological Survey Circular 409, U. S. Department of the Interior, Geological Survey, 1959.

2. A. F. Meyer, Evaporation from Lakes and Reservoirs, Minnesota Resources Commission, St. Paul, Minn., 1942.

3. R. E. Horton, "Evaporation Maps of the United States," Transactions American Geophysical Union, Vol. 24, Part 2, pp. 750-751, April 1943.

4. U. S. Geological Survey, "Water-Loss Investigations: Vol. 1 -- Lake Hefner Studies," Geological Survey Professional Paper No. 269, 1954.

5. U. S. Geological Survey, "Water-Loss Investigations: Lake Mead Studies," Geological Survey Professional Paper No. 298, 1958.

6. M. A. Kohler, T. J. Nordenson, and W. E. Fox, "Evaporation from Pans and Lakes," Research Paper No. 38, U. S. Weather Bureau, 1955.

7. S. Fortier, "Evaporation Losses in Irrigation," Engineering News, Vol. 58, No. 12, pp. 304-307, 1907.

8. L. L. Longacre, (Southern California Edison Co.), "Evaporation at High Elevations, San Joaquin River Basin, California." Report now in preparation.

9. I. A. Hunt, "Evaporation of Lake Ontario," Proceedings of the American Society of Civil Engineers, Journal of the Hydraulics Division, Vol. 85, No. HY2, Part 1, February 1959.

10. Missouri Basin Inter-Agency Committee, Adequacy of Flows in the Missouri River, Report, pp. 23-24, April 1951.

* If solar radiation, wind, dew point, and air temperature are such that water in an exposed Class A pan is warmer than the air, the coefficient is greater than 0.7, and vice versa.

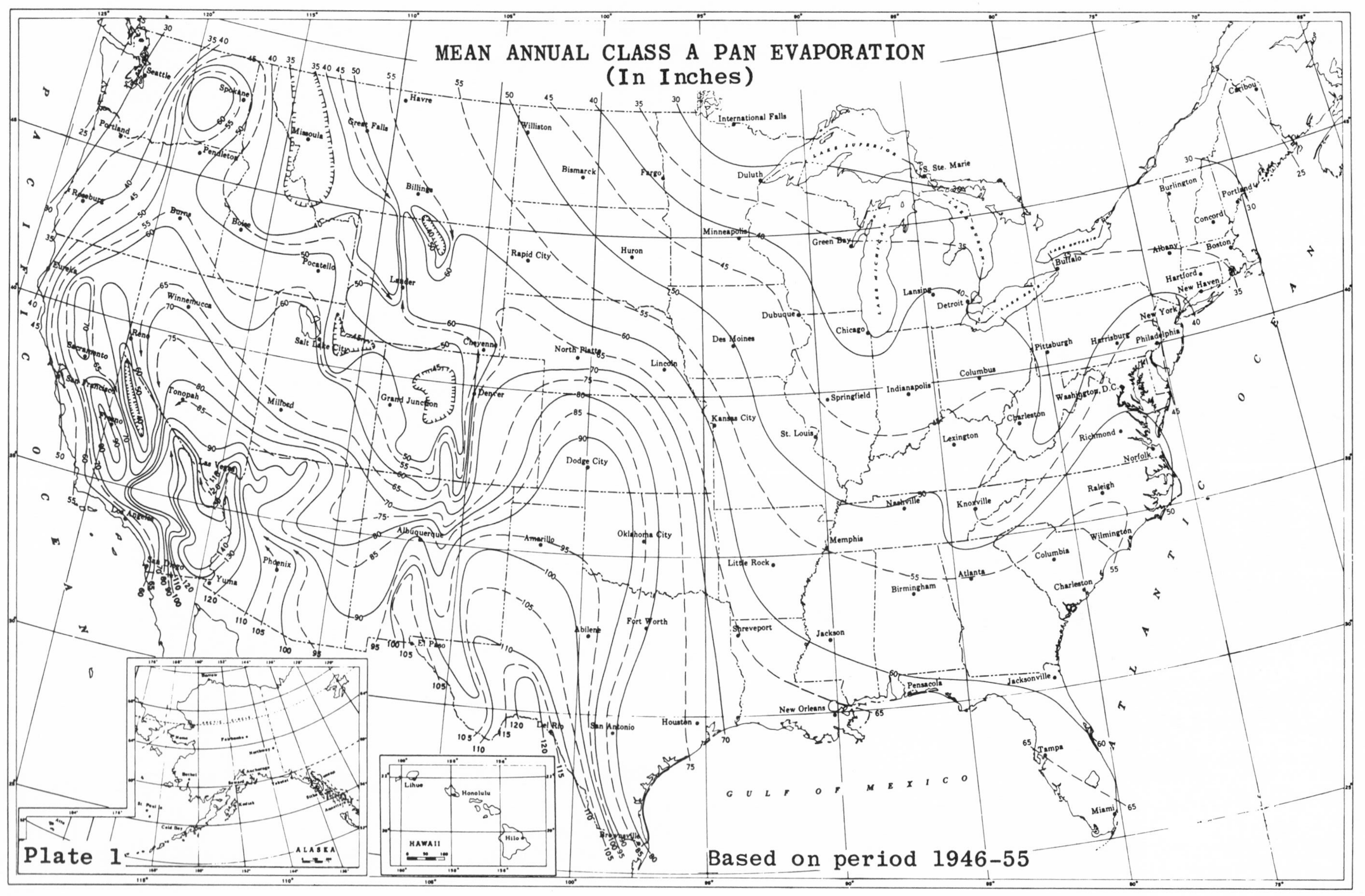
MEAN ANNUAL CLASS A PAN EVAPORATION
(In Inches)
Based on period 1946-55
Plate 1
ATLANTIC OCEAN
PACIFIC OCEAN
GULF OF MEXICO
HAWAII
ALASKA

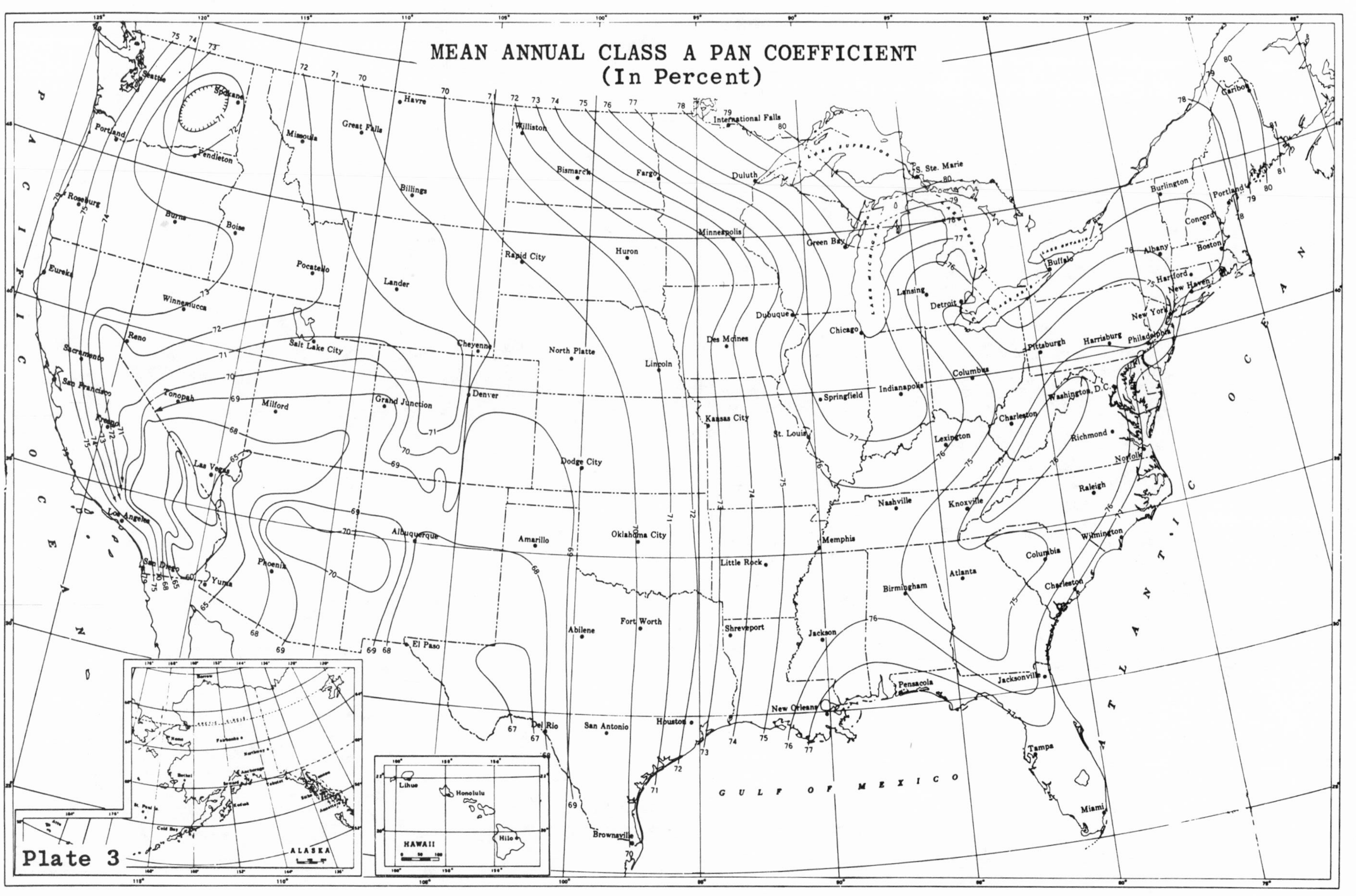
MEAN ANNUAL CLASS A PAN COEFFICIENT
(In Percent)
PACIFIC OCEAN
ATLANTIC OCEAN
GULF OF MEXICO
HAWAII
ALASKA
Plate 3

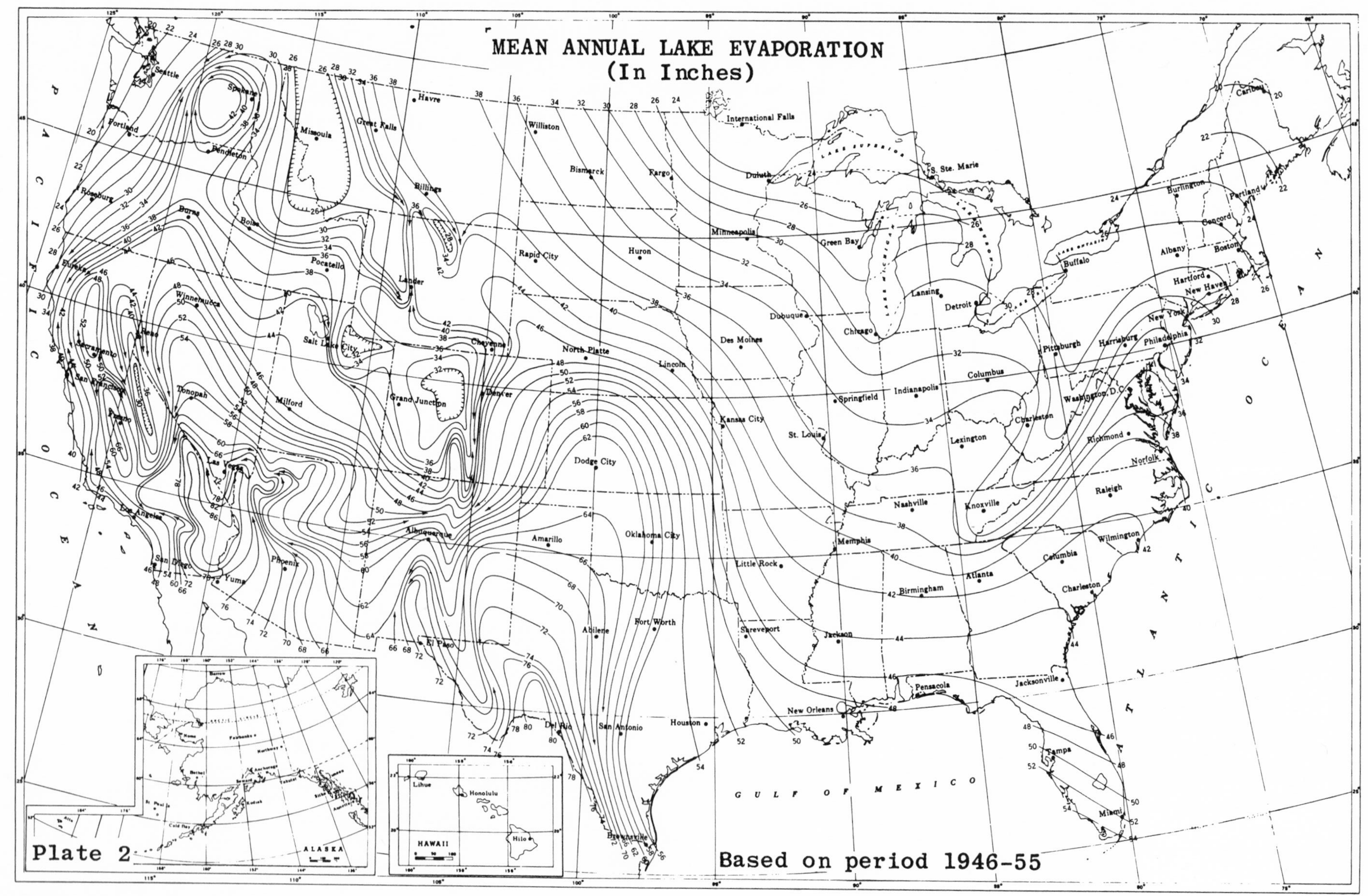
MEAN ANNUAL LAKE EVAPORATION
(In Inches)
Based on period 1946-55
Plate 2
PACIFIC OCEAN
ATLANTIC OCEAN
GULF OF MEXICO
HAWAII
ALASKA

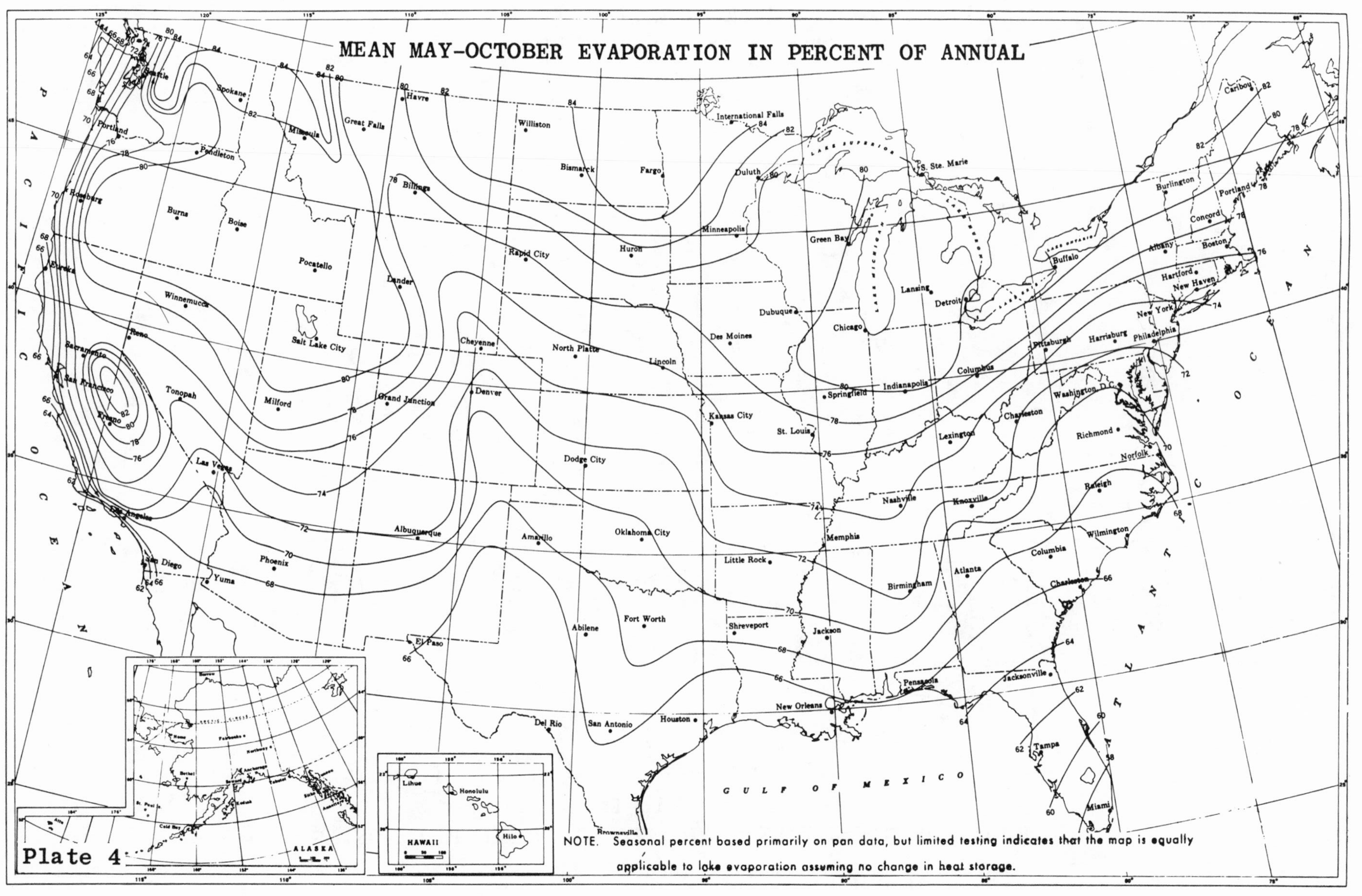
MEAN MAY-OCTOBER EVAPORATION IN PERCENT OF ANNUAL
PACIFIC OCEAN
ATLANTIC OCEAN
GULF OF MEXICO
Seattle
Spokane
Portland
Pendleton
Roseburg
Burns
Boise
Eureka
Winnemucca
Reno
Sacramento
San Francisco
Fresno
Tonopah
Las Vegas
Los Angeles
San Diego
Yuma
Phoenix
Missoula
Great Falls
Havre
Billings
Lander
Pocatello
Salt Lake City
Milford
Grand Junction
Cheyenne
Denver
Albuquerque
El Paso
Williston
Bismarck
Fargo
Rapid City
Huron
North Platte
Dodge City
Amarillo
Abilene
Del Rio
San Antonio
Houston
Brownsville
Oklahoma City
Fort Worth
International Falls
Duluth
Minneapolis
Des Moines
Lincoln
Kansas City
Little Rock
Shreveport
Jackson
New Orleans
Green Bay
Dubuque
Chicago
Springfield
St. Louis
Memphis
S. Ste. Marie
Lansing
Detroit
Indianapolis
Nashville
Birmingham
Pensacola
Columbus
Lexington
Knoxville
Atlanta
Buffalo
Pittsburgh
Charleston
Harrisburg
Philadelphia
Washington D.C.
Richmond
Norfolk
Raleigh
Wilmington
Columbia
Charleston
Jacksonville
Tampa
Miami
Burlington
Albany
Concord
Portland
Boston
Hartford
New Haven
New York
Caribou
Lake Superior
Lake Michigan
Lake Huron
Lake Erie
Lake Ontario
HAWAII
Lihue
Honolulu
Hilo
ALASKA
Plate 4
NOTE. Seasonal percent based primarily on pan data, but limited testing indicates that the map is equally applicable to lake evaporation assuming no change in heat storage.

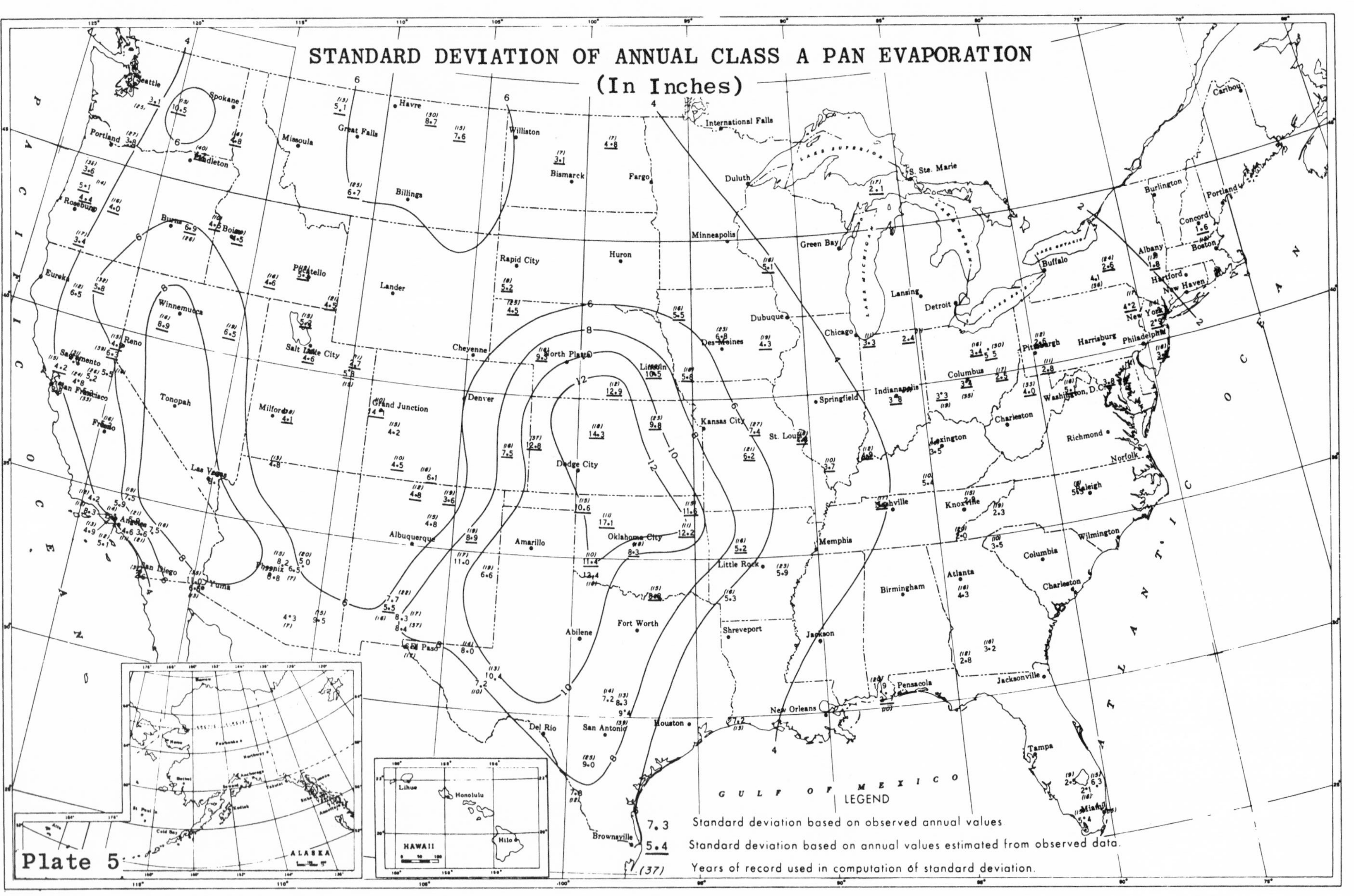
STANDARD DEVIATION OF ANNUAL CLASS A PAN EVAPORATION
(In Inches)
LEGEND
7.3 Standard deviation based on observed annual values
5.4 Standard deviation based on annual values estimated from observed data.
(37) Years of record used in computation of standard deviation.
Plate 5
PACIFIC OCEAN
ATLANTIC OCEAN
GULF OF MEXICO
HAWAII
ALASKA

SUNSHINE,
SOLAR RADIATION,
AND
SKY COVER

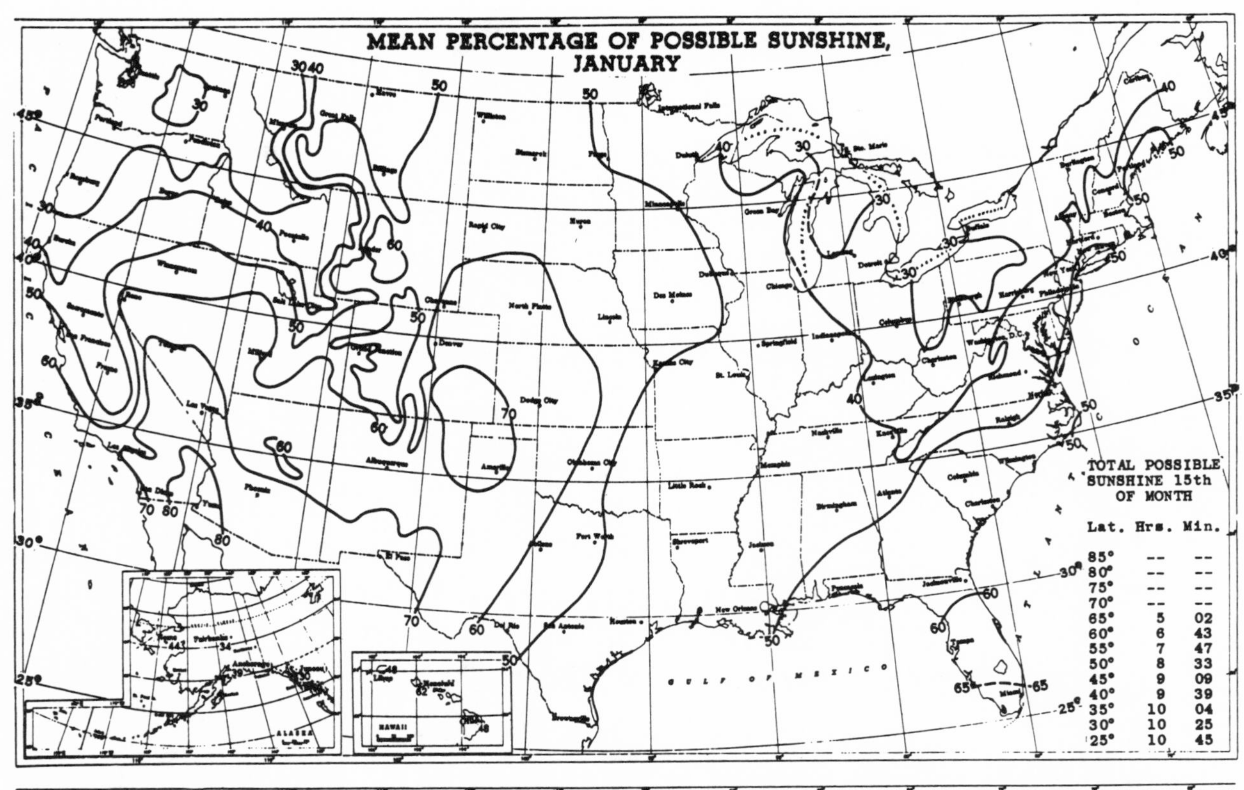
MEAN PERCENTAGE OF POSSIBLE SUNSHINE, JANUARY
TOTAL POSSIBLE SUNSHINE 15th OF MONTH
Lat. Hrs. Min.
85° -- --
80° -- --
75° -- --
70° -- --
65° 5 02
60° 6 43
55° 7 47
50° 8 33
45° 9 09
40° 9 39
35° 10 04
30° 10 25
25° 10 45
GULF OF MEXICO
ALASKA
HAWAII

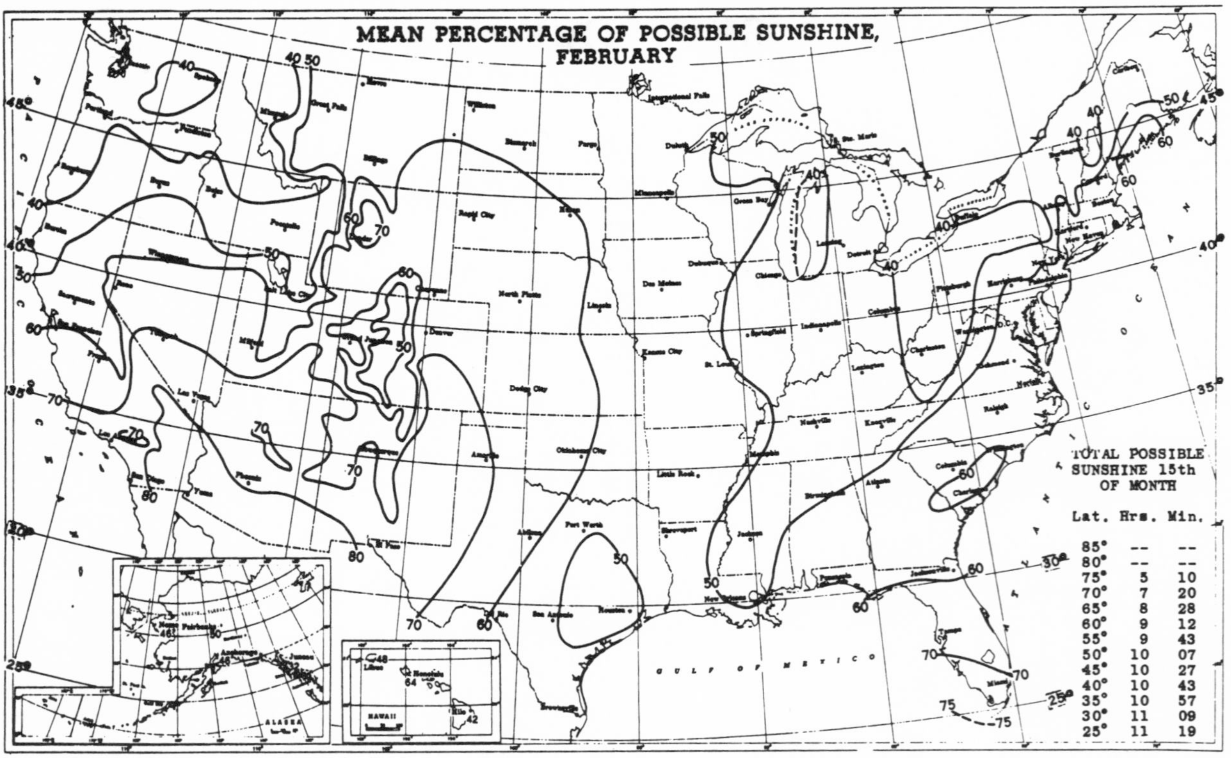
MEAN PERCENTAGE OF POSSIBLE SUNSHINE, FEBRUARY
TOTAL POSSIBLE SUNSHINE 15th OF MONTH
Lat. Hrs. Min.
85° -- --
80° -- --
75° 5 10
70° 7 20
65° 8 28
60° 9 12
55° 9 43
50° 10 07
45° 10 27
40° 10 43
35° 10 57
30° 11 09
25° 11 19
GULF OF MEXICO
ALASKA
HAWAII

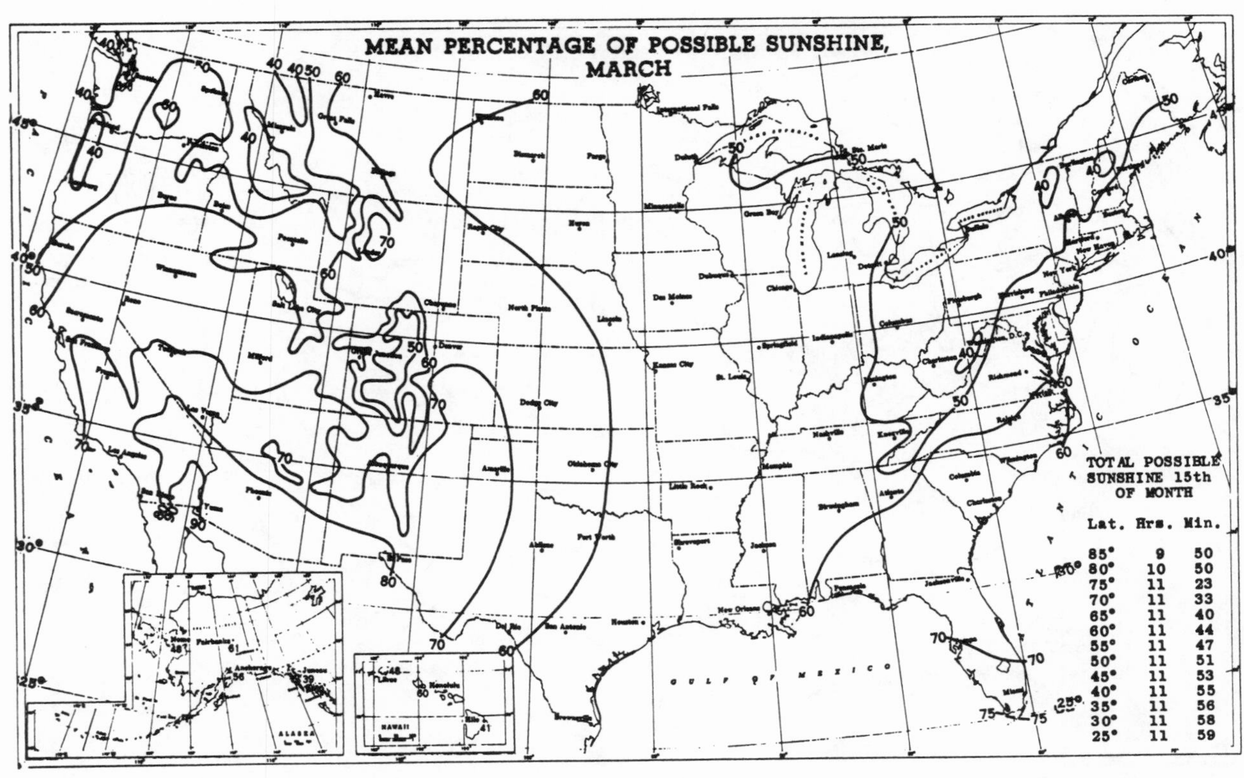
MEAN PERCENTAGE OF POSSIBLE SUNSHINE,
MARCH
GULF OF MEXICO
ALASKA
HAWAII
TOTAL POSSIBLE SUNSHINE 15th OF MONTH
Lat. Hrs. Min.
85° 9 50
80° 10 50
75° 11 23
70° 11 33
65° 11 40
60° 11 44
55° 11 47
50° 11 51
45° 11 53
40° 11 55
35° 11 56
30° 11 58
25° 11 59

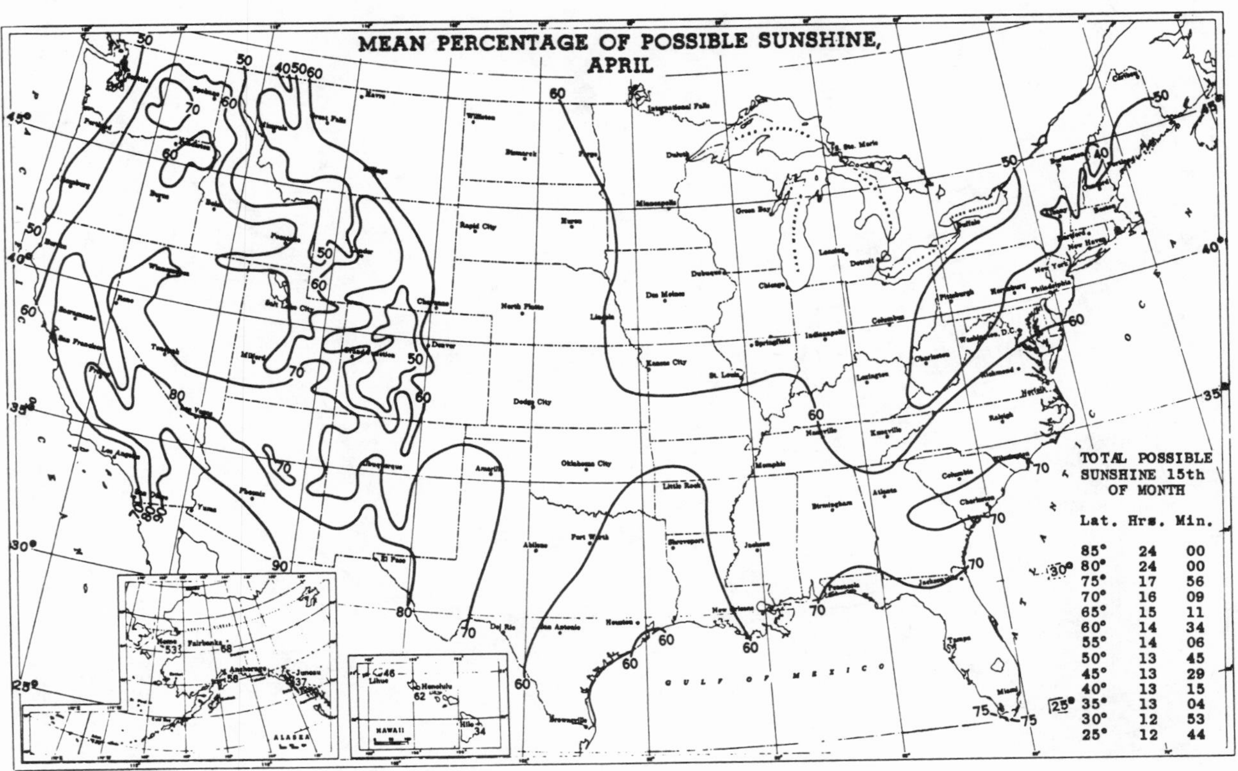
MEAN PERCENTAGE OF POSSIBLE SUNSHINE,
APRIL
GULF OF MEXICO
ALASKA
HAWAII
TOTAL POSSIBLE SUNSHINE 15th OF MONTH
Lat. Hrs. Min.
85° 24 00
80° 24 00
75° 17 56
70° 16 09
65° 15 11
60° 14 34
55° 14 06
50° 13 45
45° 13 29
40° 13 15
35° 13 04
30° 12 53
25° 12 44

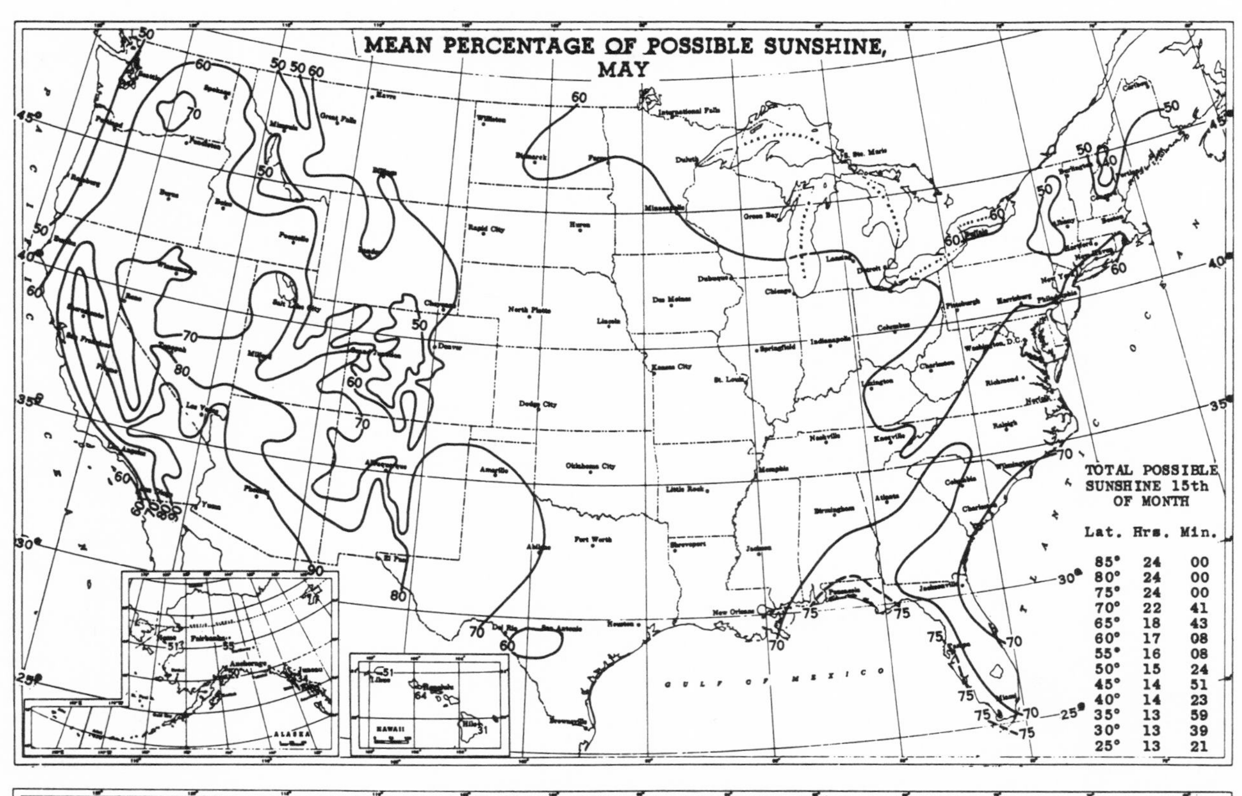
MEAN PERCENTAGE OF POSSIBLE SUNSHINE,
MAY
TOTAL POSSIBLE SUNSHINE 15th OF MONTH
Lat. Hrs. Min.
85° 24 00
80° 24 00
75° 24 00
70° 22 41
65° 18 43
60° 17 08
55° 16 08
50° 15 24
45° 14 51
40° 14 23
35° 13 59
30° 13 39
25° 13 21
GULF OF MEXICO
ALASKA
HAWAII

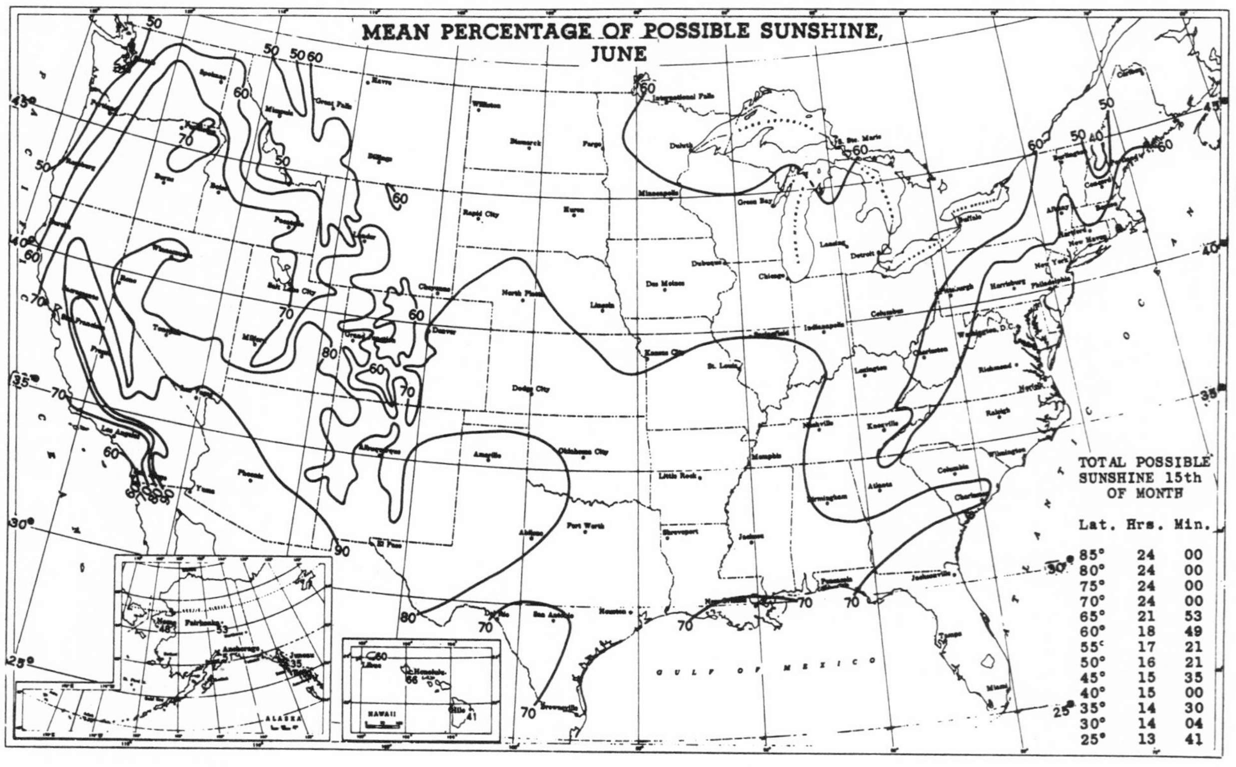
MEAN PERCENTAGE OF POSSIBLE SUNSHINE,
JUNE
TOTAL POSSIBLE SUNSHINE 15th OF MONTH
Lat. Hrs. Min.
85° 24 00
80° 24 00
75° 24 00
70° 24 00
65° 21 53
60° 18 49
55° 17 21
50° 16 21
45° 15 35
40° 15 00
35° 14 30
30° 14 04
25° 13 41
GULF OF MEXICO
ALASKA
HAWAII

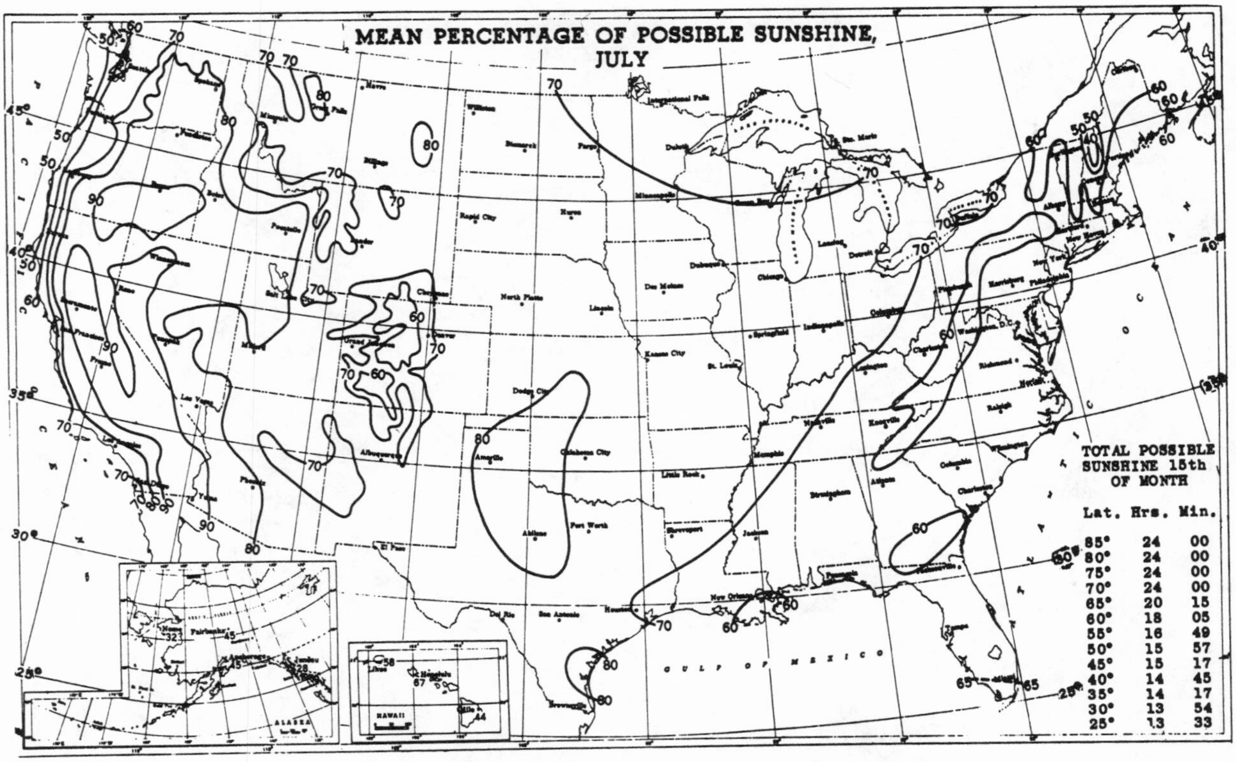

MEAN PERCENTAGE OF POSSIBLE SUNSHINE, JULY
G U L F O F M E X I C O
TOTAL POSSIBLE SUNSHINE 15th OF MONTH

Lat.	Hrs.	Min.
85°	24	00
80°	24	00
75°	24	00
70°	24	00
65°	20	15
60°	18	05
55°	16	49
50°	15	57
45°	15	17
40°	14	45
35°	14	17
30°	13	54
25°	13	33

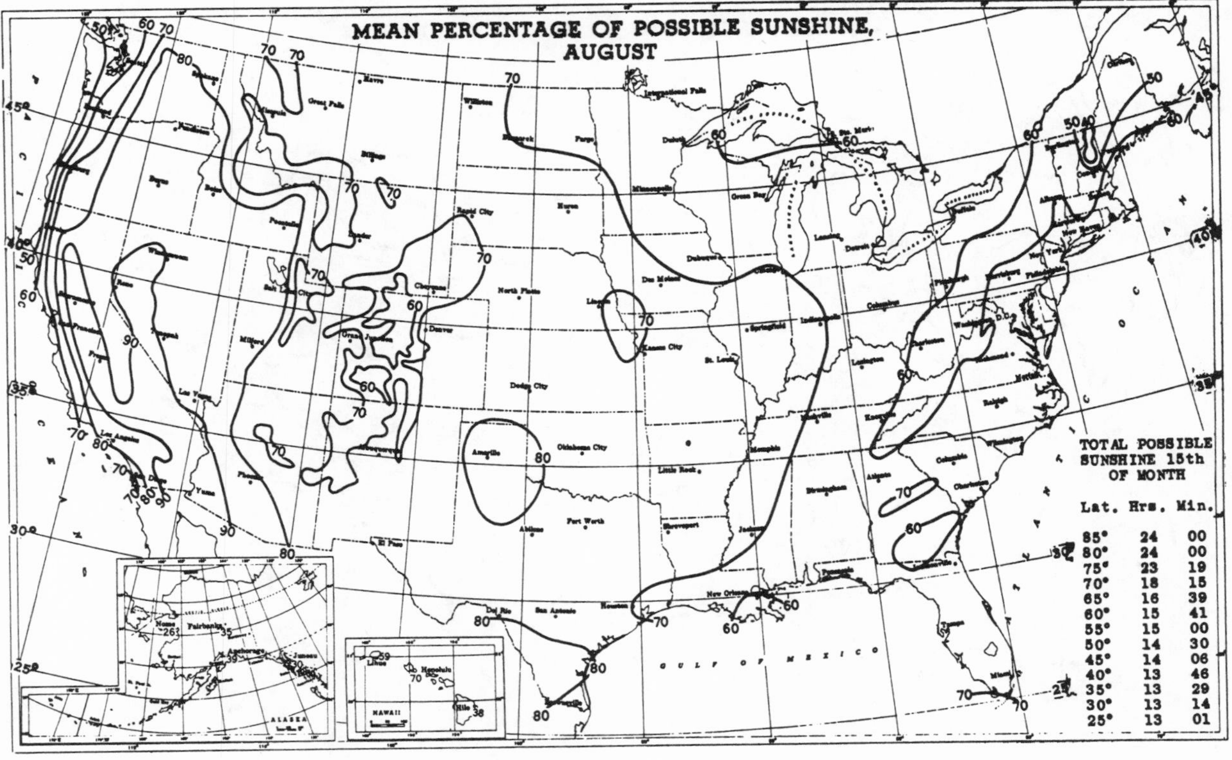

MEAN PERCENTAGE OF POSSIBLE SUNSHINE, AUGUST
G U L F O F M E X I C O
TOTAL POSSIBLE SUNSHINE 15th OF MONTH

Lat.	Hrs.	Min.
85°	24	00
80°	24	00
75°	23	19
70°	18	15
65°	16	39
60°	15	41
55°	15	00
50°	14	30
45°	14	06
40°	13	46
35°	13	29
30°	13	14
25°	13	01

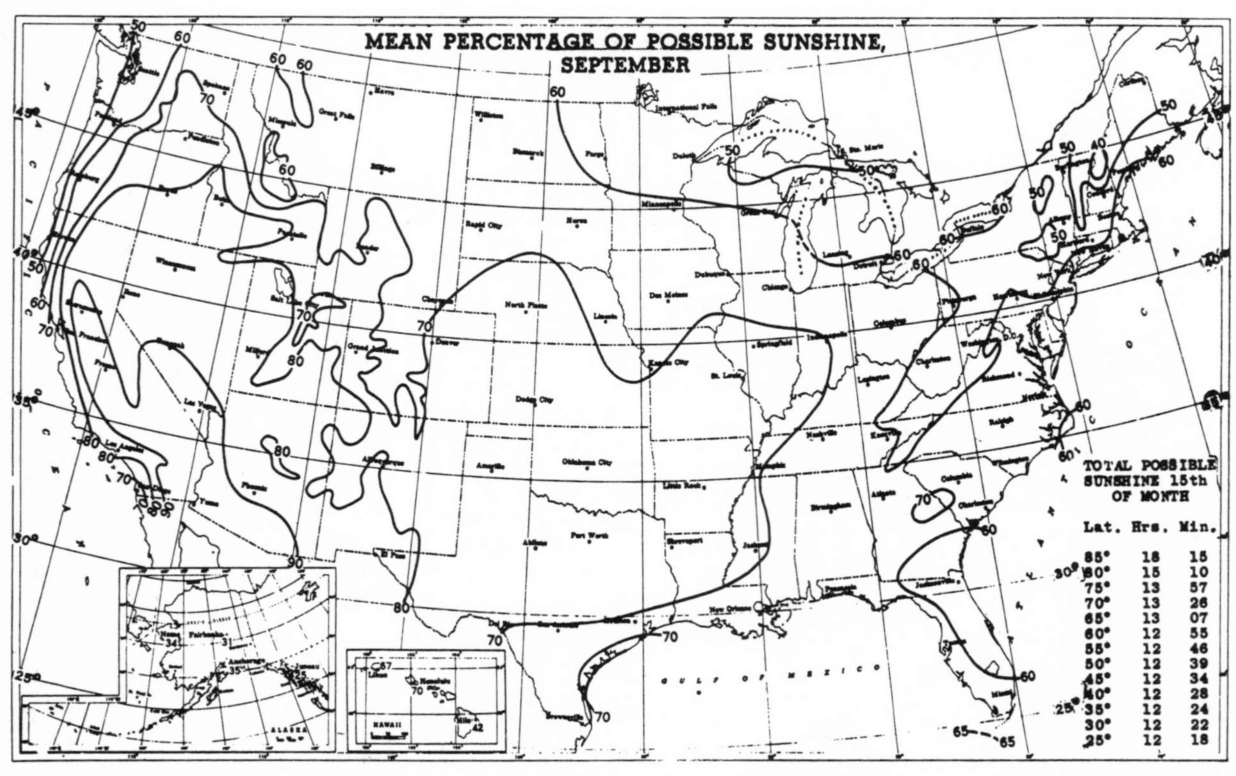
MEAN PERCENTAGE OF POSSIBLE SUNSHINE,
SEPTEMBER
GULF OF MEXICO
TOTAL POSSIBLE SUNSHINE 15th OF MONTH
Lat. Hrs. Min.
85° 18 15
80° 15 10
75° 13 57
70° 13 26
65° 13 07
60° 12 55
55° 12 46
50° 12 39
45° 12 34
40° 12 28
35° 12 24
30° 12 22
25° 12 18

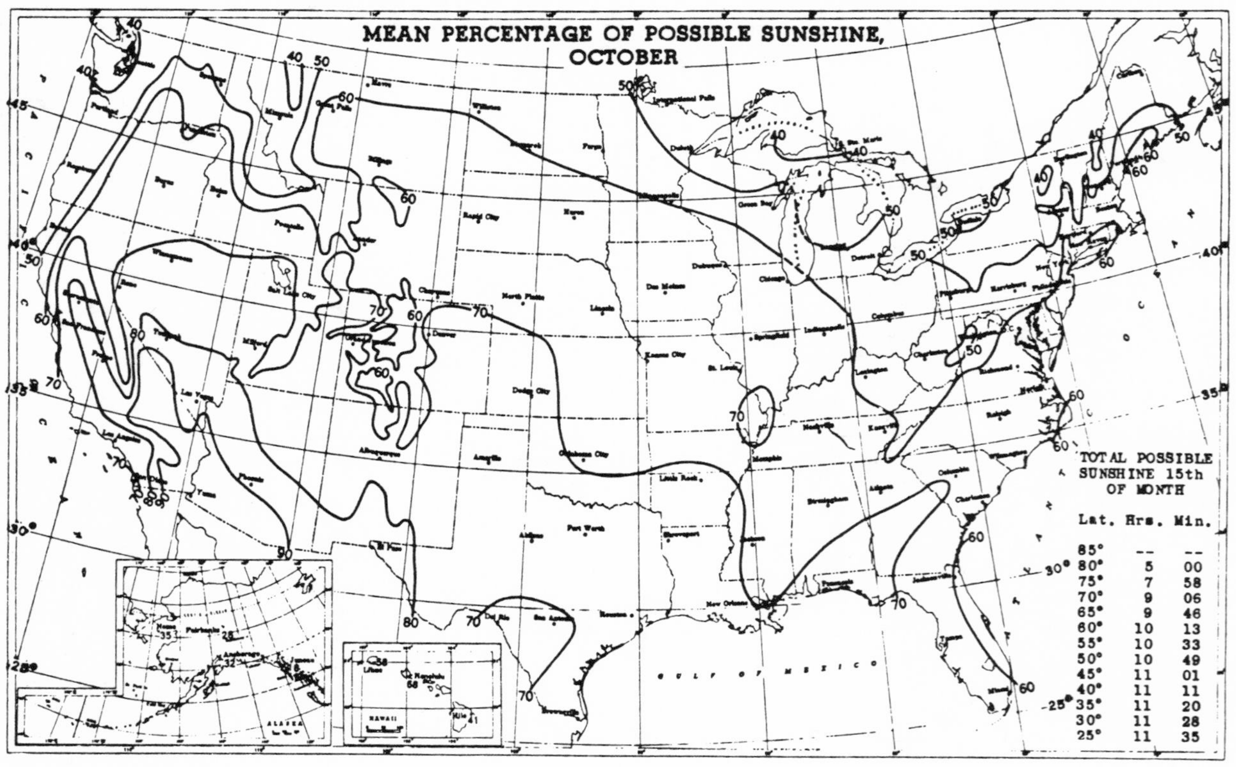
MEAN PERCENTAGE OF POSSIBLE SUNSHINE,
OCTOBER
GULF OF MEXICO
TOTAL POSSIBLE SUNSHINE 15th OF MONTH
Lat. Hrs. Min.
85° -- --
80° 5 00
75° 7 58
70° 9 06
65° 9 46
60° 10 13
55° 10 33
50° 10 49
45° 11 01
40° 11 11
35° 11 20
30° 11 28
25° 11 35

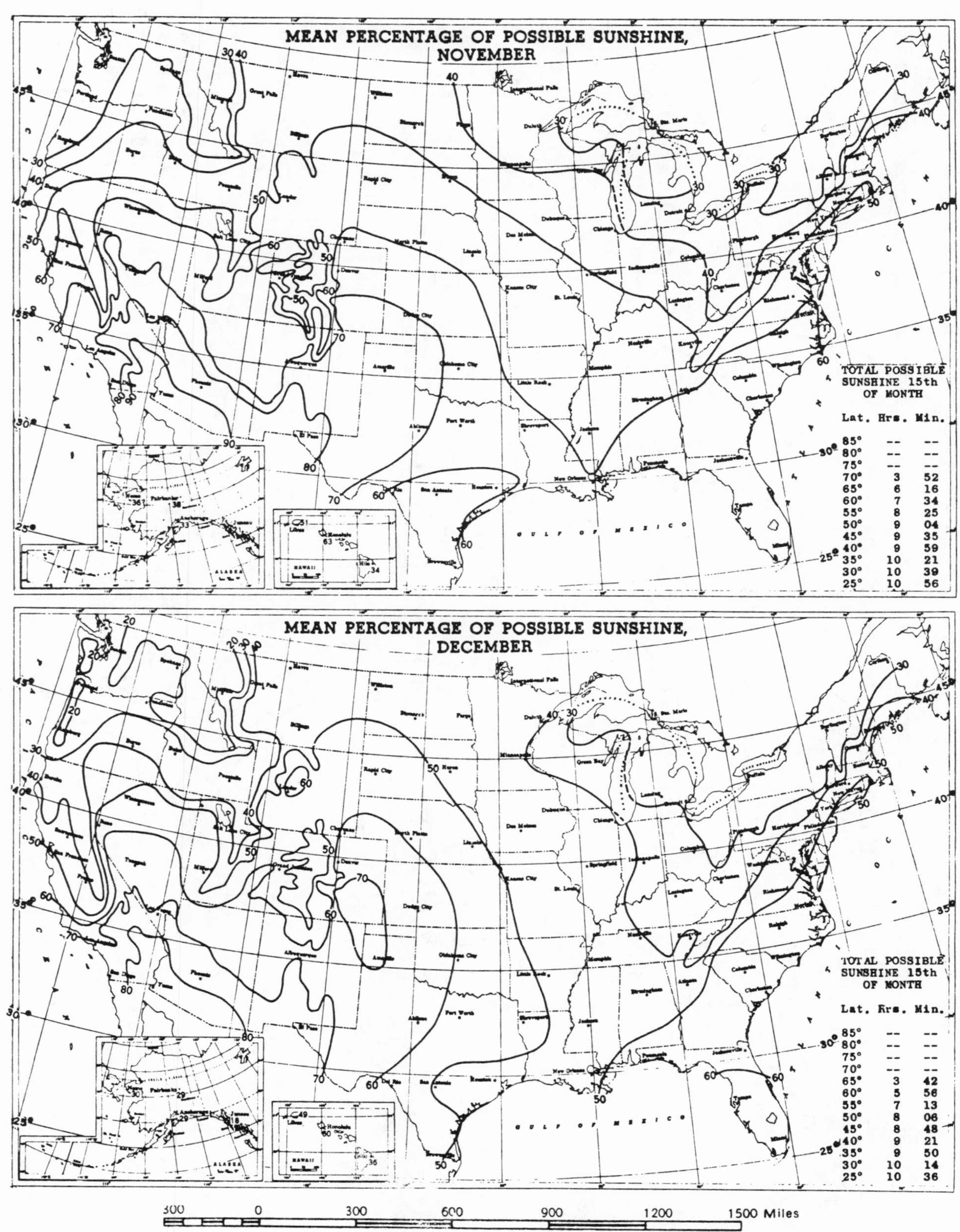

MEAN PERCENTAGE OF POSSIBLE SUNSHINE, NOVEMBER
TOTAL POSSIBLE SUNSHINE 15th OF MONTH
Lat. Hrs. Min.
85° -- --
80° -- --
75° -- --
70° 3 52
65° 6 16
60° 7 34
55° 8 25
50° 9 04
45° 9 35
40° 9 59
35° 10 21
30° 10 39
25° 10 56
GULF OF MEXICO
ALASKA
HAWAII
MEAN PERCENTAGE OF POSSIBLE SUNSHINE, DECEMBER
TOTAL POSSIBLE SUNSHINE 15th OF MONTH
Lat. Hrs. Min.
85° -- --
80° -- --
75° -- --
70° -- --
65° 3 42
60° 5 56
55° 7 13
50° 8 06
45° 8 48
40° 9 21
35° 9 50
30° 10 14
25° 10 36
GULF OF MEXICO
ALASKA
HAWAII

300 0 300 600 900 1200 1500 Miles

SCALE 1:30,000,000

ALBERS EQUAL AREA PROJECTION–STANDARD PARALLELS 29½° AND 45½°

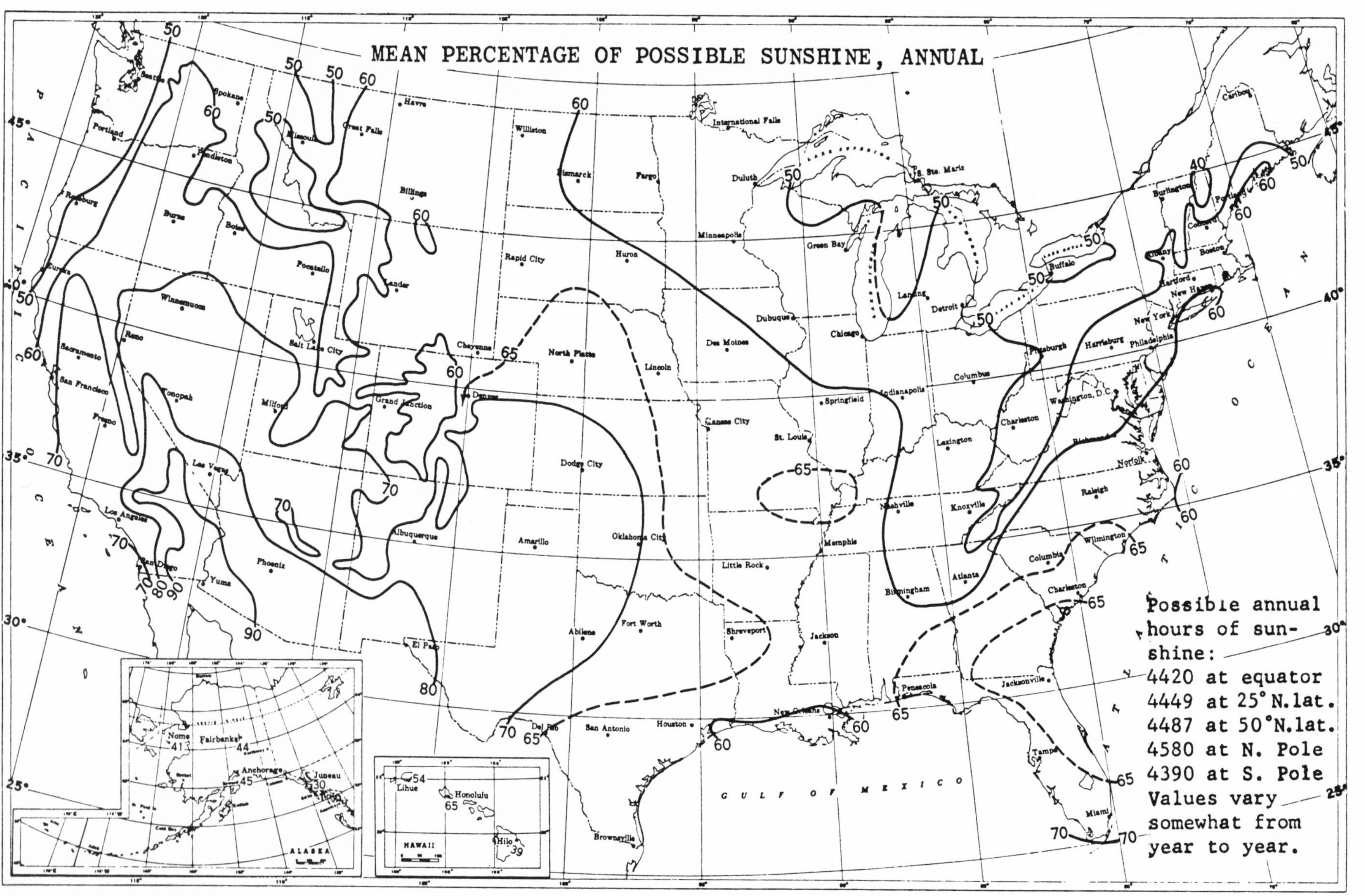
MEAN PERCENTAGE OF POSSIBLE SUNSHINE, ANNUAL
Possible annual hours of sunshine:
4420 at equator
4449 at 25°N.lat.
4487 at 50°N.lat.
4580 at N. Pole
4390 at S. Pole
Values vary somewhat from year to year.
GULF OF MEXICO
HAWAII
ALASKA

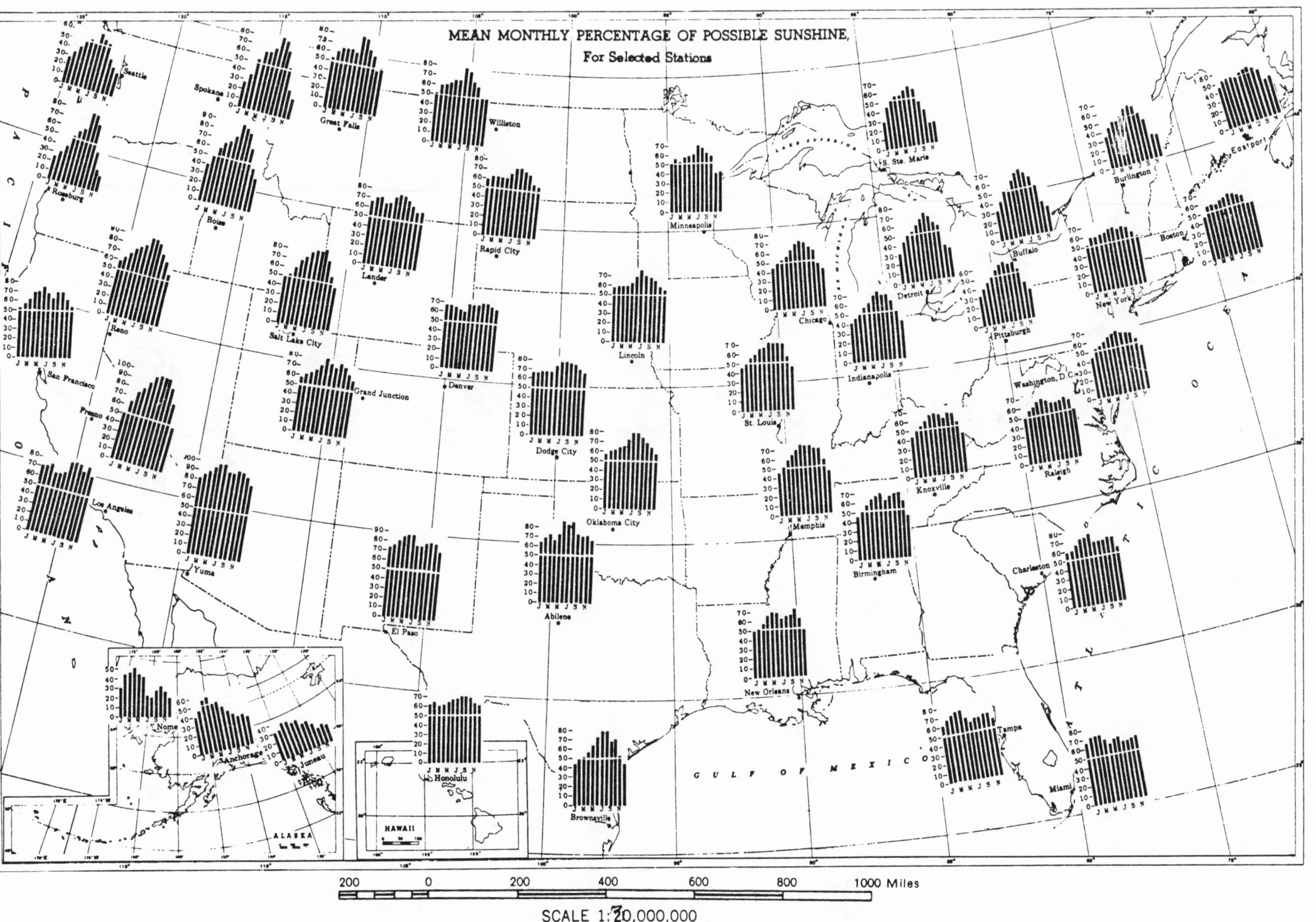
MEAN MONTHLY PERCENTAGE OF POSSIBLE SUNSHINE,
For Selected Stations
Seattle
Spokane
Great Falls
Williston
Roseburg
Boise
Rapid City
Lander
Minneapolis
S. Ste. Marie
Burlington
Eastport
Reno
Salt Lake City
Denver
Lincoln
Chicago
Detroit
Buffalo
Boston
New York
Pittsburgh
San Francisco
Fresno
Grand Junction
Dodge City
St. Louis
Indianapolis
Washington, D.C.
Raleigh
Knoxville
Los Angeles
Yuma
Oklahoma City
Memphis
Birmingham
Charleston
El Paso
Abilene
New Orleans
Tampa
Miami
Brownsville
Honolulu
HAWAII
Nome
Anchorage
Juneau
ALASKA
GULF OF MEXICO
200 0 200 400 600 800 1000 Miles
SCALE 1:20,000,000
ALBERS EQUAL AREA PROJECTION — STANDARD PARALLELS 29½° AND 45½°

MEAN PERCENTAGE OF POSSIBLE SUNSHINE

FOR SELECTED LOCATIONS

STATE AND STATION	YEARS	JAN.	FEB.	MAR.	APR.	MAY	JUNE	JULY	AUG.	SEPT.	OCT.	NOV.	DEC.	ANNUAL
ALA. BIRMINGHAM	56	43	49	56	63	66	67	62	65	66	67	58	44	59
MONTGOMERY	49	51	53	61	69	73	72	66	69	69	71	64	48	64
ALASKA. ANCHORAGE	19	39	46	56	58	50	51	45	39	35	32	33	29	45
FAIRBANKS	20	34	50	61	68	55	53	45	35	31	28	38	29	44
JUNEAU	14	30	32	39	37	34	35	28	30	25	18	21	18	30
NOME	29	44	46	48	53	51	48	32	26	34	35	36	30	41
ARIZ. PHOENIX	64	76	79	83	88	93	94	84	84	89	88	84	77	85
YUMA	52	83	87	91	94	97	98	92	91	93	93	90	83	91
ARK. LITTLE ROCK	66	44	53	57	62	67	72	71	73	71	74	58	47	62
CALIF. EUREKA	49	40	44	50	53	54	56	51	46	52	48	42	39	49
FRESNO	55	46	63	72	83	89	94	97	97	93	87	73	47	78
LOS ANGELES	63	70	69	70	67	68	69	80	81	80	76	79	72	73
RED BLUFF	39	50	60	65	75	79	86	95	94	89	77	64	50	75
SACRAMENTO	48	44	57	67	76	82	90	96	95	92	82	65	44	77
SAN DIEGO	68	68	67	68	66	60	60	67	70	70	70	76	71	68
SAN FRANCISCO	64	53	57	63	69	70	75	68	63	70	70	62	54	66
COLO. DENVER	64	67	67	65	63	61	69	68	68	71	71	67	65	67
GRAND JUNCTION	57	58	62	64	67	71	79	76	72	77	74	67	58	69
CONN. HARTFORD	48	46	55	56	54	57	60	62	60	57	55	46	46	56
D. C. WASHINGTON	66	46	53	56	57	61	64	64	62	62	61	54	47	58
FLA. APALACHICOLA	26	59	62	62	71	77	70	64	63	62	74	66	53	65
JACKSONVILLE	60	58	59	66	71	71	63	62	63	58	58	61	53	62
KEY WEST	45	68	75	78	78	76	70	69	71	65	65	69	66	71
MIAMI BEACH	48	66	72	73	73	68	62	65	67	62	62	65	65	67
TAMPA	63	63	67	71	74	75	66	61	64	64	67	67	61	68
GA. ATLANTA	65	48	53	57	65	68	68	62	63	65	67	60	47	60
HAWAII. HILO	9	48	42	41	34	31	41	44	38	42	41	34	36	39
HONOLULU	53	62	64	60	62	64	66	67	70	70	68	63	60	65
LIHUE	9	48	48	48	46	51	60	58	59	67	58	51	49	54
IDAHO. BOISE	20	40	48	59	67	68	75	89	86	81	66	46	37	66
POCATELLO	21	37	47	58	64	66	72	82	81	78	66	48	36	64
ILL. CAIRO	30	46	53	59	65	71	77	82	79	75	73	56	46	65
CHICAGO	66	44	49	53	56	63	69	73	70	65	61	47	41	59
SPRINGFIELD	59	47	51	54	58	64	69	76	72	73	64	53	45	60
IND. EVANSVILLE	48	42	49	55	61	67	73	78	76	73	67	52	42	64
FT. WAYNE	48	38	44	51	55	62	69	74	69	64	58	41	38	57
INDIANAPOLIS	63	41	47	49	55	62	68	74	70	68	64	48	39	59
IOWA. DES MOINES	66	56	56	56	59	62	66	75	70	64	64	53	48	62
DUBUQUE	54	48	52	52	58	60	63	73	67	61	55	44	40	57
SIOUX CITY	52	55	58	58	59	63	67	75	72	67	65	53	50	63
KANS. CONCORDIA	52	60	60	62	63	65	73	79	76	72	70	64	58	67
DODGE CITY	70	67	66	68	68	68	74	78	78	76	75	70	67	71
WICHITA	46	61	63	64	64	66	73	80	77	73	69	67	59	69
KY. LOUISVILLE	59	41	47	52	57	64	68	72	69	68	64	51	39	59
LA. NEW ORLEANS	69	49	50	57	63	66	64	58	60	64	70	60	46	59
SHREVEPORT	18	48	54	58	60	69	78	79	80	79	77	65	60	69
MAINE. EASTPORT	58	45	51	52	52	51	53	55	57	54	50	37	40	50
MASS. BOSTON	67	47	56	57	56	59	62	64	63	61	58	48	48	57
MICH. ALPENA	45	29	43	52	56	59	64	70	64	52	44	24	22	51
DETROIT	69	34	42	48	52	58	65	69	66	61	54	35	29	53
GRAND RAPIDS	56	26	37	48	54	60	66	72	67	58	50	31	22	49
MARQUETTE	55	31	40	47	52	53	56	63	57	47	38	24	24	47
S. STE. MARIE	60	28	44	50	54	54	59	63	58	45	36	21	22	47
MINN. DULUTH	49	47	55	60	58	58	60	68	63	53	47	36	40	55
MINNEAPOLIS	45	49	54	55	57	60	64	72	69	60	54	40	40	56
MISS. VICKSBURG	66	46	50	57	64	69	73	69	72	74	71	60	45	64
MO. KANSAS CITY	69	55	57	59	60	64	70	76	73	70	67	59	52	65
ST. LOUIS	68	48	49	56	59	64	68	72	68	67	65	54	44	61
SPRINGFIELD	45	48	54	57	60	63	69	77	72	71	65	58	48	63
MONT. HAVRE	55	49	58	61	63	63	65	78	75	64	57	48	46	62
HELENA	65	46	55	58	60	59	63	77	74	63	57	48	43	60
KALISPELL	50	28	40	49	57	58	60	77	73	61	50	28	20	53
NEBR. LINCOLN	55	57	59	60	60	63	69	76	71	67	66	59	55	64
NORTH PLATTE	53	63	63	64	62	64	72	78	74	72	70	62	58	68
NEV. ELY	21	61	64	68	65	67	79	79	81	81	73	67	62	72
LAS VEGAS	19	74	77	78	81	85	91	84	86	92	84	83	75	82
RENO	51	59	64	69	75	77	82	90	89	86	76	68	56	76
WINNEMUCCA	53	52	60	64	70	76	83	90	90	86	75	62	53	74
N. H. CONCORD	44	48	53	55	53	51	56	57	58	55	50	43	43	52
N. J. ATLANTIC CITY	62	51	57	58	59	62	65	67	66	65	54	58	52	60
N. MEX. ALBUQUERQUE	28	70	72	72	76	79	84	76	75	81	80	79	70	76
ROSWELL	47	69	72	75	77	76	80	76	75	74	74	74	69	74
N. Y. ALBANY	63	43	51	53	53	57	62	63	61	58	54	39	38	53
BINGHAMTON	63	31	39	41	44	50	56	54	51	47	43	29	26	44
BUFFALO	49	32	41	49	51	59	67	70	67	60	51	31	28	53
CANTON	43	37	47	50	48	54	61	63	61	54	45	30	31	49
NEW YORK	83	49	56	57	59	62	65	66	64	64	61	53	50	59
SYRACUSE	49	31	38	45	50	58	64	67	63	56	47	29	26	50
N. C. ASHEVILLE	57	48	53	56	61	64	63	59	59	62	64	59	48	58
RALEIGH	61	50	56	59	64	67	65	62	62	63	64	62	52	61
N. DAK. BISMARCK	65	52	58	56	57	58	61	73	69	62	59	49	48	59
DEVILS LAKE	55	53	60	59	60	59	62	71	67	59	56	44	45	58
FARGO	39	47	55	56	58	62	63	73	69	60	57	39	46	59
WILLISTON	43	51	59	60	63	66	66	78	75	65	60	48	48	63
OHIO. CINCINNATI	44	41	46	52	56	62	69	72	68	68	60	46	39	57
CLEVELAND	65	29	36	45	52	61	67	71	68	62	54	32	25	50
COLUMBUS	65	36	44	49	54	63	68	71	68	66	60	44	35	55
OKLA. OKLAHOMA CITY	62	57	60	63	64	65	74	78	78	74	68	64	57	68
OREG. BAKER	46	41	49	56	61	63	67	83	81	74	62	46	37	60
PORTLAND	69	27	34	41	49	52	55	70	65	55	42	28	23	48
ROSEBURG	29	24	32	40	51	57	59	79	77	68	42	28	18	51
PA. HARRISBURG	60	43	52	55	57	61	65	68	63	62	58	47	43	57
PHILADELPHIA	66	45	56	57	58	61	62	64	61	62	61	53	49	57
PITTSBURGH	63	32	39	45	50	57	62	64	61	62	54	39	30	51
R. I. BLOCK ISLAND	48	45	54	47	56	58	60	62	62	60	59	50	44	56
S. C. CHARLESTON	61	58	60	65	72	73	70	66	66	67	68	68	57	66
COLUMBIA	55	53	57	62	68	69	68	63	65	64	68	64	51	63
S. DAK. HURON	62	55	62	60	62	65	68	76	72	66	61	52	49	63
RAPID CITY	53	58	62	63	62	61	66	73	73	69	66	58	54	64
TENN. KNOXVILLE	62	42	49	53	59	64	66	64	59	64	64	53	41	57
MEMPHIS	55	44	51	57	64	68	74	73	74	70	69	58	45	64
NASHVILLE	63	42	47	54	60	65	69	69	68	69	65	55	42	59
TEX. ABILENE	14	64	68	73	66	73	86	83	85	73	71	72	66	73
AMARILLO	54	71	71	75	75	75	82	81	81	79	76	76	70	76
AUSTIN	33	46	50	57	60	62	72	76	79	70	70	57	49	63
BROWNSVILLE	37	44	49	51	57	65	73	78	78	67	70	54	44	61
DEL RIO	36	53	55	61	63	60	66	75	80	69	66	58	52	63
EL PASO	53	74	77	81	85	87	87	78	78	80	82	80	73	80
FT. WORTH	33	56	57	65	66	67	75	78	78	74	70	63	58	68
GALVESTON	66	50	50	55	61	69	76	72	71	70	74	62	49	63
SAN ANTONIO	57	48	51	56	58	60	69	74	75	69	67	55	49	62
UTAH. SALT LAKE CITY	22	48	53	61	68	73	78	82	82	84	73	56	49	69
VT. BURLINGTON	54	34	43	48	47	53	59	62	59	51	43	25	24	46
VA. NORFOLK	60	50	57	60	63	67	66	66	66	63	64	60	51	62
RICHMOND	56	49	55	59	63	67	66	65	62	63	64	58	50	61
WASH. NORTH HEAD	44	28	37	42	48	48	48	50	46	48	41	31	27	41
SEATTLE	26	27	34	42	48	53	48	62	56	53	36	28	24	45
SPOKANE	62	26	41	53	63	64	68	82	79	68	53	28	22	58
TATOOSH ISLAND	49	26	36	39	45	47	46	48	44	47	38	26	23	40
WALLA WALLA	44	24	35	51	63	67	72	86	84	72	59	33	20	60
YAKIMA	18	34	49	62	70	72	74	86	86	74	61	38	29	65
W. VA. ELKINS	55	33	37	42	47	55	55	56	53	55	51	41	33	48
PARKERSBURG	62	30	36	42	49	56	60	63	60	60	53	37	29	48
WIS. GREEN BAY	57	44	51	55	56	58	64	70	65	58	52	40	40	55
MADISON	59	44	49	52	53	58	64	70	66	60	56	41	38	56
MILWAUKEE	59	44	48	53	56	60	65	73	67	62	56	44	39	57
WYO. CHEYENNE	63	65	66	64	61	59	68	70	68	69	69	65	63	66
LANDER	57	66	70	71	66	65	74	76	75	72	67	61	62	69
SHERIDAN	52	56	61	62	61	61	67	76	74	67	60	53	52	64
YELLOWSTONE PARK	35	39	51	55	57	56	63	73	71	65	57	45	38	56
P. R. SAN JUAN	57	64	69	71	66	59	62	65	67	61	63	63	65	65

Based on period of record through December 1959, except in a few instances.

These charts and tabulation derived from "Normals, Means, and Extremes" table in U. S. Weather Bureau publication Local Climatological Data, except inset table on charts from U. S. Naval Observatory publication Tables of Sunrise, Sunset, and Twilight.

MEAN MONTHLY TOTAL HOURS OF SUNSHINE
JANUARY
GULF OF MEXICO
ALASKA
ARCTIC CIRCLE
HAWAII
PUERTO RICO AND VIRGIN ISLANDS

MEAN MONTHLY TOTAL HOURS OF SUNSHINE,
FEBRUARY
GULF OF MEXICO
ALASKA
ARCTIC CIRCLE
HAWAII
PUERTO RICO AND VIRGIN ISLANDS

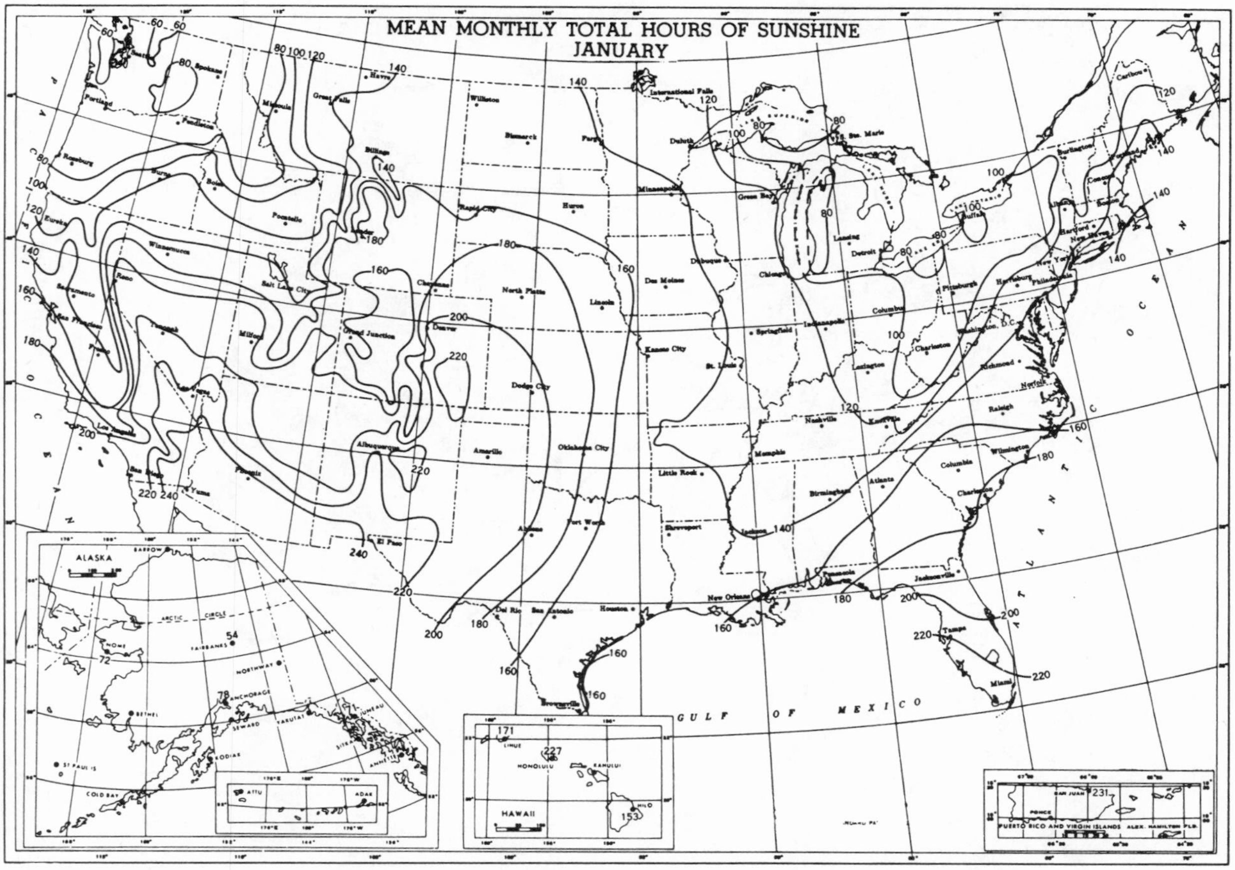

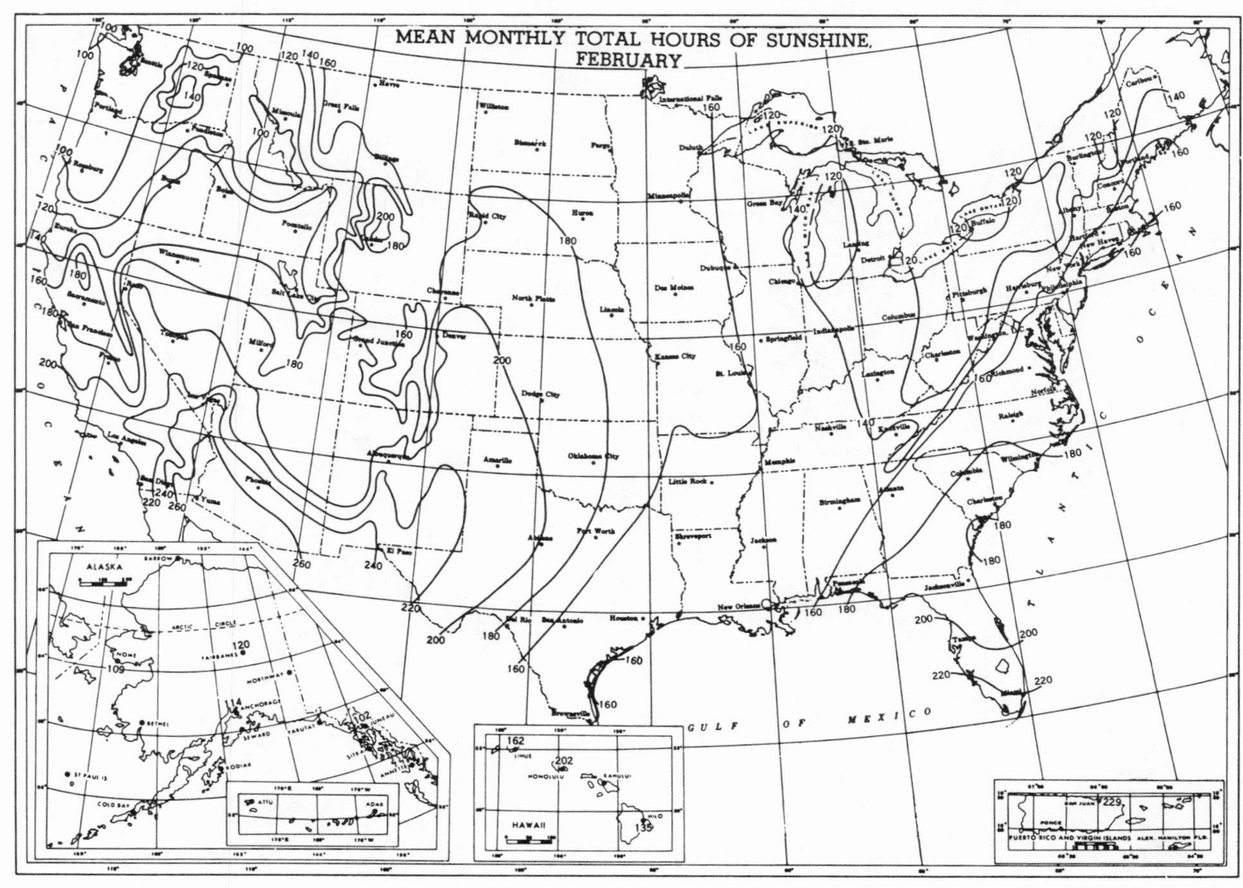

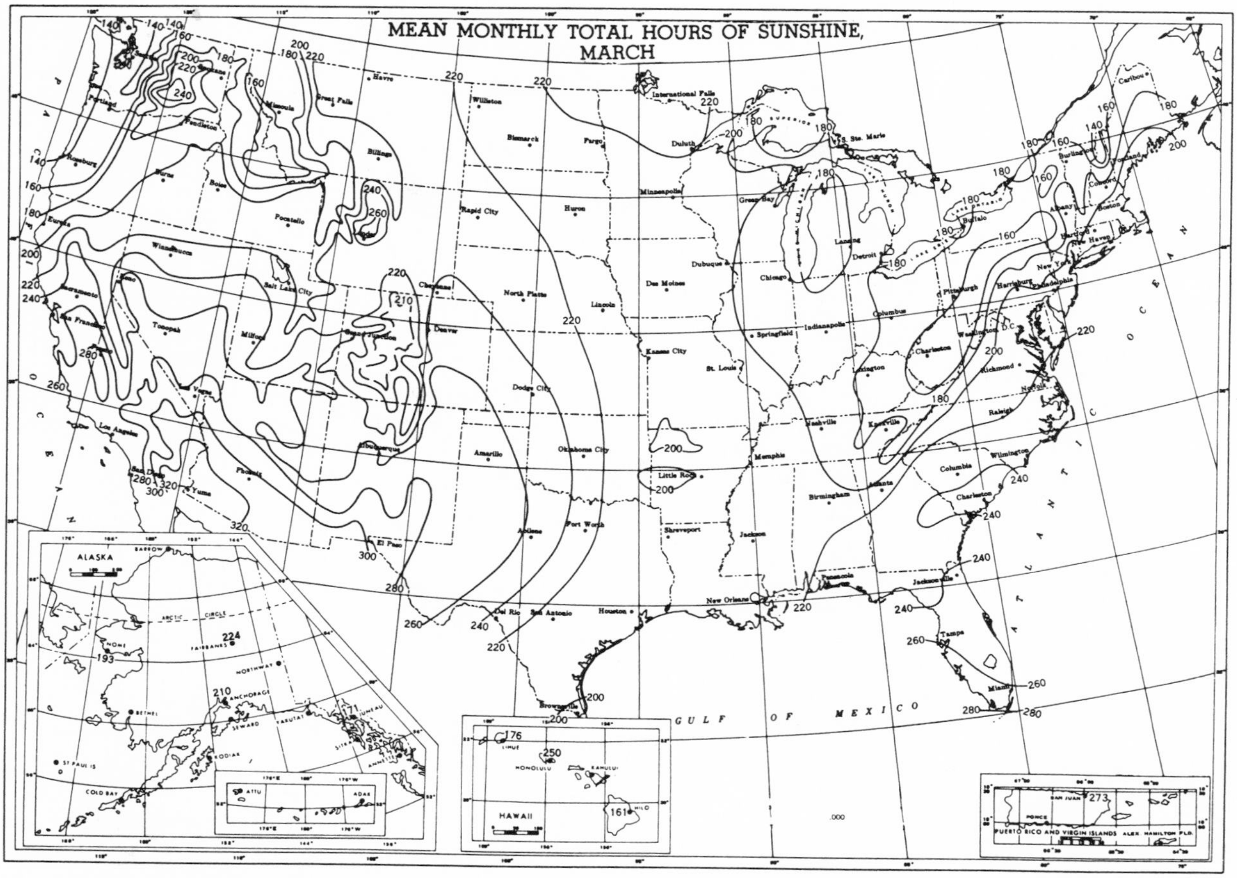

MEAN MONTHLY TOTAL HOURS OF SUNSHINE, MARCH

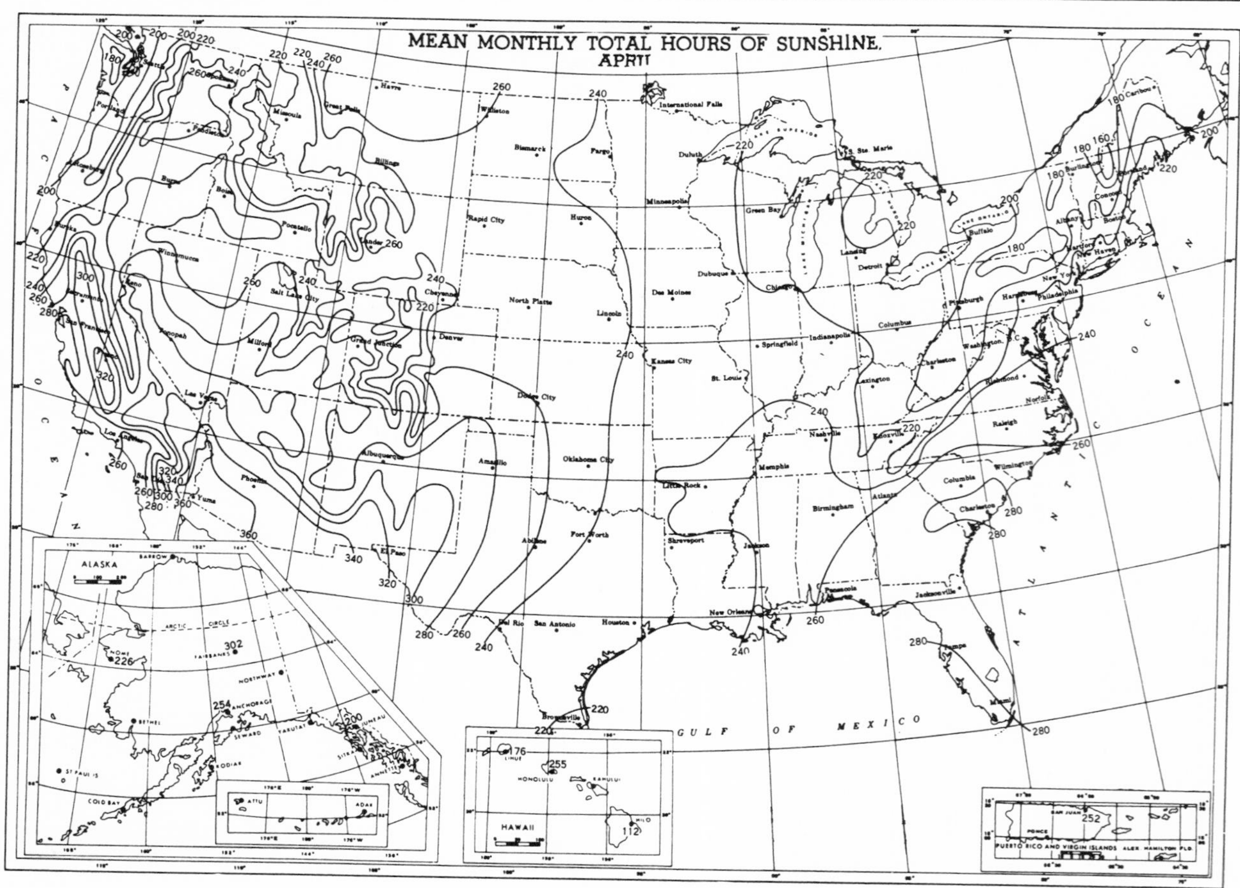

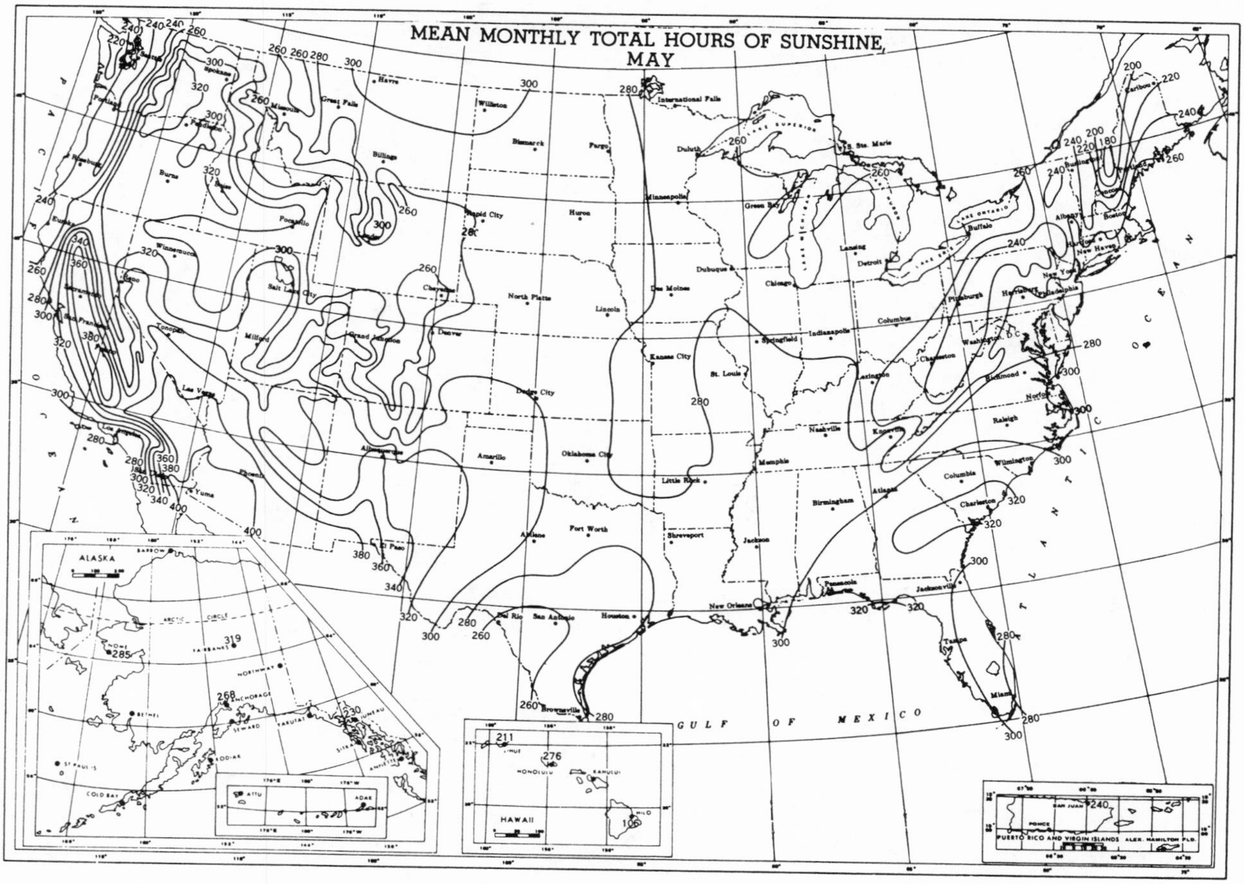
MEAN MONTHLY TOTAL HOURS OF SUNSHINE,
MAY
GULF OF MEXICO
ALASKA
HAWAII
PUERTO RICO AND VIRGIN ISLANDS

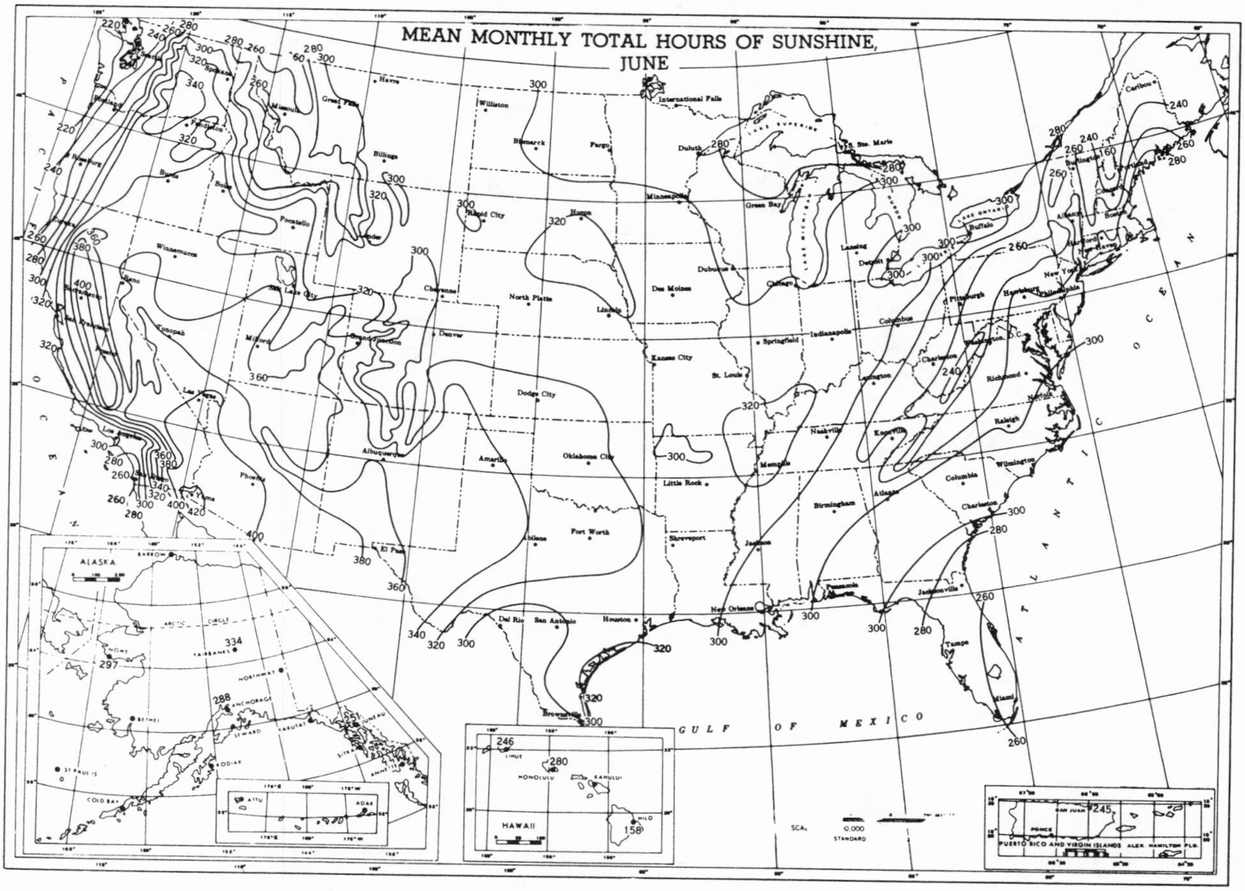
MEAN MONTHLY TOTAL HOURS OF SUNSHINE,
JUNE
GULF OF MEXICO
ALASKA
HAWAII
PUERTO RICO AND VIRGIN ISLANDS

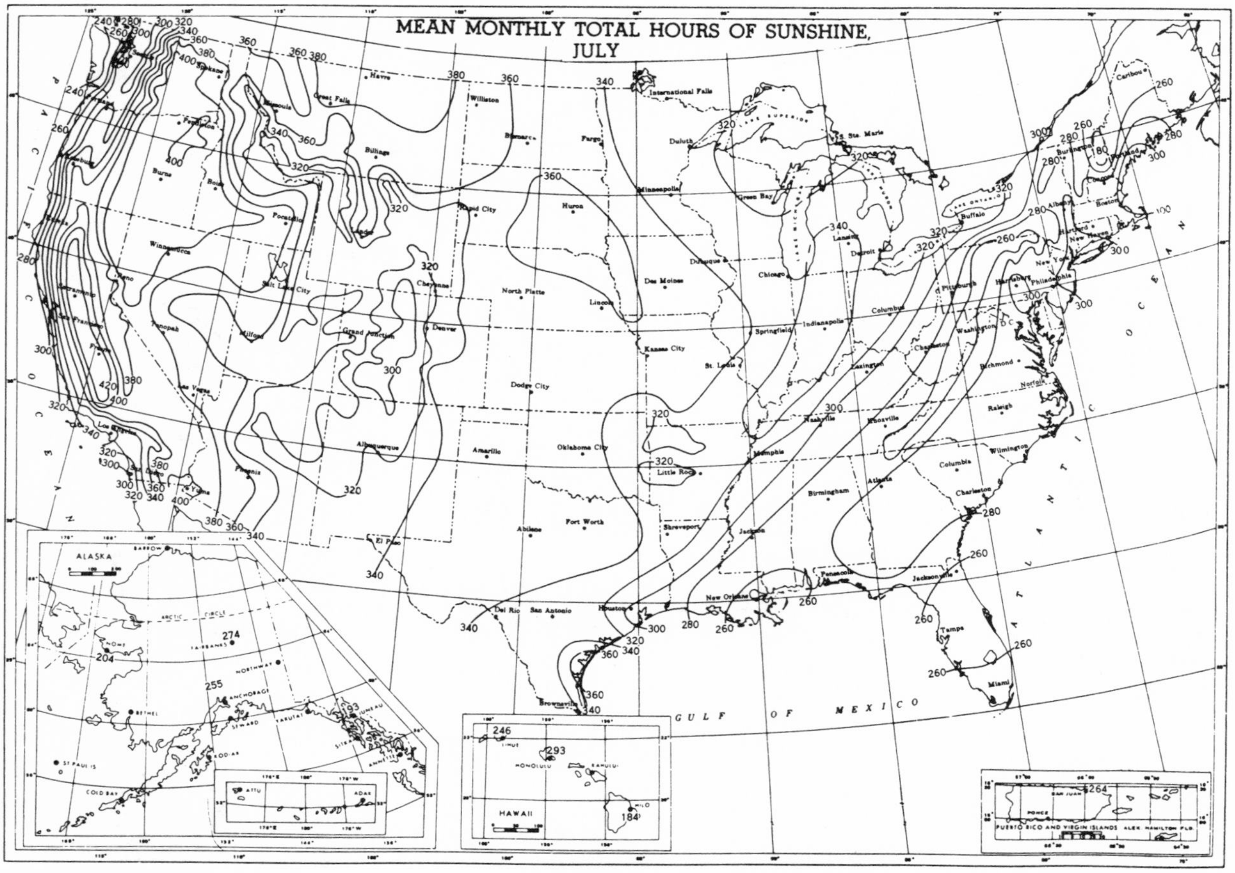
MEAN MONTHLY TOTAL HOURS OF SUNSHINE,
JULY
ALASKA
HAWAII
PUERTO RICO AND VIRGIN ISLANDS

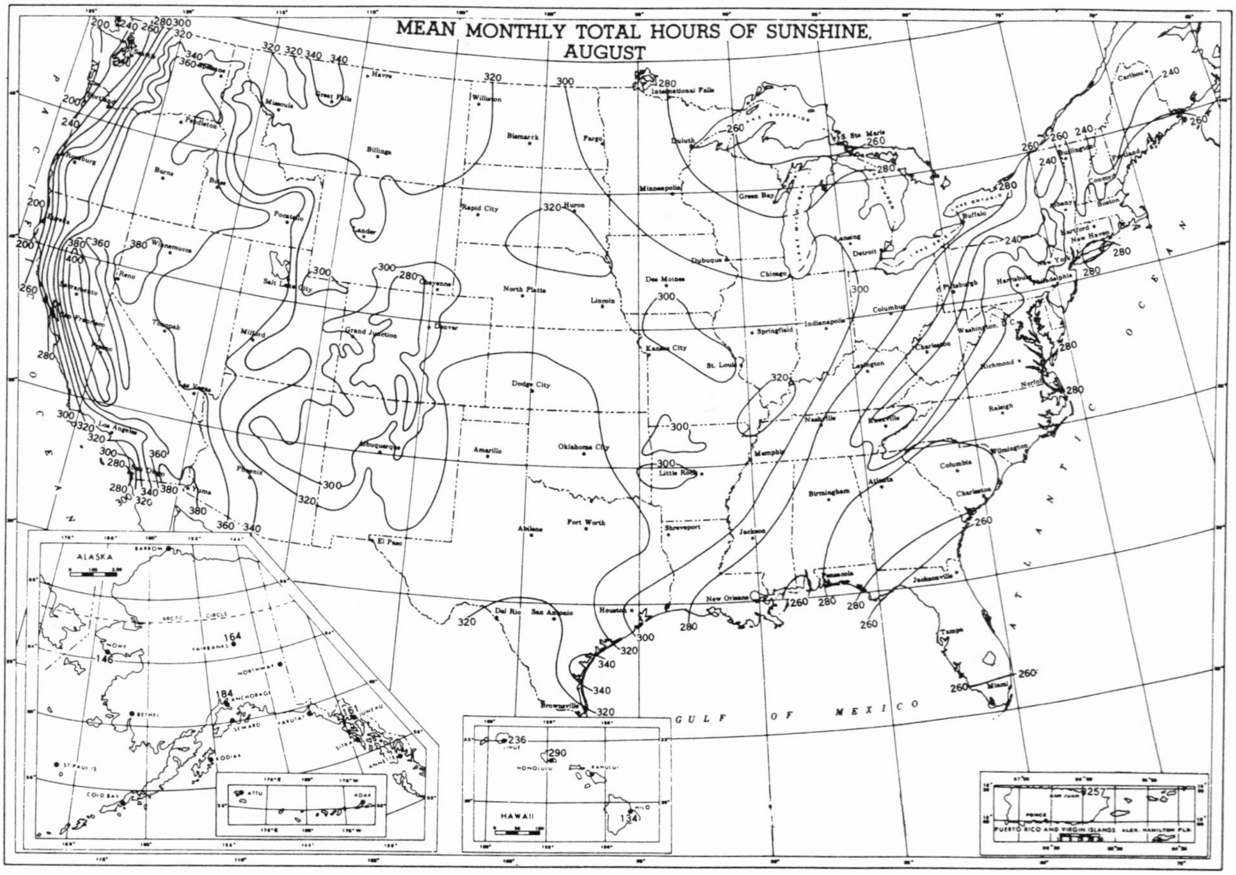
MEAN MONTHLY TOTAL HOURS OF SUNSHINE,
AUGUST
ALASKA
HAWAII
PUERTO RICO AND VIRGIN ISLANDS

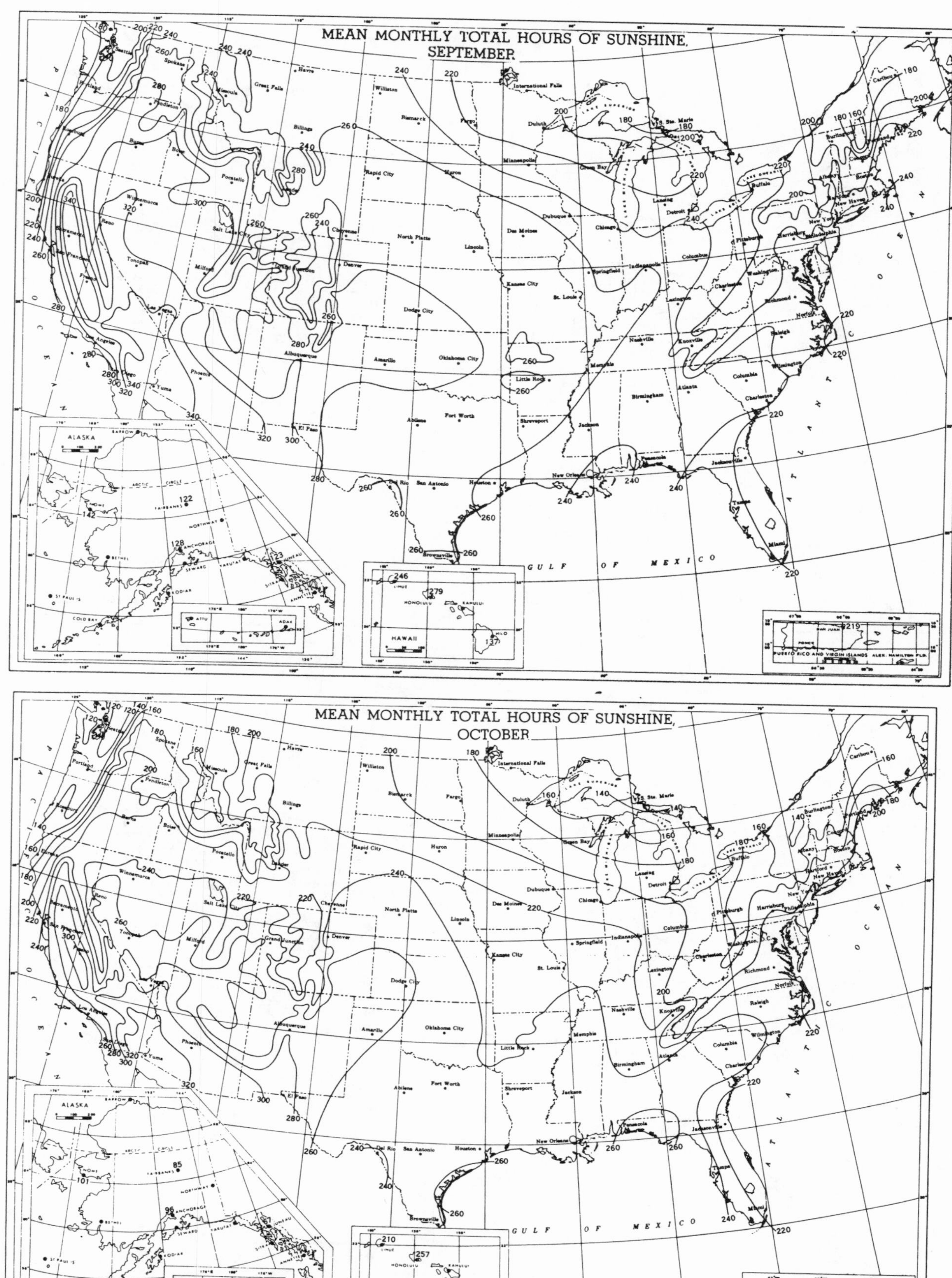
MEAN MONTHLY TOTAL HOURS OF SUNSHINE, SEPTEMBER
ALASKA
HAWAII
GULF OF MEXICO
PUERTO RICO AND VIRGIN ISLANDS ALEX. HAMILTON FLD.
MEAN MONTHLY TOTAL HOURS OF SUNSHINE, OCTOBER
ALASKA
HAWAII
GULF OF MEXICO
PUERTO RICO AND VIRGIN ISLANDS ALEX. HAMILTON FLD.

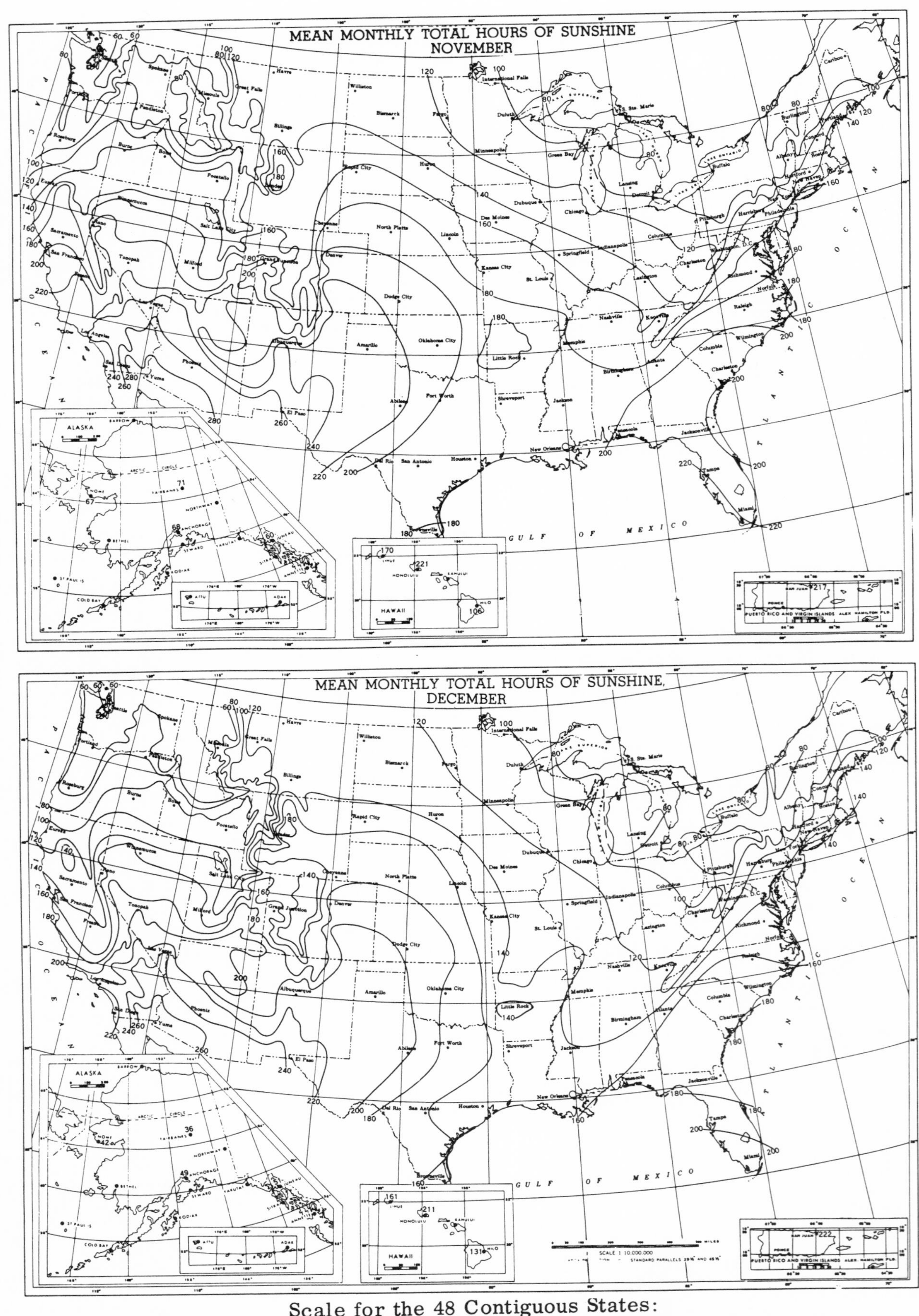

Scale for the 48 Contiguous States:

300 0 300 600 900 1200 1500 Miles

SCALE 1:30,000,000

ALBERS EQUAL AREA PROJECTION–STANDARD PARALLELS 29½° AND 45½°

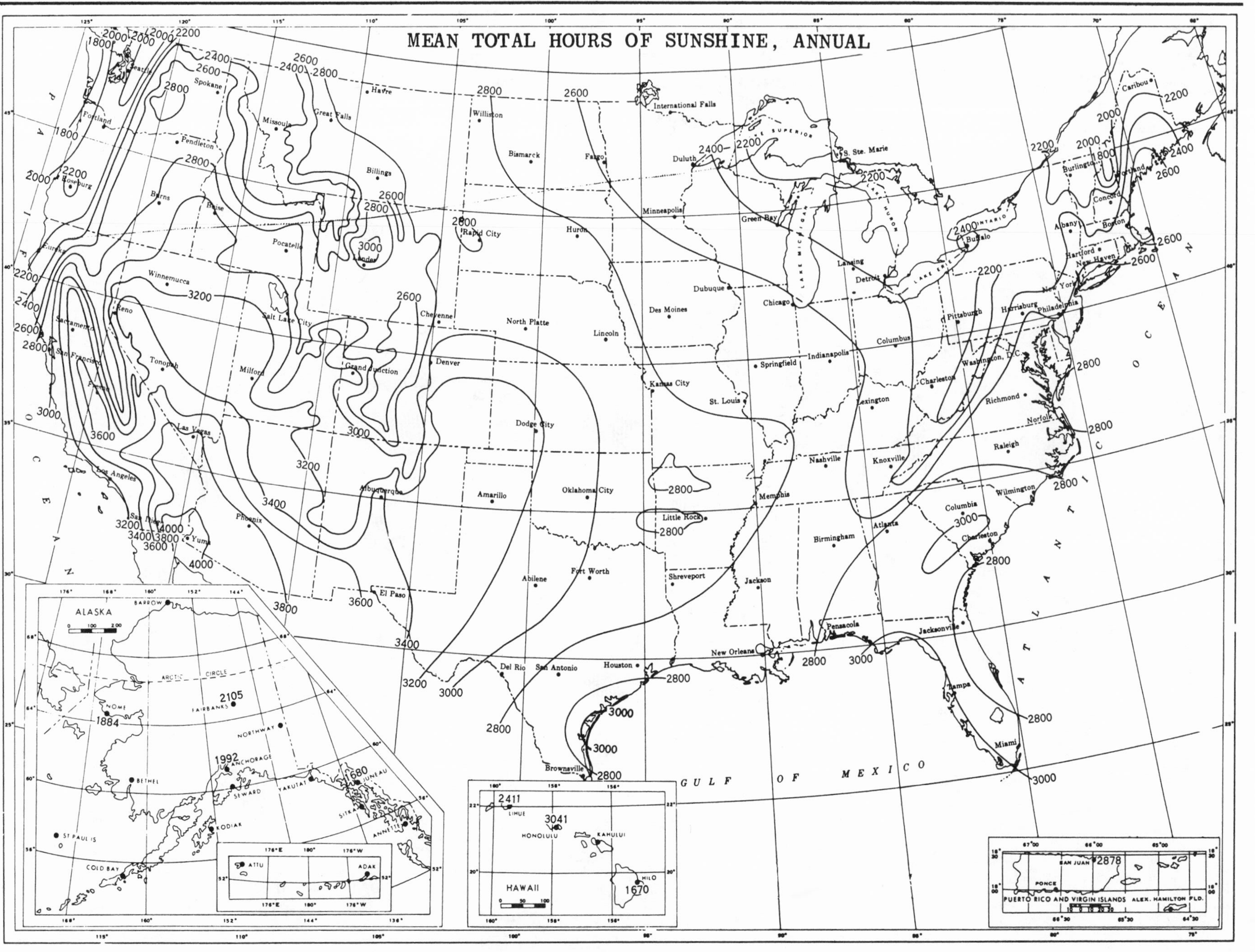
MEAN TOTAL HOURS OF SUNSHINE, ANNUAL
ALASKA
HAWAII
PUERTO RICO AND VIRGIN ISLANDS
GULF OF MEXICO
PACIFIC OCEAN
ATLANTIC OCEAN

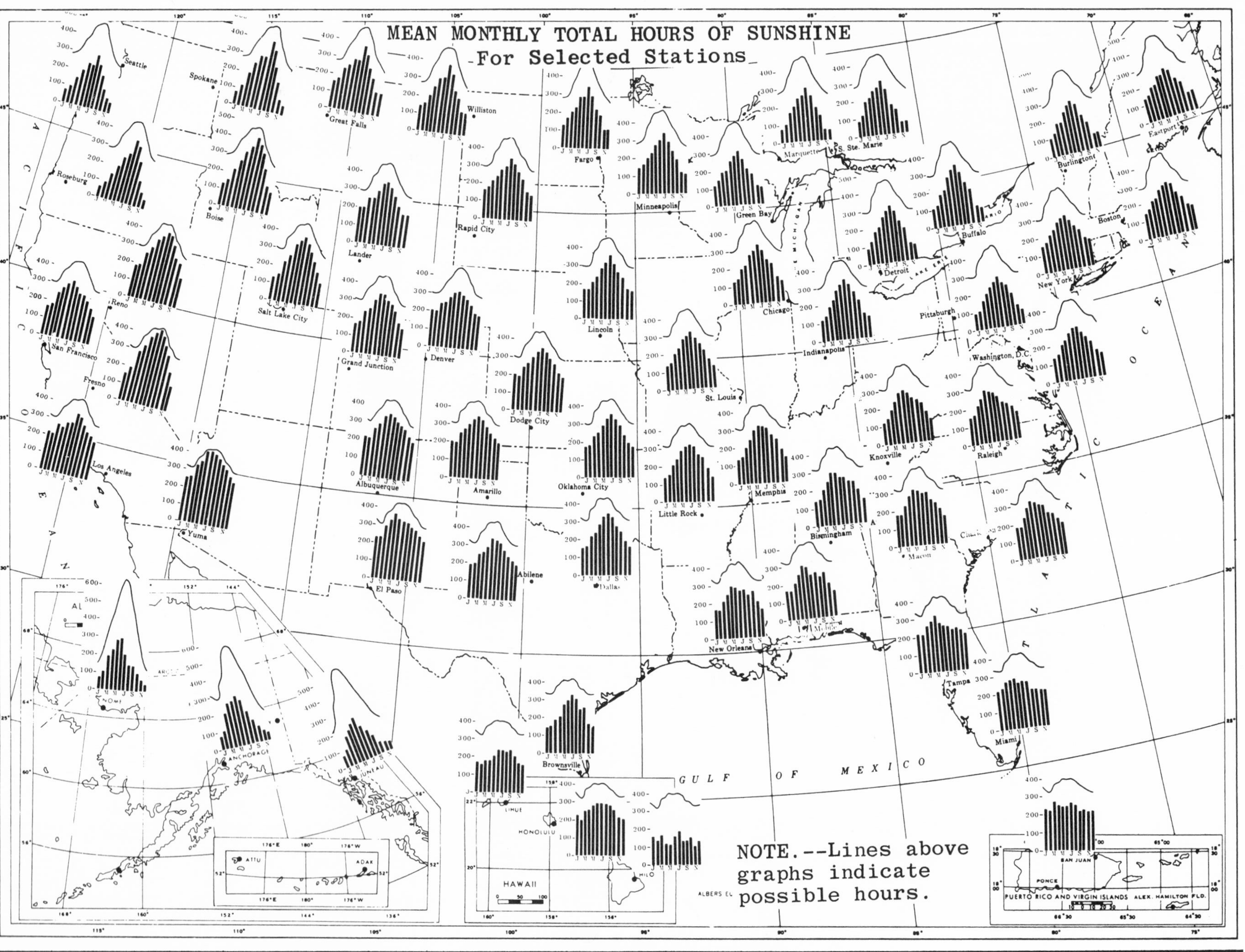
MEAN MONTHLY TOTAL HOURS OF SUNSHINE
-For Selected Stations-
Seattle
Spokane
Roseburg
Boise
Reno
San Francisco
Fresno
Los Angeles
Yuma
Salt Lake City
Great Falls
Lander
Grand Junction
Albuquerque
El Paso
Williston
Rapid City
Denver
Amarillo
Abilene
Fargo
Lincoln
Dodge City
Oklahoma City
Dallas
Minneapolis
St. Louis
Little Rock
Brownsville
Green Bay
Chicago
Memphis
New Orleans
Marquette
S. Ste. Marie
Indianapolis
Birmingham
Detroit
Knoxville
Buffalo
Pittsburgh
Raleigh
Washington, D.C.
Tampa
Miami
Burlington
New York
Boston
Eastport
GULF OF MEXICO
HAWAII
Honolulu
Hilo
Lihue
Juneau
Anchorage
Nome
Attu
Adak
PUERTO RICO AND VIRGIN ISLANDS
San Juan
Ponce
NOTE.--Lines above graphs indicate possible hours.
Scale for the 48 Contiguous States:
200 0 200 400 600 800 1000 Miles
SCALE 1:20,000,000
ALBERS EQUAL AREA PROJECTION-STANDARD PARALLELS 29½° AND 45½°

MEAN NUMBER OF HOURS OF SUNSHINE

STATE AND STATION	YEARS	JAN.	FEB.	MAR.	APR.	MAY	JUNE	JULY	AUG.	SEPT.	OCT.	NOV.	DEC.	ANNUAL
ALA. BIRMINGHAM	30	138	152	207	248	293	294	269	265	244	234	182	136	2662
MOBILE	22	157	158	212	253	301	289	249	259	235	254	195	146	2708
MONTGOMERY	30	160	168	227	267	317	311	288	290	260	250	200	156	2894
ALASKA ANCHORAGE	19	78	114	210	254	268	288	255	184	128	96	68	49	1992
FAIRBANKS	20	54	120	224	302	319	334	274	164	122	85	71	36	2105
JUNEAU	29	71	102	171	200	230	251	193	161	123	67	60	51	1680
NOME	27	72	109	193	226	285	297	204	146	142	101	67	42	1884
ARIZ. PHOENIX	30	248	244	314	346	404	404	377	351	334	307	267	236	3832
PRESCOTT	14	222	230	293	323	378	392	323	305	315	286	254	228	3549
TUCSON	13	255	266	317	350	399	394	329	329	335	317	280	258	3829
YUMA	30	258	266	337	365	419	420	404	380	351	330	285	262	4077
ARK. FT. SMITH	30	146	156	202	234	268	303	321	305	261	230	174	147	2747
LITTLE ROCK	30	143	158	213	243	291	316	321	316	265	251	181	142	2840
CALIF. EUREKA	30	120	138	180	209	247	261	244	205	195	164	127	108	2198
FRESNO	29	153	192	283	330	389	418	435	406	355	306	221	144	3632
LOS ANGELES	30	224	217	273	264	292	299	352	336	295	263	249	220	3284
RED BLUFF	15	156	186	246	302	366	396	438	407	341	277	199	154	3468
SACRAMENTO	30	134	169	255	300	367	405	437	406	347	283	197	122	3422
SAN DIEGO	30	216	212	262	242	261	253	293	277	255	234	236	217	2958
SAN FRANCISCO	30	165	182	251	281	314	330	300	272	267	243	198	156	2959
COLO. DENVER	30	207	205	247	252	281	311	321	297	274	246	200	192	3033
GRAND JUNCTION	30	169	182	243	265	314	350	349	311	291	255	198	168	3095
PUEBLO	30	224	217	261	271	299	340	349	318	290	265	225	211	3270
CONN. HARTFORD	30	141	166	206	223	267	285	299	268	220	193	137	136	2541
NEW HAVEN	30	155	178	215	234	274	291	309	284	238	215	157	154	2704
D. C. WASHINGTON	30	138	160	205	226	267	288	291	264	233	207	162	135	2576
FLA. APALACHICOLA	26	193	195	233	274	328	296	273	259	236	263	216	175	2941
JACKSONVILLE	30	192	189	241	267	296	260	255	248	199	205	191	170	2713
KEY WEST	30	229	238	285	296	307	273	277	269	236	237	226	225	3098
LAKELAND	7	204	186	222	251	285	268	252	242	203	209	212	198	2732
MIAMI	30	222	227	266	275	280	251	267	263	216	215	212	209	2903
PENSACOLA	30	175	180	232	270	311	302	278	284	249	265	206	166	2918
TAMPA	30	223	220	260	283	320	275	257	252	232	243	227	209	3001
GA. ATLANTA	25	154	165	218	266	309	304	284	285	247	241	188	160	2821
MACON	30	177	178	235	279	321	314	292	295	253	236	202	168	2950
SAVANNAH	30	175	173	229	274	307	279	267	256	212	216	197	167	2752
HAWAII HILO	7	153	135	161	112	106	158	184	134	137	153	106	131	1670
HONOLULU	30	227	202	250	255	276	280	293	290	279	257	221	211	3041
LIHUE	10	171	162	176	176	211	246	246	236	246	210	170	161	2411
IDAHO BOISE	30	116	144	218	274	322	352	412	378	311	232	143	104	3006
POCATELLO	30	111	143	211	255	300	338	380	347	296	230	145	108	2864
ILL. CAIRO	15	124	160	218	254	298	324	345	336	279	254	181	145	2918
CHICAGO	30	126	142	199	221	274	300	333	299	247	216	136	118	2611
MOLINE	18	132	139	189	214	255	279	337	300	251	214	130	123	2563
PEORIA	30	134	149	198	229	273	303	336	299	259	222	149	122	2673
SPRINGFIELD	30	127	149	193	224	282	304	346	312	266	225	152	122	2702
IND. EVANSVILLE	30	123	145	199	237	294	322	342	318	274	236	156	120	2766
FT. WAYNE	30	113	136	191	217	281	310	342	306	242	210	120	102	2570
INDIANAPOLIS	30	118	140	193	227	278	313	342	313	265	222	139	118	2668
TERRE HAUTE	24	125	148	189	231	274	302	341	305	253	235	150	122	2675
IOWA BURLINGTON	19	148	165	217	241	284	315	353	327	270	243	175	147	2885
CHARLES CITY	22	137	157	190	226	258	285	336	290	241	207	130	115	2572
DES MOINES	30	155	170	203	236	276	303	346	299	263	227	156	136	2770
SIOUX CITY	30	164	177	216	254	300	320	363	320	270	236	160	146	2926
KAN. CONCORDIA	30	180	172	214	243	281	315	348	308	249	245	189	172	2916
DODGE CITY	30	205	191	249	268	305	335	359	335	290	266	218	198	3219
TOPEKA	18	159	160	193	215	260	287	310	304	263	229	173	149	2702
WICHITA	30	187	186	233	254	291	321	350	325	277	245	206	182	3057
KY. LOUISVILLE	30	115	135	188	221	283	303	324	295	256	219	148	114	2601
LA. NEW ORLEANS	30	160	158	213	247	292	287	260	269	241	260	200	157	2744
SHREVEPORT	19	151	172	214	240	298	332	339	322	289	273	208	177	3015
MAINE EASTPORT	22	133	151	196	201	245	248	275	260	205	175	105	115	2309
PORTLAND	30	155	174	213	226	268	286	312	294	229	202	146	148	2653
MD. BALTIMORE	30	148	170	211	229	270	295	299	272	238	212	164	145	2653
MASS. BLUE HILL OBS.	10	125	136	165	182	233	248	266	241	211	181	134	135	2257
BOSTON	30	148	168	212	222	263	283	300	280	232	207	152	148	2615
NANTUCKET	22	128	156	214	227	278	284	291	279	242	208	149	129	2585
MICH. ALPENA	24	86	124	198	228	261	303	339	285	204	159	70	67	2324
DETROIT	30	90	128	180	212	263	295	321	284	226	189	98	89	2375
LANSING	30	84	119	175	215	272	305	344	294	228	182	87	73	2378
ESCANABA	30	112	148	204	226	266	283	316	267	198	162	90	94	2366
GRAND RAPIDS	30	74	117	178	218	277	308	349	304	231	188	92	70	2406
MARQUETTE	30	78	113	172	207	248	268	305	251	186	142	68	66	2104
SAULT STE. MARIE	30	83	123	187	217	252	269	309	256	165	133	61	62	2117
MINN. DULUTH	30	125	163	221	235	268	282	328	277	203	166	100	107	2475
MINNEAPOLIS	30	140	166	200	231	272	302	343	296	237	193	115	112	2607
MISS. JACKSON	12	130	147	199	244	280	287	279	287	235	223	185	150	2646
VICKSBURG	30	136	141	199	232	284	304	291	297	254	244	183	140	2705
MO. COLUMBIA	30	147	164	207	232	281	296	341	298	262	225	166	138	2757
KANSAS CITY	30	154	170	211	235	278	313	347	308	266	235	178	151	2846
ST. JOSEPH	23	154	165	211	231	274	301	347	287	260	224	168	144	2766
ST. LOUIS	30	137	152	202	235	283	301	325	289	256	223	166	125	2694
SPRINGFIELD	30	145	164	213	238	278	305	342	310	269	233	183	140	2820
MONT. BILLINGS	21	140	154	208	236	283	301	372	332	258	213	136	129	2762
GREAT FALLS	19	154	176	245	261	299	299	381	342	256	206	132	133	2884
HAVRE	30	136	174	234	268	311	312	384	339	260	202	132	122	2874
HELENA	30	138	168	215	241	292	292	342	336	258	202	137	121	2742
MISSOULA	25	85	109	167	209	261	260	378	328	246	178	90	66	2377
NEBR. LINCOLN	30	173	172	213	244	287	316	356	309	266	237	174	160	2907
NORTH PLATTE	30	181	179	221	246	282	310	343	304	264	242	184	169	2925
OMAHA	30	172	188	222	259	305	332	379	311	270	248	166	145	2997
VALENTINE	30	185	194	229	252	296	323	369	326	275	242	174	172	3037
NEV. ELY	22	186	197	262	260	300	354	359	344	303	255	204	187	3211
LAS VEGAS	8	239	251	314	336	386	411	383	364	345	301	258	250	3838
RENO	30	185	199	267	306	354	376	414	391	336	273	212	170	3483
WINNEMUCCA	30	142	155	207	255	312	346	395	375	316	242	177	139	3061
N. H. CONCORD	23	136	153	192	196	229	261	286	260	214	179	122	126	2354
MT. WASHINGTON OBS.	18	94	98	133	141	162	145	150	143	139	159	89	87	1540
N. J. ATLANTIC CITY	30	151	173	210	233	273	287	298	271	239	218	177	153	2683
TRENTON	30	145	168	203	235	277	294	309	273	239	208	160	142	2653
N. MEX. ALBUQUERQUE	30	221	218	273	299	343	365	340	317	299	279	245	219	3418
ROSWELL	21	218	223	286	306	330	333	341	313	266	266	242	216	3340
N. Y. ALBANY	30	125	151	194	213	266	301	317	286	224	192	115	112	2496
BINGHAMTON	30	94	119	151	170	226	256	266	230	184	158	92	79	2025
BUFFALO	30	110	125	180	212	274	319	338	297	239	183	97	84	2458
NEW YORK	30	154	171	213	237	268	289	302	271	235	213	169	155	2677
ROCHESTER	30	93	123	172	209	274	314	333	294	224	173	97	86	2392
SYRACUSE	30	87	115	165	197	261	295	316	276	211	163	81	74	2241
N. C. ASHEVILLE	30	146	161	211	247	289	292	268	250	235	222	179	146	2646
CAPE HATTERAS	9	152	168	206	259	293	301	286	265	214	202	169	154	2669
CHARLOTTE	30	165	177	230	267	313	316	291	277	247	243	198	167	2891
GREENSBORO	30	157	171	217	231	298	302	287	272	243	236	190	163	2767
RALEIGH	29	154	168	220	255	290	284	277	253	224	215	184	156	2680
WILMINGTON	30	179	180	237	279	314	312	286	273	237	238	206	178	2919
N. DAK. BISMARCK	30	141	170	205	236	279	294	358	307	243	198	130	125	2686
DEVILS LAKE	30	150	177	220	250	291	297	352	302	230	198	123	124	2714
FARGO	30	132	170	210	232	283	288	343	293	222	187	112	114	2586
WILLISTON	29	141	168	215	260	305	312	377	328	247	206	131	129	2819
OHIO CINCINNATI (ABBE)	30	115	137	186	222	273	309	323	295	253	205	138	118	2574
CLEVELAND	30	79	111	167	209	274	301	325	288	235	187	99	77	2352
COLUMBUS	30	112	132	177	215	270	296	323	291	250	210	131	101	2508
DAYTON	10	114	136	195	222	281	313	323	307	268	229	152	124	2664
SANDUSKY	30	100	128	183	229	285	312	343	302	248	201	111	91	2533
TOLEDO	30	93	120	170	203	263	296	331	298	241	196	106	92	2409
OKLA. OKLAHOMA CITY	29	175	182	235	253	290	329	352	331	282	243	201	175	3048
TULSA	18	152	164	200	213	244	287	314	308	281	241	207	172	2783
OREG. BAKER	22	118	143	198	251	302	313	406	368	289	215	132	100	2835
PORTLAND	30	77	97	142	203	246	249	329	275	218	134	87	65	2122
ROSEBURG	30	69	96	148	205	257	278	369	329	255	146	81	50	2283
PA. HARRISBURG	30	132	160	203	230	277	297	319	282	233	200	140	131	2604
PHILADELPHIA	30	142	166	203	231	270	281	288	253	225	205	158	142	2564
PITTSBURGH	25	89	114	163	200	239	260	283	250	234	180	114	76	2202
READING	30	133	151	195	220	259	275	293	259	219	198	144	127	2473
SCRANTON	30	108	138	178	199	251	269	290	249	213	183	120	105	2303
R. I. PROVIDENCE	30	145	168	211	221	271	285	292	267	226	207	153	143	2589
S. C. CHARLESTON	30	188	189	243	284	323	308	297	281	244	239	210	187	2993
COLUMBIA	30	173	183	233	274	312	312	291	283	243	242	202	166	2914
GREENVILLE	26	166	176	227	274	307	300	278	274	239	232	192	157	2822
S. DAK. HURON	30	153	177	213	250	295	321	367	320	260	212	142	134	2844
RAPID CITY	30	164	182	222	245	278	300	348	317	266	228	164	144	2858
TENN. CHATTANOOGA	30	126	146	187	239	290	295	278	266	247	220	169	128	2591
KNOXVILLE	30	124	144	189	237	281	288	277	248	237	213	157	120	2515
MEMPHIS	30	135	152	204	244	296	321	319	314	261	243	180	139	2808
NASHVILLE	30	123	142	196	241	285	308	292	279	250	224	168	126	2634
TEX. ABILENE	13	190	199	250	259	290	347	335	322	276	245	223	201	3137
AMARILLO	30	207	199	258	276	305	338	350	328	288	260	229	205	3243
AUSTIN	30	148	152	207	221	266	302	331	320	261	242	180	160	2790
BROWNSVILLE	30	147	152	187	210	272	297	326	311	246	252	165	151	2716
CORPUS CHRISTI	24	160	165	212	237	295	329	366	341	276	264	194	164	3003
DALLAS	30	155	159	220	238	279	326	341	325	274	240	191	163	2911
DEL RIO	27	173	173	230	237	259	279	331	319	252	240	195	178	2866
EL PASO	30	234	236	299	329	373	369	336	327	300	287	257	236	3583
GALVESTON	30	151	149	203	230	288	322	305	292	257	264	199	151	2811
HOUSTON	30	144	141	193	212	266	298	294	281	238	239	181	146	2633
PORT ARTHUR	30	153	149	209	235	292	317	285	281	252	256	191	148	2768
SAN ANTONIO	30	148	153	213	224	258	292	325	307	261	241	183	160	2765
UTAH SALT LAKE CITY	30	137	155	227	269	329	358	377	346	306	249	171	135	3059
VT. BURLINGTON	30	103	127	184	185	244	270	291	266	199	152	77	80	2178
VA. LYNCHBURG	26	153	169	216	243	288	297	288	264	235	217	177	158	2705
NORFOLK	30	156	174	223	257	304	311	296	282	237	220	182	161	2803
RICHMOND	30	144	166	211	248	280	296	286	263	230	211	176	152	2663
WASH. NORTH HEAD	22	76	97	135	182	221	214	226	186	170	123	87	66	1783
SEATTLE	30	74	99	154	201	247	234	304	248	197	122	77	62	2019
SPOKANE	30	78	120	197	262	308	309	397	350	264	177	86	57	2605
TATOOSH ISLAND	30	70	100	135	182	229	217	235	190	175	129	71	60	1793
WALLA WALLA	30	72	106	194	262	317	335	411	367	280	198	92	51	2685
W. VA. ELKINS	24	110	119	158	198	227	256	225	236	211	186	131	103	2160
PARKERSBURG	30	91	111	155	200	252	277	286	264	230	189	117	93	2265
WIS. GREEN BAY	30	121	148	194	210	251	279	314	266	213	176	110	106	2388
MADISON	30	126	147	196	214	258	285	336	288	230	198	116	108	2502
MILWAUKEE	30	116	134	191	218	267	293	340	292	235	193	125	106	2510
WYO. CHEYENNE	30	191	197	243	237	259	304	318	286	265	242	188	170	2900
LANDER	30	200	208	260	264	301	340	361	326	280	233	186	185	3144
SHERIDAN	30	160	179	226	245	286	303	367	333	266	221	153	145	2884
P. R. SAN JUAN	30	231	229	273	252	240	245	264	257	219	229	217	222	2878

The smoothed isolines on these charts and the data in this tabulation are based on Weather Bureau records from black-bulb type sunshine recorders. These values are those made during the 1931-60 period.

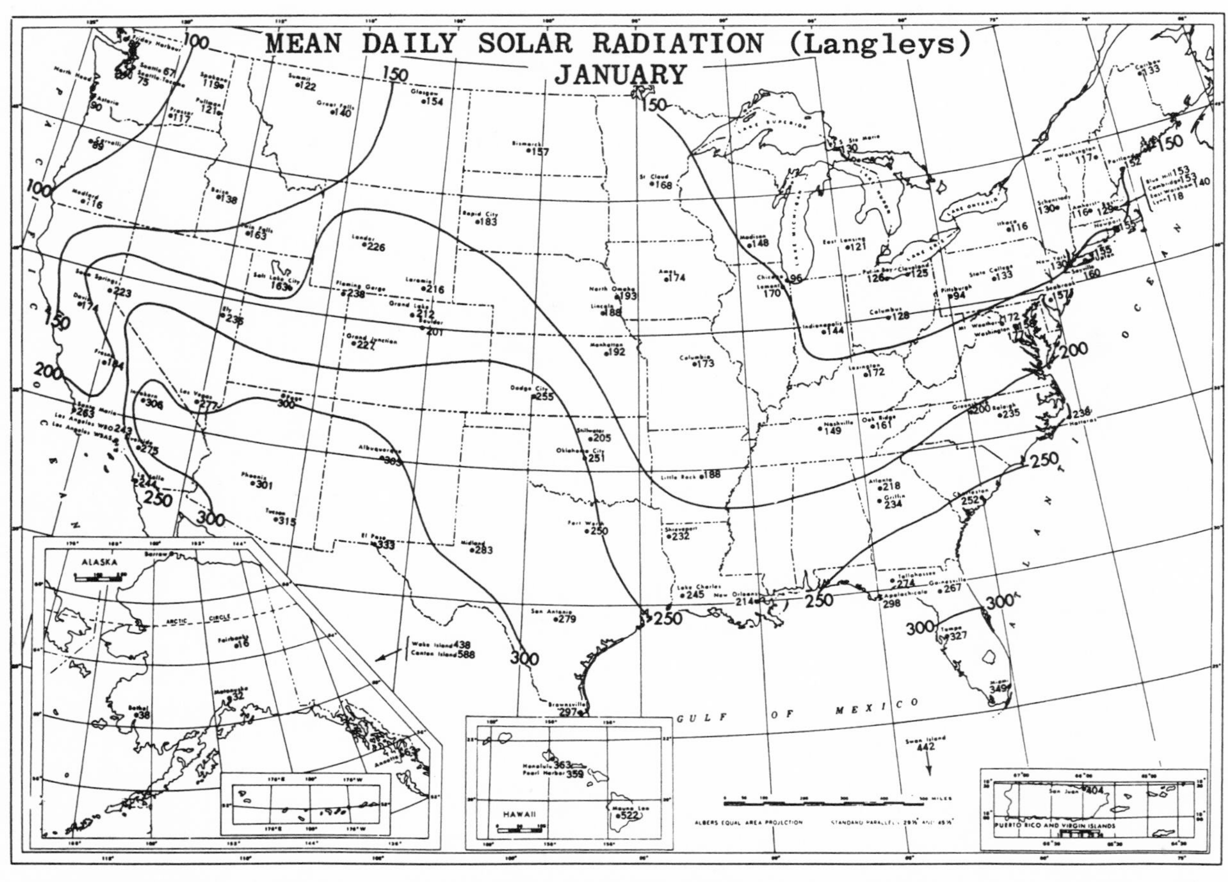
MEAN DAILY SOLAR RADIATION (Langleys)
JANUARY
ALASKA
HAWAII
PUERTO RICO AND VIRGIN ISLANDS
GULF OF MEXICO
ALBERS EQUAL AREA PROJECTION

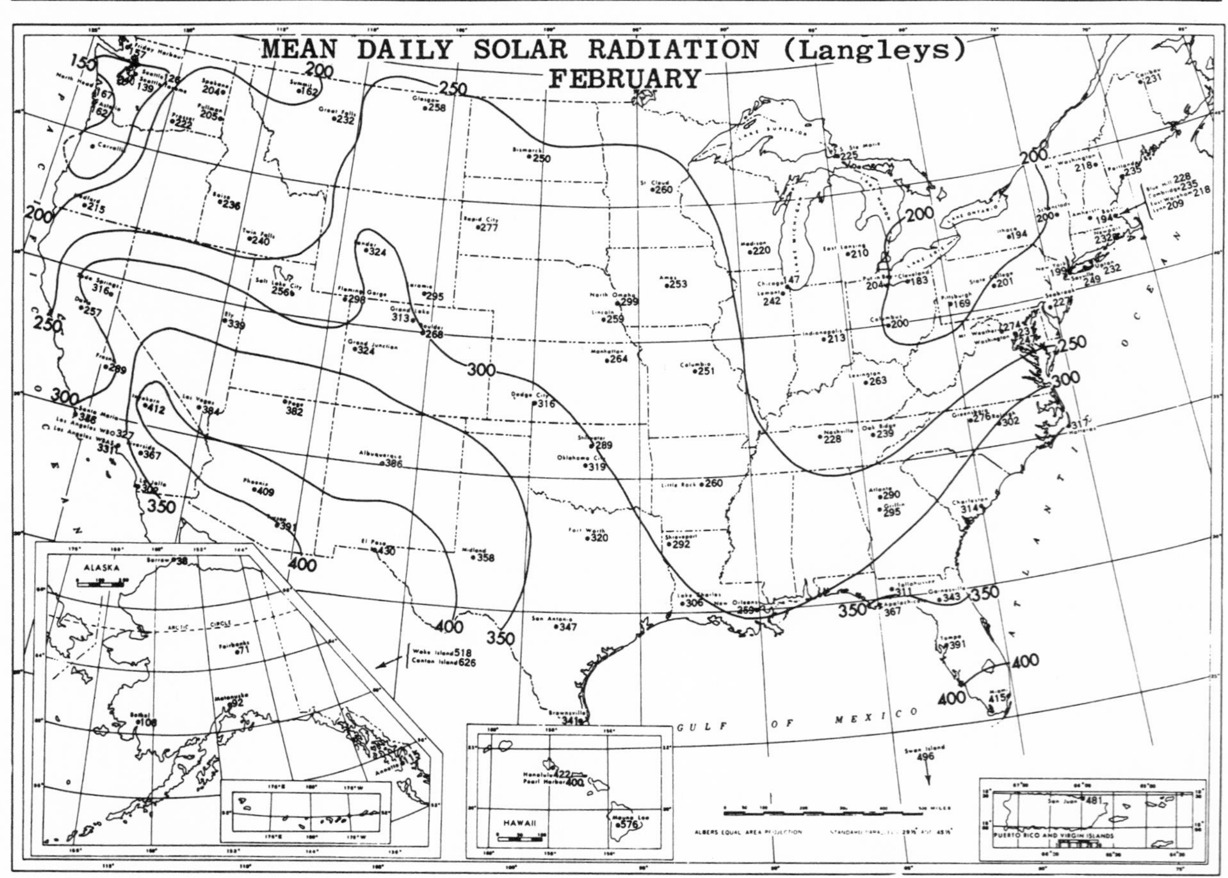
MEAN DAILY SOLAR RADIATION (Langleys)
FEBRUARY
ALASKA
HAWAII
PUERTO RICO AND VIRGIN ISLANDS
GULF OF MEXICO
ALBERS EQUAL AREA PROJECTION

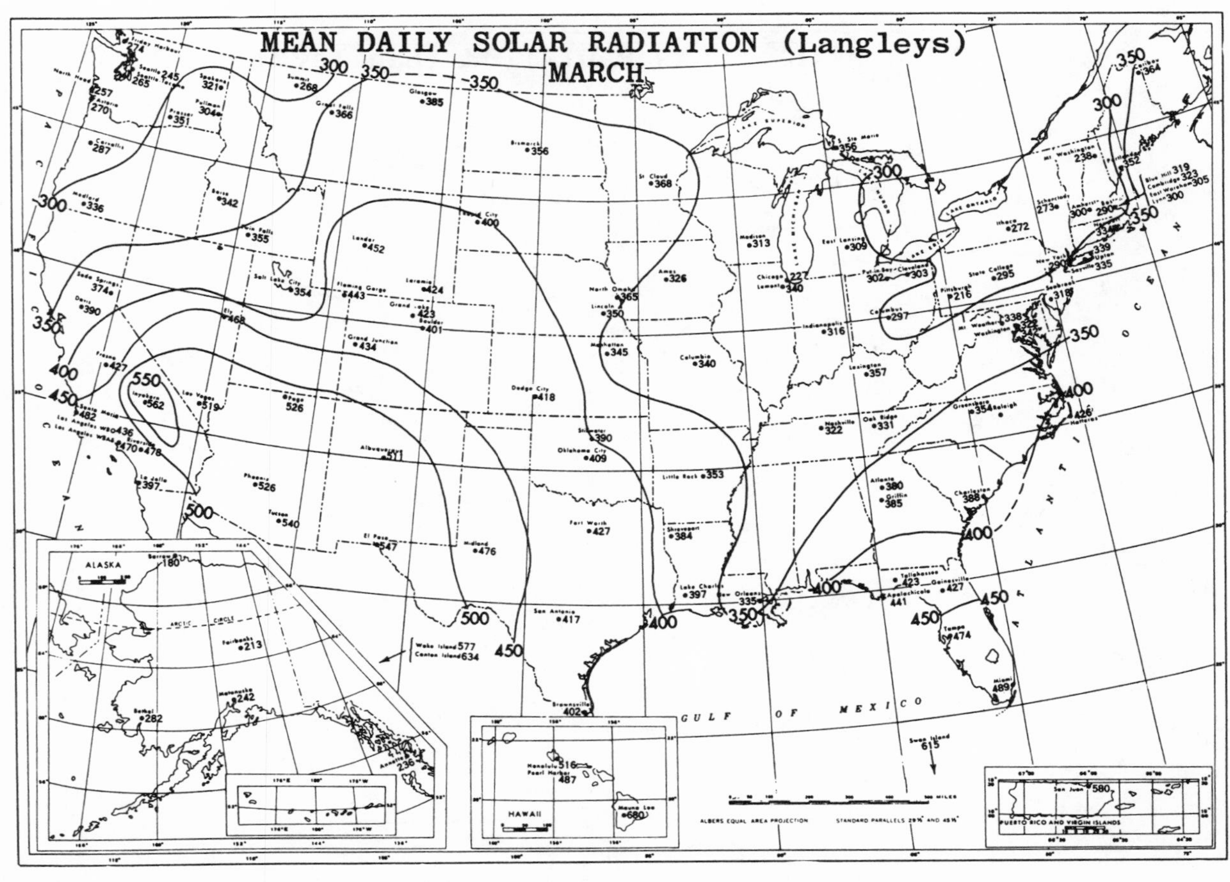
MEAN DAILY SOLAR RADIATION (Langleys)
MARCH
ALASKA
ARCTIC CIRCLE
Wake Island 577
Canton Island 634
GULF OF MEXICO
HAWAII
ALBERS EQUAL AREA PROJECTION
STANDARD PARALLELS 29½° AND 45½°
PUERTO RICO AND VIRGIN ISLANDS

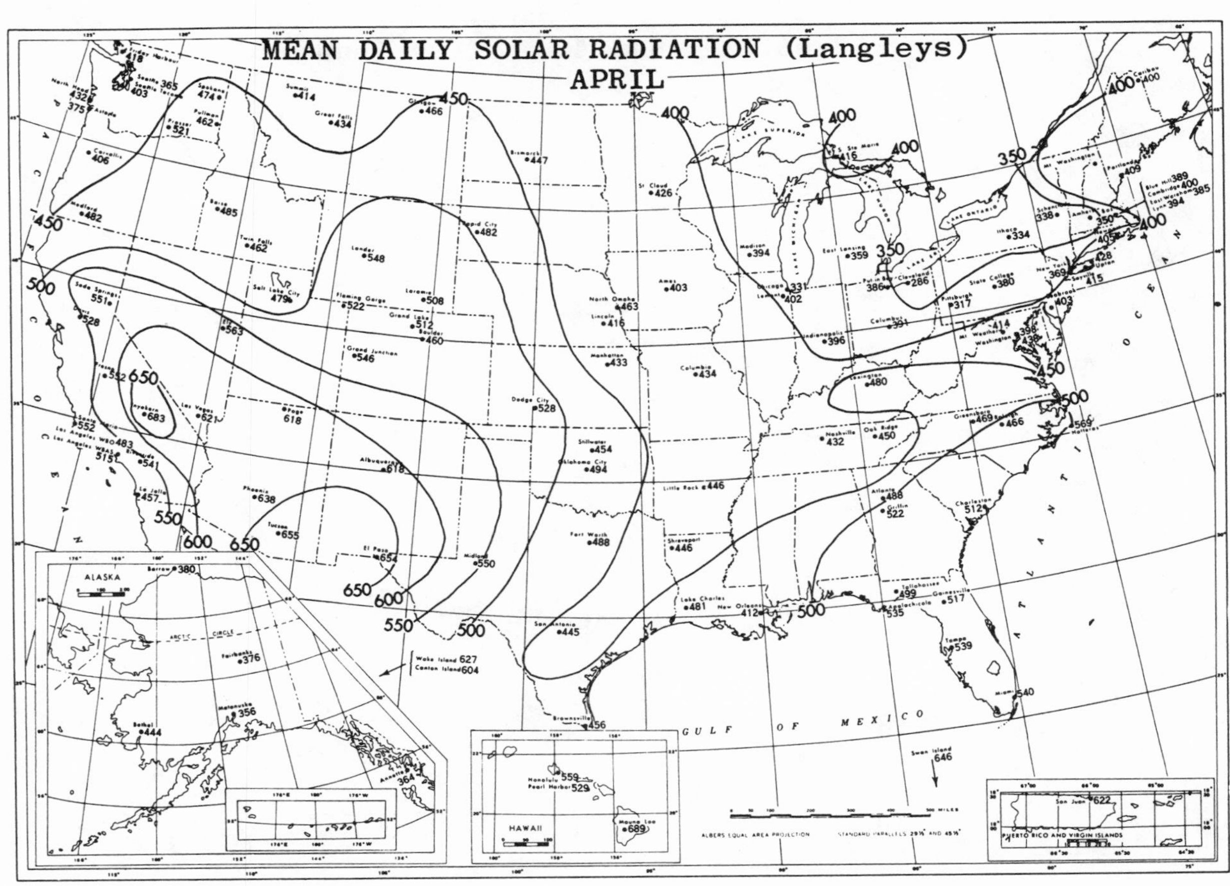
MEAN DAILY SOLAR RADIATION (Langleys)
APRIL
ALASKA
ARCTIC CIRCLE
Wake Island 627
Canton Island 604
GULF OF MEXICO
HAWAII
ALBERS EQUAL AREA PROJECTION
STANDARD PARALLELS 29½° AND 45½°
PUERTO RICO AND VIRGIN ISLANDS

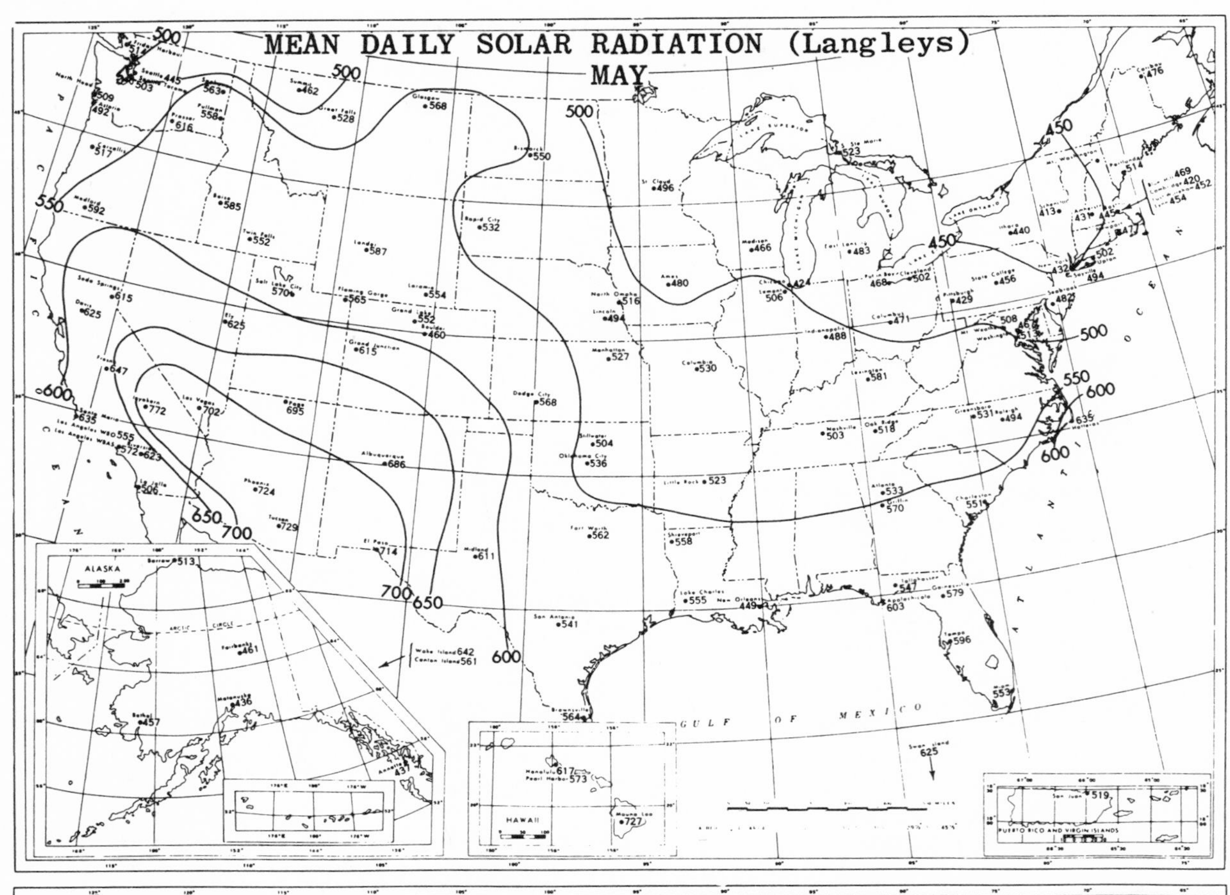
MEAN DAILY SOLAR RADIATION (Langleys)
MAY
GULF OF MEXICO
ALASKA
HAWAII
PUERTO RICO AND VIRGIN ISLANDS

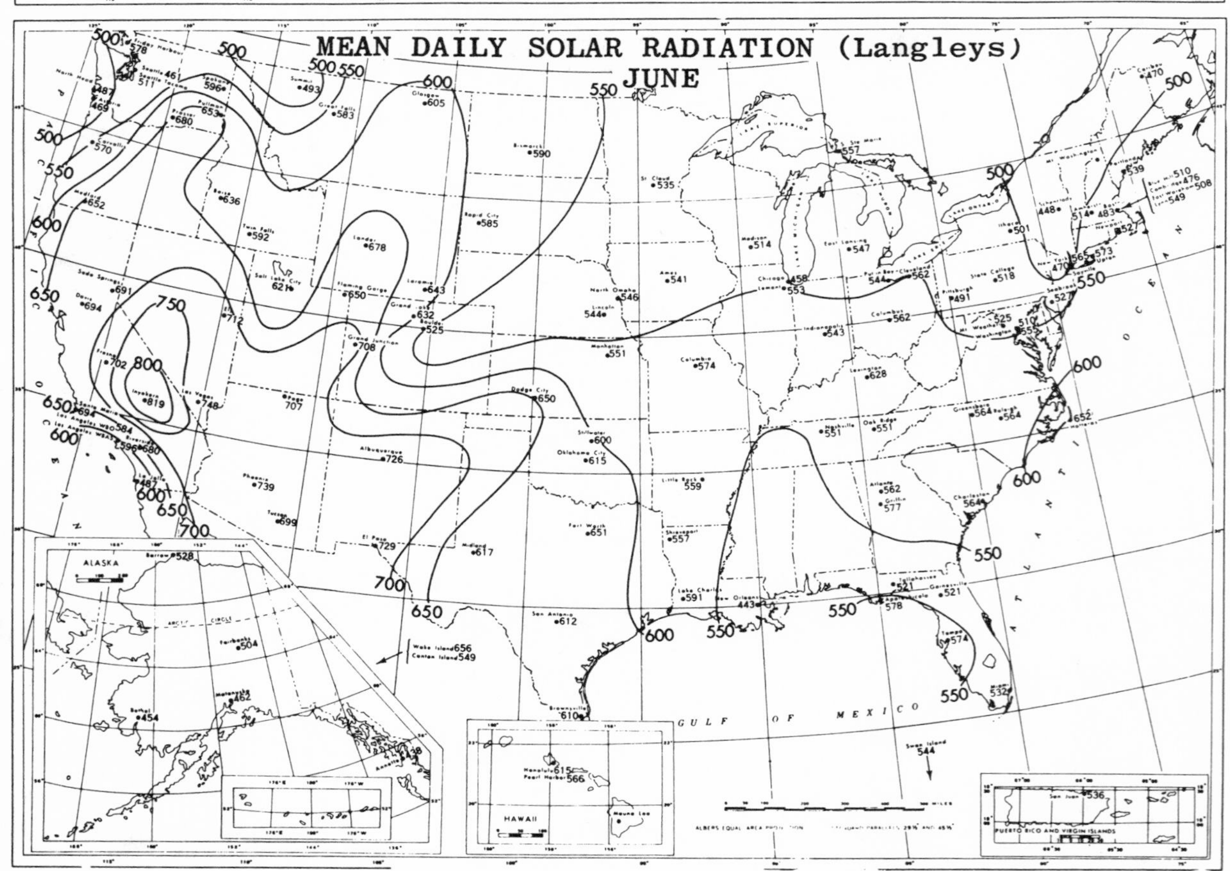
MEAN DAILY SOLAR RADIATION (Langleys)
JUNE
GULF OF MEXICO
ALASKA
HAWAII
PUERTO RICO AND VIRGIN ISLANDS

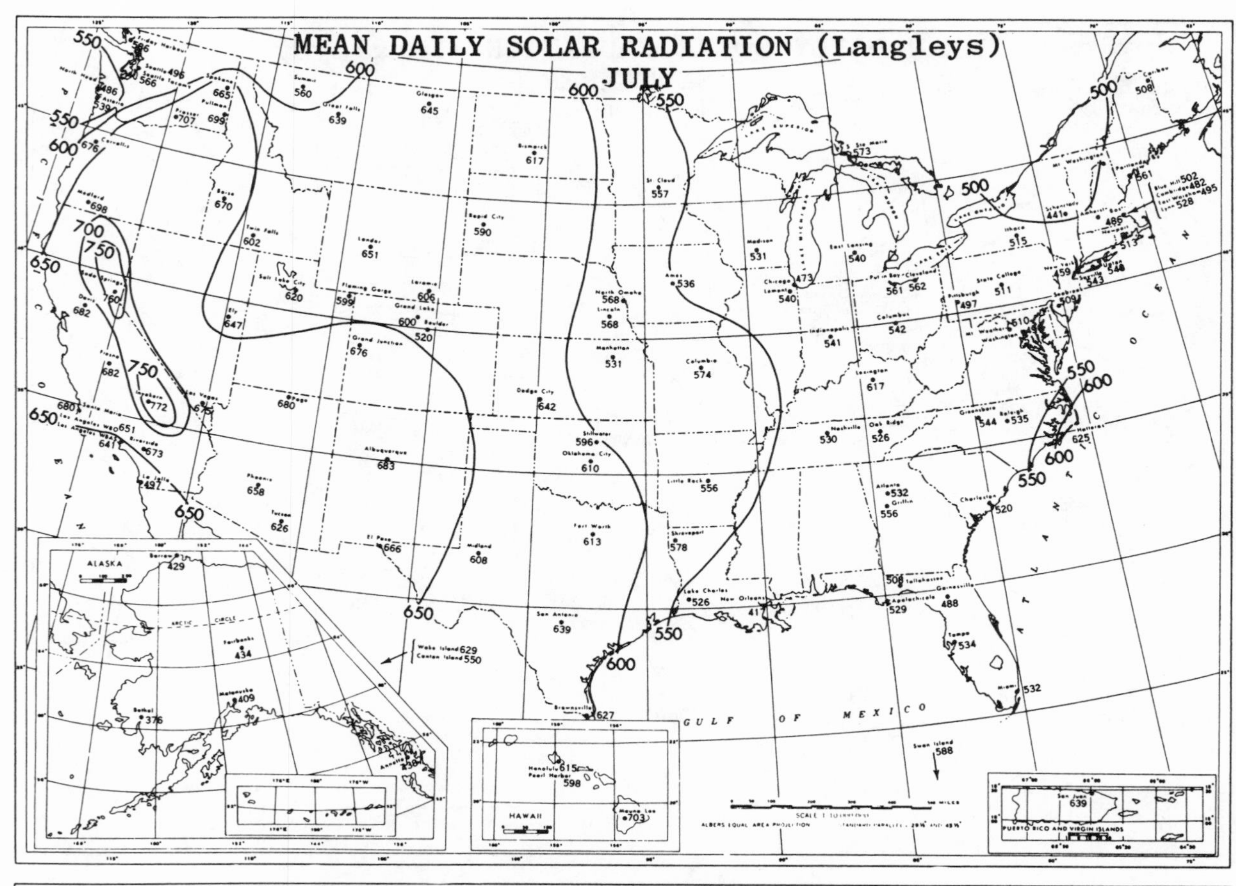
MEAN DAILY SOLAR RADIATION (Langleys)
JULY
GULF OF MEXICO
ALASKA
HAWAII
PUERTO RICO AND VIRGIN ISLANDS

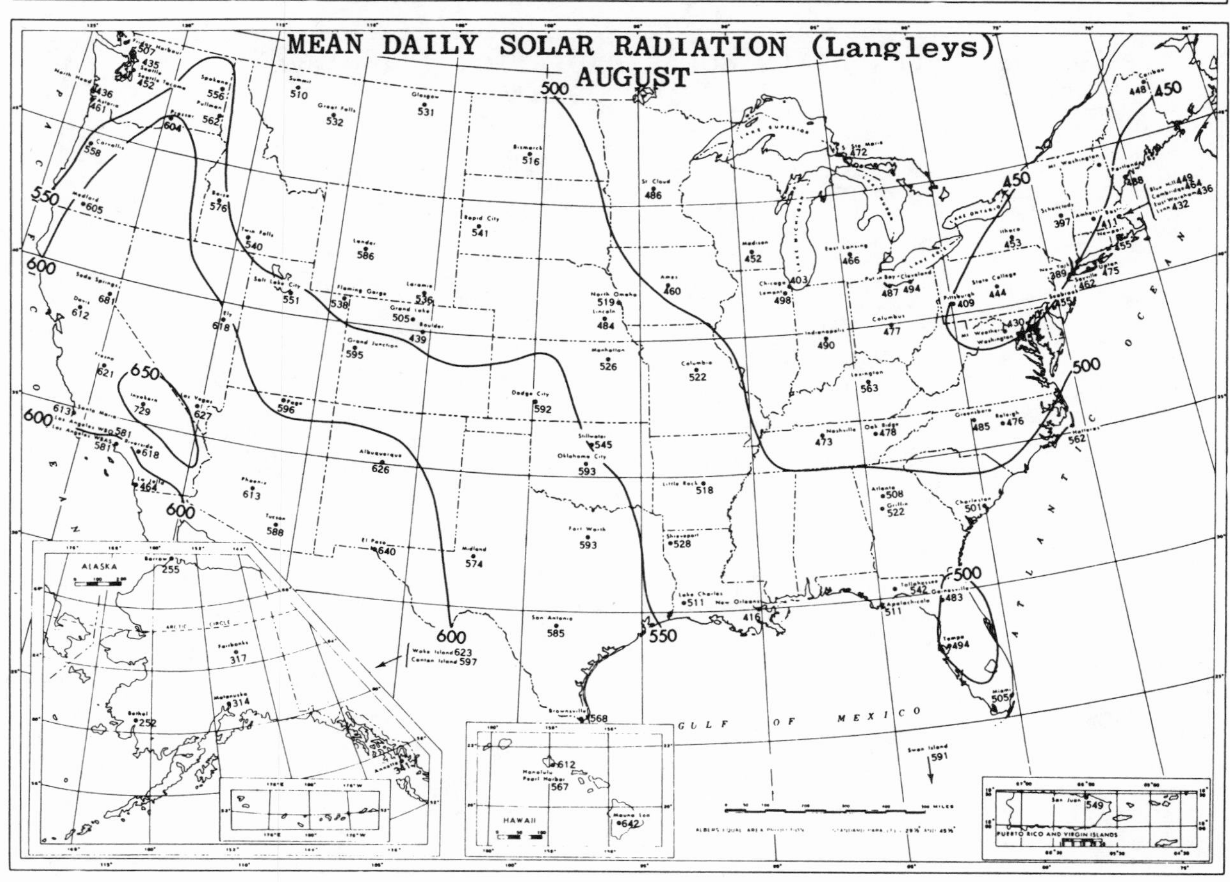
MEAN DAILY SOLAR RADIATION (Langleys)
AUGUST
GULF OF MEXICO
ALASKA
HAWAII
PUERTO RICO AND VIRGIN ISLANDS

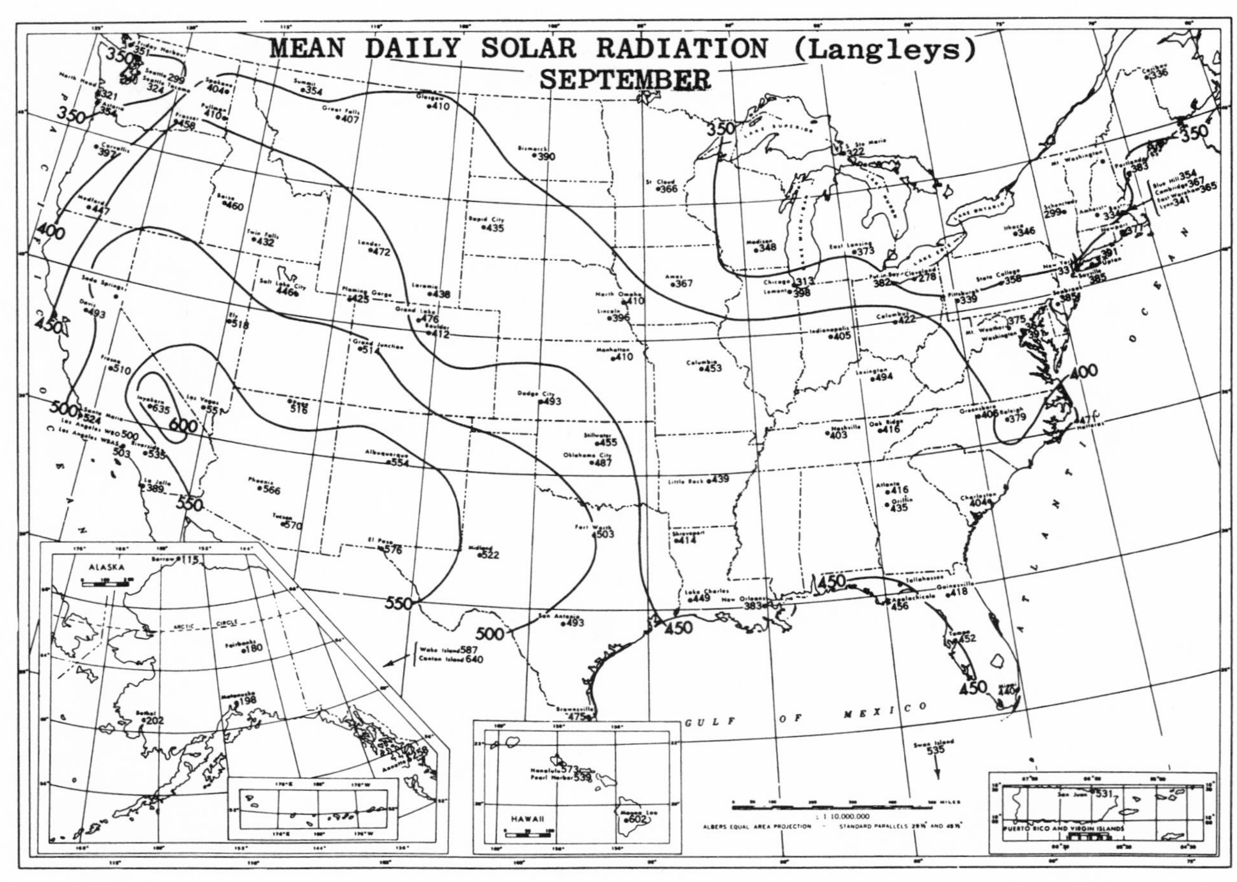
MEAN DAILY SOLAR RADIATION (Langleys)
SEPTEMBER

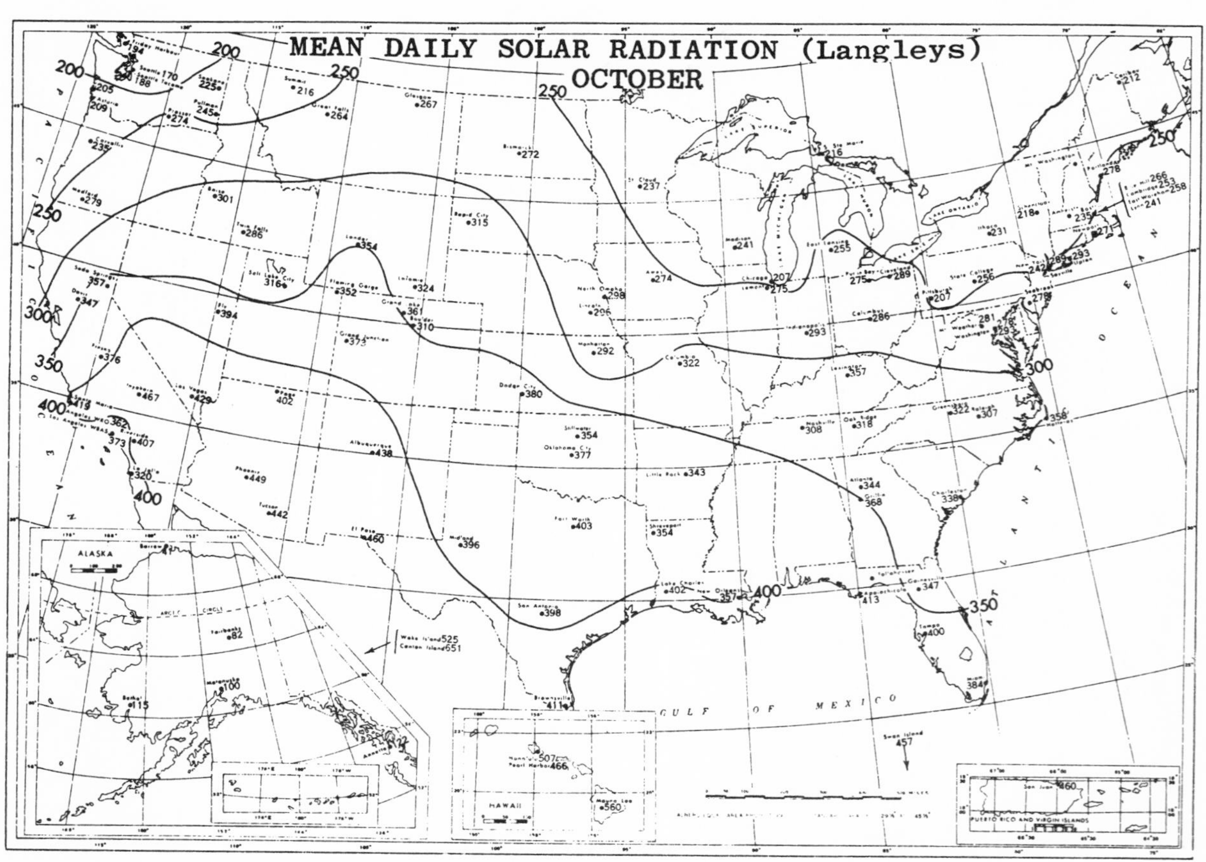
MEAN DAILY SOLAR RADIATION (Langleys)
OCTOBER

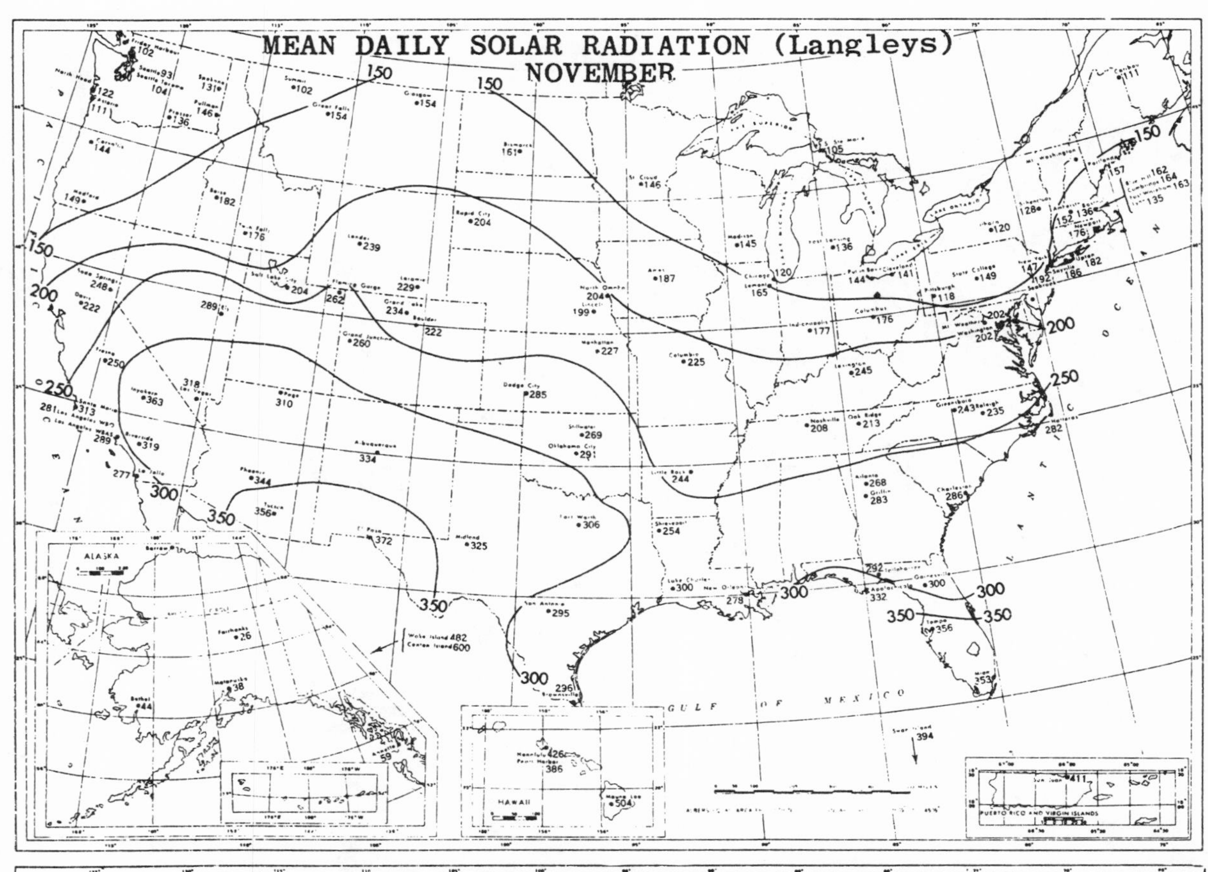

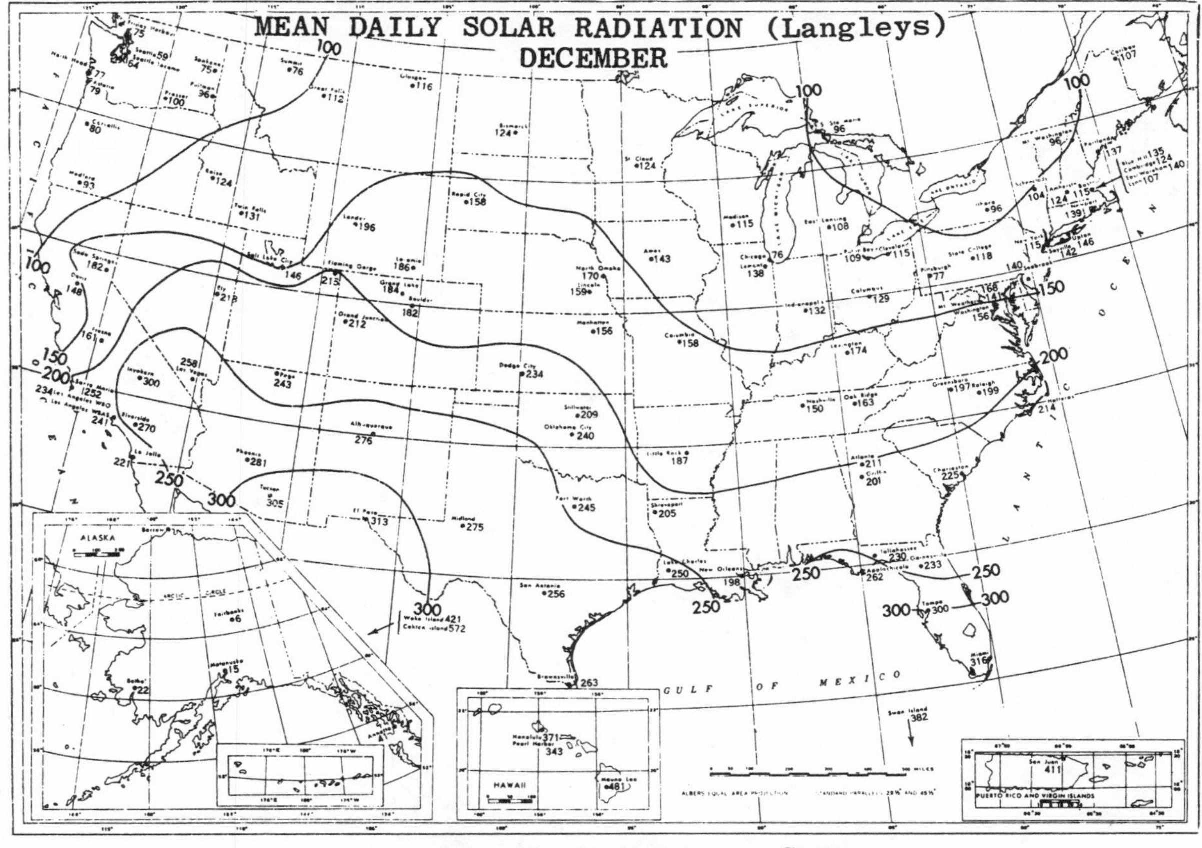

Scale for the 48 Contiguous States:

300 0 300 600 900 1200 1500 Miles

SCALE 1 : 30,000,000

ALBERS EQUAL AREA PROJECTION-STANDARD PARALLELS 29½° AND 45½°

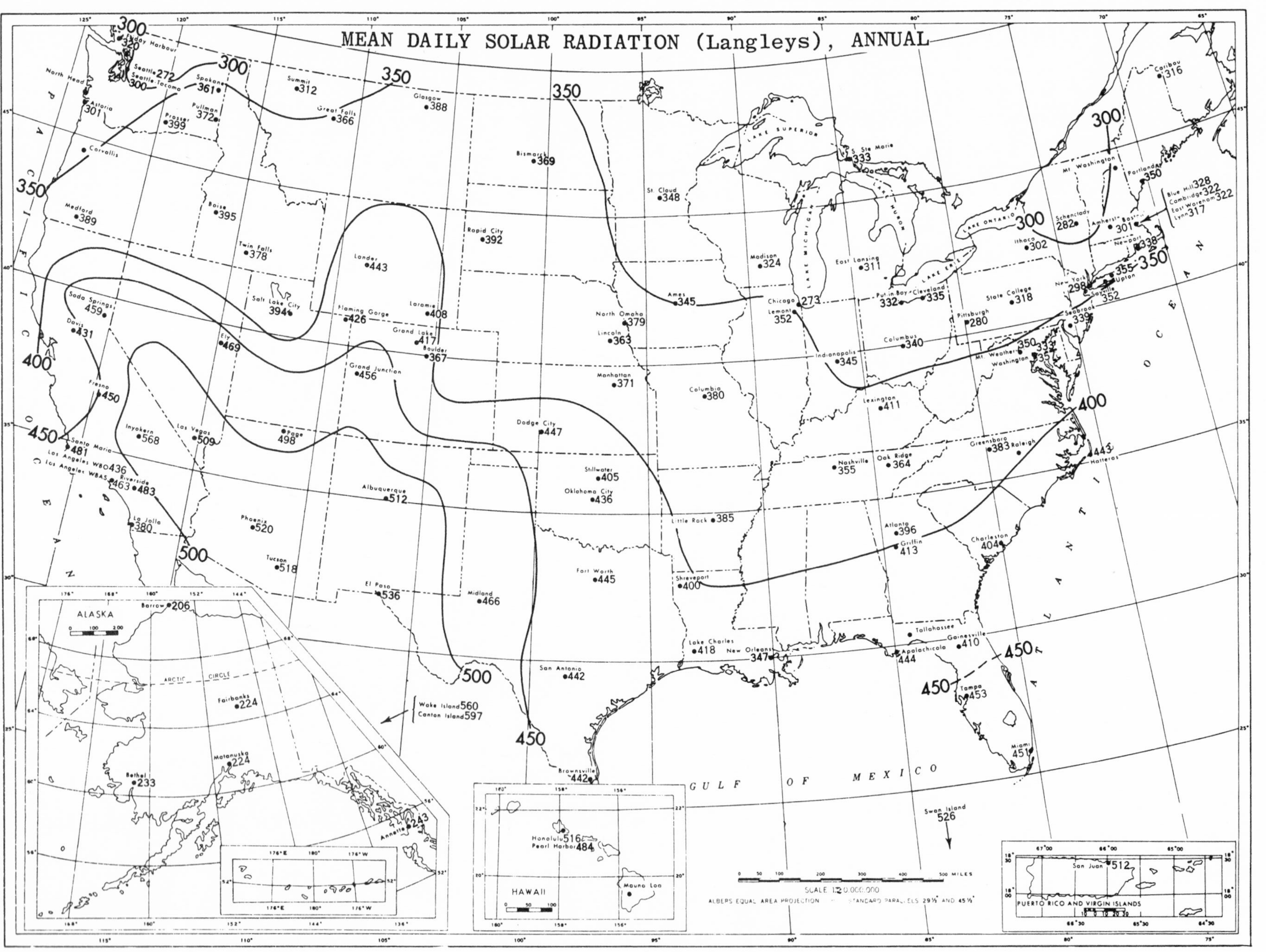

MEAN DAILY SOLAR RADIATION (Langleys), ANNUAL
Friday Harbour 320
Seattle 272
Seattle-Tacoma 300
North Head
Astoria 301
Corvallis
Medford 389
Spokane 361
Pullman 372
Prosser 399
Summit 312
Great Falls 366
Glasgow 388
Bismarck 369
Boise 395
Twin Falls 378
Salt Lake City 394
Lander 443
Rapid City 392
Flaming Gorge 426
Laramie 408
Grand Lake 417
Boulder 367
Grand Junction 456
Ely 469
Soda Springs 459
Davis 431
Fresno 450
Inyokern 568
Las Vegas 509
Page 498
Santa Maria 481
Los Angeles WBO 436
Los Angeles WBAS 463
Riverside 483
La Jolla 380
Phoenix 520
Tucson 518
Albuquerque 512
El Paso 536
Midland 466
Dodge City 447
Stillwater 405
Oklahoma City 436
Fort Worth 445
San Antonio 442
Brownsville 442
St. Cloud 348
Ames 345
North Omaha 379
Lincoln 363
Manhattan 371
Columbia 380
Little Rock 385
Shreveport 400
Lake Charles 418
New Orleans 347
Madison 324
Chicago 273
Lemont 352
S. Ste Marie 333
East Lansing 311
Put-in-Bay 332
Cleveland 335
Indianapolis 345
Columbus 340
Lexington 411
Nashville 355
Oak Ridge 364
Atlanta 396
Griffin 413
Tallahassee
Apalachicola 444
Gainesville 410
Tampa 453
Miami 451
Charleston 404
Greensboro 383
Raleigh
Hatteras 443
Pittsburgh 280
State College 318
Ithaca 302
Schenectady 282
Mt. Weather 350
Mt. Washington 337
Seabrook 339
New York 298
Sayville 352
Upton 355
Newport 338
Amherst
Boston 301
Blue Hill 328
Cambridge 322
East Wareham 322
Lynn 317
Mt. Washington
Portland 350
Caribou 316
Wake Island 560
Canton Island 597
Swan Island 526
LAKE SUPERIOR
LAKE MICHIGAN
LAKE HURON
LAKE ERIE
LAKE ONTARIO
PACIFIC OCEAN
ATLANTIC OCEAN
GULF OF MEXICO
300
350
400
450
500
SCALE 1:20,000,000
ALBERS EQUAL AREA PROJECTION STANDARD PARALLELS 29½ AND 45½
500 MILES
ALASKA
Barrow 206
ARCTIC CIRCLE
Fairbanks 224
Matanuska 224
Bethel 233
Annette 243
HAWAII
Honolulu 516
Pearl Harbor 484
Mauna Loa
San Juan 512
PUERTO RICO AND VIRGIN ISLANDS

These charts and table are based on all usuable solar radiation data, direct and diffuse, measured on a horizontal surface and published in the Monthly Weather Review and Climatological Data National Summary through 1962. All data were measured in, or were reduced to, the International Scale of Pyrheliometry, 1956.

Langley is the unit used to denote one gram calorie per square centimeter (1 langley = 1 gm. cal. cm^{-2}.

MEAN DAILY SOLAR RADIATION (Langleys) AND YEARS OF RECORD USED

STATES AND STATIONS	JAN	YRS	FEB	YRS	MAR	YRS	APR	YRS	MAY	YRS	JUNE	YRS	JULY	YRS	AUG	YRS	SEPT	YRS	OCT	YRS	NOV	YRS	DEC	YRS	ANNUAL
ALASKA, Annette	63	6	115	6	236	7	364	7	437	6	438	6	438	6	341	6	258	7	122	7	59	7	41	7	243
Barrow	#		38	8	180	8	380	8	513	8	528	8	429	9	255	10	115	10	41	10	#		#		206
Bethel	38	9	108	10	282	9	444	10	457	10	454	10	376	10	252	10	202	10	115	10	44	9	22	9	233
Fairbanks	16	25	71	27	213	25	376	28	461	28	504	29	434	28	317	29	180	29	82	30	26	26	6	26	224
Matanuska	32	6	92	6	242	4	356	7	436	7	462	6	409	6	314	6	198	6	100	6	38	6	15	7	224
ARIZ., Page	300	2	382	3	526	3	618	2	695	2	707	2	680	3	596	3	516	3	402	3	310	3	243	3	498
Phoenix	301	11	409	11	526	11	638	11	724	11	739	11	658	11	613	11	566	11	449	11	344	11	281	11	520
Tucson	315	5	391	5	540	4	655	5	729	5	699	5	626	6	588	6	570	6	442	6	356	6	305	6	518
ARK., Little Rock	188	9	260	9	353	10	446	9	523	9	559	9	556	8	518	9	439	7	343	8	244	10	187	10	385
CALIFORNIA, Davis	174	18	257	17	390	18	528	18	625	18	694	18	682	18	612	18	493	18	347	19	222	19	148	19	431
Fresno	184	31	289	31	427	31	552	31	647	31	702	32	682	32	621	31	510	31	376	32	250	31	161	32	450
Inyokern (China Lake)	306	11	412	11	562	11	683	11	772	11	819	11	772	11	729	10	635	8	467	9	363	11	300	12	568
LaJolla	244	19	302	18	397	19	457	20	506	19	487	21	497	22	464	22	389	22	320	21	277	20	221	20	380
Los Angeles WBAS	248	10	331	10	470	10	515	10	572	9	596	9	641	9	581	10	503	10	373	10	289	10	241	10	463
Los Angeles WBO	243	9	327	9	436	9	483	9	555	9	584	9	651	9	581	10	500	10	362	10	281	10	234	10	436
Riverside ‡	275	8	367	8	478	9	541	9	623	9	680	9	673	9	618	9	535	9	407	9	319	9	270	9	483
Santa Maria	263	11	346	11	482	11	552	10	635	11	694	11	680	11	613	11	524	11	419	11	313	11	252	11	481
Soda Springs	223	4	316	3	374	4	551	4	615	3	691	4	760	3	681	3	510	3	357	4	248	4	182	3	459
COLO., Boulder	201	5	268	4	401	4	460	4	460	4	525	5	520	5	439	5	412	4	310	4	222	4	182	4	367
Grand Junction	227	9	324	9	434	8	546	8	615	9	708	8	676	8	595	8	514	8	373	10	260	10	212	10	456
Grand Lake (Granby)	212	6	313	7	423	7	512	8	552	8	632	8	600	8	505	7	476	6	361	7	234	6	184	7	417
D C., Washington (C.O.)	174	3	266	3	344	2	411	2	551	2	494	2	536	2	446	3	375	3	299	3	211	3	166	3	356
American University	158	39	231	39	322	39	398	39	467	39	510	39	496	39	440	38	364	38	278	38	192	39	141	39	333
Silver Hill	177	7	247	6	342	7	438	7	513	7	555	7	511	7	457	7	391	8	293	8	202	7	156	6	357
FLA., Apalachicola	298	10	367	10	441	10	535	10	603	9	578	9	529	9	511	9	456	9	413	10	332	10	262	10	444
Belle Isle	297	10	330	10	412	10	463	10	483	10	464	10	488	11	461	10	400	10	366	11	313	11	291	10	397
Gainesville	267	11	343	10	427	12	517	12	579	12	521	10	488	10	483	8	418	9	347	8	300	10	233	10	410
Miami Airport	349	10	415	9	489	9	540	10	553	10	532	10	532	10	505	10	440	10	384	10	353	10	316	10	451
Tallahassee	274	2	311	2	423	3	499	3	547	3	521	3	508	3	542	2	*		*		292	2	230	2	---
Tampa	327	8	391	8	474	8	539	8	596	8	574	9	534	9	494	9	452	9	400	9	356	9	300	9	453
GA., Atlanta	218	11	290	11	380	11	488	11	533	11	562	11	532	10	508	10	416	10	344	11	268	11	211	11	396
Griffin	234	9	295	9	385	10	522	11	570	11	577	11	556	11	522	11	435	11	368	11	283	11	201	11	413
HAWAII, Honolulu	363	4	422	4	516	4	559	5	617	5	615	5	615	5	612	5	573	5	507	5	426	5	371	5	516
Mauna Loa Obs.	522	2	576	2	680	2	689	3	727	3	*		703	3	642	2	602	2	560	2	504	2	481	3	---
Pearl Harbor	359	5	400	4	487	4	529	5	573	5	566	5	598	5	567	5	539	5	466	5	386	5	343	5	484
IDAHO, Boise	138	10	236	9	342	9	485	9	585	10	636	9	670	10	576	10	460	10	301	11	182	11	124	11	395
Twin Falls	163	20	240	20	355	20	462	21	552	20	592	18	602	20	540	20	432	19	286	20	176	20	131	19	378
ILL., Chicago	96	19	147	19	227	19	331	19	424	19	458	18	473	19	403	18	313	19	207	20	120	20	76	20	273
Lemont	170	6	242	6	340	6	402	6	506	6	553	6	540	6	498	6	398	5	275	5	165	5	138	5	352
IND., Indianapolis	144	10	213	10	316	10	396	10	488	9	543	11	541	10	490	11	405	11	293	11	177	11	132	11	345
IOWA, Ames	174	5	253	5	326	5	403	5	480	5	541	5	436	6	460	6	367	6	274	7	187	7	143	7	345
KANS., Dodge City	255	7	316	7	418	7	528	7	568	7	650	7	642	8	592	9	493	9	380	9	285	10	234	10	447
Manhattan	192	3	264	3	345	3	433	3	527	4	551	4	531	4	526	4	410	4	292	4	227	4	156	4	371
KY., Lexington	172	9	263	9	357	10	480	10	581	10	628	9	617	10	563	10	494	10	357	9	245	9	174	11	411
LA., Lake Charles	245	11	306	11	397	11	481	11	555	11	591	11	526	11	511	11	449	11	402	11	300	10	250	10	418
New Orleans	214	14	259	14	335	15	412	16	449	14	443	13	417	15	416	15	383	15	357	13	278	13	198	14	347
Shreveport	232	3	292	3	384	3	446	4	558	4	557	4	578	4	528	4	414	4	354	4	254	4	205	4	400
MAINE, Caribou	133	8	231	9	364	8	400	10	476	10	470	10	508	11	448	11	336	11	212	11	111	11	107	9	316
Portland	152	7	235	8	352	7	409	8	514	9	539	9	561	9	488	8	383	7	278	9	157	8	137	9	350
MASS., Amherst	116	2	*		300	2	*		431	2	514	2	*		---		---		---		152	2	124	2	---
Blue Hill	153	27	228	27	319	26	389	26	469	27	510	27	502	26	449	27	354	28	266	28	162	28	135	28	328
Boston	129	16	194	17	290	17	350	17	445	16	483	16	486	16	411	16	334	17	235	16	136	16	115	15	301
Cambridge	153	4	235	3	323	3	400	3	420	3	476	3	482	4	464	4	367	4	253	4	164	4	124	4	322
East Wareham	140	13	218	13	305	12	385	14	452	14	508	14	495	14	436	14	365	13	258	14	163	14	140	13	322
Lynn	118	2	209	2	300	2	394	2	454	2	549	4	528	4	432	3	341	2	241	3	135	3	107	3	317
MICH., East Lansing	121	10	210	11	309	11	359	11	483	10	547	11	540	11	466	11	373	11	255	11	136	11	108	11	311
Sault Ste. Marie	130	10	225	9	356	10	416	10	523	10	557	11	573	11	472	10	322	10	216	9	105	9	96	9	333
MINN., St. Cloud	168	8	260	8	368	8	426	8	496	8	535	8	557	9	486	8	366	8	237	7	146	8	124	8	348
MO., Columbia (C. O.)	173	10	251	10	340	11	434	11	530	11	574	11	574	10	522	10	453	10	322	10	225	10	158	9	380
University of Missouri	166	5	248	6	324	6	429	6	501	6	560	6	583	6	509	6	417	6	324	5	177	5	146	5	365

MEAN DAILY SOLAR RADIATION (Langleys) AND YEARS OF RECORD USED

STATES AND STATIONS	JAN	YRS	FEB	YRS	MAR	YRS	APR	YRS	MAY	YRS	JUNE	YRS	JULY	YRS	AUG	YRS	SEPT	YRS	OCT	YRS	NOV	YRS	DEC	YRS	ANNUAL
MONT., Glasgow	154	6	258	8	385	7	466	8	568	8	605	8	645	9	531	10	410	10	267	8	154	8	116	7	388
Great Falls	140	8	232	9	366	9	434	8	528	8	583	8	639	9	532	9	407	10	264	10	154	10	112	10	366
Summit	122	3	162	2	268	3	414	3	462	3	493	3	560	2	510	2	354	2	216	2	102	2	76	2	312
NEBR., Lincoln	188	39	259	39	350	39	416	39	494	40	544	38	568	38	484	38	396	38	296	36	199	40	159	39	363
North Omaha	193	3	299	3	365	3	463	3	516	3	546	4	568	4	519	4	410	4	298	4	204	4	170	4	379
NEV., Ely	236	7	339	9	468	9	563	9	625	10	712	10	647	11	618	11	518	11	394	10	289	10	218	10	469
Las Vegas	277	11	384	11	519	11	621	11	702	11	748	10	675	11	627	11	551	11	429	11	318	11	258	11	509
N. J., Seabrook	157	8	227	8	318	8	403	8	482	9	527	8	509	8	455	9	385	9	278	7	192	8	140	8	339
N. H., Mt. Washington	117	2	218	2	238	2	*		*		*		---		---		*		*		*		96	2	---
N. Mex., Albuquerque	303	13	386	13	511	13	618	13	686	13	726	13	683	12	626	13	554	14	438	15	334	15	276	14	512
N. Y., Ithaca	116	22	194	21	272	23	334	23	440	24	501	23	515	23	453	23	346	21	231	22	120	23	96	23	302
N. Y. Central Park	130	34	199	34	290	33	369	35	432	35	470	34	459	35	389	35	331	36	242	36	147	36	115	35	298
Sayville	160	11	249	11	335	10	415	10	494	10	565	10	543	10	462	10	385	10	289	10	186	10	142	11	352
Schenectady	130	8	200	9	273	9	338	9	413	9	448	8	441	8	397	8	299	8	218	8	128	8	104	8	282
Upton	155	8	232	8	339	8	428	8	502	8	573	8	543	7	475	7	391	7	293	6	182	7	146	7	355
N. C., Greensboro	200	7	276	9	354	9	469	9	531	10	564	10	544	10	485	10	406	10	322	10	243	10	197	8	383
Hatteras	238	10	317	9	426	8	569	9	635	10	652	10	625	10	562	11	471	11	358	11	282	11	214	11	443
Raleigh	235	3	302	2	*		466	3	494	2	564	2	535	3	476	3	379	3	307	3	235	3	199	3	---
N. D., Bismarck	157	7	250	8	356	6	447	8	550	8	590	9	617	10	516	11	390	11	272	11	161	10	124	10	369
OHIO., Cleveland	125	6	183	6	303	7	286	8	502	8	562	8	562	8	494	8	278	8	289	9	141	9	115	7	335
Columbus	128	7	200	7	297	7	391	7	471	6	562	4	542	5	477	4	422	4	286	4	176	4	129	5	340
Put-in-Bay	126	10	204	9	302	10	386	11	468	11	544	11	561	10	487	10	382	11	275	11	144	11	109	11	332
OKLA., Oklahoma City	251	10	319	10	409	9	494	10	536	10	615	7	610	8	593	8	487	9	377	10	291	9	240	9	436
Stillwater	205	8	289	8	390	9	454	9	504	9	600	10	596	10	545	10	455	11	354	10	269	9	209	8	405
OREG., Astoria	90	7	162	8	270	8	375	8	492	8	469	8	539	8	461	7	354	7	209	8	111	8	79	8	301
Corvallis	89	2	*		287	3	406	3	517	3	570	3	676	4	558	4	397	4	235	4	144	4	80	4	---
Medford	116	11	215	11	336	11	482	11	592	11	652	11	698	10	605	11	447	11	279	11	149	11	93	11	389
PA., Pittsburgh	94	6	169	5	216	6	317	6	429	6	491	6	497	7	409	6	339	6	207	5	118	6	77	5	280
State College	133	19	201	19	295	20	380	20	456	20	518	20	511	20	444	20	358	20	256	20	149	20	118	20	318
R. I., Newport	155	23	232	22	334	23	405	23	477	23	527	24	513	24	455	24	377	24	271	24	176	24	139	24	338
S. C., Charleston	252	11	314	11	388	11	512	11	551	11	564	11	520	11	501	11	404	11	338	11	286	11	225	11	404
S. D., Rapid City	183	11	277	11	400	11	482	11	532	11	585	11	590	11	541	11	435	11	315	10	204	10	158	10	392
TENN., Nashville	149	18	228	19	322	19	432	19	503	18	551	18	530	17	473	17	403	17	308	19	208	18	150	19	355
Oak Ridge	161	11	239	11	331	11	450	11	518	11	551	11	526	11	478	11	416	11	318	11	213	10	163	11	364
TEXAS, Brownsville	297	10	341	10	402	10	456	11	564	10	610	9	627	8	568	11	475	11	411	11	296	11	263	10	442
El Paso	333	11	430	11	547	10	654	11	714	11	729	11	666	11	640	10	576	11	460	11	372	11	313	11	536
Ft. Worth	250	11	320	11	427	11	488	11	562	11	651	11	613	11	593	11	503	11	403	11	306	11	245	9	445
Midland	283	7	358	8	476	9	550	8	611	8	617	8	608	7	574	8	522	9	396	9	325	8	275	8	466
San Antonio	279	9	347	9	417	9	445	9	541	9	612	9	639	9	585	9	493	10	398	10	295	10	256	8	442
UTAH, Flaming Gorge	238	2	298	2	443	2	522	2	565	2	650	2	599	3	538	3	425	3	352	3	262	3	215	3	426
Salt Lake City	163	8	256	8	354	8	479	8	570	7	621	7	620	6	551	7	446	8	316	8	204	8	146	9	394
VA., Mt. Weather	172	2	274	2	338	2	414	2	508	2	525	3	510	3	430	3	375	3	281	2	202	2	168	2	350
WASH., North Head	*		167	2	257	3	432	2	509	3	487	3	486	3	436	3	321	3	205	3	122	3	77	3	---
Friday Harbor	87	8	157	7	274	8	418	8	514	9	578	10	586	10	507	11	351	8	194	10	102	10	75	8	320
Prosser	117	4	222	4	351	4	521	5	616	4	680	4	707	4	604	4	458	4	274	4	136	4	100	4	399
Pullman	121	4	205	2	304	2	462	2	558	4	653	5	699	5	562	4	410	4	245	5	146	5	96	5	372
University of Washington	67	9	126	9	245	10	364	9	445	10	461	10	496	11	435	10	299	8	170	9	93	9	59	9	272
Seattle-Tacoma	75	9	139	9	265	9	403	9	503	9	511	9	566	9	452	10	324	10	188	10	104	9	64	10	300
Spokane	119	8	204	8	321	8	474	9	563	9	596	9	665	9	556	9	404	10	225	9	131	9	75	7	361
WIS., Madison †	148	46	220	46	313	45	394	47	466	47	514	47	531	47	452	47	348	47	241	47	145	44	115	46	324
WYO., Lander	226	8	324	9	452	9	548	11	587	11	678	11	651	11	586	10	472	8	354	9	239	9	196	9	443
Laramie	216	3	295	3	424	3	508	3	554	3	643	3	606	3	536	3	438	3	324	3	229	3	186	4	408
ISLAND STATIONS																									
Canton Island	588	9	626	7	634	7	604	9	561	9	549	8	550	9	597	9	640	9	651	9	600	8	572	8	597
San Juan, P. R.	404	5	481	4	580	4	622	4	519	5	536	6	639	5	549	6	531	6	460	6	411	6	411	6	512
Swan Island	442	6	496	7	615	6	646	6	625	6	544	8	588	8	591	7	535	8	457	7	394	8	382	8	526
Wake Island	438	7	518	7	577	7	627	7	642	8	656	6	629	7	623	7	587	6	525	7	482	7	421	7	560

NOTES:
- * Denotes only one year of data for the month -- no means computed.
- --- No data for the month (or incomplete data for the year).
- \# Barrow is in darkness during the winter months.
- † Madison data after 1957 not used due to exposure influences.
- ‡ Riverside data prior to March 1952 not used-instrumental discrepancies.

Langley is the unit used to denote one gram calorie per square centimeter.

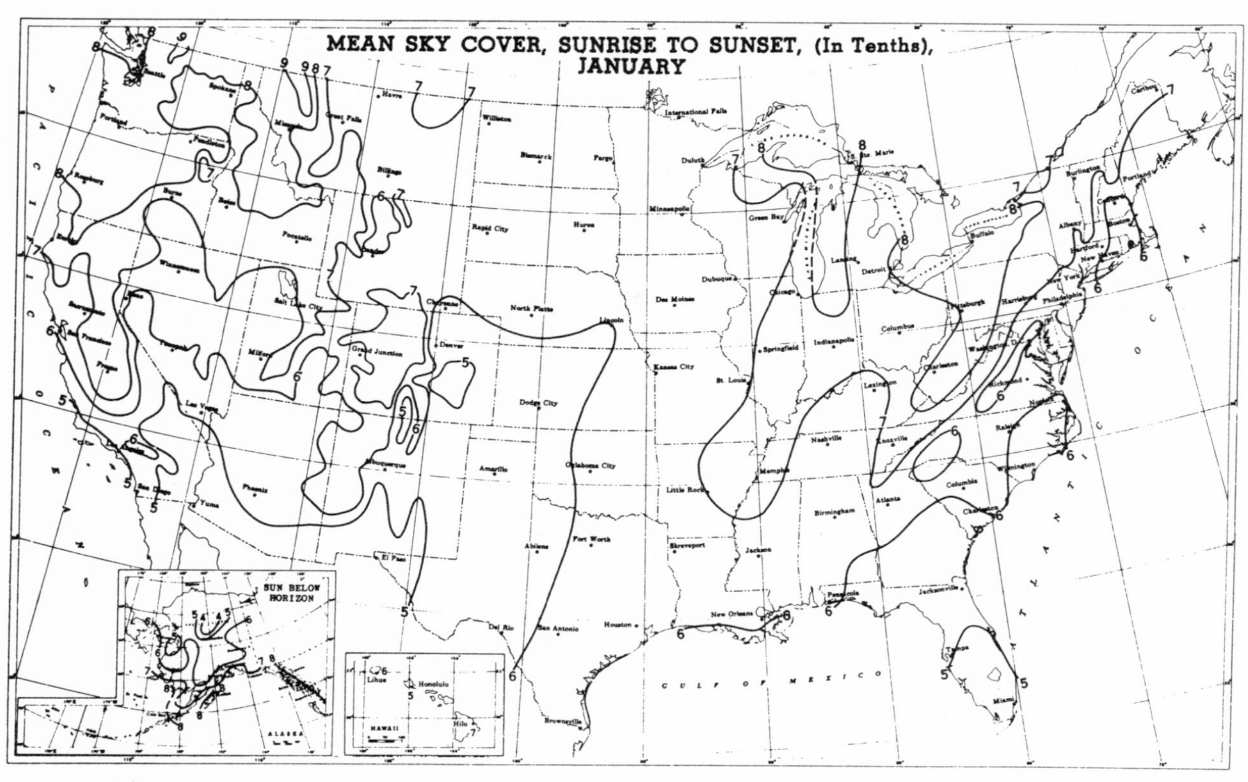
MEAN SKY COVER, SUNRISE TO SUNSET, (In Tenths),
JANUARY
SUN BELOW HORIZON
ALASKA
HAWAII
Honolulu
Lihue
Hilo
G U L F O F M E X I C O

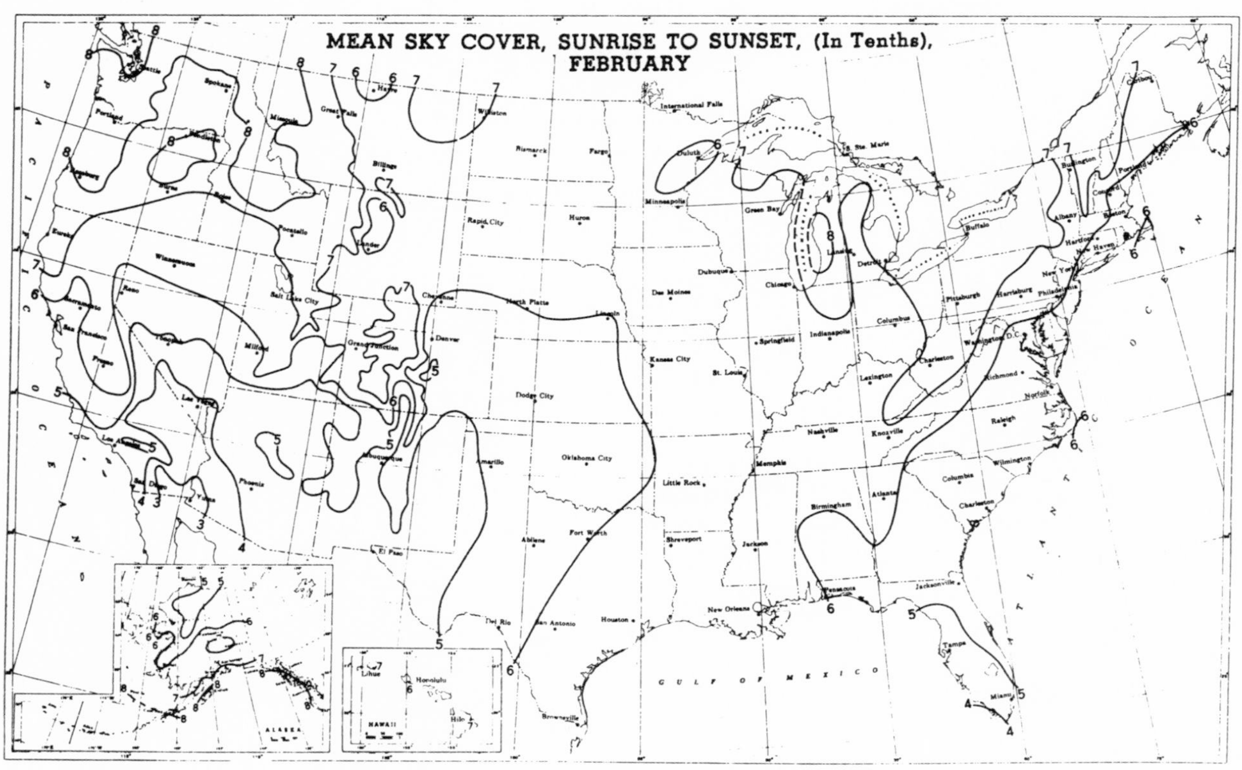
MEAN SKY COVER, SUNRISE TO SUNSET, (In Tenths),
FEBRUARY
ALASKA
HAWAII
Honolulu
Lihue
Hilo
G U L F O F M E X I C O

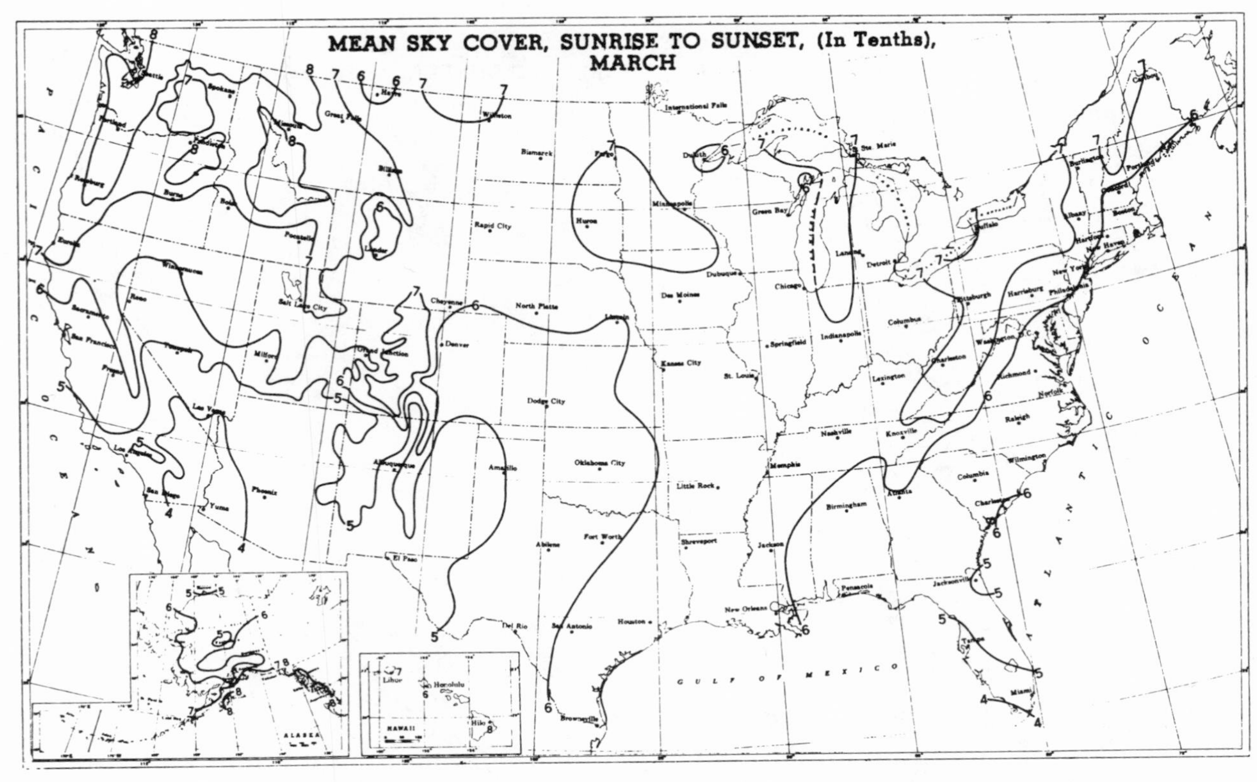
MEAN SKY COVER, SUNRISE TO SUNSET, (In Tenths),
MARCH
GULF OF MEXICO
ALASKA
HAWAII

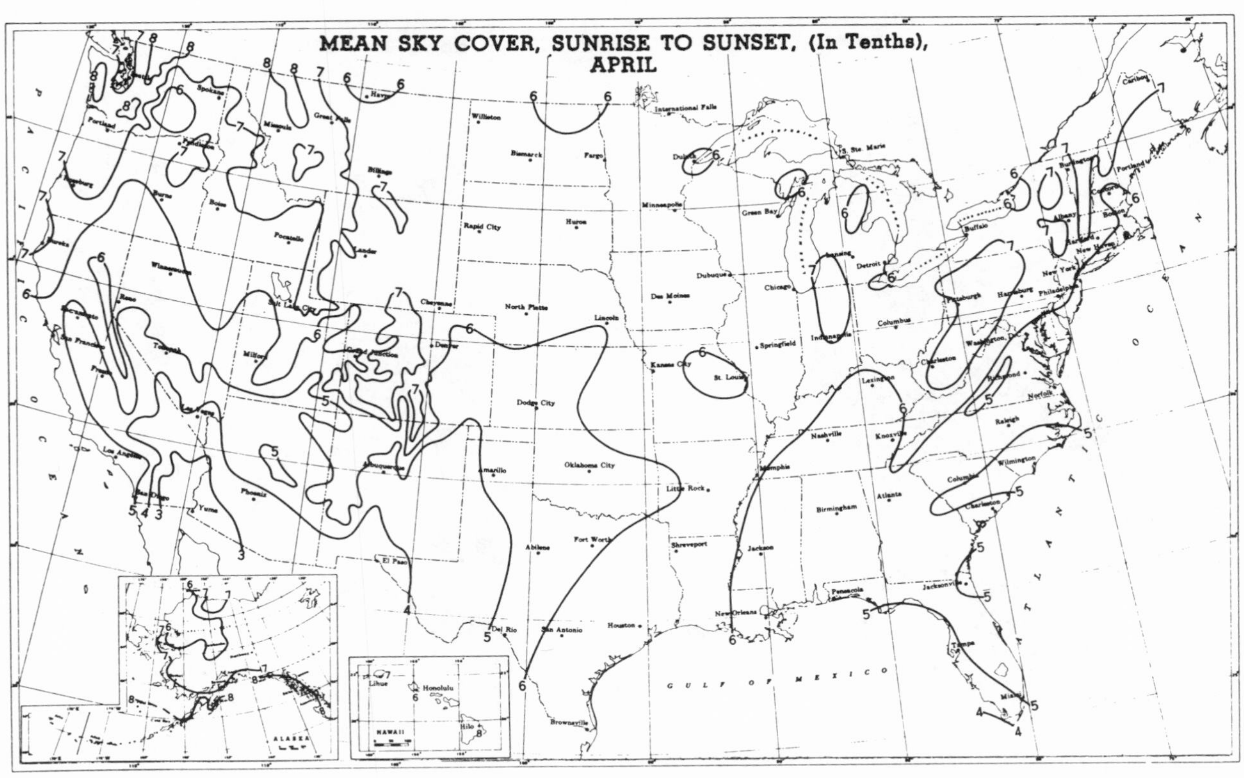
MEAN SKY COVER, SUNRISE TO SUNSET, (In Tenths),
APRIL
GULF OF MEXICO
ALASKA
HAWAII

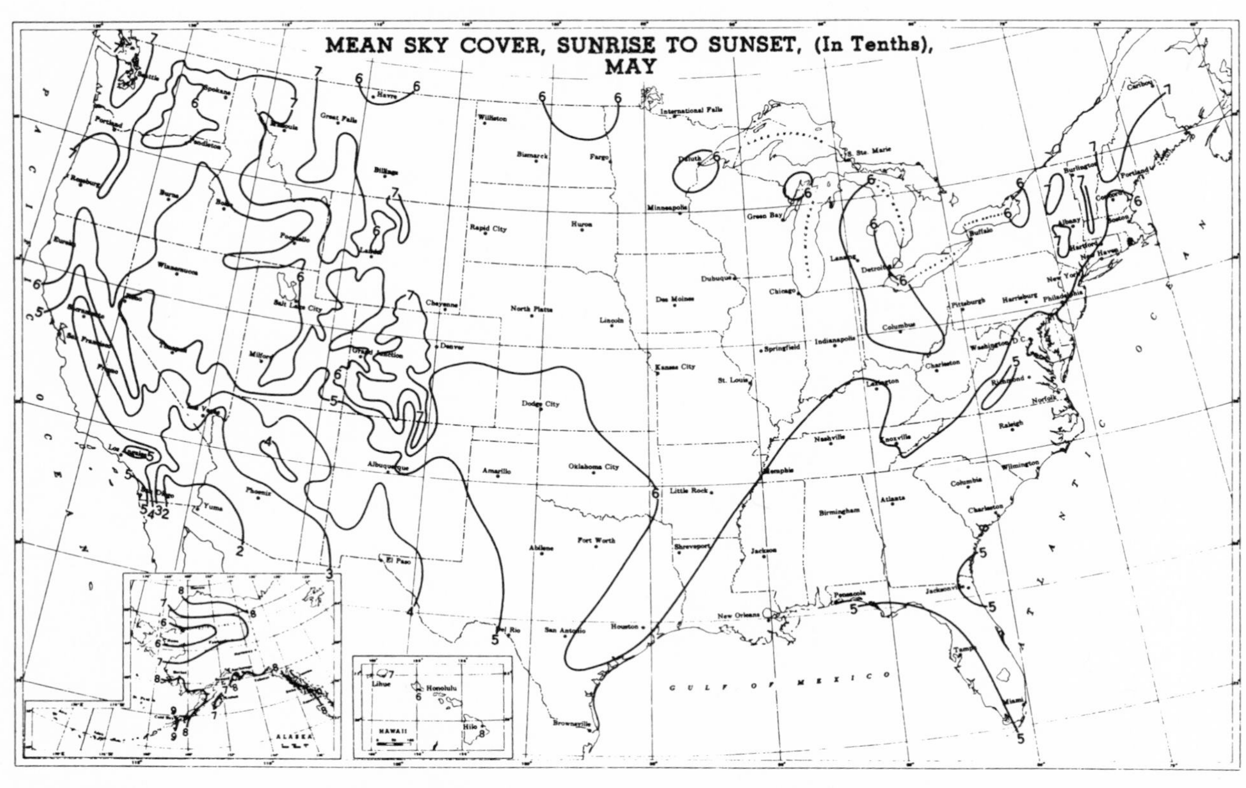
MEAN SKY COVER, SUNRISE TO SUNSET, (In Tenths),
MAY
GULF OF MEXICO
ALASKA
HAWAII

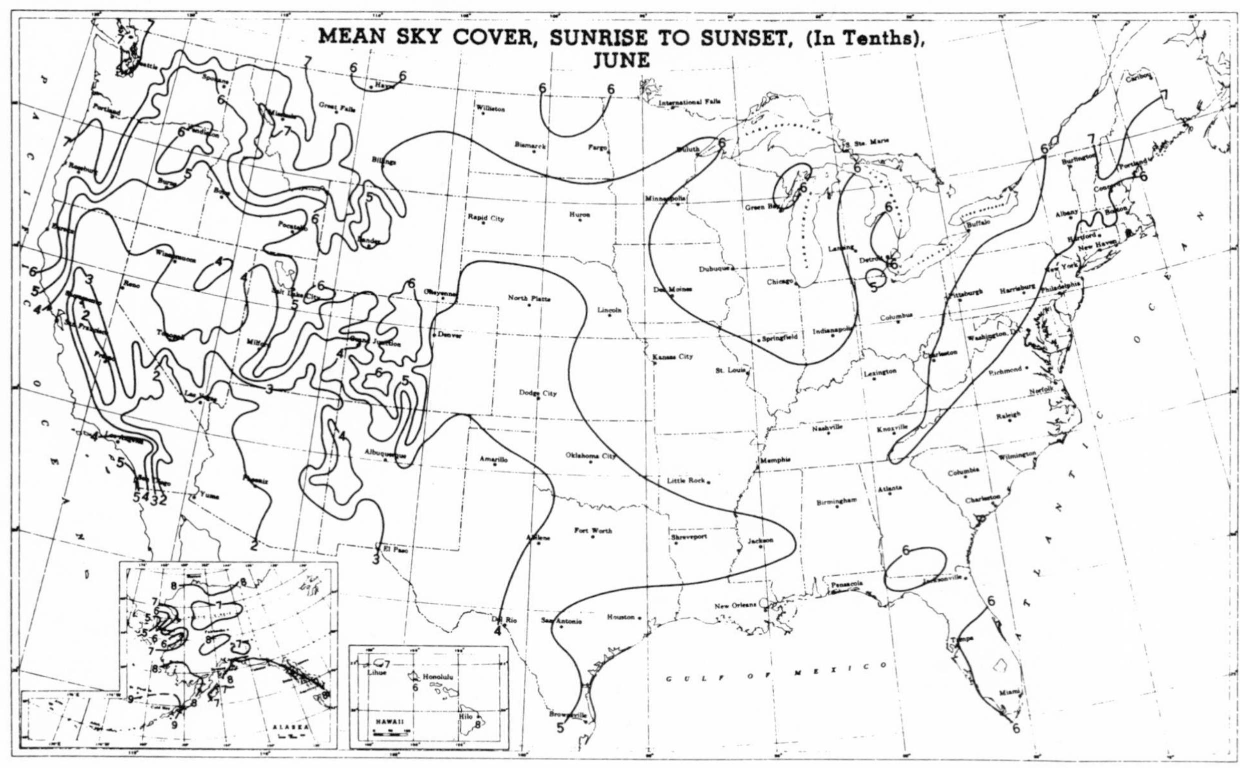
MEAN SKY COVER, SUNRISE TO SUNSET, (In Tenths),
JUNE
GULF OF MEXICO
ALASKA
HAWAII

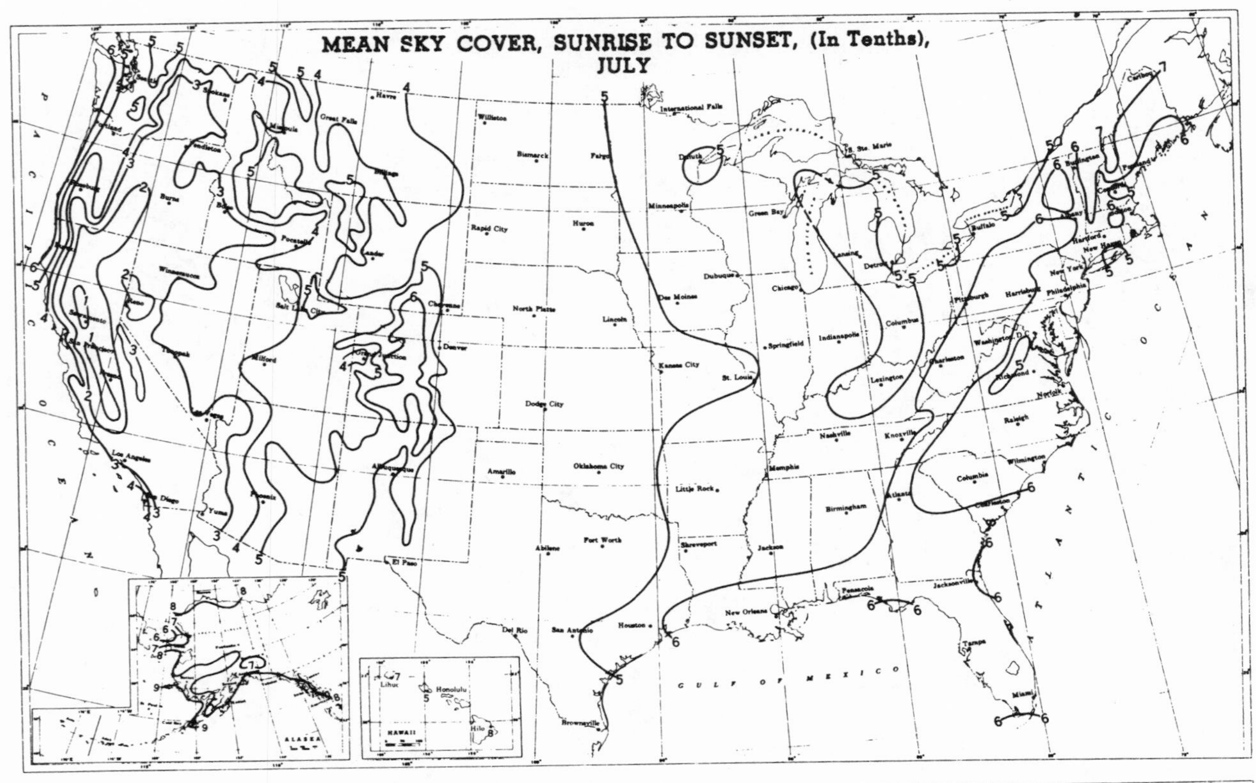
MEAN SKY COVER, SUNRISE TO SUNSET, (In Tenths),
JULY
International Falls
Bismarck
Minneapolis
Rapid City
North Platte
Lincoln
Des Moines
Dubuque
Chicago
Green Bay
Kansas City
St. Louis
Springfield
Indianapolis
Columbus
Lexington
Nashville
Knoxville
Memphis
Little Rock
Birmingham
Atlanta
Columbia
Wilmington
Raleigh
Norfolk
Richmond
Pittsburgh
Harrisburg
Philadelphia
New York
Buffalo
Dodge City
Amarillo
Oklahoma City
Fort Worth
Abilene
Shreveport
Jackson
New Orleans
Pensacola
Jacksonville
Houston
San Antonio
Del Rio
El Paso
Brownsville
Tampa
Miami
Albuquerque
Denver
Cheyenne
Los Angeles
Phoenix
Yuma
Salt Lake City
Pocatello
Great Falls
Havre
Williston
Fargo
Huron
Duluth
Sault Ste Marie
G U L F O F M E X I C O
ALASKA
HAWAII
Lihue
Honolulu
Hilo

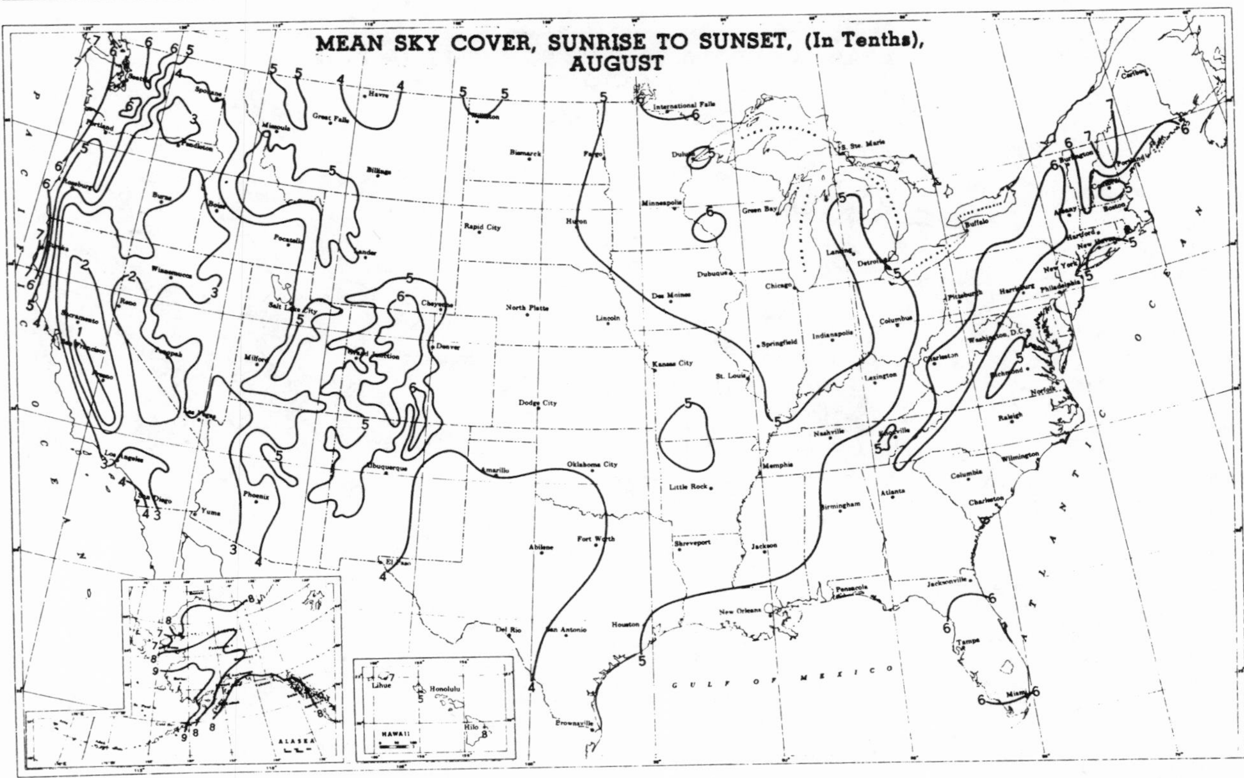
MEAN SKY COVER, SUNRISE TO SUNSET, (In Tenths),
AUGUST
International Falls
Bismarck
Minneapolis
Rapid City
North Platte
Lincoln
Des Moines
Dubuque
Chicago
Green Bay
Kansas City
St. Louis
Springfield
Indianapolis
Columbus
Lexington
Nashville
Memphis
Little Rock
Birmingham
Atlanta
Columbia
Charleston
Wilmington
Raleigh
Norfolk
Richmond
Washington, D.C.
Pittsburgh
Harrisburg
Philadelphia
New York
Buffalo
Dodge City
Amarillo
Oklahoma City
Fort Worth
Abilene
Shreveport
Jackson
New Orleans
Pensacola
Jacksonville
Houston
San Antonio
Del Rio
Brownsville
Tampa
Miami
Albuquerque
Denver
Cheyenne
Los Angeles
San Diego
Phoenix
Yuma
Salt Lake City
Pocatello
Great Falls
Havre
Billings
Fargo
Huron
Duluth
Sault Ste Marie
Portland
Burns
Reno
Sacramento
Winnemucca
Milford
G U L F O F M E X I C O
ALASKA
HAWAII
Lihue
Honolulu
Hilo

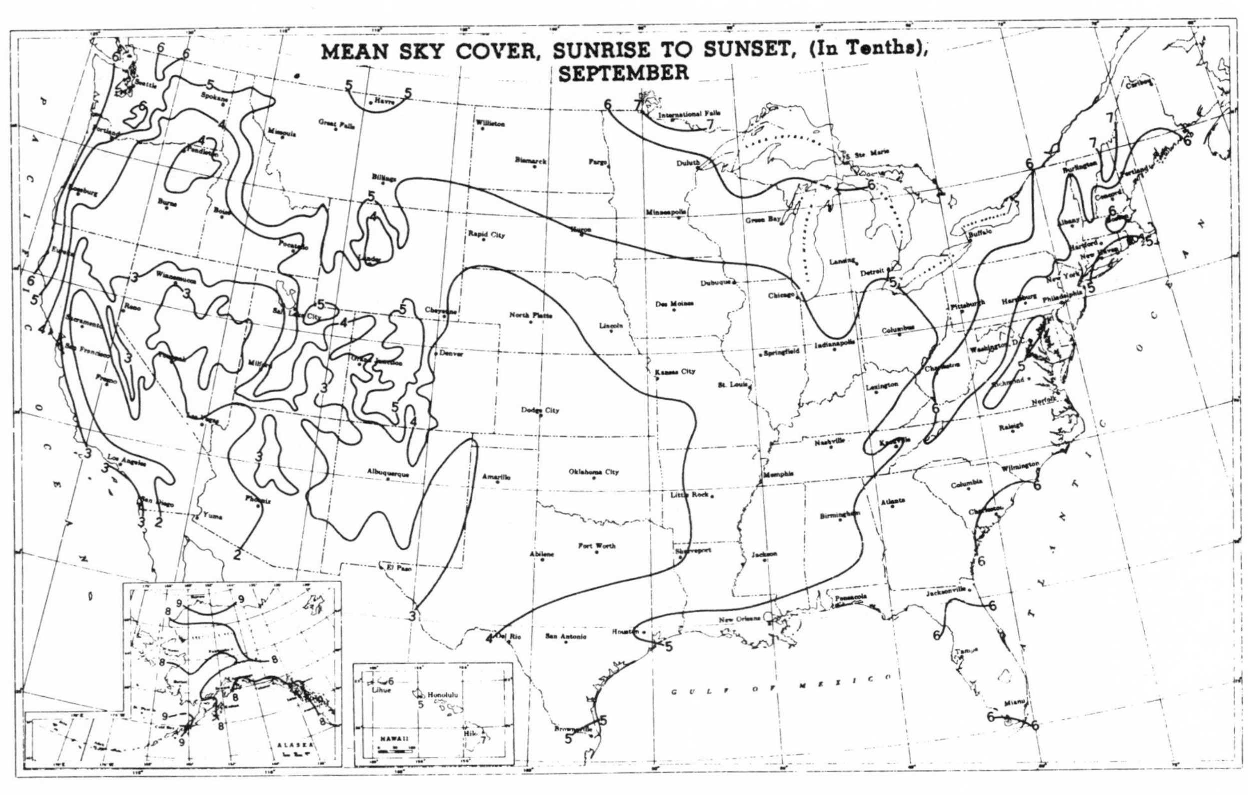
MEAN SKY COVER, SUNRISE TO SUNSET, (In Tenths),
SEPTEMBER

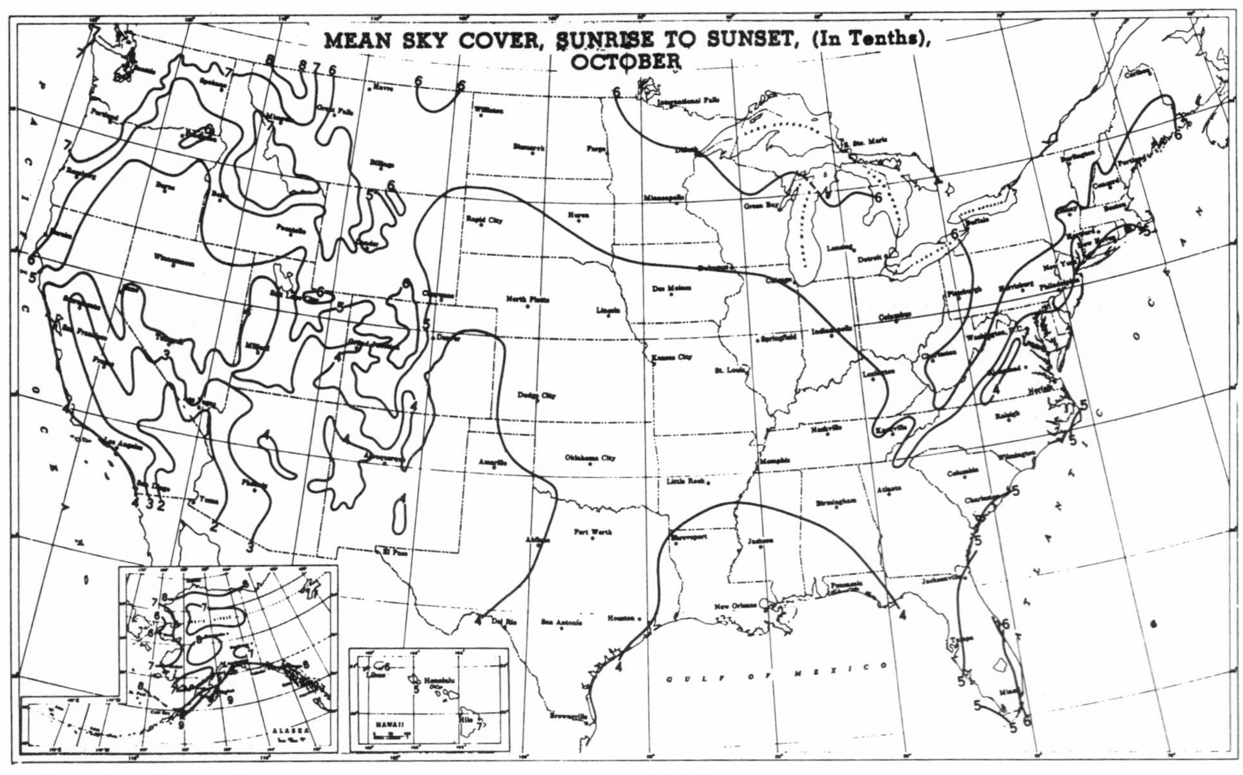
MEAN SKY COVER, SUNRISE TO SUNSET, (In Tenths),
OCTOBER

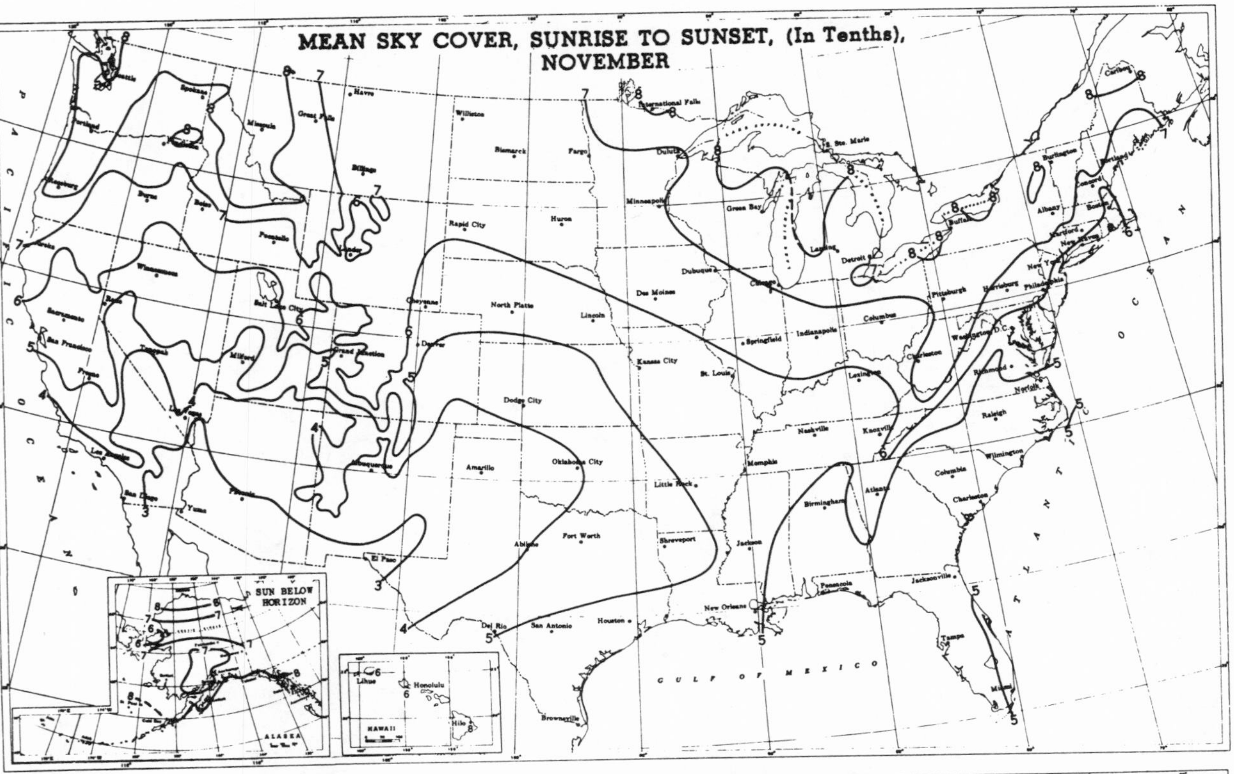

MEAN SKY COVER, SUNRISE TO SUNSET, (In Tenths),
NOVEMBER
SUN BELOW HORIZON
ALASKA
HAWAII
GULF OF MEXICO

MEAN SKY COVER, SUNRISE TO SUNSET, (In Tenths),
DECEMBER
SUN BELOW HORIZON
HAWAII
GULF OF MEXICO

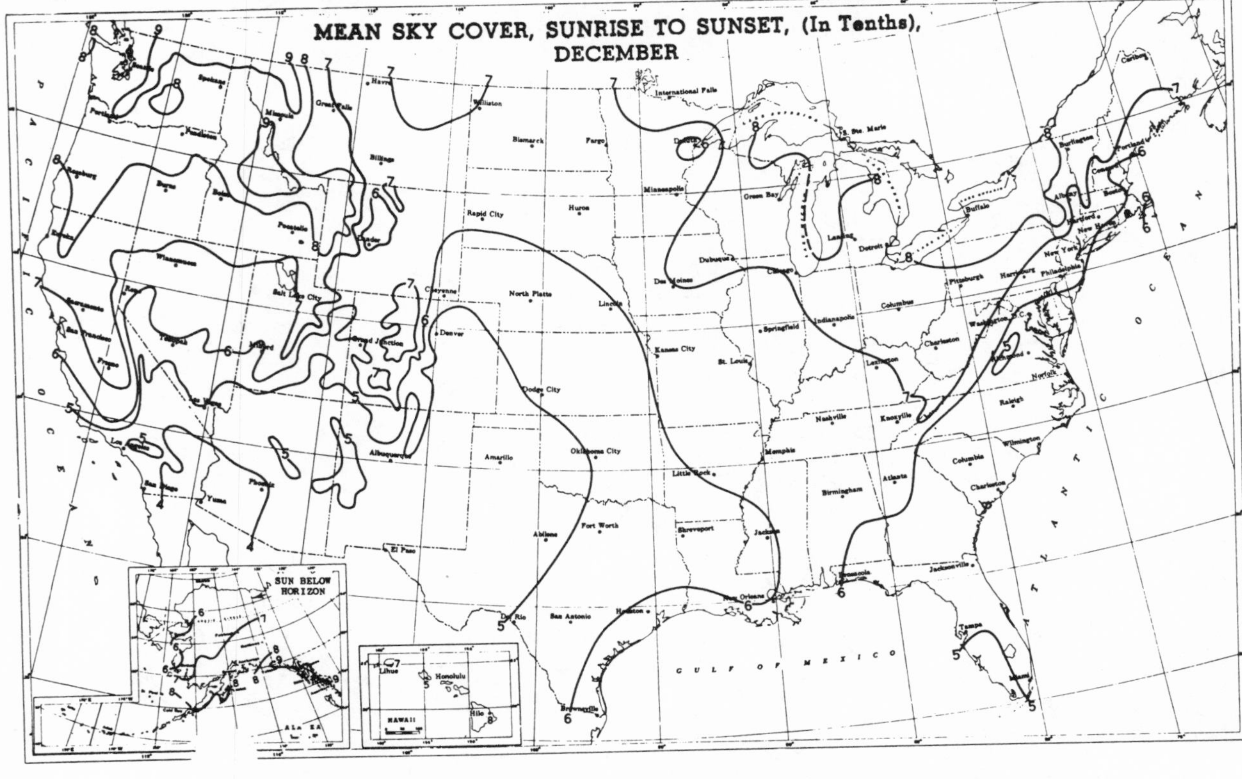

300 0 300 600 900 1200 1500 Miles

SCALE 1:30,000,000

ALBERS EQUAL AREA PROJECTION–STANDARD PARALLELS 29½° AND 45½°

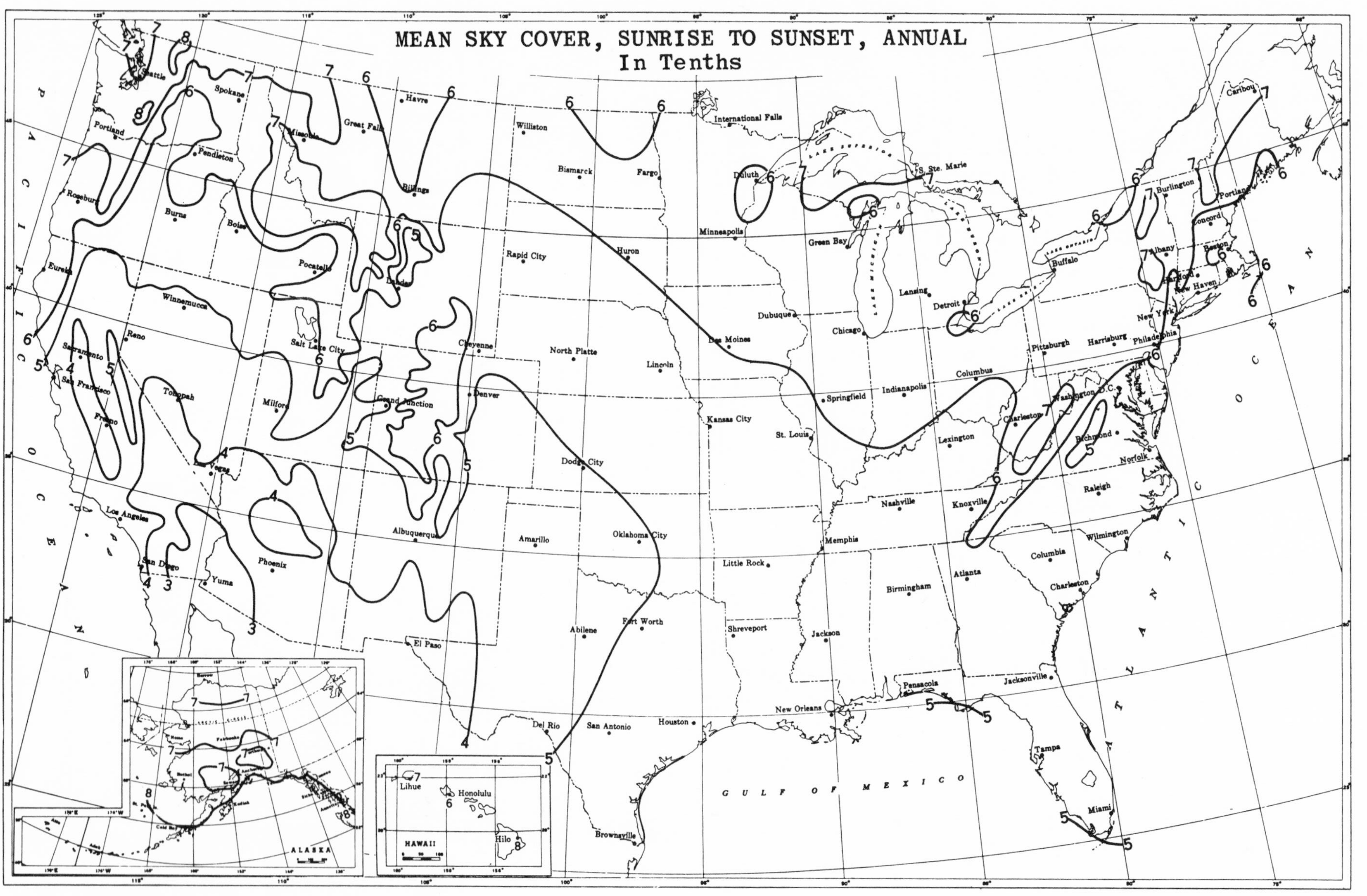
MEAN SKY COVER, SUNRISE TO SUNSET, ANNUAL
In Tenths

MEAN MONTHLY SKY COVER, SUNRISE TO SUNSET, In Tenths For Selected Stations

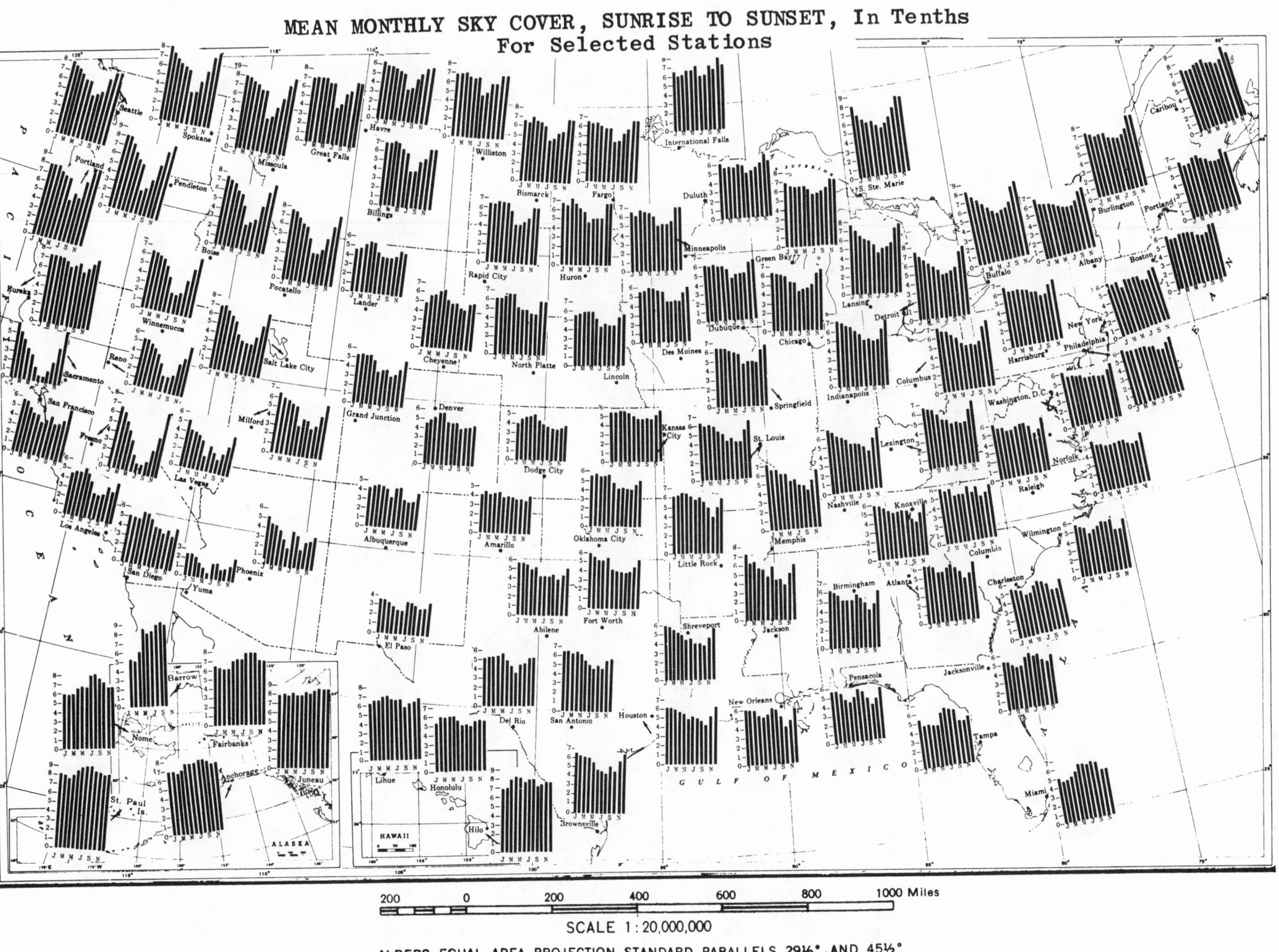

200 0 200 400 600 800 1000 Miles

SCALE 1:20,000,000

ALBERS EQUAL AREA PROJECTION-STANDARD PARALLELS 29½° AND 45½°

MEAN SKY COVER, SUNRISE TO SUNSET, (In Tenths),

STATE AND STATION	YEARS	JAN.	FEB.	MAR.	APR.	MAY	JUNE	JULY	AUG.	SEPT.	OCT.	NOV.	DEC.	ANNUAL
ALA. BIRMINGHAM	54	6.2	5.9	5.6	5.2	5.2	5.2	5.7	5.3	4.8	4.1	4.7	6.1	5.3
MOBILE	11	6.5	6.6	6.0	5.7	5.6	5.7	6.8	5.4	5.8	4.3	4.9	6.1	5.8
MONTGOMERY	83	5.9	5.7	5.2	4.8	4.7	5.1	5.5	5.2	4.5	3.9	4.5	5.7	5.1
ALASKA. ANCHORAGE	19	6.8	6.8	6.6	6.9	7.6	7.5	7.6	7.8	7.8	7.6	7.3	7.4	7.3
ANNETTE	12	7.7	8.0	7.7	7.6	7.9	8.0	8.0	7.8	7.7	8.6	8.5	8.6	8.0
BARROW	18	*	5.3	5.0	6.0	8.5	8.0	8.2	8.9	9.2	8.8	*	*	-
BARTER ISLAND	9	*	4.7	5.8	6.5	8.5	8.1	7.9	8.3	8.6	8.5	*	*	-
BETHEL	16	6.3	6.2	6.5	6.9	7.9	8.1	8.5	9.0	8.2	7.7	7.2	6.7	7.4
CORDOVA	14	7.2	7.1	7.1	7.4	8.5	8.3	8.6	8.1	8.2	8.0	7.9	7.8	7.9
FAIRBANKS	16	6.5	6.3	6.2	6.2	7.0	7.3	7.3	7.8	7.9	7.9	7.1	7.1	7.1
JUNEAU	16	7.9	8.0	7.8	8.2	7.8	7.8	8.3	8.0	8.4	8.6	8.5	8.5	8.2
KOTZEBUE	16	5.6	5.5	6.0	5.8	6.4	6.9	7.7	8.3	7.5	7.2	6.6	5.9	6.6
MCGRATH	16	6.6	6.3	6.5	6.1	7.5	7.8	7.8	8.5	8.1	8.4	7.4	6.8	7.3
NOME	25	5.9	5.8	5.9	6.5	6.7	6.6	7.9	8.1	7.5	7.1	6.5	6.6	6.8
ST. PAUL ISLAND	30	8.1	8.0	7.7	8.1	8.7	8.9	9.1	9.1	8.6	8.5	8.3	8.4	8.5
ARIZ. FLAGSTAFF	10	5.9	4.9	5.0	4.9	4.1	2.4	5.4	5.3	2.9	3.1	3.4	4.4	4.3
PHOENIX	14	5.1	4.5	4.3	3.8	2.9	2.0	3.8	3.5	2.0	2.8	3.1	3.6	3.5
TUCSON	16	5.0	4.5	4.6	3.7	2.8	2.2	5.5	4.7	2.7	3.1	3.0	4.4	3.9
YUMA	80	2.4	2.5	2.2	1.6	1.1	0.6	1.7	1.8	1.1	1.3	1.7	2.5	1.7
ARK. FT. SMITH	67	5.4	5.5	5.4	5.4	5.2	4.6	4.4	4.2	3.9	3.3	4.5	5.3	4.8
LITTLE ROCK	55	6.2	6.3	6.5	6.4	6.2	5.7	6.0	5.4	4.8	3.9	5.0	5.9	5.7
CALIF. BAKERSFIELD	14	6.2	5.5	5.4	4.4	3.2	1.3	1.1	1.1	1.3	2.7	4.1	6.3	3.6
BISHOP	12	5.3	4.8	4.7	4.7	4.1	2.3	2.4	1.9	2.0	2.8	3.5	4.9	3.6
EUREKA	17	7.3	7.3	7.3	7.2	6.8	6.4	6.6	6.9	6.0	6.5	7.0	7.6	6.9
FRESNO	22	6.7	6.1	5.3	4.4	3.4	1.9	1.1	1.1	1.5	2.8	4.4	7.0	3.8
LOS ANGELES	19	4.6	4.8	4.8	5.3	4.8	4.1	2.8	2.7	2.8	3.9	3.4	4.5	4.0
RED BLUFF	15	6.6	6.3	6.1	5.4	4.6	3.2	1.3	1.6	2.2	4.0	5.5	6.9	4.5
SACRAMENTO	50	6.0	5.2	4.4	3.5	2.7	1.6	0.7	0.7	1.4	2.4	4.0	5.9	3.2
SAN DIEGO	19	5.1	5.0	4.9	5.7	5.5	5.1	4.4	4.3	3.7	4.3	3.7	4.7	4.7
SAN FRANCISCO	49	5.7	5.5	5.0	4.5	4.4	3.7	4.2	4.6	3.8	4.0	4.5	5.6	4.6
COLO. COLORADO SPRINGS	10	4.9	5.1	5.7	6.0	6.0	4.5	4.9	4.7	3.5	3.7	4.7	4.5	4.9
DENVER	48	4.5	4.8	5.3	5.7	5.7	4.7	4.6	4.7	4.0	4.1	4.2	4.4	4.7
GRAND JUNCTION	57	5.3	5.4	5.3	5.4	4.8	3.6	3.9	4.0	3.3	3.5	4.2	5.1	4.5
CONN. HARTFORD	51	6.2	5.7	5.8	6.1	6.0	5.8	5.7	5.6	5.4	5.3	6.2	6.2	5.8
D.C. WASHINGTON	70	6.1	5.6	5.6	5.4	5.4	5.1	5.1	5.1	4.8	4.7	5.4	5.8	5.3
FLA. APALACHICOLA	26	5.2	5.4	5.5	4.8	4.3	5.2	6.0	5.7	5.6	3.9	4.4	5.7	5.1
JACKSONVILLE	68	5.3	5.2	4.9	4.6	4.7	5.6	5.9	5.6	5.7	5.1	4.7	5.4	5.2
KEY WEST	58	4.3	3.8	3.7	3.8	4.4	5.4	5.3	5.3	5.5	4.9	4.2	4.3	4.6
MIAMI	48	5.0	4.6	4.7	4.9	5.4	6.2	6.1	5.8	6.2	5.8	5.2	5.2	5.4
TAMPA	34	5.3	4.7	4.8	4.7	4.6	5.8	6.3	6.1	6.0	4.7	4.6	5.2	5.2
GA. ATLANTA	25	6.3	6.2	6.1	5.5	5.5	5.8	6.3	5.7	5.3	4.5	5.1	6.2	5.7
HAWAII. HILO	13	6.8	7.0	7.8	8.3	8.2	7.6	7.8	7.9	7.2	7.4	7.6	7.6	7.6
HONOLULU	13	5.3	6.0	5.9	6.0	6.0	5.5	5.0	5.1	4.8	5.4	5.5	5.4	5.5
LIHUE	10	6.2	6.5	6.5	6.9	7.0	6.6	6.5	6.6	5.7	6.2	6.4	6.6	6.5
IDAHO. BOISE	20	7.5	7.2	6.8	6.2	5.8	4.7	2.7	3.1	3.5	5.0	6.6	7.6	5.6
POCATELLO	21	7.7	7.5	6.8	6.3	6.0	4.8	3.4	3.4	3.8	5.1	6.5	7.7	5.8
ILL. CAIRO	66	6.5	6.3	6.3	6.0	5.9	5.6	5.2	5.0	4.6	4.4	5.5	6.3	5.6
CHICAGO	84	6.4	6.2	6.1	5.8	5.3	5.2	4.3	4.6	4.7	5.0	6.3	6.6	5.5
SPRINGFIELD	31	6.4	6.4	6.4	6.1	5.9	5.7	4.7	4.9	4.9	4.8	6.1	6.6	5.7
IND. FT. WAYNE	44	7.1	7.0	6.6	6.4	5.8	5.6	4.7	5.0	5.0	5.4	6.7	7.4	6.1
INDIANAPOLIS	51	7.1	6.8	6.6	6.4	6.0	5.6	5.0	5.1	4.8	5.0	6.3	7.1	6.0
IOWA. DES MOINES	73	5.4	5.5	5.8	5.7	5.6	5.3	4.3	4.4	4.5	4.4	5.3	5.8	5.2
DUBUQUE	62	6.1	6.0	6.2	5.9	5.9	5.6	4.7	5.0	5.1	5.2	6.3	6.5	5.7
SIOUX CITY	66	5.6	5.8	6.1	5.9	5.9	5.3	4.3	4.6	4.5	4.6	5.7	5.9	5.4
KANS. CONCORDIA	16	5.5	5.8	5.9	6.0	6.0	5.0	4.5	4.2	4.0	4.0	4.8	5.3	5.1
DODGE CITY	67	4.0	4.5	4.7	4.8	4.9	4.3	3.9	3.7	3.5	3.4	3.6	3.9	4.1
GOODLAND	11	5.8	5.9	6.0	6.0	6.2	4.4	4.4	4.4	3.6	3.9	4.8	5.1	5.0
WICHITA	66	5.0	5.1	5.2	5.3	5.3	4.6	3.9	3.8	3.9	4.0	4.4	5.0	4.6
KY. LOUISVILLE	54	6.8	6.3	6.1	6.0	5.5	5.3	4.8	4.8	4.5	4.6	5.7	6.7	5.6
LA. NEW ORLEANS	44	6.0	6.0	5.8	5.3	5.1	5.4	6.1	5.8	5.2	4.0	4.7	5.9	5.4
SHREVEPORT	83	5.8	5.7	5.4	5.1	4.8	4.4	4.5	4.0	4.0	3.7	4.7	5.6	4.8
MAINE. CARIBOU	14	7.2	6.8	6.8	7.3	7.4	7.5	7.0	6.8	6.6	6.8	8.1	7.3	7.1
EASTPORT	61	6.9	6.5	6.6	6.7	6.9	6.9	6.5	6.3	6.3	6.6	7.6	7.2	6.8
MASS. BOSTON	64	6.0	5.6	5.6	5.8	5.8	5.7	5.6	5.3	5.1	5.3	6.0	6.0	5.7
MICH. ALPENA	45	7.7	6.9	6.4	6.0	5.8	5.5	4.8	5.1	5.8	6.3	7.9	8.0	6.4
DETROIT	63	7.4	6.9	6.5	6.1	5.7	5.2	4.7	4.8	5.0	5.4	7.1	7.6	6.0
GRAND RAPIDS	51	8.2	7.5	6.8	6.3	5.9	5.5	4.7	5.0	5.4	6.0	7.5	8.3	6.4
MARQUETTE	66	7.7	7.3	6.9	6.3	6.3	5.9	5.5	5.8	6.3	7.0	8.0	8.0	6.8
SAULT STE. MARIE	64	7.9	6.8	6.3	5.9	6.0	5.5	5.2	5.6	6.3	7.2	8.4	8.3	6.6
MINN. DULUTH	64	6.1	5.5	5.6	5.5	5.8	5.6	4.8	5.0	5.4	5.9	6.8	6.1	5.7
MINNEAPOLIS	24	6.5	6.2	6.7	6.5	6.4	6.0	4.9	5.1	5.1	5.4	6.9	6.9	6.1
MISS. MERIDIAN	22	6.1	6.0	5.5	5.1	4.9	4.9	5.6	4.9	4.5	3.9	4.6	6.1	5.2
VICKSBURG	55	5.7	5.8	5.6	5.3	5.1	4.8	5.5	4.9	4.1	3.4	4.5	5.6	5.0
MO. KANSAS CITY	66	5.4	5.4	5.5	5.5	5.3	4.9	4.1	4.1	4.2	4.1	5.0	5.5	4.9
ST. LOUIS	24	6.3	5.9	6.0	5.7	5.3	5.0	4.2	4.5	4.1	4.1	5.4	6.4	5.2
SPRINGFIELD	50	5.7	5.3	5.2	5.0	4.6	4.1	3.6	3.9	3.8	4.0	4.8	5.6	4.6
MONT. HAVRE	56	6.3	5.9	5.9	5.7	5.5	5.3	3.6	3.8	4.8	5.4	6.2	6.3	5.4
HELENA	65	6.8	6.6	6.6	6.5	6.5	6.0	4.0	4.2	5.1	5.7	6.6	6.8	6.0
KALISPELL	50	7.7	6.7	6.2	5.7	5.6	5.3	3.5	3.9	4.9	5.7	7.6	8.0	5.9
MILES CITY	10	6.8	6.9	7.0	6.8	6.3	5.7	3.9	4.1	5.2	5.3	6.6	6.5	5.9
NEBR. LINCOLN	60	5.5	5.7	5.8	5.9	5.9	5.2	4.3	4.6	4.4	4.4	5.3	5.6	5.2
NORTH PLATTE	11	6.2	6.2	6.4	6.6	6.6	5.3	4.9	4.9	4.5	4.4	5.8	5.8	5.6
NEV. ELY	16	6.2	6.1	6.2	6.0	5.9	4.0	3.9	3.4	3.0	4.1	5.4	6.0	5.0
LAS VEGAS	11	5.2	4.2	4.1	3.8	3.3	1.6	2.8	2.3	1.5	2.3	3.2	4.4	3.2
RENO	35	5.3	5.3	4.9	4.6	4.1	3.1	1.8	1.8	2.1	3.4	4.3	5.5	3.9
WINNEMUCCA	24	6.4	6.3	6.0	5.8	5.1	4.0	2.3	2.2	2.4	4.0	5.3	6.6	4.7

MEAN SKY COVER, SUNRISE TO SUNSET, (In Tenths),

STATE AND STATION	YEARS	JAN.	FEB.	MAR.	APR.	MAY	JUNE	JULY	AUG.	SEPT.	OCT.	NOV.	DEC.	ANNUAL
N.H. CONCORD	41	5.6	5.0	4.9	5.2	5.2	4.8	4.7	4.8	5.0	5.3	6.0	5.9	5.2
N.J. ATLANTIC CITY	59	6.2	5.6	5.7	5.8	5.6	5.5	5.3	5.2	5.1	4.8	5.3	6.0	5.5
N.MEX. ALBUQUERQUE	25	4.3	4.5	4.6	4.5	4.3	3.4	4.4	4.5	3.2	2.9	3.1	4.0	4.0
RATON	8	5.1	5.4	5.2	5.9	6.1	4.5	5.3	5.4	3.9	4.0	4.4	4.2	5.0
ROSWELL	38	4.1	4.0	4.0	3.8	3.9	3.2	3.6	3.6	3.4	3.3	3.3	3.9	3.7
N.Y. ALBANY	65	6.6	6.0	5.9	5.8	5.6	5.3	5.1	5.0	5.0	5.4	6.4	6.7	5.7
BINGHAMTON	58	7.5	7.2	7.1	6.9	6.5	6.2	6.1	6.0	6.0	6.3	7.5	7.9	6.8
BUFFALO	57	8.1	7.5	7.0	6.5	6.0	5.5	5.2	5.4	5.6	6.2	7.7	8.2	6.6
CANTON	43	6.9	6.3	6.0	6.1	5.7	5.0	4.8	5.0	5.5	6.1	7.5	7.4	6.0
NEW YORK	69	6.2	5.8	5.9	5.2	5.8	5.7	5.6	5.2	5.0	5.0	5.8	6.0	5.7
SYRACUSE	55	7.9	7.5	7.1	6.6	6.0	5.7	5.4	5.6	5.8	6.3	7.7	8.1	6.6
N.C. ASHEVILLE	56	6.0	5.7	5.6	5.4	5.3	5.5	5.9	5.7	5.2	4.4	4.8	5.8	5.5
HATTERAS	24	6.1	5.9	5.9	5.1	5.3	5.4	5.8	5.6	5.7	5.1	5.1	5.9	5.6
RALEIGH	54	6.0	5.5	5.4	5.1	5.1	5.4	5.6	5.5	5.2	4.4	4.7	5.8	5.3
WILMINGTON	19	5.7	5.4	5.3	4.6	4.9	5.3	5.6	5.4	5.6	4.4	4.5	5.4	5.2
N.DAK. BISMARCK	15	6.7	6.6	6.9	6.6	6.4	6.2	4.5	4.8	5.4	5.6	6.6	6.6	6.1
DEVILS LAKE	55	6.1	6.0	6.1	5.8	5.8	5.6	4.5	4.7	5.3	5.6	6.6	6.3	5.7
FARGO	40	6.5	6.3	6.4	6.0	5.7	5.7	4.4	4.6	5.3	5.7	6.7	6.7	5.8
WILLISTON	18	6.8	6.9	7.1	6.6	6.4	6.6	4.7	5.0	5.9	6.0	6.9	6.9	6.3
OHIO. CINCINNATI	24	7.4	6.8	6.8	6.5	6.2	5.8	5.1	5.1	4.8	5.0	6.4	7.3	6.1
CLEVELAND	65	7.7	7.4	6.7	6.1	5.6	5.2	4.6	4.8	5.1	5.7	7.4	8.0	6.2
COLUMBUS	64	7.1	6.7	6.6	6.2	5.6	5.2	4.8	4.8	4.7	5.0	6.5	7.2	5.9
OKLA. OKLAHOMA CITY	23	5.5	5.7	5.4	5.5	5.6	4.7	4.2	4.3	4.1	4.1	4.6	5.2	4.9
OREG. BAKER	58	6.9	6.7	6.3	6.9	5.7	5.0	2.8	2.8	3.8	4.5	6.0	6.7	5.3
MEDFORD	29	8.2	7.6	7.2	6.6	5.8	4.9	2.1	2.2	3.3	5.6	7.4	8.6	5.8
PORTLAND	13	7.9	7.9	8.0	7.4	7.1	6.9	4.6	5.4	5.1	7.0	8.2	8.6	7.0
ROSEBURG	7	8.9	8.5	7.9	7.2	6.4	6.4	2.8	3.7	4.9	6.9	8.4	8.9	6.7
PA. HARRISBURG	58	6.6	6.1	6.1	6.2	6.0	5.9	5.6	5.6	5.1	5.2	6.2	6.5	5.9
PHILADELPHIA	63	6.1	5.8	5.8	5.8	5.8	5.8	5.7	5.6	5.2	4.9	5.6	6.0	5.7
PITTSBURGH	83	7.4	7.0	6.7	6.3	5.7	5.5	5.3	5.2	5.0	5.4	6.7	7.5	6.1
R.I. BLOCK ISLAND	27	6.0	5.5	5.5	5.7	5.7	5.3	5.2	5.0	4.8	4.6	5.7	5.9	5.4
PROVIDENCE	49	5.9	5.4	5.4	5.7	5.6	5.4	5.4	5.1	5.0	4.7	5.5	5.7	5.4
S.C. CHARLESTON	59	5.4	5.3	5.1	4.5	4.6	5.3	5.7	5.5	5.1	4.3	4.3	5.3	5.0
COLUMBIA	15	6.0	5.7	5.8	5.1	5.2	5.1	5.8	5.1	5.6	4.6	4.8	5.7	5.4
S.DAK. HURON	16	6.6	6.4	7.2	6.6	6.2	5.7	4.6	4.9	5.0	5.1	6.6	6.7	6.0
RAPID CITY	17	6.4	6.4	6.6	6.6	6.5	5.7	4.2	4.3	4.6	4.9	6.2	6.2	5.7
TENN. KNOXVILLE	78	5.9	5.8	5.5	5.2	4.9	5.0	5.1	5.0	4.1	3.7	4.7	5.7	5.0
MEMPHIS	14	7.2	6.5	6.5	6.0	6.0	5.2	5.3	4.7	4.5	4.1	5.2	6.2	5.6
NASHVILLE	36	6.8	6.5	6.1	5.8	5.6	5.2	5.2	5.0	4.7	4.3	5.5	6.5	5.6
TEX. ABILENE	25	5.6	5.6	5.4	4.9	5.2	4.2	4.2	4.2	4.4	3.8	4.3	5.1	4.7
AMARILLO	64	4.2	4.5	4.2	4.3	4.4	3.8	4.0	3.8	3.7	3.7	3.6	4.2	4.0
AUSTIN	29	6.2	6.1	5.7	5.6	5.5	4.7	4.5	4.0	4.5	4.2	5.1	5.8	5.2
BROWNSVILLE	32	6.5	6.2	6.1	5.9	5.4	4.7	4.5	4.3	5.0	4.4	5.7	6.4	5.4
CORPUS CHRISTI	23	6.5	6.7	6.3	6.2	5.8	4.6	4.5	4.2	4.7	3.9	5.2	6.4	5.4
DEL RIO	49	5.2	5.3	5.3	5.4	5.7	4.9	4.2	3.5	4.4	4.6	5.1	5.2	4.9
EL PASO	64	3.7	3.6	3.4	2.9	2.6	2.4	4.0	3.8	3.1	2.7	2.8	3.5	3.2
FT. WORTH	25	5.9	5.6	5.2	5.2	5.4	4.2	4.0	3.8	3.7	3.9	4.4	5.3	4.7
GALVESTON	52	5.9	5.9	5.7	5.4	4.8	4.1	4.7	4.5	4.3	3.6	4.7	5.9	5.0
HOUSTON	46	6.1	6.1	5.9	5.7	5.4	4.8	5.2	5.0	4.7	4.2	5.2	6.2	5.4
MIDLAND	11	5.4	5.3	5.1	4.9	4.7	3.8	4.6	3.9	3.5	3.6	3.5	4.3	4.4
SAN ANTONIO	17	6.4	6.5	6.3	6.4	6.2	5.4	5.0	4.6	4.9	4.6	5.4	5.7	5.6
UTAH. MODENA	46	5.2	5.5	5.0	5.1	4.4	2.8	3.7	3.5	2.8	3.6	4.1	5.0	4.2
SALT LAKE CITY	24	6.9	7.0	6.5	6.1	5.4	4.2	3.5	3.4	3.4	4.3	5.6	6.9	5.3
VT. BURLINGTON	49	7.2	6.9	6.6	6.7	6.7	6.1	5.8	5.7	6.0	6.5	7.9	7.8	6.7
VA. NORFOLK	40	5.6	5.2	5.1	4.8	4.8	4.9	5.0	5.0	4.6	4.4	4.6	5.4	5.0
RICHMOND	56	5.9	5.2	5.3	5.1	5.0	5.2	5.2	5.2	4.9	4.3	4.8	5.5	5.1
ROANOKE	11	6.7	6.3	6.4	6.3	6.3	5.8	6.0	5.9	5.5	5.2	5.3	6.2	6.0
WASH. NORTH HEAD	56	7.8	7.3	7.1	6.9	6.9	6.8	6.4	6.5	6.2	6.8	7.6	7.8	7.0
SEATTLE	24	8.0	7.7	7.4	6.9	6.4	6.4	4.9	5.3	5.6	7.2	8.0	8.1	6.8
SPOKANE	67	8.1	7.4	6.8	6.4	6.2	5.8	3.5	3.6	4.8	6.0	7.7	8.2	6.2
TATOOSH ISLAND	49	7.9	7.3	7.3	7.2	7.1	7.1	6.9	7.0	6.5	6.9	8.0	7.9	7.3
WALLA WALLA	50	8.1	7.4	6.3	5.4	5.0	4.5	2.4	2.7	4.0	5.2	7.4	8.4	5.6
YAKIMA	27	7.7	7.1	6.5	6.1	5.6	5.0	2.6	2.8	3.9	5.4	7.1	7.8	5.6
W.VA. ELKINS	27	7.9	7.6	7.4	7.0	6.8	6.7	6.5	6.4	6.1	6.1	6.9	7.6	6.9
PARKERSBURG	68	7.3	6.8	6.4	6.0	5.5	5.2	5.0	5.1	4.8	5.1	6.6	7.3	5.9
WIS. GREEN BAY	69	6.8	6.5	6.4	6.4	6.4	6.2	5.6	5.7	5.9	6.3	7.2	7.1	6.4
MILWAUKEE	58	6.5	6.3	6.3	6.0	5.7	5.5	4.5	4.8	5.1	5.5	6.5	6.7	5.8
WYO. CHEYENNE	56	4.8	5.6	5.7	6.2	6.3	5.3	4.8	5.0	4.5	4.4	5.0	5.1	5.2
LANDER	64	4.7	4.8	5.3	5.6	5.5	4.5	4.1	4.0	4.0	4.3	4.9	4.7	4.7
SHERIDAN	15	6.5	6.7	6.9	6.6	6.6	5.8	3.9	4.1	4.7	5.1	6.4	6.6	5.8
YELLOWSTONE	35	6.9	6.4	6.4	6.1	6.3	5.4	4.3	4.3	4.9	5.4	6.4	6.8	5.8
P.R. SAN JUAN	57	4.9	4.5	4.5	5.0	5.8	5.8	5.5	5.1	5.8	5.4	5.2	5.1	5.2
VIRGIN ISLANDS. ST. CROIX	6	5.6	5.8	5.7	5.9	7.2	6.9	7.0	6.1	7.0	6.5	6.0	5.8	6.3

* SUN BELOW HORIZON.

BASED ON PERIOD OF RECORD THROUGH DECEMBER 1959, EXCEPT IN A FEW INSTANCES. VALUES ARE IN TENTHS; (10.0 WOULD BE COMPLETE SKY COVERAGE). DERIVED FROM "NORMALS, MEANS, AND EXTREMES" TABLE IN U. S. WEATHER BUREAU PUBLICATION LOCAL CLIMATOLOGICAL DATA.

WIND

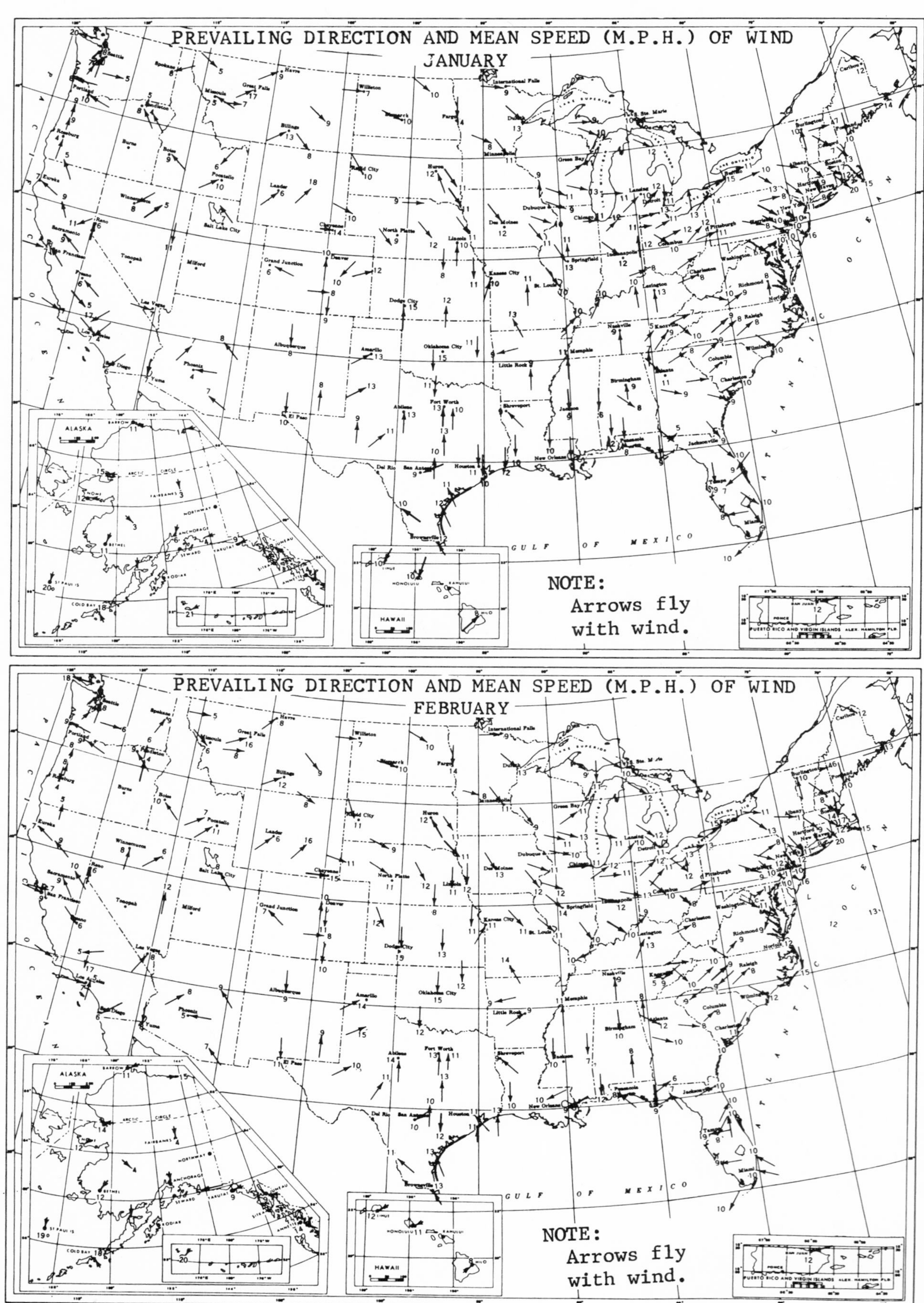
PREVAILING DIRECTION AND MEAN SPEED (M.P.H.) OF WIND
JANUARY
GULF OF MEXICO
ALASKA
HAWAII
NOTE:
Arrows fly
with wind.
PREVAILING DIRECTION AND MEAN SPEED (M.P.H.) OF WIND
FEBRUARY
GULF OF MEXICO
ALASKA
HAWAII
NOTE:
Arrows fly
with wind.

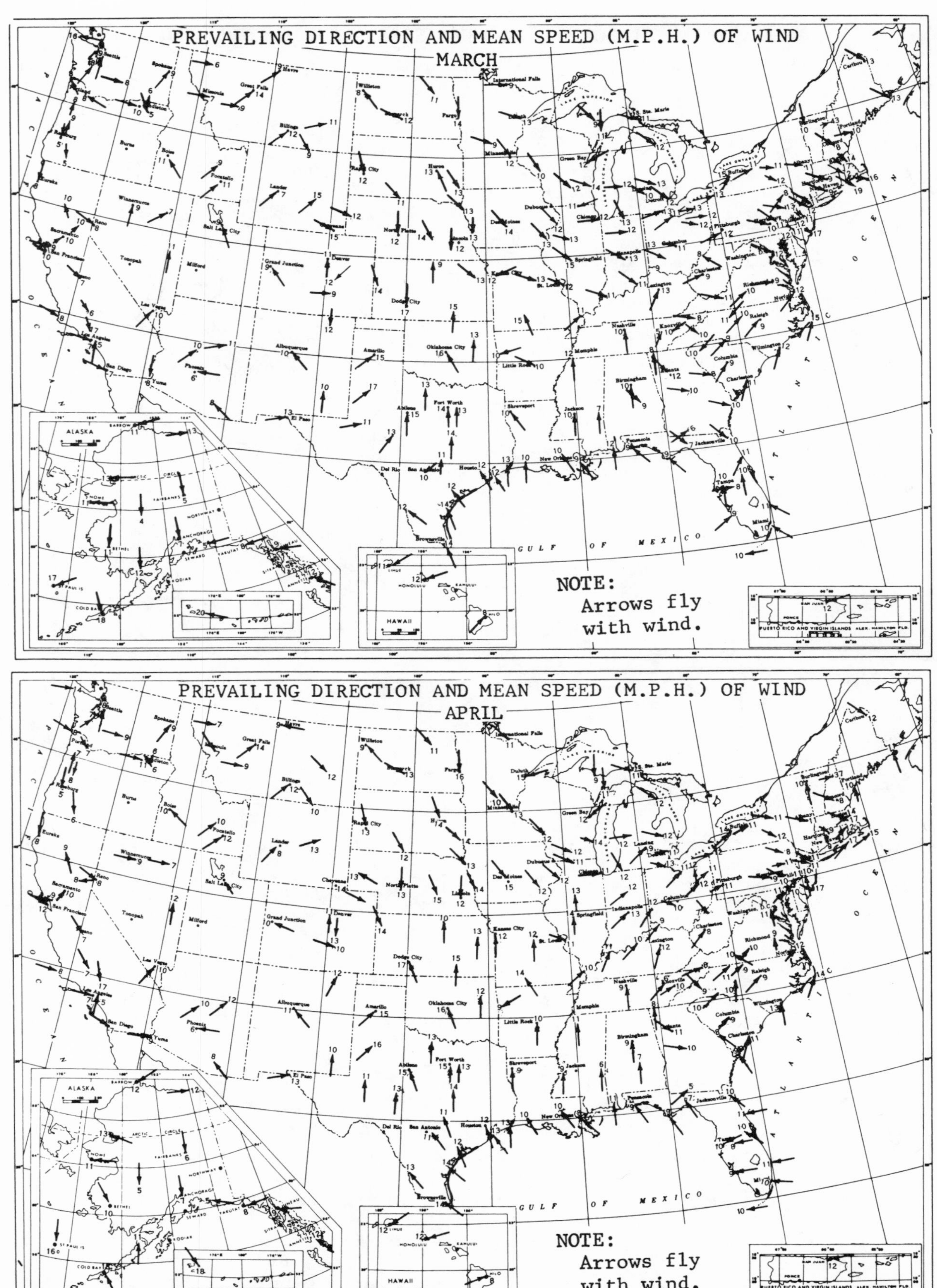
PREVAILING DIRECTION AND MEAN SPEED (M.P.H.) OF WIND
MARCH
NOTE:
Arrows fly
with wind.
PREVAILING DIRECTION AND MEAN SPEED (M.P.H.) OF WIND
APRIL
NOTE:
Arrows fly
with wind.

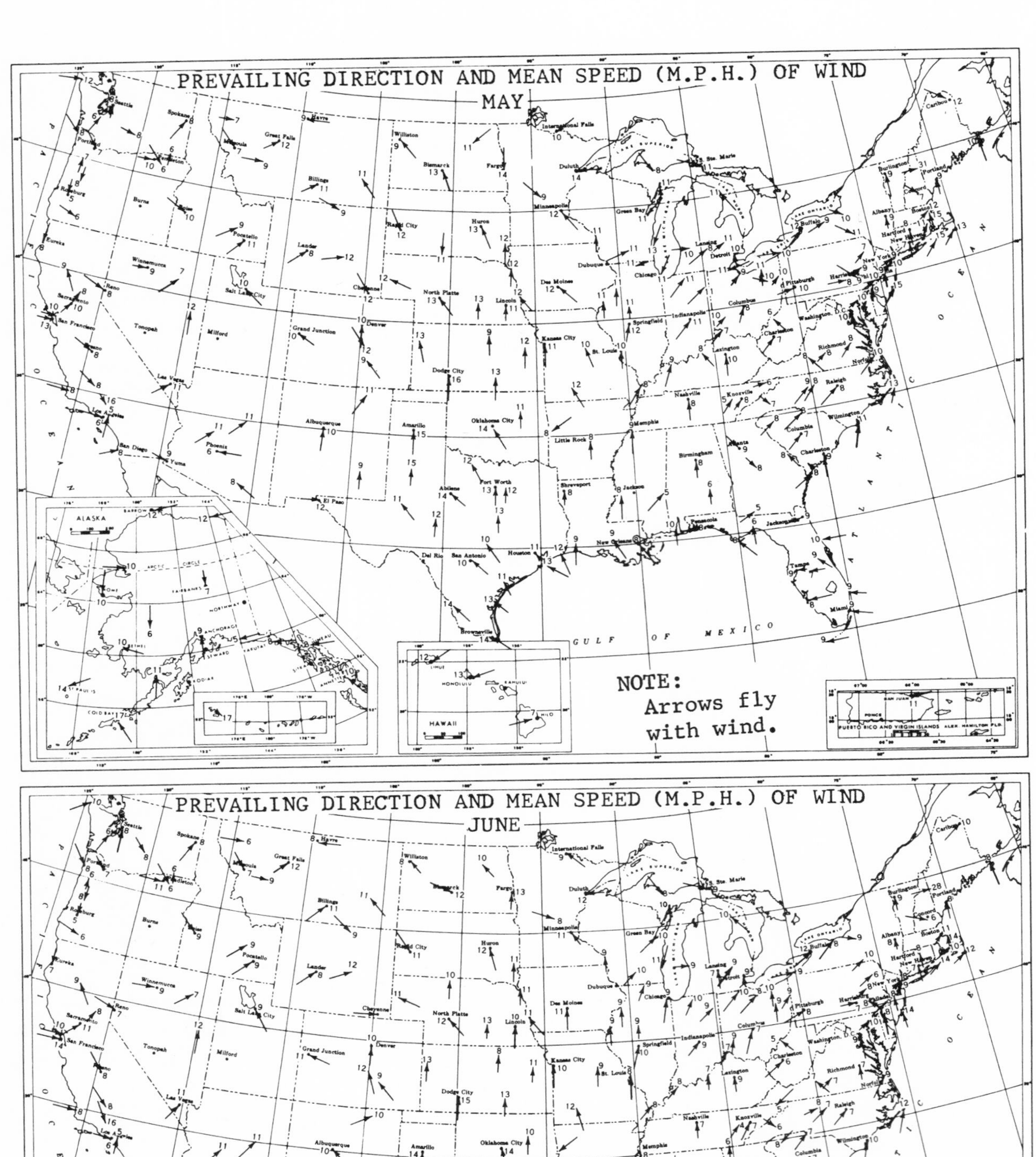
PREVAILING DIRECTION AND MEAN SPEED (M.P.H.) OF WIND
MAY
NOTE:
Arrows fly
with wind.
GULF OF MEXICO
ALASKA
HAWAII
PUERTO RICO AND VIRGIN ISLANDS
PREVAILING DIRECTION AND MEAN SPEED (M.P.H.) OF WIND
JUNE
NOTE:
Arrows fly
with wind.
GULF OF MEXICO
ALASKA
HAWAII
PUERTO RICO AND VIRGIN ISLANDS

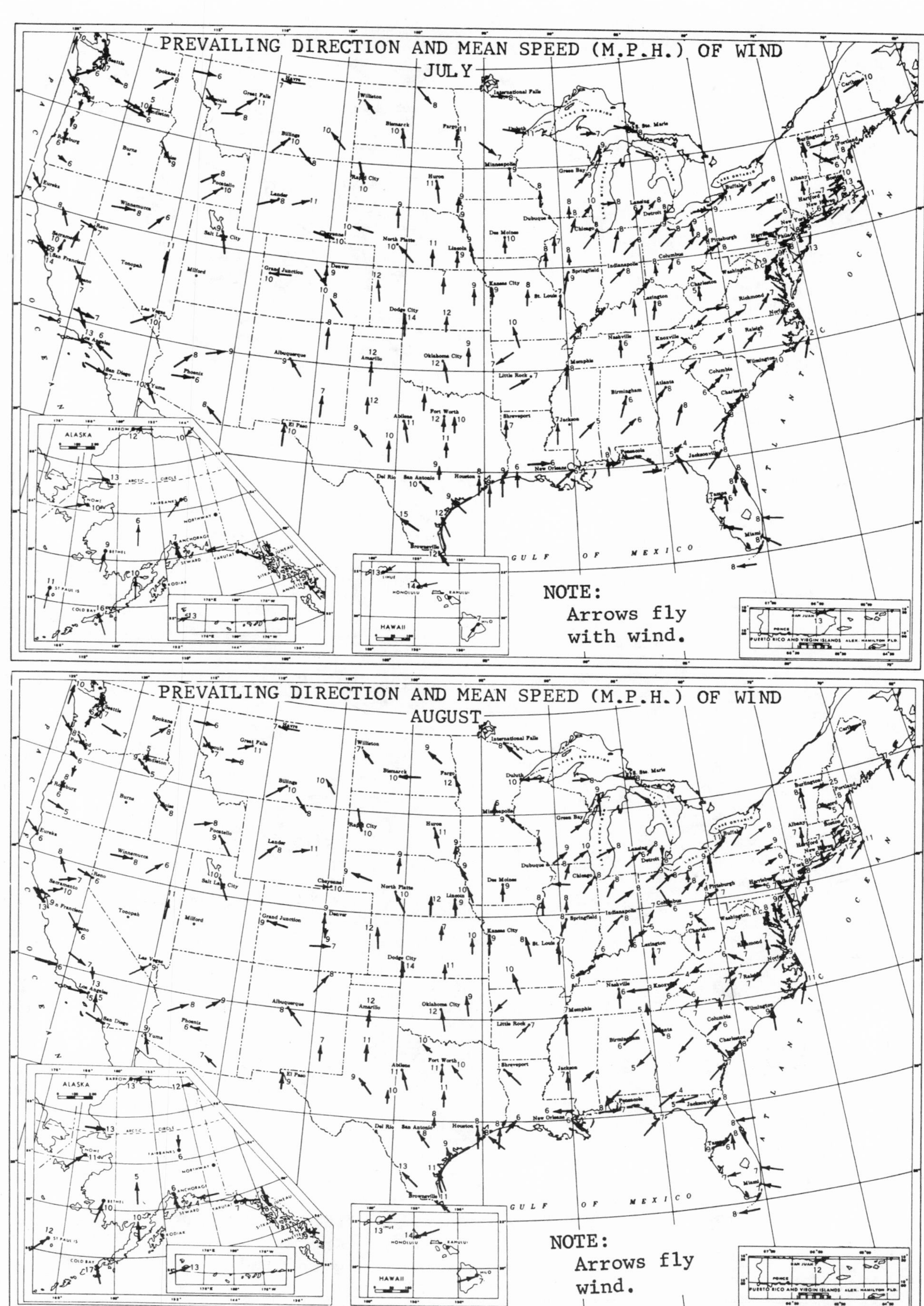
PREVAILING DIRECTION AND MEAN SPEED (M.P.H.) OF WIND
JULY
NOTE:
Arrows fly with wind.
GULF OF MEXICO
ALASKA
HAWAII
PUERTO RICO AND VIRGIN ISLANDS
PREVAILING DIRECTION AND MEAN SPEED (M.P.H.) OF WIND
AUGUST
NOTE:
Arrows fly wind.

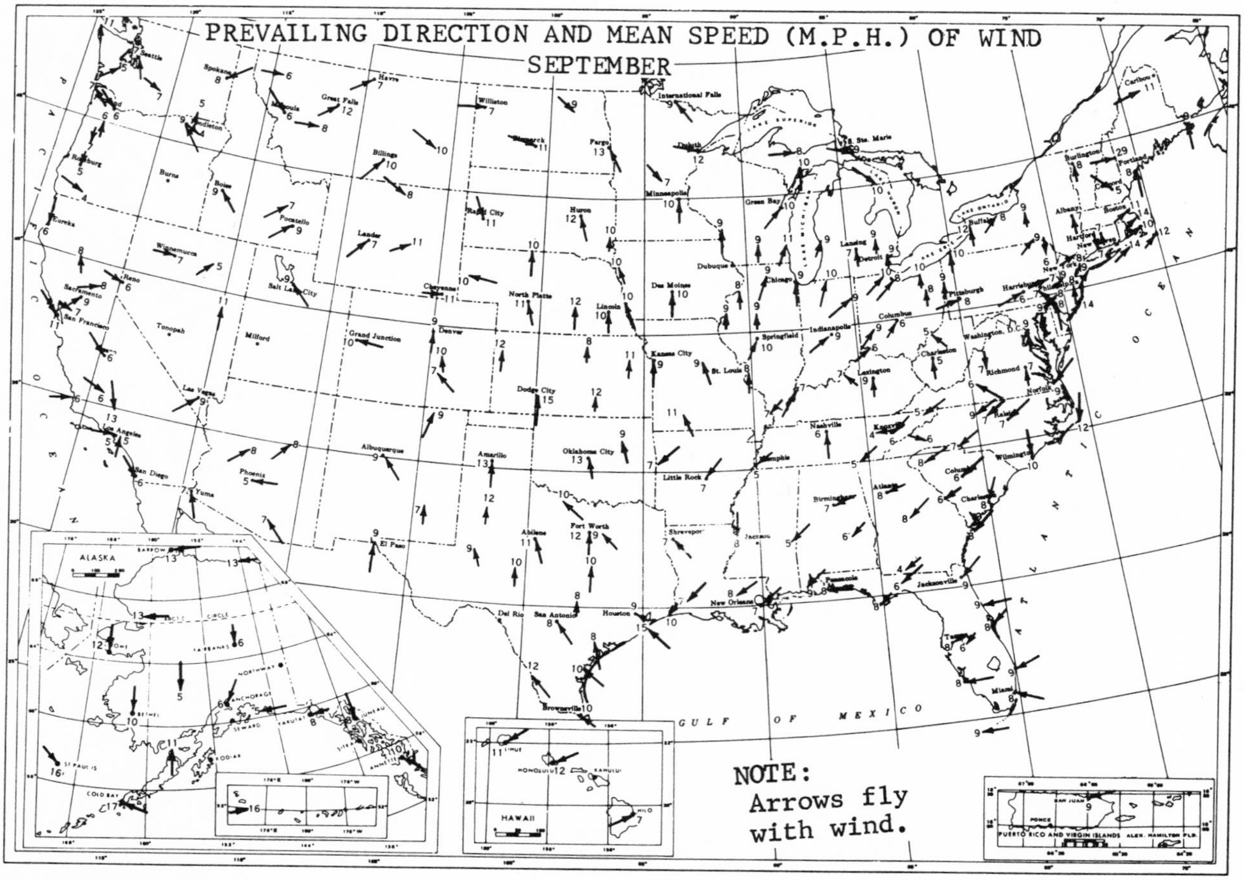

PREVAILING DIRECTION AND MEAN SPEED (M.P.H.) OF WIND

SEPTEMBER

NOTE:
Arrows fly with wind.

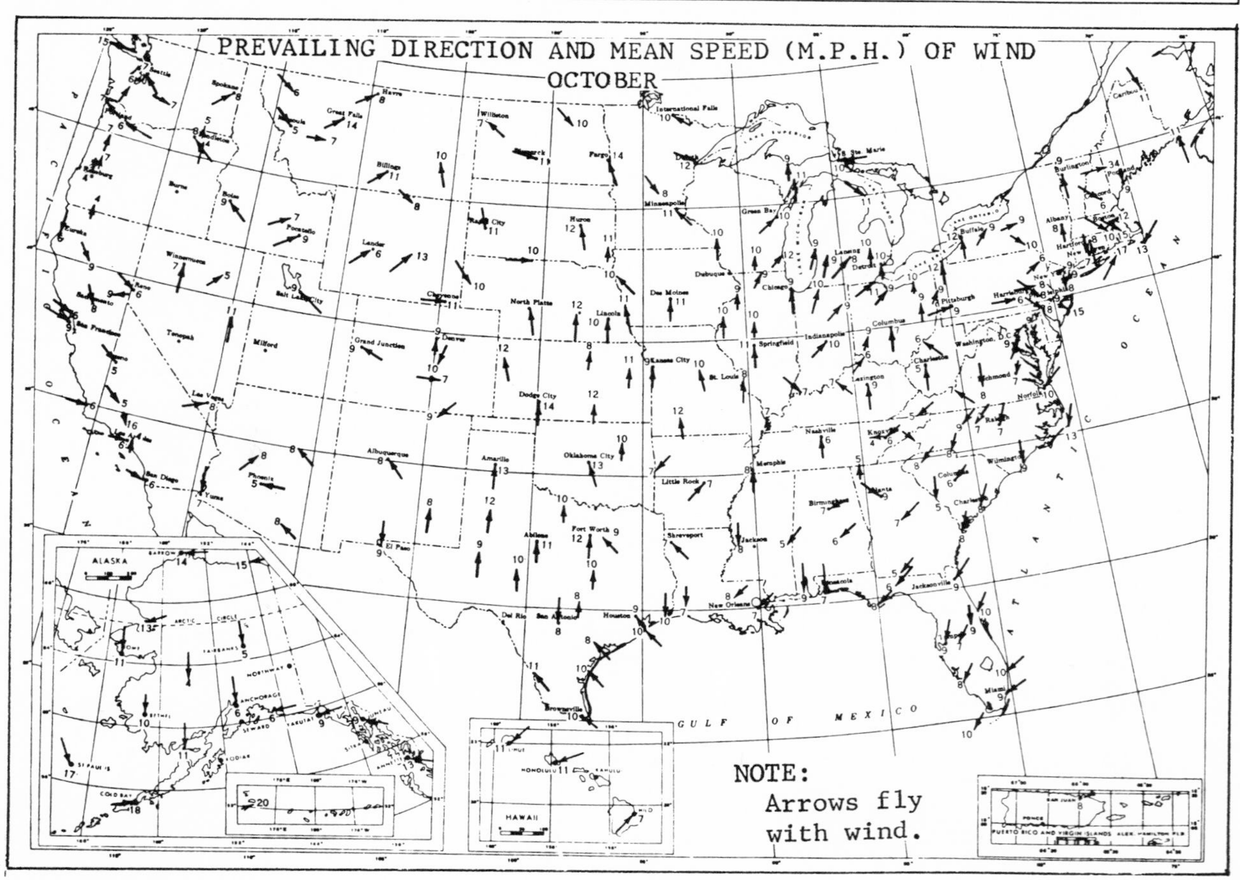

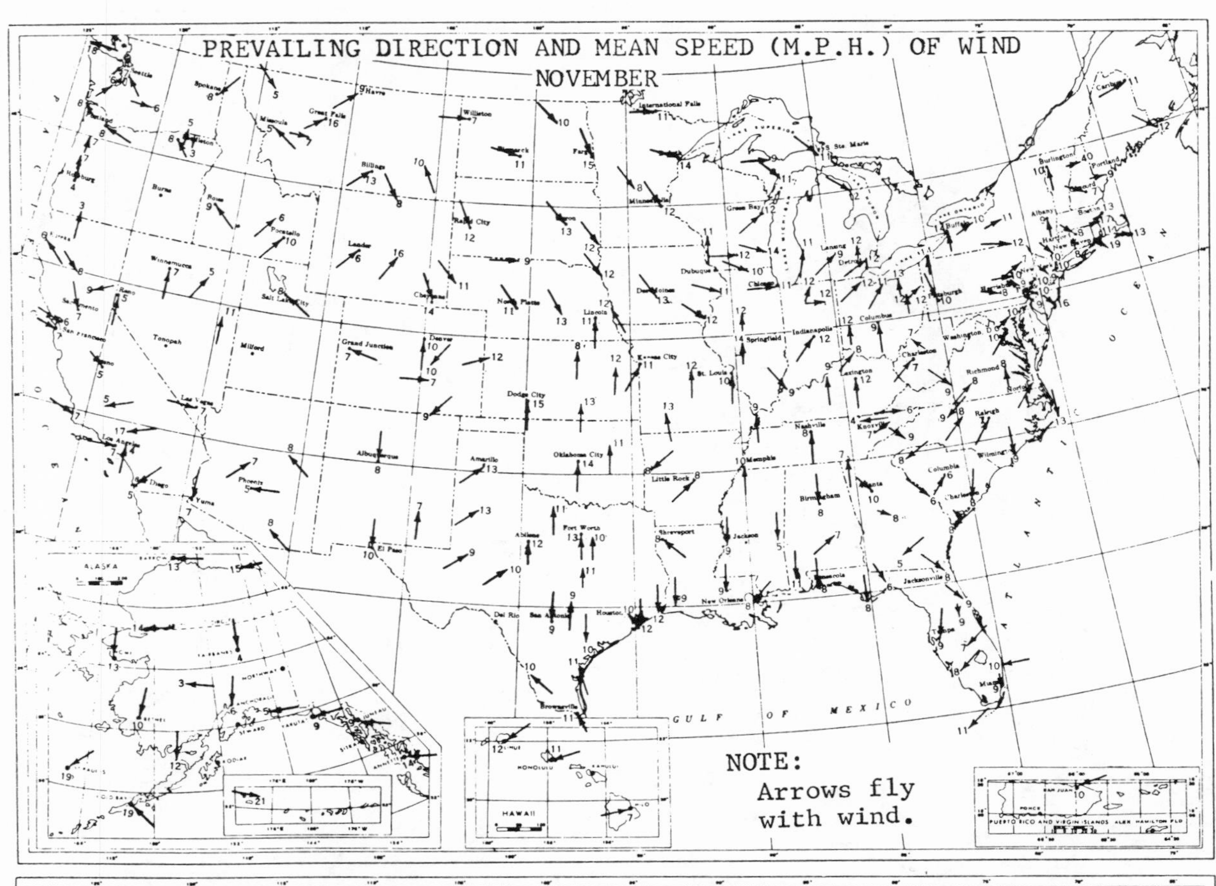
PREVAILING DIRECTION AND MEAN SPEED (M.P.H.) OF WIND
NOVEMBER
GULF OF MEXICO
HAWAII
ALASKA
NOTE:
Arrows fly
with wind.

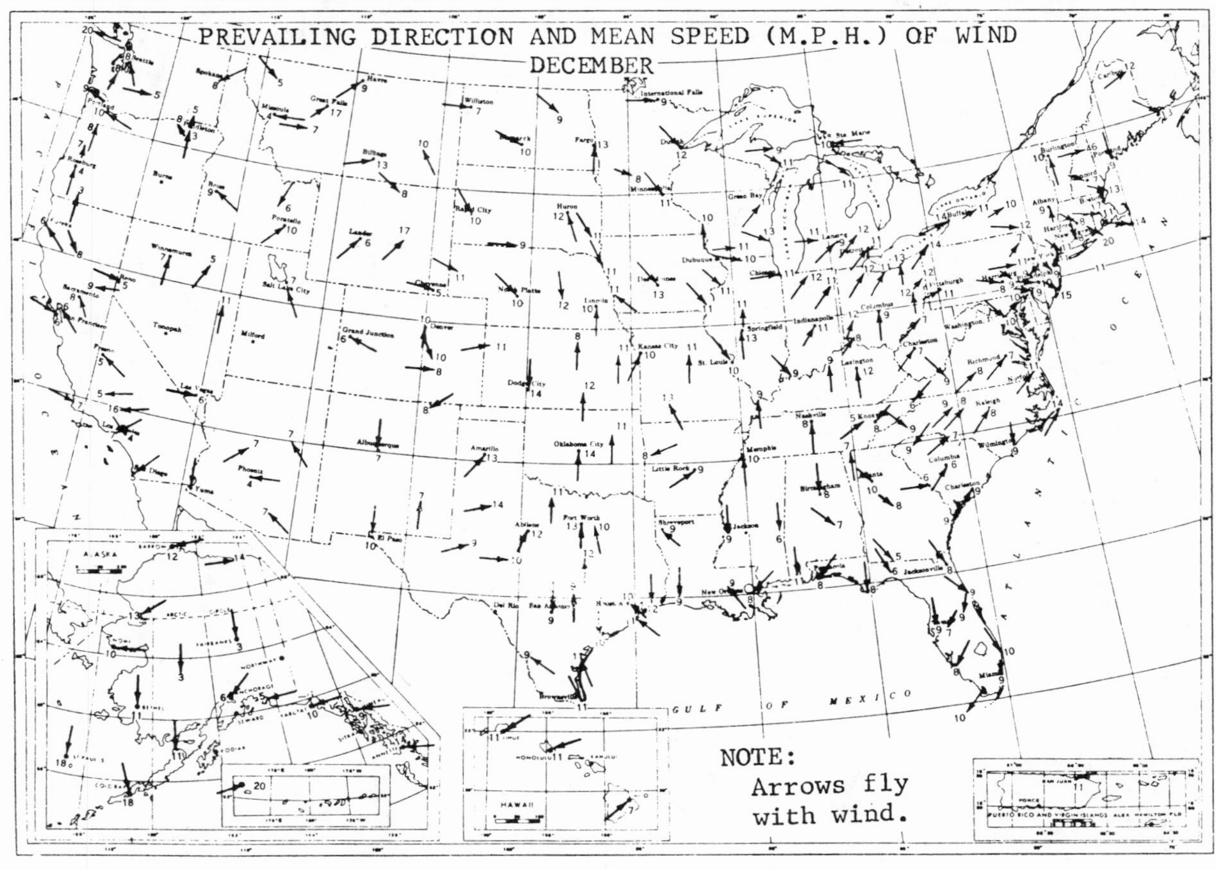
PREVAILING DIRECTION AND MEAN SPEED (M.P.H.) OF WIND
DECEMBER
GULF OF MEXICO
HAWAII
ALASKA
NOTE:
Arrows fly
with wind.

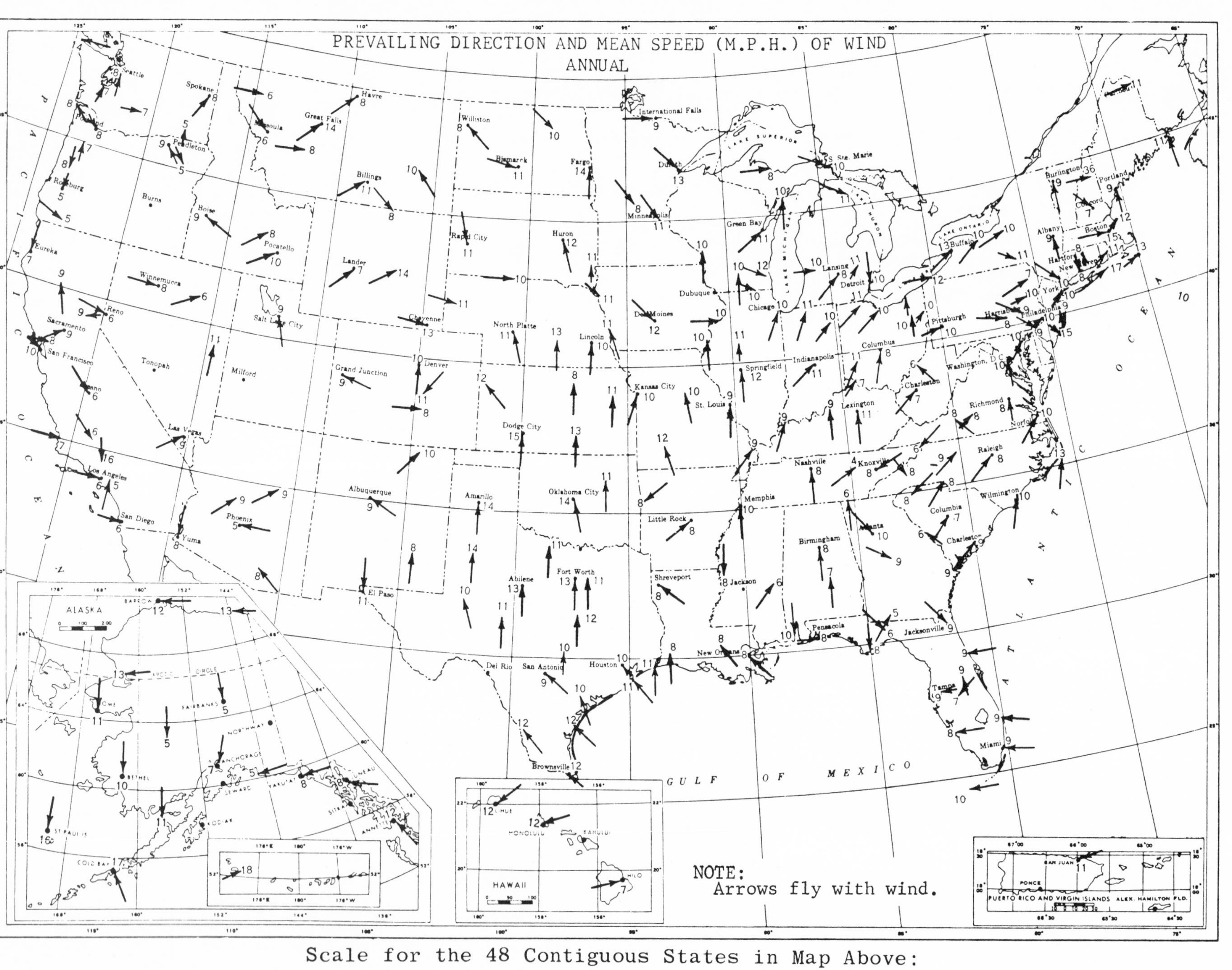

Scale for the 48 Contiguous States in Map Above:

200 0 200 400 600 800 1000 Miles

SCALE 1:20,000,000

ALBERS EQUAL AREA PROJECTION-STANDARD PARALLELS 29½° AND 45½°

FASTEST MILE AND DIRECTION OF WIND

STATE & STATION	YRS	JAN	FEB	MAR	APR	MAY	JUNE	JULY	AUG	SEPT	OCT	NOV	DEC.	YEAR
ALA. BIRMINGHAM	50	49 SW	59 SE	65 SW	56 SW	65 NW	56 SW	57 SW	50 NW	50 SE	43 W	52 N	45 SE	65 SW
MOBILE	28	45 SW	44 E	48 SE	56 SE	44 E	47 E	81 E	49 NE	75 N	87 E	47 E	42 E	87 E
MONTGOMERY	50	47 SW	47 W	60 SW	54 SE	50 W	51 NW	51 E	43 SW	40 NE	40 W	46 S	46 SW	60 SW
ALASKA ANCHORAGE	10	60 NE	62 S	49 N	66 S	31 S	33 S	32 S	45 S	49 SE	59 S	66 NE	56 SW	66 NE
BARROW	30	56 E	58 W	48 -	52 -	43 SW	38 W	56 SW	47 E	56 W	51 -	63 W	70 W	70 W
FAIRBANKS	11	41 SW	57 W	60 SW	35 SW	42 SE	40 SW	50 SW	34 W	49 SW	50 SW	50 SW	45 WSW	60 SW
JUNEAU	15	68 SE	66 SE	49 SE	57 SE	47 SE	45 SE	32 SE	33 SE	48 SE	61 SE	59 SE	50 SE	68 SE
ST. PAULS ISLAND	14	70 SE	69 NE	59 SW	60 NE	46 NW	38 NE	35 SE	52 SW	51 NE	60 S	114 NE	84 SE	114 NE
NOME	21	75 E	74 W	72 E	50 E	68 NE	37 S	36 SE	45 S	52 SE	73 SW	73 SW	54 NE	75 E
SHEMYA	16	99 WSW	81 SW	79 SSE	73 SSE	62 WSW	55 SW	50 WSW	55 S	71 SW	79 WNW	39 W	139 SW	139 SW
ARIZ. PHOENIX	24	41 WNW	49 SSE	50 WNW	45 NW	59 SSE	46 NE	71 N	60 SSW	75 SW	48 SSW	45 SSW	68 W	75 SW
PRESCOTT	17	47 SW	65 SW	65 SW	65 SW	60 SE	54 SW	54 NW	44 SW	56 SW	61 S	72 SW	52 W	72 SW
TUCSON	14	40 E	59 E	41 SE	46 SE	40 SW	50 SE	54 S	54 E	54 SE	47 SE	55 E	44 W	59 E
YUMA	51	36 W	42 NW	56 NW	47 NW	38 W	41 E	52 NE	54 S	42 SE	39 NW	47 N	47 W	56 NW
ARK. FORT SMITH	16	42 SW	56 W	41 SW	42 NW	57 NW	58 N	37 W	44 NE	52 NE	54 W	56 W	45 SW	58 N
LITTLE ROCK	20	44 S	46 WSW	56 SE	65 NW	61 NW	60 NE	56 NW	54 NW	50 NW	58 SSW	49 SW	42 SE	65 NW
CALIF. BAKERSFIELD	18	47 ESE	50 ESE	35 WNW	60 NNW	40 WNW	37 NW	28 NW	30 WNW	32 NNW	32 E	45 WNW	40 SSE	60 NNW
BLUE CANYON	4	46 WSW	64 SSW	67 ENE	42 ENE	37 SW	49 ENE	32 E	30 ENE	49 E	70 NE	47 ENE	51 ENE	70 NE
BURBANK	5	40 NE	36 NNW	32 NW	38 NNW	28 N	25 NE	19 S	20 SE	19 ESE	30 NNW	38 NNW	45 NE	45 NE
EUREKA	51	54 S	48 SW	48 SW	49 N	40 NW	39 NW	35 N	34 N	44 N	43 SW	43 S	56 S	56 S
FRESNO	12	32 SE	38 W	31 SE	36 NW	38 N	34 W	25 NE	31 S	29 SW	40 NE	27 SE	43 NW	43 NW
LOS ANGELES	21	49 N	40 NW	40 NW	40 NW	39 NW	32 N	21 W	24 E	27 NW	43 N	42 N	44 SE	49 N
RED BLUFF	17	59 SE	61 SE	63 SE	50 SE	40 SE	38 NW	34 SE	30 SW	43 W	56 SE	47 SE	60 SE	63 SE
SACRAMENTO	85	60 SE	58 SE	66 S	45 SW	40 SE	47 SW	36 SW	38 SW	37 NW	68 SE	70 SE	70 SE	70 SE
SANDBERG	7	85 NW	91 NW	90 NNW	77 NW	77 NNW	77 NW	56 NNW	52 S	68 NNW	87 NW	81 NNW	97 E	97 E
SAN DIEGO	18	39 SW	35 SW	46 SW	37 S	27 SW	26 S	18 SW	23 SW	25 W	31 N	51 SE	34 S	51 SE
SAN FRANCISCO	24	60 SW	62 SW	52 NW	48 W	56 W	48 W	50 W	45 WNW	48 S	51 SW	55 SSW	53 S	62 SW
COLO. COLO. SPRINGS	7	74 NW	74 SW	84 WNW	59 W	68 SSW	90 SSW	70 SE	54 W	74 SW	62 NW	80 W	80 WNW	90 SSW
DENVER	89	51 SW	50 W	53 NW	56 NW	65 W	56 SE	57 NE	47 N	50 NW	47 S	58 NE	51 NW	65 W
GRAND JUNCTION	64	54 S	50 NW	65 S	59 W	61 SW	66 S	56 SE	56 NW	61 S	61 NW	56 W	48 NW	66 S
PUEBLO	21	80 W	56 N	68 N	72 N	66 NW	63 N	61 W	54 NE	56 N	65 N	61 N	57 N	80 W
CONN. BRIDGEPORT	5	40 NW	47 WNW	45 WNW	45 SSW	32 W	40 WSW	40 WSW	32 S	30 SSE	38 ESE	62 E	35 S	62 E
HARTFORD	45	51 SW	50 W	57 SE	47 NE	41 NW	52 N	42 SW	56 NE	62 N	57 S	70 E	54 NW	70 E
NEW HAVEN	18	38 S	38 SE	42 E	40 SE	31 SW	32 SW	33 NW	38 NE	41 N	44 E	57 E	45 SW	57 E
D. C. WASHINGTON	13	56 NW	57 SW	60 E	56 N	48 S	57 NW	54 E	49 NE	56 SE	78 SE	60 E	62 SW	78 SE
WASHINGTON U	87	47 NW	47 NW	56 NW	50 NW	47 NW	47 NW	54 NE	49 NW	62 SE	54 NW	47 SW	40 NW	62 SE
FLA. APALACHICOLA	29	48 E	40 SE	54 E	51 SE	47 SE	47 NE	63 N	59 NE	67 E	56 NW	47 SE	42 SE	67 E
DAYTONA BEACH	12	38 W	40 S	40 SW	46 S	32 S	35 WSW	40 NNW	50 ESE	45 ESE	59 SE	23 ENE	40 NNW	59 SE
FORT MYERS	10	39 SSE	40 SSE	40 NW	40 SSE	30 SW	45 SSE	45 S	47 WNW	*90 E	*58 ENE	30 NW	30 SSW	*90 E
JACKSONVILLE	90	51 SW	76 W	66 SW	54 NW	57 SE	59 SW	53 W	51 SW	76 SE	72 E	60 S	54 SW	76 W,SE
KEY WEST	50	71 N	63 WNW	52 W	50 W	51 E	48 S	52 ESE	56 S	122 NW	66 SE	76 NW	49 NW	122 NW
FLA. (CON'T) MIAMI	51	50 SW	68 SE	53 S	70 N	48 SW	48 NE	53 SE	62 SW	132 E	122 S	94 SE	52 S	132 E
PENSACOLA	22	35 NW	42 SE	49 NW	41 NW	40 NW	33 SW	35 SE	57 W	91 SE	32 NW	35 SW	36 SE	91 SE
TAMPA	51	47 SE	59 SW	56 SW	50 SW	56 NE	56 NE	62 SE	57 S	84 SE	68 NE	40 NW	51 NW	84 SE
GA. ATLANTA	50	54 NW	59 W	66 NW	68 SE	59 NW	70 NE	56 SE	49 NW	49 N	47 NW	46 NE	63 W	70 NE
AUGUSTA	40	46 -	43 -	37 -	41 -	46 SSW	43 -	48 -	46 -	44 -	37 -	40 W	39 -	48 -
MACON	62	47 SE	60 NE	60 S	57 SE	65 SE	56 NW	59 SW	70 S	43 SE	36 N	54 SW	38 W	70 S
SAVANNAH	51	57 SW	62 SW	60 S	64 SW	69 W	66 E	54 W	90 N	56 NW	85 NE	57 SE	51 SW	90 N
HAWAII HONOLULU	27	67 SW	63 W	59 SE	40 NE	35 E	39 E	34 E	52 SE	35 NE	41 NE	65 NE	59 NE	67 SW
LIHUE	11	36 SW	34 E	36 N	29 NE	49 NE	30 NE	31 NE	73 NE	30 NE	32 E	43 S	58 NE	73 NE
IDAHO BOISE	22	50 SE	56 W	52 W	50 W	50 W	50 SW	61 W	50 SE	50 SE	56 SE	57 NW	56 NW	61 W
IDAHO FALLS (46W)	11	39 WSW	36 SW	51 WSW	39 WSW	41 SW	36 SW	35 WSW	33 SW	38 WSW	44 WSW	40 SW	43 SW	51 WSW
POCATELLO	49	61 SE	54 W	72 W	61 S	61 W	50 W	47 S	56 SW	57 W	54 W	67 W	56 SE	72 W
ILL. CAIRO	71	56 SW	56 SW	59 SW	59 SW	62 N	52 N	59 N	56 N	37 NW	40 NW	50 SW	43 S	62 N
CHICAGO	90	58 NE	87 NE	76 NE	69 SW	69 SW	62 NW	69 SW	62 SW	69 SW	62 SW	76 SE	66 SW	87 NE
MOLINE	33	56 SW	58 NW	66 SW	73 SW	68 SW	77 SW	65 W	61 SW	66 NW	56 SW	60 W	61 SW	77 SW
PEORIA	50	54 S	52 W	56 S	66 SW	61 SW	66 N	75 NW	65 W	60 W	60 SW	56 SW	58 SW	75 NW
SPRINGFIELD	83	65 SW	63 SE	66 SW	56 W	58 W	75 SW	59 NW	58 W	52 SW	44 S	68 SW	66 SW	75 SW
IND. EVANSVILLE	50	51 SW	76 W	62 SW	61 SW	56 SW	113 S	60 W	80 SW	51 W	41 SW	62 SW	54 N	113 S
FORT WAYNE	50	59 SW	52 W	65 S	60 W	54 SW	65 SE	61 NW	47 N	56 W	41 NW	57 SW	52 SW	65 SE
INDIANAPOLIS	90	90 W	65 SW	68 NW	60 W	68 W	111 NW	72 NE	68 W	62 W	56 NW	59 SW	53 W	111 NW
SOUTH BEND	11	42 W	47 SSW	47 SW	42 SW	40 WNW	50 W	45 NNW	63 NW	35 WSW	38 WSW	45 S	43 SW	63 NW
IOWA BURLINGTON	20	49 W	46 NW	56 SW	73 W	68 SW	70 NW	56 NW	73 N	45 W	63 W	47 W	72 SW	73 N
DES MOINES	30	66 NW	56 W	70 NW	63 NW	71 SW	76 NW	66 NW	60 SSE	55 NW	56 W	72 W	61 SW	76 NW
DUBUQUE	76	31 NW	30 NW	34 SW	32 NW	32 SE	47 NW	44 NW	36 NE	35 NE	30 NW	34 NW	34 NW	47 NW
SIOUX CITY	50	59 NW	62 NW	69 NW	68 W	80 W	91 W	61 N	91 SW	66 S	70 W	59 NW	56 NW	91 W,SW
KANS. CONCORDIA	50	37 N	44 NW	54 NW	47 NW	40 NW	65 W	50 NW	47 NE	34 SW	41 NW	47 N	40 NW	65 W
DODGE CITY	18	56 NW	72 NW	74 NW	70 N	71 W	73 S	78 SW	72 NW	66 N	63 SW	59 NW	57 NW	78 SW
TOPEKA	50	43 SW	50 N	66 SW	63 SE	67 SE	57 NW	81 N	57 N	57 N	63 NE	49 S	61 S	81 N
WICHITA	73	51 NW	56 NW	82 NW	68 NW	70 SW	73 NE	100 N	59 NW	59 NW	52 NE	72 S	52 S	100 N
KY. LEXINGTON	45	52 W	47 SW	53 SW	50 SW	50 SW	47 SW	49 NW	56 S	53 W	40 NW	52 NW	47 S	56 S
LOUISVILLE	49	53 S	62 W	62 SW	66 SW	68 NW	66 NW	57 SW	63 NW	57 NW	50 NW	60 SE	61 SW	68 NW
LA. LAKE CHARLES	21	46 SSE	42 SSE	55 NW	46 SSE	50 WNW	69 SE	45 S	43 SSE	50 NE	42 NNE	46 WSW	50 S	69 SE
NEW ORLEANS	47	41 NE	44 E	43 NW	35 SE	43 SE	40 NE	37 NW	41 SE	98 SE	31 SE	28 NW	31 SW	98 SE
SHREVEPORT	82	47 NW	45 NW	57 S	52 N	47 NW	50 E	46 NW	52 W	59 E	47 NW	50 SW	52 NW	59 E
MAINE CARIBOU	8	76 NW	52 W	63 W	46 NW	53 WNW	50 WNW	66 N	45 WSW	45 W	48 NW	48 W	55 WNW	76 NW
PORTLAND	62	50 S	58 N	76 NE	57 S	64 NW	48 NW	51 S	69 E	62 SE	47 SE	76 NE	62 SE	76 NE
MD. BALTIMORE	39	56 NW	65 NW	66 W	65 W	57 NE	61 N	61 NE	60 SW	54 SW	49 NW	57 S	56 NW	66 W
MASS. BLUE HILL	53	72 SW	74 NNW	69 SE	57 W	59 S	76 WNW	67 NW	93 SE	121 S	63 SE	70 S	82 WNW	121 S
BOSTON	43	66 SW	58 NE	73 E	63 NE	55 W	46 NW	52 NW	52 NE	87 S	63 NW	80 SE	73 W	87 S
NANTUCKET	75	69 NE	71 NE	91 E	61 NE	52 NE	68 NE	62 NE	79 SW	73 SE	69 E	70 NW	62 NE	91 E
WORCESTER	4	60 WSW	76 NW	76 WNW	54 NE	48 W	39 WSW	43 NW	34 W	34 SW	43 WSW	54 SSW	51 SW	76 WNW
MICH. ALPENA	45	56 E	50 NW	56 E	52 W	47 NW	50 NW	49 S	52 S	49 SE	50 NW	61 SW	49 NW	61 SW
MICH. (CON'T) DETROIT	28	57 SW	49 W	68 SW	56 N	61 NW	56 SW	77 NW	50 NW	52 SW	56 S	66 W	59 NW	77 NW
ESCANABA	30	52 NE	56 N	68 N	47 N	57 NW	50 SE	49 N	43 NW	47 NW	52 NE	56 N	47 NW	68 N
FLINT	19	76 SW	65 SSW	63 SW	67 SSW	81 NW	68 S	44 NNW	60 WSW	46 NNW	52 WSW	56 WSW	46 SW	81 NW
GRAND RAPIDS	58	72 SW	65 SW	69 SW	79 SW	73 SW	68 S	62 SW	51 SW	65 SW	69 SW	80 SW	68 SW	80 SW
LANSING	22	46 S	41 SW	38 SW	38 S	49 NW	38 NW	42 NW	29 NW	37 NW	44 SW	50 SW	33 W	50 SW
MARQUETTE	49	52 -	56 SW	57 SW	47 SW	91 S	59 SW	61 S	52 SW	45 S	66 SW	50 SW	57 S	91 S
S. STE. MARIE	71	50 NW	72 NW	60 SE	72 SW	49 SE	54 W	56 W	56 SW	62 SW	62 W	66 W	58 NW	72 SW
MINN. DULUTH	35	65 NW	67 E	75 NW	75 NE	61 W	59 NW	72 W	68 N	60 NW	61 S	68 E	72 NW	75 NE
INTERN'L FALLS	9	32 SW	35 W	42 WNW	52 SW	52 SSW	36 S	46 WNW	36 NW	35 SW	29 NW	35 W	32 W	52 SW
MINNEAPOLIS	50	47 NW	52 NW	56 SW	58 N	56 NW	63 NW	92 W	57 NW	50 NW	73 S	60 SW	52 W	92 W
MISS. JACKSON	14	56 NW	45 NW	68 S	49 S	56 NW	52 SE	54 N	51 NW	54 NE	39 NW	51 S	59 NW	68 S
MERIDIAN	55	42 -	41 -	47 -	44 -	40 -	40 -	40 -	36 -	32 -	32 -	31 -	38 -	47 -
VICKSBURG	50	38 NW	47 NW	40 NW	56 W	49 NW	38 SW	46 N	41 SE	47 S	35 SW	47 SW	49 SW	56 W
MO. COLUMBIA	50	56 NW	45 NW	56 SW	57 NW	58 SW	58 NW	61 NW	56 NW	63 NW	49 NW	49 NW	56 NW	63 NW
KANSAS	71	66 NE	49 NW	70 SW	61 W	68 SW	68 NW	70 NW	72 NW	50 NW	56 W	56 SW	57 SW	72 NW
ST. JOSEPH	48	46 NW	46 W	64 S	61 NW	63 W	60 NW	52 W	57 NW	43 SW	42 NW	56 NW	59 S	64 S
ST. LOUIS U	45	59 SW	57 SW	82 SW	61 SW	61 SW	60 SW	56 SW	60 SW	73 W	62 SW	65 SE	56 SW	82 SW
SPRINGFIELD	16	55 SW	59 S	65 SW	60 NW	66 W	57 N	56 NW	59 W	54 S	49 S	59 SE	55 W	66 W
MONT. BILLINGS	18	66 W	68 NW	61 NW	72 NW	68 -	65 SE	73 N	66 N	61 NW	68 NW	63 NW	66 NW	73 N
GREAT FALLS	18	65 SW	72 W	73 W	70 W	65 SW	70 W	73 W	71 SW	73 NW	73 W	73 SW	82 SW	82 SW
HAVRE	48	52 SW	56 SW	47 SW	59 SW	59 NW	54 W	53 -	50 SW	70 SW	53 -	54 SW	47 -	70 SW
HELENA	50	73 SW	73 W	61 SW	52 W	62 S	59 SW	60 SW	65 S	54 NW	62 W	59 SW	59 NW	73 W
MISSOULA	26	52 S	53 E	53 NE	51 NW	57 SW	51 S	72 SE	58 SW	41 W	51 SW	42 NW	56 W	72 SE
NEBR. LINCOLN U	50	49 NW	56 NW	63 NW	56 NW	60 W	66 NW	65 NW	62 NW	50 N	44 S	52 NW	54 NW	66 NW
NORTH PLATTE	50	57 N	68 N	72 N	63 NW	70 S	72 N	60 NW	56 SW	53 N	72 SW	64 NW	59 NW	72 N
OMAHA	89	60 NW	59 N	73 NW	65 NW	73 NW	72 N	109 N	66 N	51 N	59 SW	58 NW	52 NW	109 N
SCOTTSBLUFF	10	40 NW	60 WNW	48 W	46 NW	80 NW	80 WNW	52 WSW	41 NNW	46 NW	44 WNW	52 NW	47 NW	80 WNW
VALENTINE	50	45 N	56 N	49 NW	55 N	62 NW	56 W	56 SW	65 NW	57 S	49 SE	56 W	50 N	65 NW
NEV. ELY	18	66 SE	56 S	59 SE	59 S	74 S	63 SW	50 S	57 E	57 S	65 S	51 S	61 SE	74 S
RENO	56	66 SW	66 SW	68 S	66 SW	63 SW	66 S	43 W	57 S	48 S	66 S	59 SW	72 S	72 S
WINNEMUCCA	46	56 W	59 W	63 W	49 S	61 N	53 W	56 W	51 W	57 W	54 NE	59 SW	69 E	69 E
N. H. CONCORD	50	43 NW	42 N	71 NE	52 NW	48 NW	38 SW	37 NW	56 E	61 SE	39 NW	72 NE	51 NE	72 NE
MT. WASHINGTON	29	170 W	144 SE	180 W	231 SE	164 W	136 NW	110 -	142 ENE	157 SE	161 W	160 SE	175 W	231 SE
N. J. ATLANTIC CITY	37	80 E	77 SE	72 SE	80 NE	59 NE	57 NE	61 SW	76 E	91 NE	72 NE	72 E	70 NE	91 NE
NEWARK	10	46 WSW	50 WSW	55 WSW	52 WSW	40 W	55 ENE	45 S	46 E	43 NW	48 ESE	82 E	42 SSW	82 E
TRENTON	49	62 NE	60 W	60 NW	54 NE	56 W	57 NW	73 NE	47 N	56 NW	60 NW	64 E	56 W	73 NE
N. MEX. ALBUQUERQUE	30	61 E	68 NW	80 NW	72 S	72 W	82 SE	68 E	61 SE	62 SE	66 N	57 NW	90 SE	90 SE
ROSWELL	57	67 NW	70 NW	66 NW	75 W	72 N	66 W	66 SW	72 W	54 NW	66 W	65 W	72 W	75 W
N. Y. ALBANY	50	57 W	71 NW	55 W	49 W	50 W	45 W	43 NW	38 S	48 S	45 NW	70 E	54 W	71 NW
BINGHAMTON	18	59 NW	66 NW	61 NW	52 W	54 W	60 NW	58 W	58 N	42 NW	72 S	57 SE	59 NW	72 S
BUFFALO	50	91 SW	73 SW	84 SW	73 W	63 SW	73 W	62 SW	60 W	68 SW	71 W	76 SW	85 SW	91 SW
NEW YORK CITY	46	76 SW	91 SW	91 SW	95 NW	74 W	94 W	95 NW	74 NW	99 N	113 SE	87 W	91 NW	113 SE

FASTEST MILE AND DIRECTION OF WIND -- Cont.

STATE & STATION	YRS	JAN	FEB	MAR	APR	MAY	JUNE	JULY	AUG	SEPT	OCT	NOV	DEC	YEAR
N.Y. (CON'T) ROCHESTER	51	73 W	66 W	65 W	59 W	63 SW	61 SW	56 W	59 NE	59 SW	65 SW	59 E	57 W	73 W
SYRACUSE	50	56 SW	61 W	59 SW	57 S	49 NW	59 NW	52 NW	49 NW	47 SW	63 SE	59 E	69 SW	69 SW
N. C. ASHEVILLE	50	52 -	43 NW	44 SE	47 SE	49 -	49 -	40 -	34 -	43 -	44 NW	40 -	42 NW	52 -
CAPE HATTERAS	50	76 SW	68 S	68 W	61 SSW	60 SW	35 NNE	77 NW	77 SE	110 W	56 SE	65 N	56 NW	110 W
CHARLOTTE	50	57 SW	54 SW	47 W	53 NW	48 NW	57 NW	50 NW	54 NW	47 NW	50 SW	47 NW	57 NE	57 SW,NW,NE
GREENSBORO	33	40 NW	51 W	54 SW	42 W	59 SW	56 W	63 N	45 W	40 N	43 N	40 SW	45 SSW	63 N
RALEIGH	41	50 SW	66 SW	59 SW	56 NW	65 SE	50 W	47 NW	50 NW	38 NE	52 W	47 SE	52 W	66 SW
WILMINGTON	49	57 S	66 SW	56 SW	56 NE	50 NW	54 NW	42 NW	72 NE	88 N	82 SE	42 SW	47 SW	88 N
WINSTON-SALEM	8	40 SSE	46 WSW	45 W	43 W	48 NW	29 NE	48 SSW	46 NE	42 E	42 NE	42 NW	43 NE	48 NW,SSW
N. DAK. BISMARCK	87	70 NW	54 NW	65 NW	63 W	66 NW	66 NW	72 W	72 S	66 NW	61 W	67 NW	61 NW	72 W
DEVILS LAKE	57	41 NW	54 N	54 N	47 N	52 NW	50 W	56 NW	47 NW	46 NW	47 N	57 N	42 NW	57 N
FARGO	41	57 W	56 W	56 N	65 N	72 NW	115 NW	60 SE	71 NW	88 N	57 NW	66 N	58 N	115 NW
WILLISTON	34	38 W	37 SE	43 W	41 NW	45 W	56 E	60 W	46 W	42 NW	40 W	53 NW	47 NW	60 W
OHIO CINCINNATI	40	49 SW	49 SW	49 SW	47 SW	36 W	40 W	43 SW	38 W	38 SW	35 SW	47 SW	41 SW	49 SW
CLEVELAND	31	71 W	65 SW	74 W	65 NW	78 SW	68 NW	66 W	61 W	56 S	68 W	67 SW	59 W	78 SW
COLUMBUS	59	63 NW	58 SW	68 SW	76 W	56 NW	62 NW	84 NW	78 NW	51 SW	60 S	61 N	58 W	84 NW
DAYTON	47	60 NW	72 SW	75 W	72 SW	60 SW	78 NW	74 NW	70 W	65 W	56 SW	68 SW	70 W	78 NW
SANDUSKY	84	56 SW	64 NE	65 SW	75 SW	48 SW	77 NW	69 N	63 NE	52 NW	54 N	68 SW	56 NW	77 NW
TOLEDO	50	66 SW	69 SW	87 SW	72 SW	58 SW	56 W	76 SW	76 W	62 S	57 SW	76 SW	69 SW	87 SW
OKLA. OKLA. CITY	47	63 W	61 W	61 S	75 NW	72 SE	87 NNW	73 NW	56 S	54 NW	65 S	66 S	56 NW	87 NNW
TULSA	19	55 SW	50 N	51 SW	70 NW	75 SW	65 NW	56 N	56 N	48 E	45 NW	56 SW	56 SW	75 SW
OREG. PORTLAND	50	54 S	61 SW	50 S	60 S	42 SW	40 SW	31 S	29 SW	38 SW	50 SW	56 SW	57 S	61 SW
ROSEBURG	9	34 SW	38 SW	27 SW	29 SW	22 S	22 N	25 NW	25 N	25 N	33 S	31 SW	31 S	38 SW
PA. ERIE	42	56 SE	57 W	60 SW	56 W	51 SW	47 SW	51 W	58 SE	56 SW	51 SE	60 SE	53 SW	60 SW,SE
HARRISBURG	50	47 W	60 -	68 W	56 SW	46 -	46 -	47 NW	45 -	40 -	50 E	58 SE	61 NW	68 W
PHILADELPHIA	50	62 N	59 NW	61 S	54 S	73 NW	73 NW	88 N	67 E	49 NE	73 S	60 SW	47 NW	88 N
PITTSBURGH	79	67 NW	58 NW	72 W	60 W	73 NW	58 NW	64 SW	57 NW	54 NW	56 NW	49 NW	60 NW	73 NW
READING	49	84 SE	79 NW	65 NW	80 E	95 NW	62 NW	66 W	56 E	58 NW	72 SE	69 E	64 S	95 NW
SCRANTON	17	42 W	60 W	42 W	47 NW	40 NW	43 W	56 NW	50 NE	42 NW	50 SE	45 S	49 SW	60 W
R. I. BLOCK ISLAND	70	69 NW	66 SE	90 E	72 NE	72 NE	72 NE	51 N	82 N	91 SE	65 E	90 NE	77 NW	91 SE
PROVIDENCE	49	70 NW	63 SW	69 NW	65 NW	56 N	60 NW	56 NW	49 NW	95 SW	60 W	60 NW	85 W	95 SW
S. C. CHARLESTON	47	61 -	52 NW	72 -	65 E	68 -	54 NW	60 -	73 -	76 -	56 -	49 -	73 NE	76 -
COLUMBIA	41	54 SW	47 W	54 SW	47 SW	47 SW	41 N	51 SW	49 S	35 W	38 NE	43 SW	44 W	54 SW
GREENVILLE	18	73 NE	56 NE	63 NW	66 NW	65 SW	52 N	60 N	70 SE	47 NE	79 N	52 SW	50 SW	79 N
S. DAK. HURON	29	57 NW	56 NW	68 NW	73 SE	70 NW	65 SE	77 NW	70 S	64 NW	72 W	73 NW	56 NW	77 NW
RAPID CITY	50	73 NW	75 NW	72 NW	72 NW	67 NW	72 SW	70 SW	72 SW	73 NW	72 NW	59 NW	65 NW	75 NW
TENN. CHATTANOOGA	83	59 SW	63 W	82 W	57 W	63 NW	67 NW	48 NW	62 W	57 SW	35 NW	45 S	46 SW	82 W
KNOXVILLE	50	60 SW	61 SW	61 SW	71 SW	59 SW	65 W	73 SW	50 S	56 W	42 SW	43 SW	52 SW	73 SW
MEMPHIS	50	56 -	51 -	54 S	51 -	57 -	51 -	54 -	47 -	51 -	47 W	43 -	56 -	57 -
NASHVILLE	50	56 SW	57 -	70 NW	61 W	57 -	73 NW	59 -	49 -	47 -	51 -	58 -	47 W	73 NW
TEXAS ABILENE	49	52 NW	60 N	71 S	61 S	73 N	109 NW	63 NE	51 NW	55 SW	49 S	50 N	56 N	109 NW
AMARILLO	70	62 NE	70 NW	72 W	74 SW	84 SW	65 S	66 SW	65 E	68 NE	68 S	56 NW	62 NE	84 SW
AUSTIN	35	44 N	57 N	48 NW	44 W	53 NW	50 N	43 S	49 NE	43 SW	52 S	48 NW	49 NW	57 N
BROWNSVILLE	39	51 S	56 NW	47 S	52 SE	57 SE	52 E	41 NE	57 NW	106 NW	47 NE	42 NW	50 NW	106 NW
CORPUS CHRISTI	75	59 NE	50 SE	56 S	54 N	69 NE	55 NE	56 S	100 E	110 NE	47 N	62 NE	47 NW	110 NE
DALLAS	48	66 SW	61 SW	59 W	58 N	65 S	65 N	77 N	56 NW	48 W	61 NW	56 W	47 S	77 N
EL PASO	83	61 SW	69 W	70 SW	66 NW	70 NW	68 N	65 N	63 N	58 SW	58 W	57 W	66 W	70 NW
FORT WORTH	38	61 SW	61 W	59 W	53 NW	68 W	57 NW	56 NW	49 N	49 N	56 W	53 NW	53 N	68 W
GALVESTON	86	53 S	60 N	50 SE	56 N	60 W	62 SE	68 NW	91 E	87 NE	66 SE	54 N	50 NW	91 E
HOUSTON	52	50 NW	53 SE	84 NW	63 NE	60 SE	54 SE	73 NW	80 SE	66 E	70 E	46 NW	49 SE	84 NW
LAREDO	11	45 WNW	39 NNW	43 NE	42 NW	40 ESE	61 NE	55 ESE	40 ESE	46 NE	45 E	37 NE	34 SE	61 NE
LUBBOCK	12	53 W	58 WSW	69 NNW	58 WSW	70 N	63 NE	64 WSW	46 SSE	45 N	65 WSW	59 WSW	58 WSW	70 N
MIDLAND	7	41 NNW	67 WSW	48 WSW	38 SW	52 SSW	58 NNE	29 N	30 NE	40 N	32 NNE	32 N	37 SSE	67 WSW
PORT ARTHUR	24	68 SW	56 SW	57 E	56 NW	57 NW	72 NW	65 SE	91 NE	73 SE	70 SE	56 SW	69 S	91 NE
SAN ANGELO	10	44 W	48 WNW	48 NW	69 WNW	58 NNW	57 NNE	40 SSE	37 NNE	44 E	60 WNW	66 WNW	40 WNW	69 WNW
SAN ANTONIO	50	56 NW	56 N	63 W	61 W	73 NW	59 E	54 SE	74 NE	49 NE	57 NW	57 W	51 NW	74 NE
VICTORIA	11	55 NNW	56 NW	57 W	58 NW	60 NW	50 NW	52 ENE	72 NW	57 SSE	53 NW	62 NW	54 N	72 NW
WACO	18	45 NW	46 NW	58 NNW	52 W	56 N	64 W	49 SW	46 WNW	48 NW	54 N	50 NW	45 NW	64 W
WICHITA FALLS	16	49 NW	68 WSW	59 W	64 W	62 W	92 W	50 S	55 NNW	56 WNW	60 WNW	70 WNW	64 WNW	92 W
UTAH SALT LAKE CITY	50	52 SE	56 SE	71 NW	56 NW	57 NW	54 W	53 W	58 SW	61 W	67 NW	63 NW	54 S	71 NW
VT. BURLINGTON	55	62 S	57 S	66 S	63 S	49 S	50 SW	42 S	54 N	56 S	70 SE	72 SE	57 S	72 SE
VA. LYNCHBURG	50	57 SW	48 SW	48 NW	62 W	56 N	47 N	47 N	46 NE	47 SW	62 NE	46 SW	46 NW	62 W
NORFOLK	50	62 SW	62 SW	63 W	62 N	70 W	80 W	63 SW	70 NE	75 NE	78 S	53 NW	62 W	80 W
RICHMOND	50	62 SW	51 NW	51 W	62 NW	56 SW	53 NW	56 NW	50 SE	45 SE	68 SE	47 SW	46 SW	68 SE
WASH. SEATTLE	28	63 SW	64 S	60 SW	65 S	45 SW	54 SW	38 SW	35 SW	55 SW	63 S	57 SW	60 SW	65 S
SPOKANE	49	56 SW	54 SW	56 W	43 W	40 W	47 SW	46 S	38 SW	38 SW	56 SW	54 SW	56 SW	56 W,SW
STAMPEDE PASS	9	94 W	91 SW	78 SW	61 E	61 W	56 W	42 W	33 E	56 W	61 W	62 SW	64 W	94 W
TATOOSH ISLAND	59	87 S	84 NE	91 E	73 SW	66 S	70 S	53 S	59 SW	68 NE	73 S	94 S	85 S	94 S
WALLA WALLA	46	42 SW	47 W	62 W	41 NW	32 W	37 S	36 W	26 W	51 W	54 SE	67 SW	47 W	67 SW
YAKIMA	18	32 NW	31 NW	34 SW	34 W	30 SW	30 W	29 NW	25 NW	35 SW	29 SW	29 NW	32 S	35 SW
W.VA. PARKERSBURG	73	45 NW	45 NW	47 SE	47 SW	43 NW	49 NW	62 NW	37 N	61 NW	38 NW	66 NW	35 W	66 NW
WIS. GREEN BAY	49	61 W	66 W	68 W	57 NE	109 SW	73 SW	70 NE	56 SW	66 W	66 SW	67 W	61 SW	109 SW
LA CROSSE	10	35 NNW	36 WNW	40 NNW	50 SSW	58 E	60 NNW	36 N	46 N	36 SSW	38 WNW	46 S	43 NNW	60 NNW
MADISON	50	68 E	57 W	70 SW	73 SW	77 SW	59 W	72 NW	47 W	52 W	73 SW	56 SE	65 SW	77 SW
MILWAUKEE	50	62 W	58 NE	73 SW	66 SW	72 SW	66 W	59 W	56 N	62 S	60 S	72 W	62 SW	73 SW
WYO. CHEYENNE	50	75 NW	69 NW	73 W	67 NW	69 NW	63 NW	57 W	46 NW	54 W	61 W	66 W	71 W	75 NW
LANDER	50	65 SW	77 SW	62 SW	72 SW	56 SW	61 SW	57 W	56 SW	70 W	70 SW	75 N	72 SW	77 SW
SHERIDAN	50	73 NW	70 W	65 SW	66 NW	71 NW	66 SW	73 W	72 NW	66 NW	66 NW	84 SW	63 NW	84 SW
P. R. SAN JUAN	60	46 NE	43 N	41 E	35 E	45 N	40 SE	62 E	80 NE	*149 NE	43 N	40 NE	44 E	*149 NE

Direction indicates the direction from which wind was blowing at time of fastest mile.

U City   * Estimates

Prevailing means most frequently observed. Arrows fly with wind.

Charts and tabulations based on "Normals, Means, and Extremes" tables in U. S. Weather Bureau publication, Local Climatological Data. Use with caution because of the effects of local topography, particularly in mountainous terrain

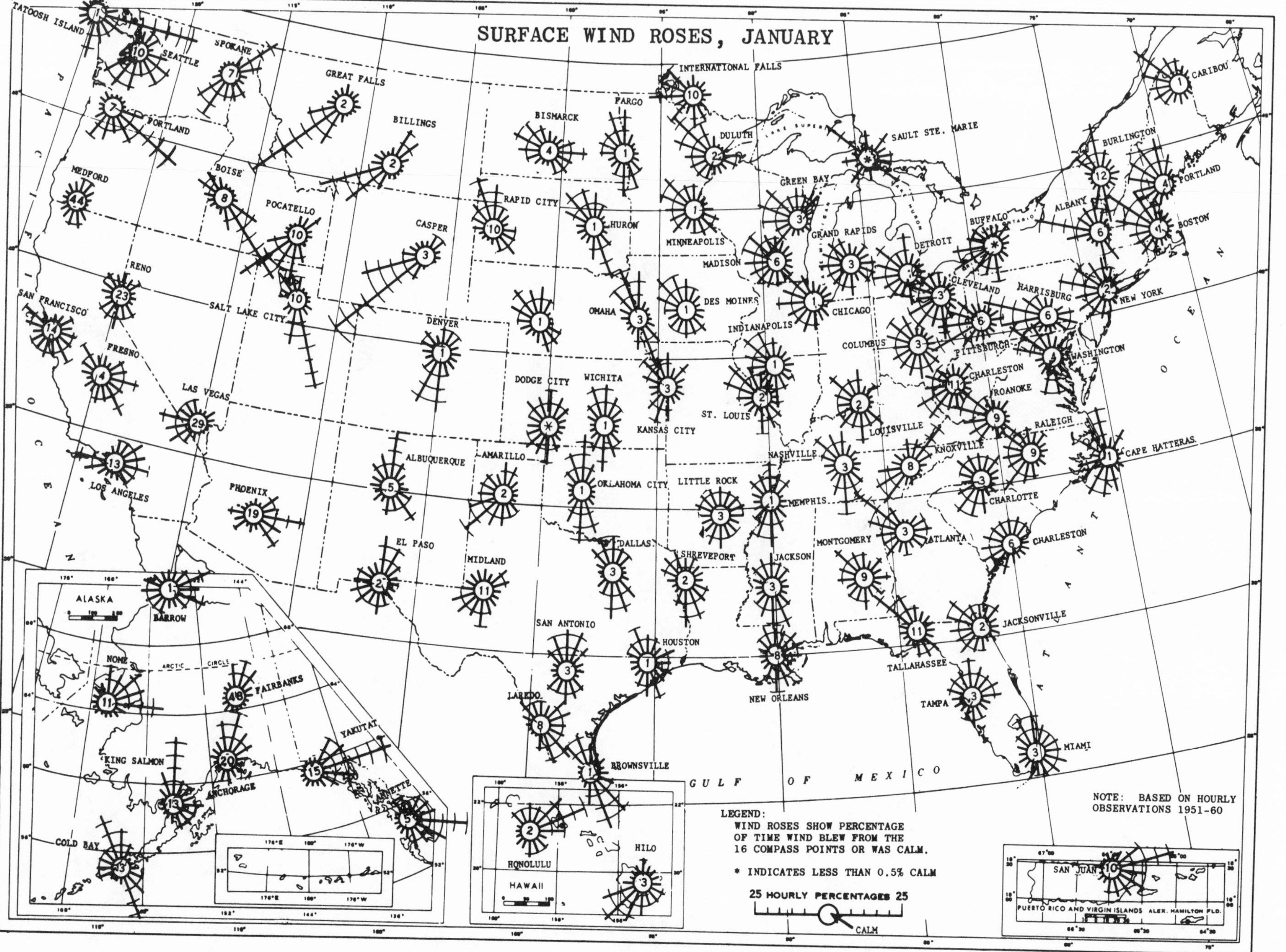
SURFACE WIND ROSES, JANUARY
NOTE: BASED ON HOURLY OBSERVATIONS 1951-60
LEGEND:
WIND ROSES SHOW PERCENTAGE OF TIME WIND BLEW FROM THE 16 COMPASS POINTS OR WAS CALM.
* INDICATES LESS THAN 0.5% CALM
25 HOURLY PERCENTAGES 25
CALM
PACIFIC OCEAN
ATLANTIC OCEAN
GULF OF MEXICO
TATOOSH ISLAND
SEATTLE
SPOKANE
GREAT FALLS
PORTLAND
BILLINGS
BISMARCK
FARGO
INTERNATIONAL FALLS
DULUTH
SAULT STE. MARIE
CARIBOU
BURLINGTON
PORTLAND
MEDFORD
BOISE
POCATELLO
CASPER
RAPID CITY
HURON
MINNEAPOLIS
GREEN BAY
GRAND RAPIDS
DETROIT
BUFFALO
ALBANY
BOSTON
RENO
SALT LAKE CITY
MADISON
SAN FRANCISCO
DENVER
OMAHA
DES MOINES
CHICAGO
CLEVELAND
HARRISBURG
NEW YORK
FRESNO
INDIANAPOLIS
COLUMBUS
PITTSBURGH
WASHINGTON
LAS VEGAS
DODGE CITY
WICHITA
ST. LOUIS
CHARLESTON
ROANOKE
KANSAS CITY
LOUISVILLE
RALEIGH
KNOXVILLE
NASHVILLE
CAPE HATTERAS
LOS ANGELES
ALBUQUERQUE
AMARILLO
OKLAHOMA CITY
LITTLE ROCK
MEMPHIS
CHARLOTTE
PHOENIX
DALLAS
SHREVEPORT
MONTGOMERY
ATLANTA
CHARLESTON
EL PASO
MIDLAND
JACKSON
JACKSONVILLE
SAN ANTONIO
HOUSTON
TALLAHASSEE
NEW ORLEANS
LAREDO
TAMPA
MIAMI
BROWNSVILLE
ALASKA
BARROW
ARCTIC CIRCLE
NOME
FAIRBANKS
KING SALMON
ANCHORAGE
YAKUTAT
ANNETTE
COLD BAY
HONOLULU
HAWAII
HILO
SAN JUAN
PUERTO RICO AND VIRGIN ISLANDS ALEX. HAMILTON FLD.

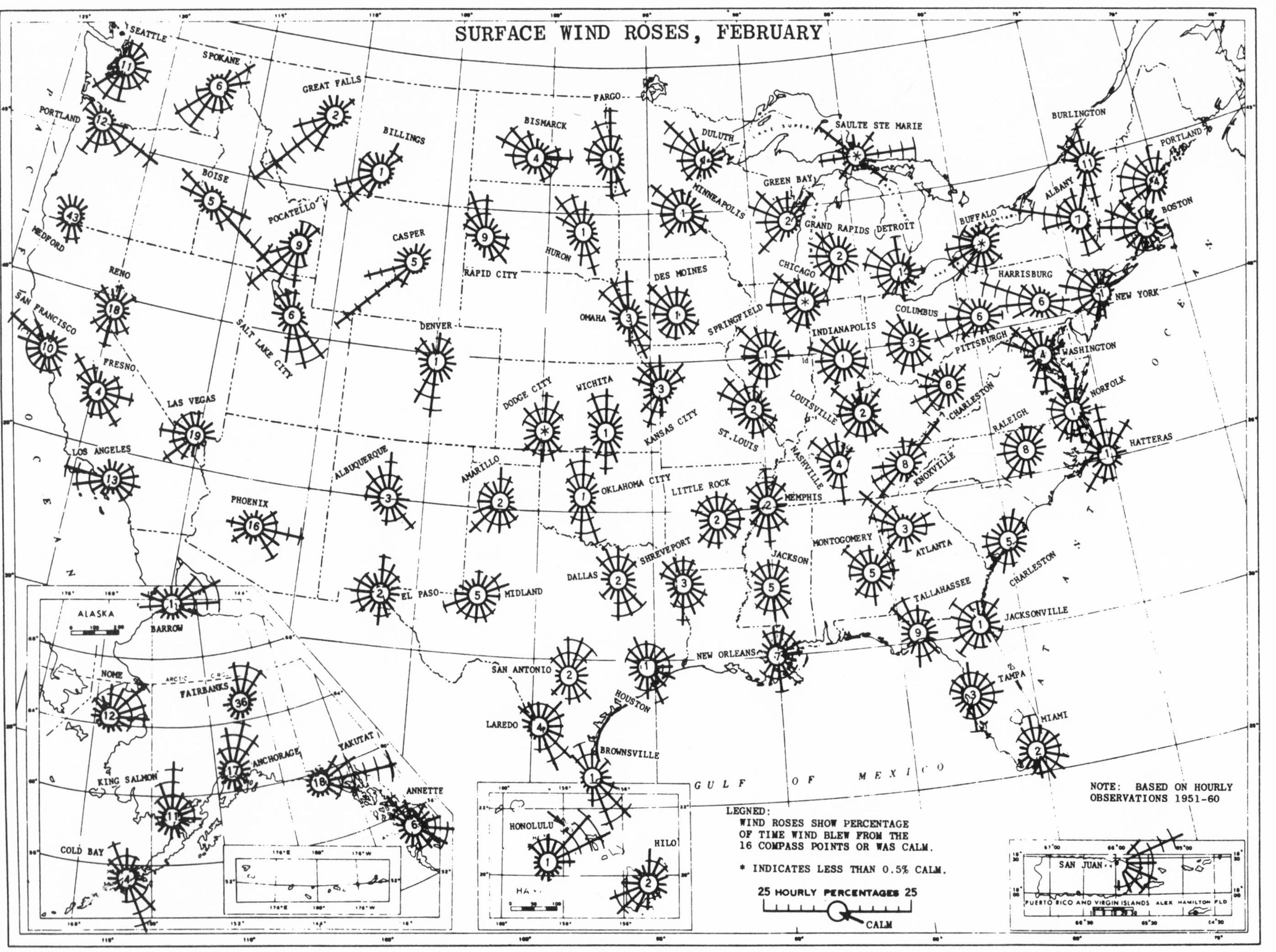
SURFACE WIND ROSES, FEBRUARY
NOTE: BASED ON HOURLY OBSERVATIONS 1951-60
LEGNED:
WIND ROSES SHOW PERCENTAGE OF TIME WIND BLEW FROM THE 16 COMPASS POINTS OR WAS CALM.
* INDICATES LESS THAN 0.5% CALM.
25 HOURLY PERCENTAGES 25
CALM
SEATTLE
SPOKANE
GREAT FALLS
PORTLAND
BILLINGS
BISMARCK
FARGO
DULUTH
SAULTE STE MARIE
BURLINGTON
PORTLAND
BOISE
MEDFORD
POCATELLO
CASPER
RAPID CITY
HURON
MINNEAPOLIS
GREEN BAY
GRAND RAPIDS
DETROIT
BUFFALO
ALBANY
BOSTON
RENO
SAN FRANCISCO
SALT LAKE CITY
DENVER
OMAHA
DES MOINES
CHICAGO
SPRINGFIELD
INDIANAPOLIS
COLUMBUS
HARRISBURG
NEW YORK
PITTSBURGH
WASHINGTON
FRESNO
LAS VEGAS
DODGE CITY
WICHITA
KANSAS CITY
ST. LOUIS
LOUISVILLE
CHARLESTON
NORFOLK
RALEIGH
HATTERAS
LOS ANGELES
ALBUQUERQUE
AMARILLO
OKLAHOMA CITY
LITTLE ROCK
MEMPHIS
NASHVILLE
KNOXVILLE
PHOENIX
SHREVEPORT
MONTGOMERY
ATLANTA
CHARLESTON
DALLAS
JACKSON
EL PASO
MIDLAND
TALLAHASSEE
JACKSONVILLE
SAN ANTONIO
NEW ORLEANS
TAMPA
HOUSTON
LAREDO
BROWNSVILLE
MIAMI
GULF OF MEXICO
ALASKA
BARROW
NOME
FAIRBANKS
ANCHORAGE
YAKUTAT
KING SALMON
ANNETTE
COLD BAY
HONOLULU
HILO
SAN JUAN
PUERTO RICO AND VIRGIN ISLANDS ALEX HAMILTON FLD

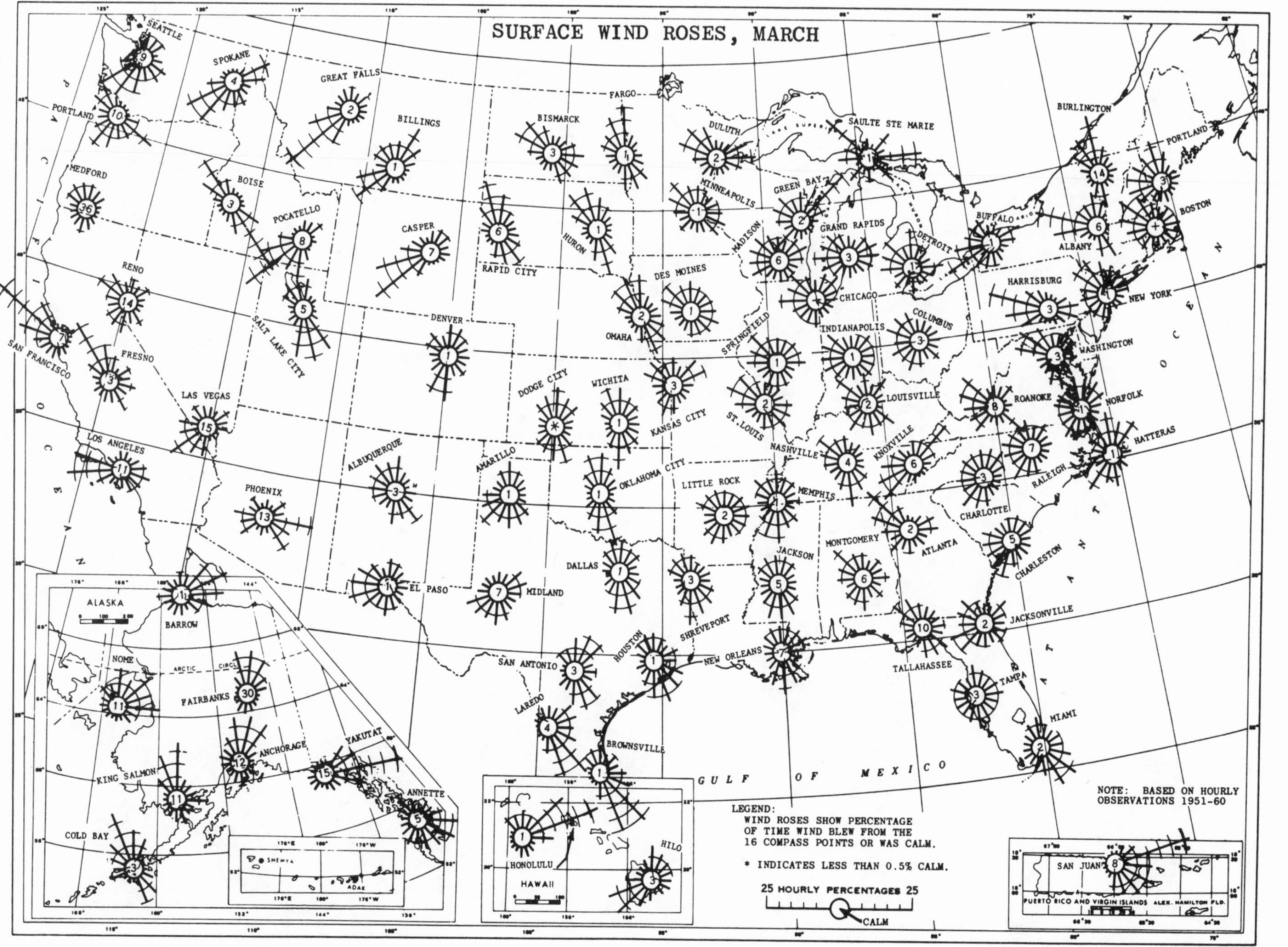

SURFACE WIND ROSES, MARCH
NOTE: BASED ON HOURLY OBSERVATIONS 1951-60
LEGEND: WIND ROSES SHOW PERCENTAGE OF TIME WIND BLEW FROM THE 16 COMPASS POINTS OR WAS CALM.
* INDICATES LESS THAN 0.5% CALM.
25 HOURLY PERCENTAGES 25
CALM
PACIFIC OCEAN
ATLANTIC OCEAN
GULF OF MEXICO
PUERTO RICO AND VIRGIN ISLANDS
SAN JUAN
ALEX. HAMILTON FLD.
HAWAII
HILO
HONOLULU
ALASKA
ARCTIC CIRCLE
BARROW
NOME
FAIRBANKS
ANCHORAGE
YAKUTAT
ANNETTE
KING SALMON
COLD BAY
ADAK
SHEMYA
SEATTLE
PORTLAND
MEDFORD
SPOKANE
BOISE
GREAT FALLS
POCATELLO
RENO
FRESNO
SAN FRANCISCO
LOS ANGELES
LAS VEGAS
SALT LAKE CITY
BILLINGS
CASPER
PHOENIX
ALBUQUERQUE
EL PASO
DENVER
AMARILLO
MIDLAND
RAPID CITY
BISMARCK
FARGO
HURON
DODGE CITY
WICHITA
OKLAHOMA CITY
DALLAS
SAN ANTONIO
LAREDO
BROWNSVILLE
HOUSTON
OMAHA
KANSAS CITY
DES MOINES
SHREVEPORT
NEW ORLEANS
LITTLE ROCK
JACKSON
DULUTH
MINNEAPOLIS
MADISON
SPRINGFIELD
ST. LOUIS
MEMPHIS
NASHVILLE
GREEN BAY
SAULTE STE MARIE
LAKE SUPERIOR
GRAND RAPIDS
CHICAGO
INDIANAPOLIS
LOUISVILLE
KNOXVILLE
MONTGOMERY
ATLANTA
DETROIT
COLUMBUS
BUFFALO
CHARLOTTE
TALLAHASSEE
JACKSONVILLE
TAMPA
MIAMI
CHARLESTON
RALEIGH
ROANOKE
HARRISBURG
WASHINGTON
NORFOLK
HATTERAS
NEW YORK
ALBANY
BURLINGTON
PORTLAND
BOSTON

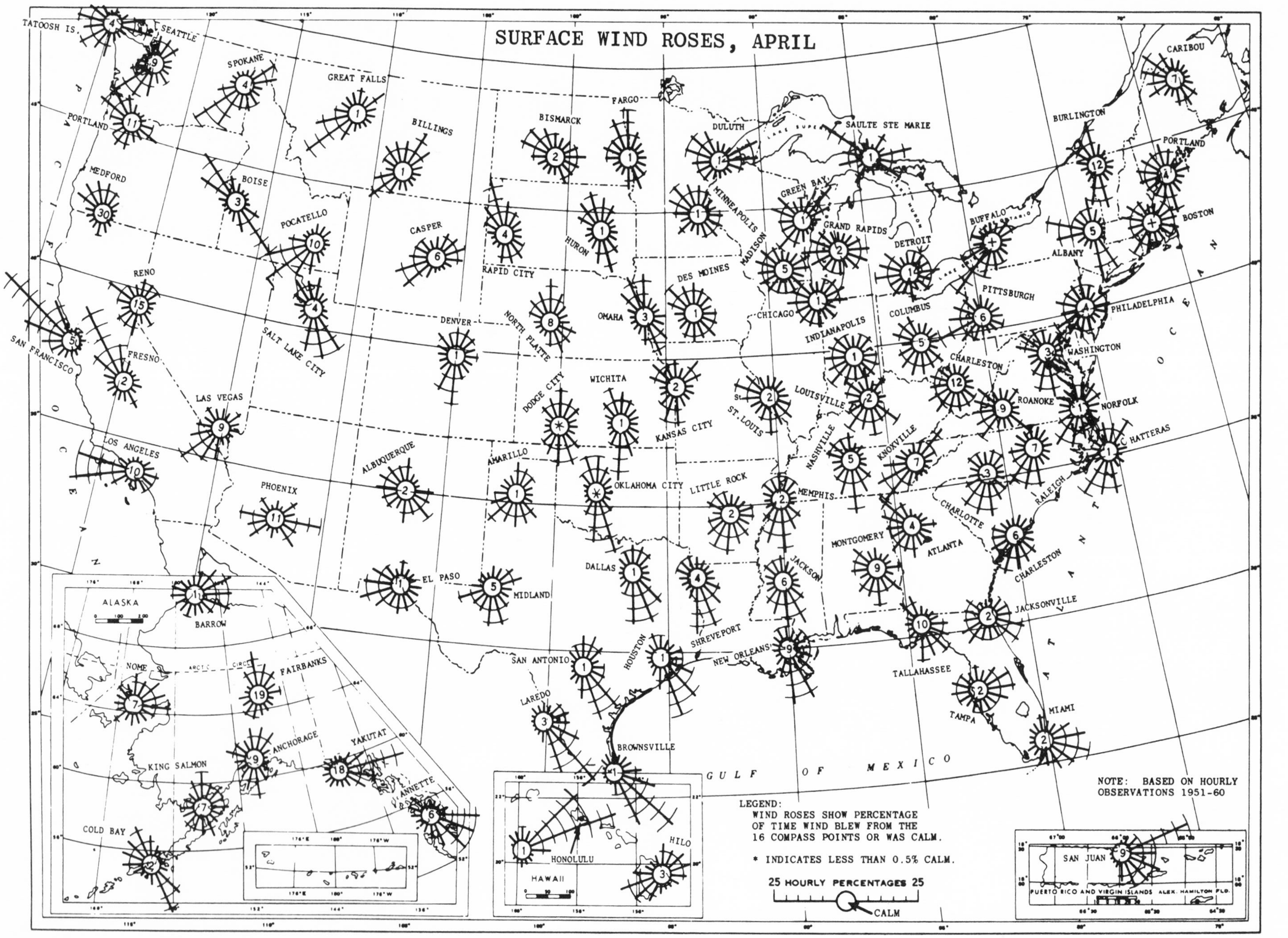

SURFACE WIND ROSES, APRIL
NOTE: BASED ON HOURLY OBSERVATIONS 1951-60
LEGEND:
WIND ROSES SHOW PERCENTAGE OF TIME WIND BLEW FROM THE 16 COMPASS POINTS OR WAS CALM.
* INDICATES LESS THAN 0.5% CALM.
25 HOURLY PERCENTAGES 25
CALM
TATOOSH IS.
SEATTLE
SPOKANE
GREAT FALLS
FARGO
DULUTH
SAULTE STE MARIE
BURLINGTON
CARIBOU
PORTLAND
BILLINGS
BISMARCK
MINNEAPOLIS
GREEN BAY
BOSTON
MEDFORD
BOISE
POCATELLO
CASPER
HURON
MADISON
GRAND RAPIDS
BUFFALO
ALBANY
RAPID CITY
DES MOINES
DETROIT
RENO
SAN FRANCISCO
FRESNO
SALT LAKE CITY
DENVER
NORTH PLATTE
OMAHA
CHICAGO
INDIANAPOLIS
COLUMBUS
PITTSBURGH
PHILADELPHIA
WASHINGTON
CHARLESTON
LAS VEGAS
DODGE CITY
WICHITA
KANSAS CITY
ST. LOUIS
LOUISVILLE
ROANOKE
NORFOLK
HATTERAS
LOS ANGELES
AMARILLO
ALBUQUERQUE
NASHVILLE
KNOXVILLE
PHOENIX
OKLAHOMA CITY
LITTLE ROCK
MEMPHIS
RALEIGH
CHARLOTTE
ATLANTA
MONTGOMERY
DALLAS
EL PASO
MIDLAND
JACKSON
SHREVEPORT
CHARLESTON
JACKSONVILLE
SAN ANTONIO
HOUSTON
NEW ORLEANS
TALLAHASSEE
TAMPA
MIAMI
LAREDO
BROWNSVILLE
GULF OF MEXICO
PACIFIC OCEAN
ATLANTIC OCEAN
ALASKA
BARROW
NOME
ARCTIC CIRCLE
FAIRBANKS
ANCHORAGE
YAKUTAT
KING SALMON
ANNETTE
COLD BAY
HAWAII
HONOLULU
HILO
SAN JUAN
PUERTO RICO AND VIRGIN ISLANDS
ALEX. HAMILTON FLD.

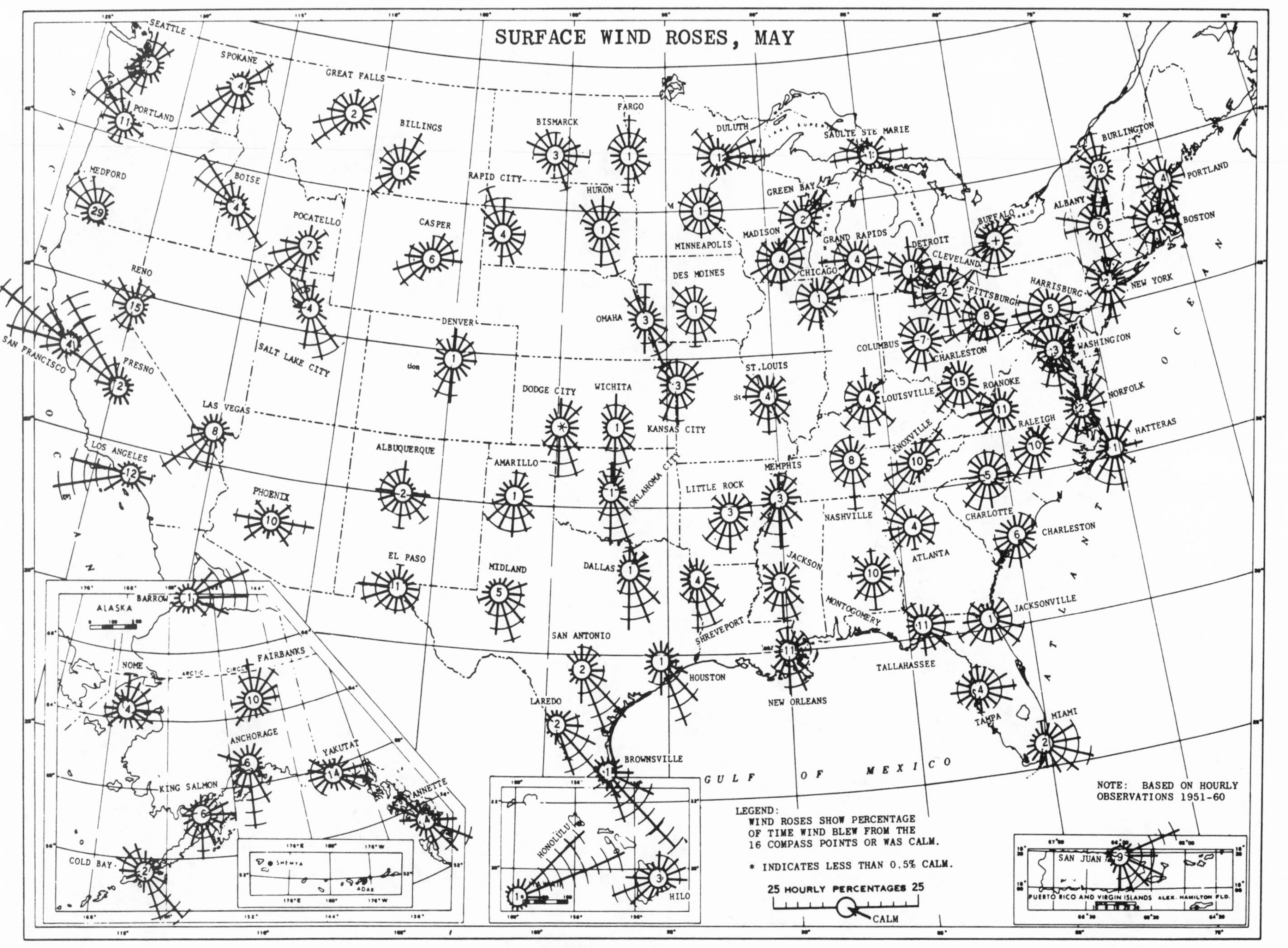
SURFACE WIND ROSES, MAY
NOTE: BASED ON HOURLY OBSERVATIONS 1951-60
LEGEND:
WIND ROSES SHOW PERCENTAGE OF TIME WIND BLEW FROM THE 16 COMPASS POINTS OR WAS CALM.
* INDICATES LESS THAN 0.5% CALM.
25 HOURLY PERCENTAGES 25
CALM
SEATTLE
SPOKANE
GREAT FALLS
PORTLAND
BILLINGS
BISMARCK
FARGO
DULUTH
SAULTE STE MARIE
BURLINGTON
PORTLAND
MEDFORD
BOISE
RAPID CITY
HURON
MINNEAPOLIS
GREEN BAY
ALBANY
BOSTON
POCATELLO
CASPER
MADISON
GRAND RAPIDS
DETROIT
BUFFALO
CLEVELAND
NEW YORK
RENO
DES MOINES
CHICAGO
PITTSBURGH
HARRISBURG
SAN FRANCISCO
SALT LAKE CITY
DENVER
OMAHA
COLUMBUS
CHARLESTON
WASHINGTON
FRESNO
ST. LOUIS
LOUISVILLE
ROANOKE
NORFOLK
LAS VEGAS
DODGE CITY
WICHITA
KANSAS CITY
KNOXVILLE
RALEIGH
HATTERAS
LOS ANGELES
ALBUQUERQUE
AMARILLO
OKLAHOMA CITY
MEMPHIS
LITTLE ROCK
NASHVILLE
CHARLOTTE
CHARLESTON
PHOENIX
EL PASO
MIDLAND
DALLAS
JACKSON
ATLANTA
SHREVEPORT
MONTGOMERY
JACKSONVILLE
SAN ANTONIO
HOUSTON
NEW ORLEANS
TALLAHASSEE
LAREDO
TAMPA
MIAMI
BROWNSVILLE
GULF OF MEXICO
PACIFIC OCEAN
ATLANTIC OCEAN
ALASKA
BARROW
NOME
ARCTIC CIRCLE
FAIRBANKS
ANCHORAGE
YAKUTAT
ANNETTE
KING SALMON
COLD BAY
SHEMYA
ADAK
HONOLULU
HILO
SAN JUAN
PUERTO RICO AND VIRGIN ISLANDS
ALEX. HAMILTON FLD.

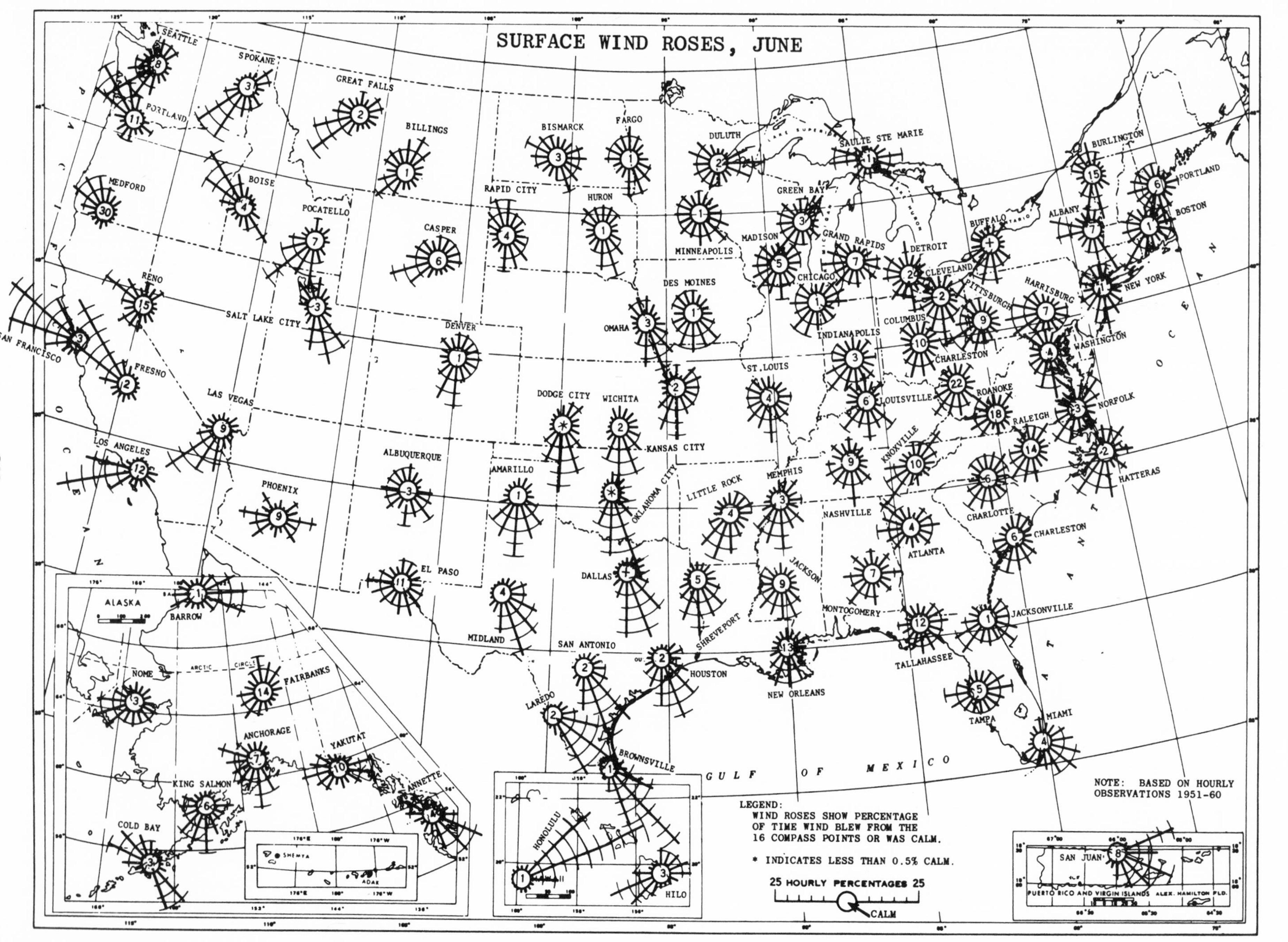
SURFACE WIND ROSES, JUNE
NOTE: BASED ON HOURLY OBSERVATIONS 1951-60
LEGEND:
WIND ROSES SHOW PERCENTAGE OF TIME WIND BLEW FROM THE 16 COMPASS POINTS OR WAS CALM.
* INDICATES LESS THAN 0.5% CALM.
25 HOURLY PERCENTAGES 25
CALM
PACIFIC OCEAN
ATLANTIC OCEAN
GULF OF MEXICO
SEATTLE
PORTLAND
SPOKANE
GREAT FALLS
BILLINGS
BISMARCK
FARGO
DULUTH
SAULTE STE MARIE
BURLINGTON
PORTLAND
MEDFORD
BOISE
POCATELLO
CASPER
RAPID CITY
HURON
MINNEAPOLIS
GREEN BAY
MADISON
GRAND RAPIDS
DETROIT
BUFFALO
ALBANY
BOSTON
RENO
SALT LAKE CITY
SAN FRANCISCO
FRESNO
DENVER
DES MOINES
OMAHA
CHICAGO
CLEVELAND
PITTSBURGH
HARRISBURG
NEW YORK
COLUMBUS
INDIANAPOLIS
WASHINGTON
CHARLESTON
LAS VEGAS
DODGE CITY
WICHITA
KANSAS CITY
ST. LOUIS
LOUISVILLE
ROANOKE
NORFOLK
LOS ANGELES
ALBUQUERQUE
AMARILLO
OKLAHOMA CITY
LITTLE ROCK
MEMPHIS
KNOXVILLE
RALEIGH
HATTERAS
PHOENIX
NASHVILLE
CHARLOTTE
CHARLESTON
EL PASO
DALLAS
JACKSON
ATLANTA
MIDLAND
SHREVEPORT
MONTGOMERY
JACKSONVILLE
SAN ANTONIO
HOUSTON
NEW ORLEANS
TALLAHASSEE
LAREDO
BROWNSVILLE
TAMPA
MIAMI
ALASKA
BARROW
ARCTIC CIRCLE
NOME
FAIRBANKS
ANCHORAGE
YAKUTAT
KING SALMON
ANNETTE
COLD BAY
SHEMYA
ADAK
HONOLULU
HILO
SAN JUAN
PUERTO RICO AND VIRGIN ISLANDS
ALEX. HAMILTON FLD.

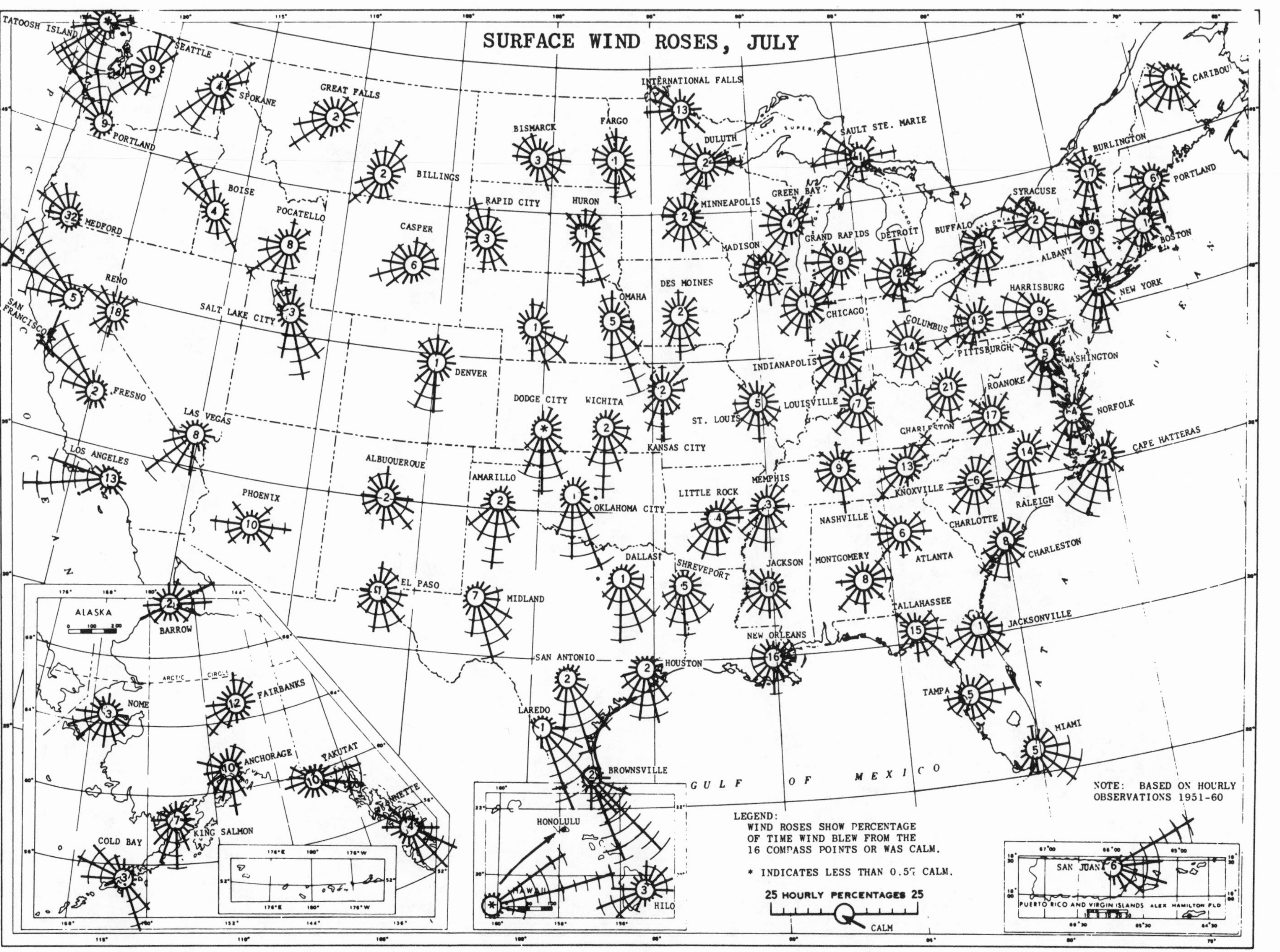

SURFACE WIND ROSES, JULY
TATOOSH ISLAND
SEATTLE
SPOKANE
GREAT FALLS
PORTLAND
BILLINGS
BISMARCK
FARGO
INTERNATIONAL FALLS
DULUTH
SAULT STE. MARIE
CARIBOU
BURLINGTON
PORTLAND
BOISE
MEDFORD
POCATELLO
CASPER
RAPID CITY
HURON
MINNEAPOLIS
GREEN BAY
MADISON
GRAND RAPIDS
DETROIT
BUFFALO
SYRACUSE
ALBANY
BOSTON
NEW YORK
HARRISBURG
RENO
SAN FRANCISCO
SALT LAKE CITY
DES MOINES
OMAHA
CHICAGO
COLUMBUS
PITTSBURGH
WASHINGTON
FRESNO
DENVER
INDIANAPOLIS
LOUISVILLE
ROANOKE
NORFOLK
LAS VEGAS
DODGE CITY
WICHITA
ST. LOUIS
KANSAS CITY
CHARLESTON
CAPE HATTERAS
LOS ANGELES
ALBUQUERQUE
AMARILLO
MEMPHIS
KNOXVILLE
RALEIGH
PHOENIX
OKLAHOMA CITY
LITTLE ROCK
NASHVILLE
CHARLOTTE
ATLANTA
CHARLESTON
EL PASO
MIDLAND
DALLAS
SHREVEPORT
JACKSON
MONTGOMERY
TALLAHASSEE
JACKSONVILLE
NEW ORLEANS
SAN ANTONIO
HOUSTON
TAMPA
LAREDO
BROWNSVILLE
MIAMI
GULF OF MEXICO
ALASKA
BARROW
ARCTIC CIRCLE
NOME
FAIRBANKS
ANCHORAGE
YAKUTAT
ANNETTE
KING SALMON
COLD BAY
HONOLULU
HAWAII
HILO
SAN JUAN
PUERTO RICO AND VIRGIN ISLANDS
LEGEND:
WIND ROSES SHOW PERCENTAGE
OF TIME WIND BLEW FROM THE
16 COMPASS POINTS OR WAS CALM.
* INDICATES LESS THAN 0.5% CALM.
25 HOURLY PERCENTAGES 25
CALM
NOTE: BASED ON HOURLY
OBSERVATIONS 1951-60

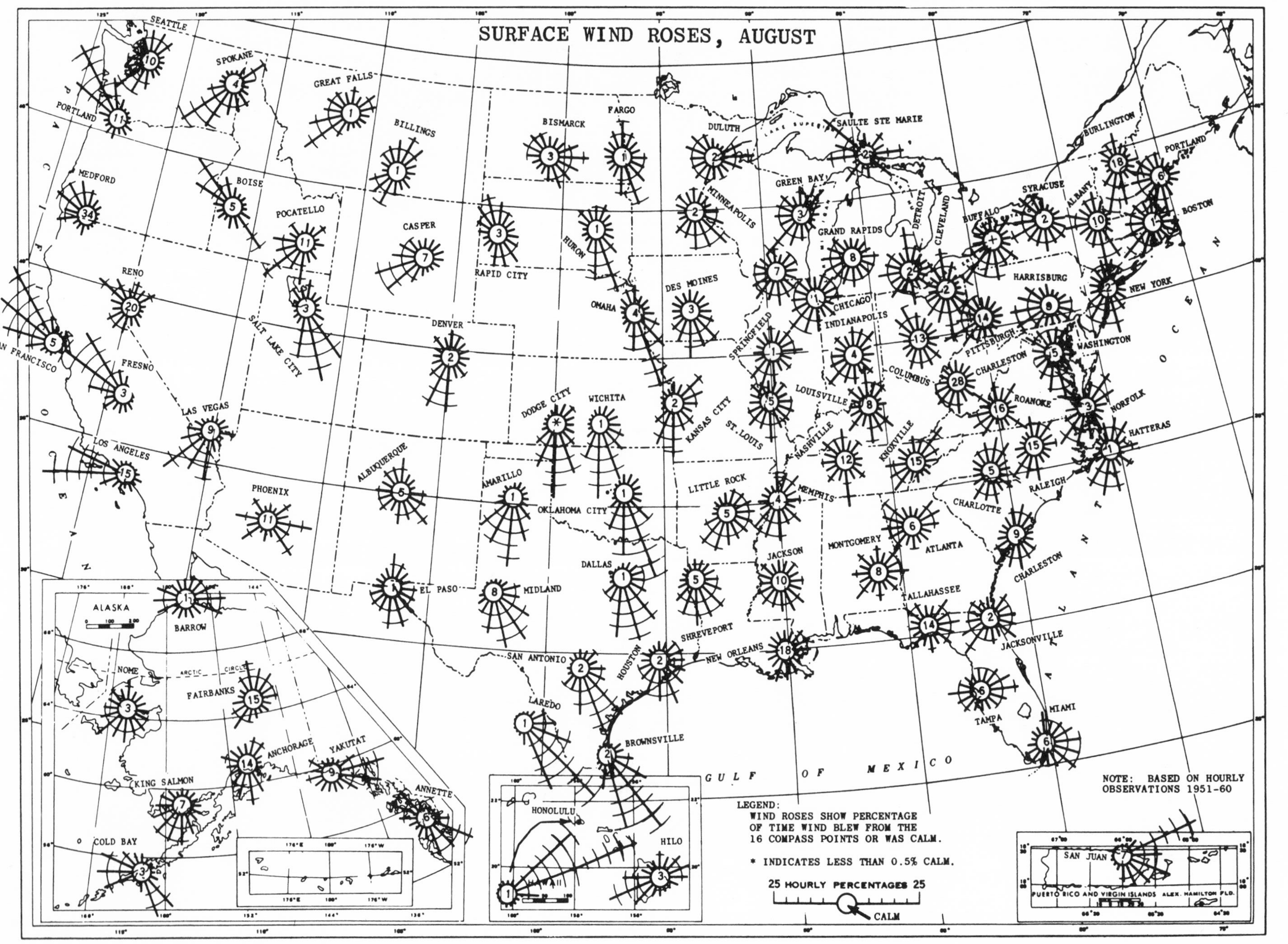
SURFACE WIND ROSES, AUGUST
NOTE: BASED ON HOURLY OBSERVATIONS 1951-60
LEGEND:
WIND ROSES SHOW PERCENTAGE OF TIME WIND BLEW FROM THE 16 COMPASS POINTS OR WAS CALM.
* INDICATES LESS THAN 0.5% CALM.
25 HOURLY PERCENTAGES 25
CALM
PACIFIC OCEAN
ATLANTIC OCEAN
GULF OF MEXICO
SEATTLE
PORTLAND
SPOKANE
GREAT FALLS
BILLINGS
BISMARCK
FARGO
DULUTH
SAULTE STE MARIE
BURLINGTON
PORTLAND
BOSTON
MEDFORD
BOISE
POCATELLO
CASPER
RAPID CITY
HURON
MINNEAPOLIS
GREEN BAY
GRAND RAPIDS
DETROIT
CLEVELAND
BUFFALO
SYRACUSE
ALBANY
RENO
SAN FRANCISCO
FRESNO
SALT LAKE CITY
DENVER
OMAHA
DES MOINES
SPRINGFIELD
CHICAGO
INDIANAPOLIS
COLUMBUS
PITTSBURGH
HARRISBURG
NEW YORK
WASHINGTON
LAS VEGAS
LOS ANGELES
DODGE CITY
WICHITA
KANSAS CITY
ST. LOUIS
LOUISVILLE
NASHVILLE
KNOXVILLE
CHARLESTON
ROANOKE
NORFOLK
HATTERAS
PHOENIX
ALBUQUERQUE
AMARILLO
OKLAHOMA CITY
LITTLE ROCK
MEMPHIS
CHARLOTTE
RALEIGH
EL PASO
MIDLAND
DALLAS
JACKSON
MONTGOMERY
ATLANTA
CHARLESTON
SHREVEPORT
TALLAHASSEE
JACKSONVILLE
SAN ANTONIO
HOUSTON
NEW ORLEANS
LAREDO
BROWNSVILLE
TAMPA
MIAMI
ALASKA
BARROW
NOME
ARCTIC CIRCLE
FAIRBANKS
ANCHORAGE
YAKUTAT
KING SALMON
ANNETTE
COLD BAY
HONOLULU
HAWAII
HILO
SAN JUAN
PUERTO RICO AND VIRGIN ISLANDS
ALEX. HAMILTON FLD.

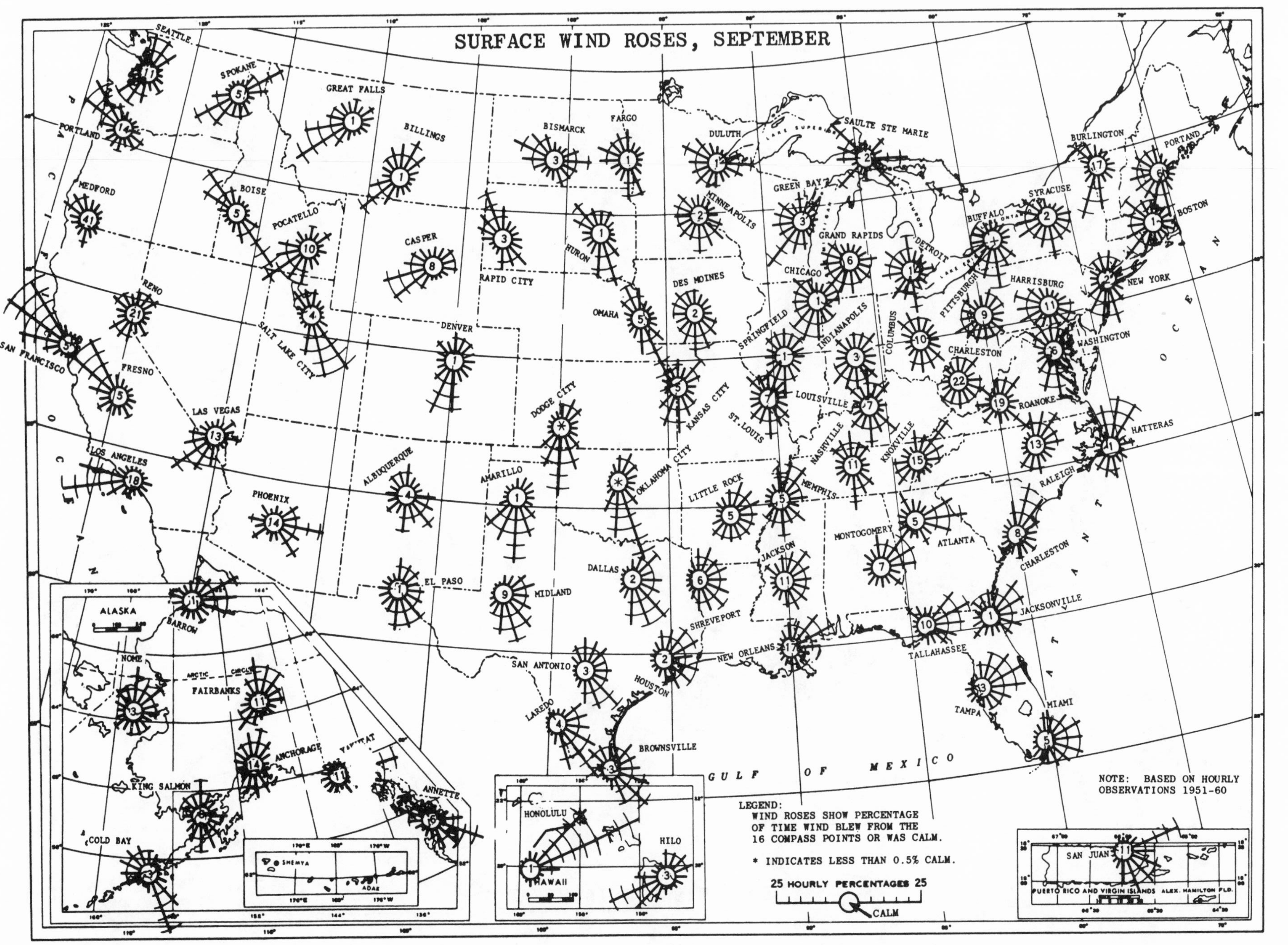

SURFACE WIND ROSES, SEPTEMBER
SEATTLE
SPOKANE
GREAT FALLS
BILLINGS
BISMARCK
FARGO
DULUTH
SAULTE STE MARIE
BURLINGTON
PORTLAND
PORTLAND
MEDFORD
BOISE
POCATELLO
CASPER
RAPID CITY
HURON
MINNEAPOLIS
GREEN BAY
GRAND RAPIDS
DETROIT
BUFFALO
SYRACUSE
BOSTON
RENO
SAN FRANCISCO
FRESNO
SALT LAKE CITY
DENVER
OMAHA
DES MOINES
CHICAGO
SPRINGFIELD
INDIANAPOLIS
COLUMBUS
PITTSBURGH
HARRISBURG
NEW YORK
WASHINGTON
CHARLESTON
LAS VEGAS
LOS ANGELES
PHOENIX
ALBUQUERQUE
AMARILLO
DODGE CITY
OKLAHOMA CITY
KANSAS CITY
ST. LOUIS
LOUISVILLE
NASHVILLE
KNOXVILLE
ROANOKE
HATTERAS
RALEIGH
LITTLE ROCK
MEMPHIS
ATLANTA
MONTGOMERY
CHARLESTON
JACKSONVILLE
EL PASO
MIDLAND
DALLAS
SHREVEPORT
JACKSON
NEW ORLEANS
TALLAHASSEE
TAMPA
MIAMI
SAN ANTONIO
HOUSTON
LAREDO
BROWNSVILLE
GULF OF MEXICO
ALASKA
BARROW
NOME
ARCTIC CIRCLE
FAIRBANKS
ANCHORAGE
YAKUTAT
KING SALMON
ANNETTE
COLD BAY
SHEMYA
ADAK
HONOLULU
HAWAII
HILO
SAN JUAN
PUERTO RICO AND VIRGIN ISLANDS
ALEX. HAMILTON FLD.
PACIFIC OCEAN
ATLANTIC OCEAN
NOTE: BASED ON HOURLY OBSERVATIONS 1951-60
LEGEND:
WIND ROSES SHOW PERCENTAGE OF TIME WIND BLEW FROM THE 16 COMPASS POINTS OR WAS CALM.
* INDICATES LESS THAN 0.5% CALM.
25 HOURLY PERCENTAGES 25
CALM

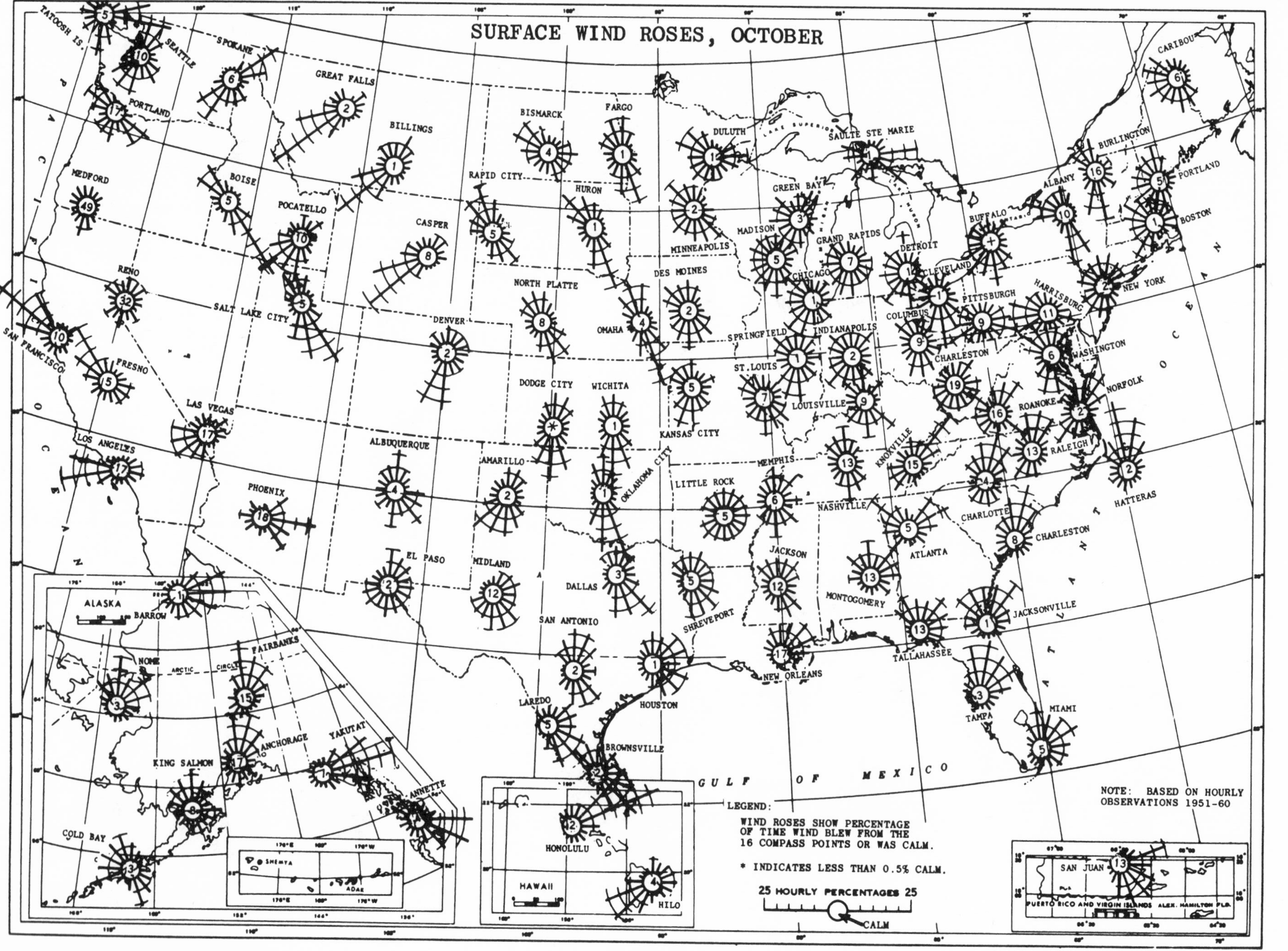
SURFACE WIND ROSES, OCTOBER
NOTE: BASED ON HOURLY OBSERVATIONS 1951-60
LEGEND:
WIND ROSES SHOW PERCENTAGE OF TIME WIND BLEW FROM THE 16 COMPASS POINTS OR WAS CALM.
* INDICATES LESS THAN 0.5% CALM.
25 HOURLY PERCENTAGES 25
CALM
TATOOSH IS.
SEATTLE
SPOKANE
GREAT FALLS
PORTLAND
BILLINGS
BISMARCK
FARGO
DULUTH
SAULTE STE MARIE
CARIBOU
BURLINGTON
PORTLAND
MEDFORD
BOISE
RAPID CITY
HURON
GREEN BAY
ALBANY
BOSTON
POCATELLO
CASPER
MINNEAPOLIS
MADISON
GRAND RAPIDS
DETROIT
BUFFALO
RENO
NORTH PLATTE
DES MOINES
CHICAGO
CLEVELAND
NEW YORK
SALT LAKE CITY
DENVER
OMAHA
SPRINGFIELD
INDIANAPOLIS
COLUMBUS
PITTSBURGH
HARRISBURG
SAN FRANCISCO
FRESNO
ST. LOUIS
CHARLESTON
WASHINGTON
DODGE CITY
WICHITA
LOUISVILLE
NORFOLK
LAS VEGAS
KANSAS CITY
ROANOKE
LOS ANGELES
ALBUQUERQUE
AMARILLO
OKLAHOMA CITY
MEMPHIS
KNOXVILLE
RALEIGH
LITTLE ROCK
PHOENIX
NASHVILLE
CHARLOTTE
HATTERAS
EL PASO
MIDLAND
DALLAS
JACKSON
ATLANTA
CHARLESTON
MONTGOMERY
SHREVEPORT
SAN ANTONIO
JACKSONVILLE
TALLAHASSEE
NEW ORLEANS
HOUSTON
LAREDO
TAMPA
MIAMI
BROWNSVILLE
GULF OF MEXICO
PACIFIC OCEAN
ATLANTIC OCEAN
LAKE SUPERIOR
ALASKA
BARROW
NOME
ARCTIC CIRCLE
FAIRBANKS
ANCHORAGE
YAKUTAT
KING SALMON
ANNETTE
COLD BAY
SHEMYA
ADAK
HONOLULU
HAWAII
HILO
SAN JUAN
PUERTO RICO AND VIRGIN ISLANDS
ALEX. HAMILTON FLD.

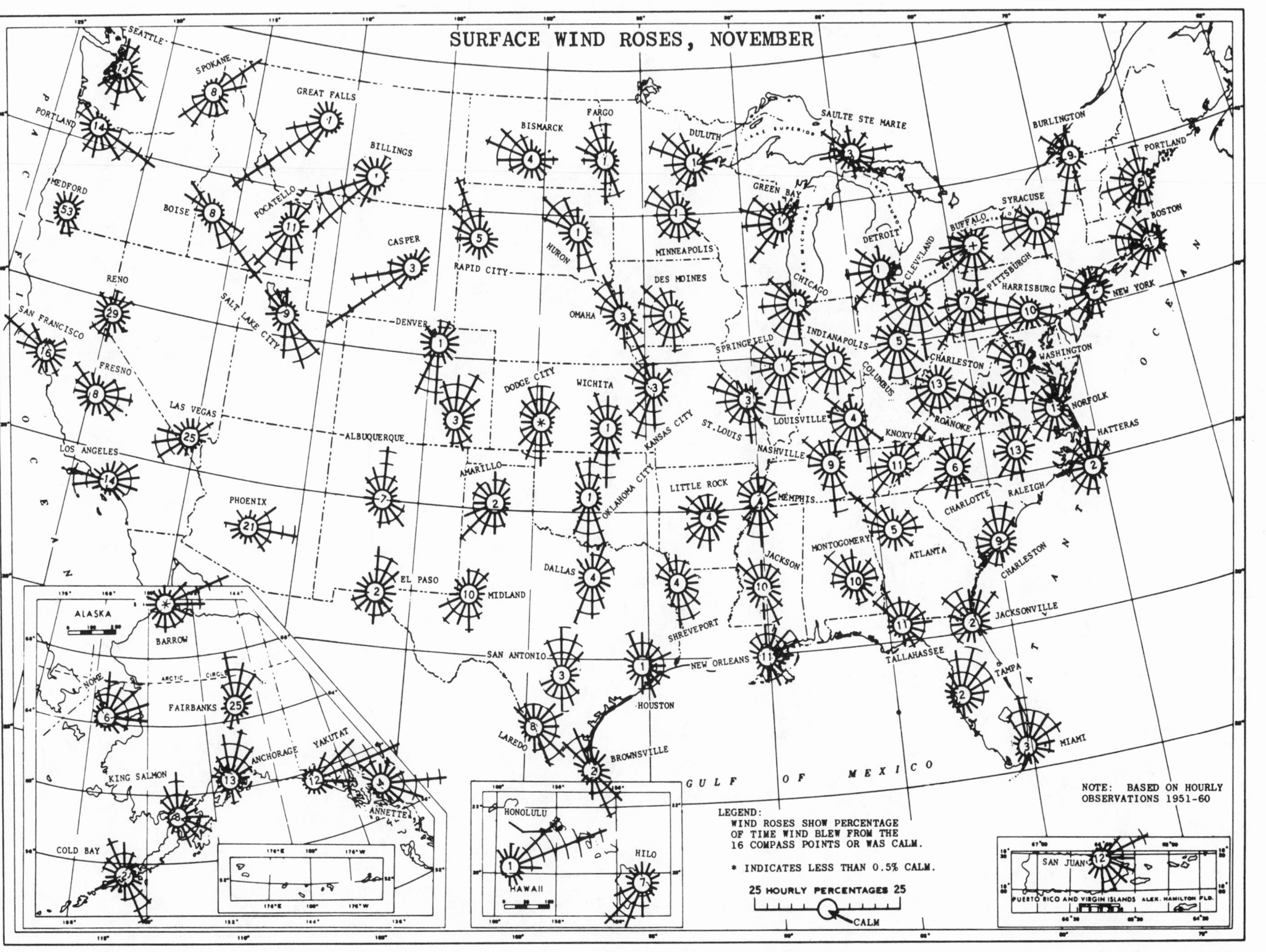
SURFACE WIND ROSES, NOVEMBER
SEATTLE
SPOKANE
GREAT FALLS
PORTLAND
BISMARCK
FARGO
DULUTH
SAULTE STE MARIE
BURLINGTON
PORTLAND
BILLINGS
MEDFORD
BOISE
POCATELLO
GREEN BAY
SYRACUSE
BOSTON
CASPER
HURON
MINNEAPOLIS
DETROIT
BUFFALO
CLEVELAND
PITTSBURGH
HARRISBURG
NEW YORK
RAPID CITY
RENO
SALT LAKE CITY
DES MOINES
CHICAGO
OMAHA
SAN FRANCISCO
DENVER
SPRINGFIELD
INDIANAPOLIS
COLUMBUS
CHARLESTON
WASHINGTON
FRESNO
DODGE CITY
WICHITA
LAS VEGAS
ALBUQUERQUE
KANSAS CITY
ST. LOUIS
LOUISVILLE
KNOXVILLE
ROANOKE
NORFOLK
HATTERAS
LOS ANGELES
AMARILLO
OKLAHOMA CITY
NASHVILLE
LITTLE ROCK
MEMPHIS
CHARLOTTE
RALEIGH
PHOENIX
MONTGOMERY
ATLANTA
JACKSON
CHARLESTON
DALLAS
EL PASO
MIDLAND
SHREVEPORT
JACKSONVILLE
NEW ORLEANS
TALLAHASSEE
SAN ANTONIO
HOUSTON
TAMPA
LAREDO
BROWNSVILLE
MIAMI
PACIFIC OCEAN
ATLANTIC OCEAN
GULF OF MEXICO
ALASKA
BARROW
ARCTIC CIRCLE
NOME
FAIRBANKS
ANCHORAGE
YAKUTAT
KING SALMON
ANNETTE
COLD BAY
HONOLULU
HAWAII
HILO
NOTE: BASED ON HOURLY OBSERVATIONS 1951-60
LEGEND:
WIND ROSES SHOW PERCENTAGE OF TIME WIND BLEW FROM THE 16 COMPASS POINTS OR WAS CALM.
* INDICATES LESS THAN 0.5% CALM.
25 HOURLY PERCENTAGES 25
CALM
SAN JUAN
PUERTO RICO AND VIRGIN ISLANDS
ALEX. HAMILTON FLD.

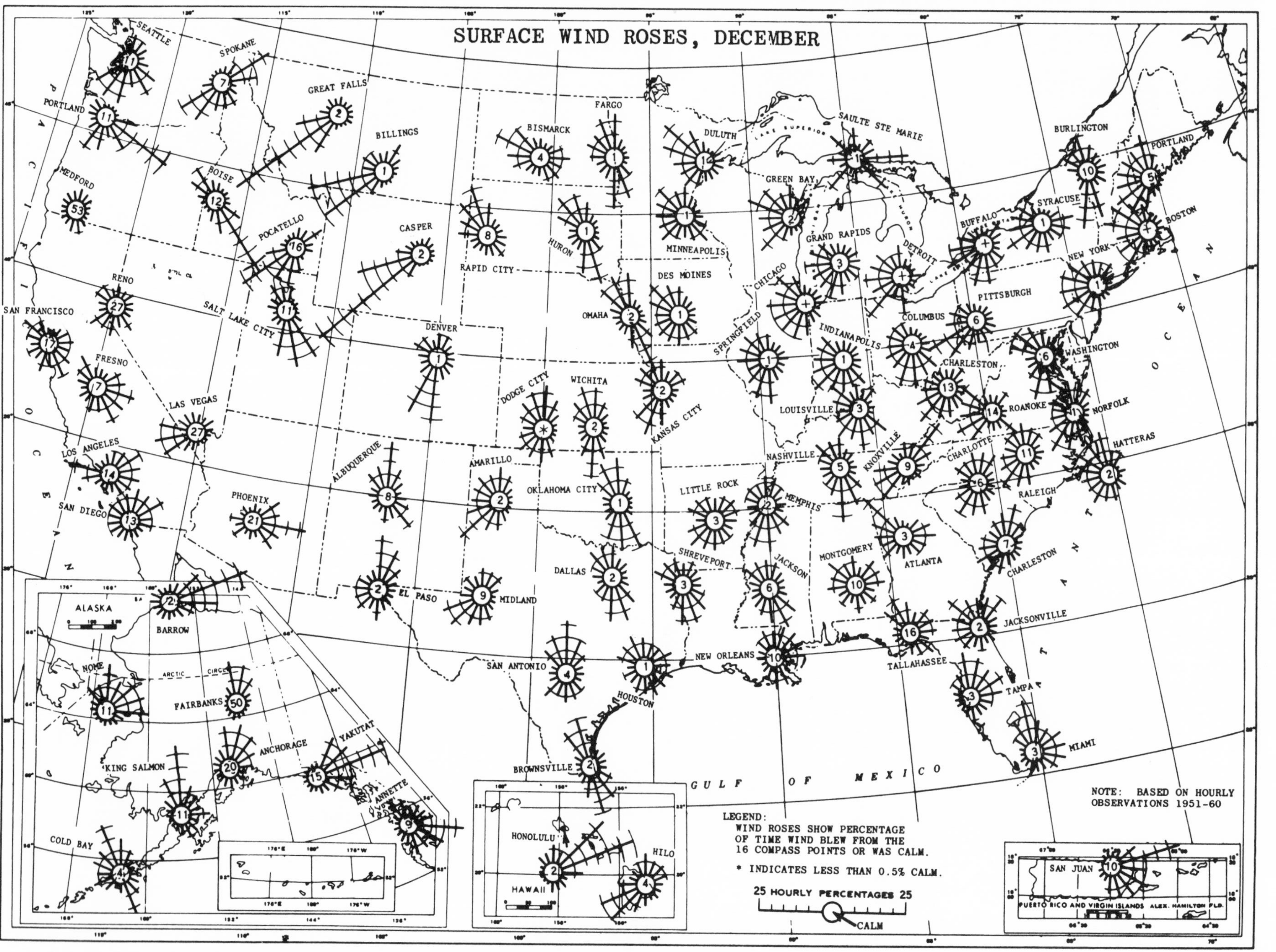
SURFACE WIND ROSES, DECEMBER
NOTE: BASED ON HOURLY OBSERVATIONS 1951-60
LEGEND:
WIND ROSES SHOW PERCENTAGE OF TIME WIND BLEW FROM THE 16 COMPASS POINTS OR WAS CALM.
* INDICATES LESS THAN 0.5% CALM.
25 HOURLY PERCENTAGES 25
CALM
PACIFIC OCEAN
ATLANTIC OCEAN
GULF OF MEXICO
SEATTLE
SPOKANE
GREAT FALLS
PORTLAND
BILLINGS
BISMARCK
FARGO
DULUTH
SAULTE STE MARIE
BURLINGTON
PORTLAND
MEDFORD
BOISE
CASPER
HURON
MINNEAPOLIS
GREEN BAY
GRAND RAPIDS
SYRACUSE
BUFFALO
BOSTON
POCATELLO
RAPID CITY
DES MOINES
DETROIT
NEW YORK
RENO
SAN FRANCISCO
SALT LAKE CITY
OMAHA
CHICAGO
PITTSBURGH
DENVER
SPRINGFIELD
INDIANAPOLIS
COLUMBUS
WASHINGTON
FRESNO
DODGE CITY
WICHITA
KANSAS CITY
LOUISVILLE
CHARLESTON
ROANOKE
NORFOLK
LAS VEGAS
LOS ANGELES
ALBUQUERQUE
AMARILLO
NASHVILLE
KNOXVILLE
CHARLOTTE
HATTERAS
OKLAHOMA CITY
LITTLE ROCK
MEMPHIS
RALEIGH
SAN DIEGO
PHOENIX
SHREVEPORT
JACKSON
MONTGOMERY
ATLANTA
CHARLESTON
DALLAS
EL PASO
MIDLAND
JACKSONVILLE
SAN ANTONIO
NEW ORLEANS
TALLAHASSEE
HOUSTON
TAMPA
MIAMI
BROWNSVILLE
ALASKA
BARROW
NOME
ARCTIC CIRCLE
FAIRBANKS
ANCHORAGE
YAKUTAT
KING SALMON
ANNETTE
COLD BAY
HONOLULU
HAWAII
HILO
SAN JUAN
PUERTO RICO AND VIRGIN ISLANDS
ALEX. HAMILTON FLD.

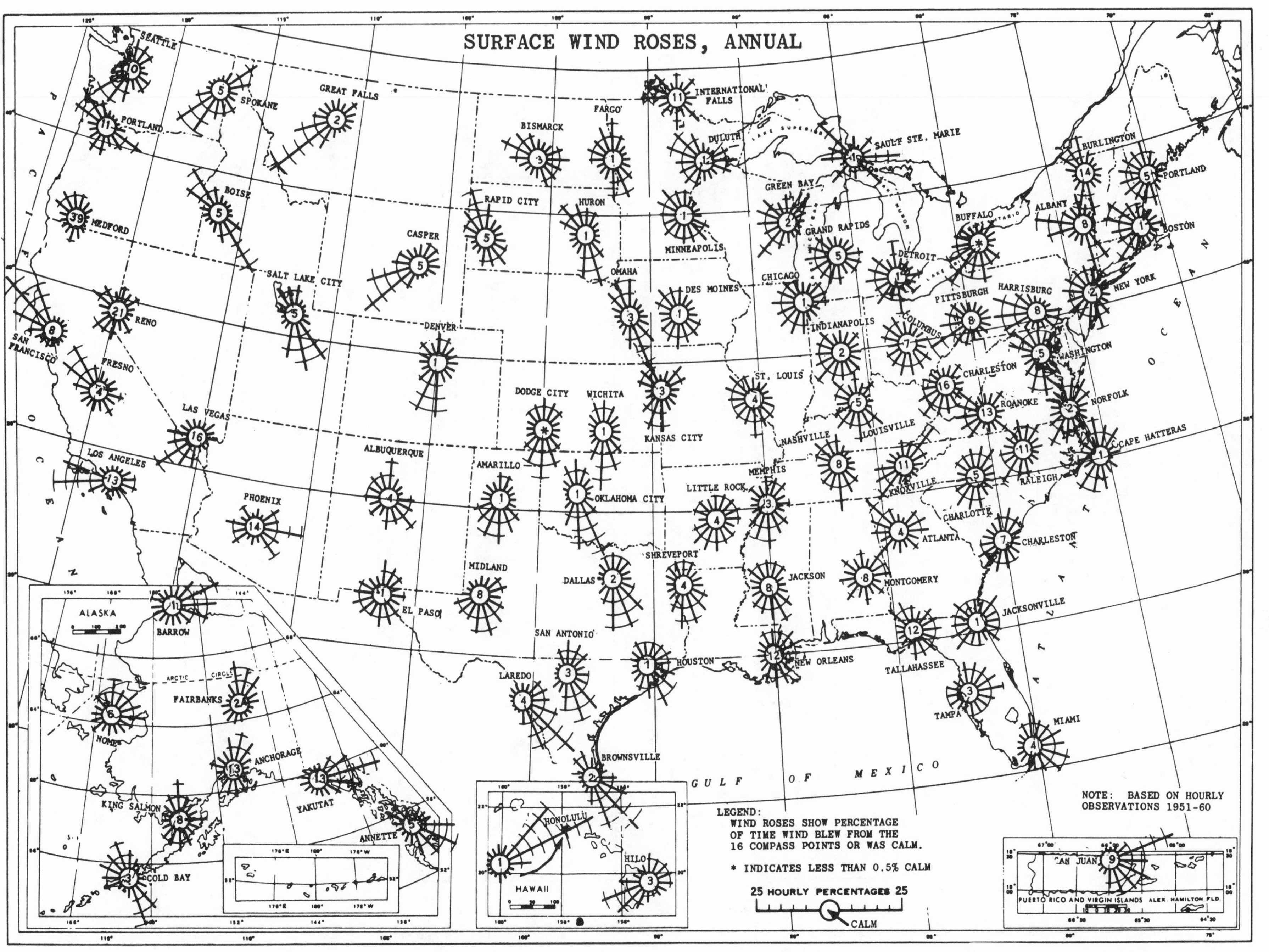

SURFACE WIND ROSES, ANNUAL
NOTE: BASED ON HOURLY OBSERVATIONS 1951-60
LEGEND:
WIND ROSES SHOW PERCENTAGE OF TIME WIND BLEW FROM THE 16 COMPASS POINTS OR WAS CALM.
* INDICATES LESS THAN 0.5% CALM
25 HOURLY PERCENTAGES 25
CALM
PACIFIC OCEAN
ATLANTIC OCEAN
GULF OF MEXICO
SEATTLE
PORTLAND
MEDFORD
SAN FRANCISCO
FRESNO
RENO
LOS ANGELES
LAS VEGAS
SPOKANE
BOISE
GREAT FALLS
SALT LAKE CITY
PHOENIX
CASPER
DENVER
ALBUQUERQUE
EL PASO
BISMARCK
RAPID CITY
FARGO
HURON
DODGE CITY
AMARILLO
MIDLAND
WICHITA
OKLAHOMA CITY
DALLAS
SAN ANTONIO
LAREDO
BROWNSVILLE
HOUSTON
SHREVEPORT
LITTLE ROCK
KANSAS CITY
OMAHA
DES MOINES
MINNEAPOLIS
DULUTH
INTERNATIONAL FALLS
GREEN BAY
CHICAGO
ST. LOUIS
MEMPHIS
JACKSON
NEW ORLEANS
NASHVILLE
LOUISVILLE
INDIANAPOLIS
GRAND RAPIDS
SAULT STE. MARIE
DETROIT
COLUMBUS
KNOXVILLE
ATLANTA
MONTGOMERY
TALLAHASSEE
CHARLOTTE
CHARLESTON
JACKSONVILLE
TAMPA
MIAMI
RALEIGH
ROANOKE
PITTSBURGH
BUFFALO
HARRISBURG
WASHINGTON
NORFOLK
CAPE HATTERAS
NEW YORK
ALBANY
BURLINGTON
BOSTON
PORTLAND
SAN JUAN
PUERTO RICO AND VIRGIN ISLANDS
ALEX. HAMILTON FLD.
HAWAII
HONOLULU
HILO
ALASKA
BARROW
FAIRBANKS
NOME
ANCHORAGE
KING SALMON
COLD BAY
YAKUTAT
ANNETTE
ARCTIC CIRCLE

ANNUAL PERCENTAGE FREQUENCY OF WIND BY SPEED GROUPS AND THE MEAN SPEED

STATE AND STATION	0 - 3 m.p.h.	4 - 7 m.p.h.	8 - 12 m.p.h.	13 - 18 m.p.h.	19 - 24 m.p.h.	25 - 31 m.p.h.	32 - 38 m.p.h.	39 - 46 m.p.h.	47 m.p.h. and over	Mean speed m.p.h.
ALA. Birmingham	27	22	30	17	3	1	*	*	*	7.9
Mobile	7	28	38	20	6	1	*	*	*	10.0
Montgomery	31	29	27	12	2	*	*			6.9
ALASKA, Anchorage	28	35	25	11	2	*	*	*		6.8
Cold Bay	4	9	18	27	21	14	5	2	*	17.4
Fairbanks	40	35	19	5	1	*	*	*		5.2
King Salmon	11	20	30	24	10	4	1	*	*	11.4
ARIZ. Phoenix	38	36	20	5	1	*	*	*		5.4
Tucson	18	35	30	14	3	1	*	*		8.1
ARK. Little Rock	12	30	39	16	2	*	*	*		8.7
CALIF. Bakersfield	35	30	24	10	1	*	*			5.8
Burbank	52	26	18	4	1	*	*	*		4.5
Fresno	30	41	22	7	1	*	*			6.1
Los Angeles	28	33	27	11	1	*	*			6.8
Oakland	26	28	28	16	2	1	*	*	*	7.5
Sacramento	15	28	31	18	5	1	*	*	*	9.3
San Diego	28	38	28	6	*	*	*			6.3
San Francisco	16	21	26	22	11	3	*	*	*	10.6
COLO. Colorado Springs	9	27	38	19	6	2	*	*	*	10.0
Denver	11	27	34	22	5	2	*	*	*	10.0
CONN. Hartford	13	26	32	24	6	1	*	*	*	9.8
D.C. Washington	11	26	35	22	5	1	*	*	*	9.7
DEL. Wilmington	15	31	30	19	4	1	*	*	*	8.8
FLA. Jacksonville	10	33	35	18	3	*	*	*		8.9
Miami	14	30	34	20	2	*	*	*	*	8.8
Orlando	18	28	32	17	4	*	*	*		8.6
Tallahassee	33	36	23	7	*	*				6.1
Tampa	9	31	40	16	2	*	*	*	*	8.8
West Palm Beach	9	22	36	27	6	1	*	*		10.5
GA. Atlanta	13	24	36	21	6	1	*	*		9.7
Augusta	36	29	25	9	1	*	*			6.3
Macon	10	26	46	16	2	*	*	*	*	8.9
Savannah	12	34	37	14	3	*	*	*	*	8.4
HAWAII, Hilo	7	34	43	15	2	*	*			8.7
Honolulu	9	17	27	32	12	2	*	*	*	12.1
IDAHO, Boise	15	30	32	18	4	1	*	*		8.9
ILL. Chicago (O'Hare)	8	22	33	27	8	2	*	*	*	11.2
Chicago (Midway)	7	26	36	25	5	1	*	*	*	10.2
Moline	14	23	32	24	7	2	*	*	*	10.0
Springfield	7	22	28	27	12	3	1	*	*	12.0
IND. Evansville	19	23	32	21	5	1	*	*		9.1
Fort Wayne	9	23	33	25	8	2	*	*	*	10.9
Indianapolis	9	22	34	26	7	2	*	*	*	10.8
South Bend	7	21	35	30	7	1	*	*		10.9
IOWA, Des Moines	3	17	38	29	10	3	1	*	*	12.1
Sioux City	10	20	31	25	10	4	1	*	*	11.7
KANS. Topeka	11	19	30	27	10	2	*	*	*	11.2
Wichita	4	12	30	31	16	5	1	*	*	13.7
KY. Lexington	8	25	39	22	6	1	*	*		10.1
Louisville	17	28	31	20	3	1	*	*		8.8
LA. Baton Rouge	17	29	34	17	3	*	*	*		8.3
Lake Charles	19	31	29	17	4	1	*	*	*	8.5
New Orleans	16	27	32	19	5	1	*	*	*	9.0
Shreveport	12	26	37	21	4	1	*	*	*	9.5
MAINE, Portland	10	30	33	22	4	1	*	*	*	9.6
MD. Baltimore	7	24	39	22	6	2	*	*	*	10.4
MASS. Boston	3	12	33	35	12	4	1	*	*	13.3
MICH. Detroit (City AP)	8	23	37	26	5	1	*	*	*	10.3
Flint	16	26	32	22	3	1	*	*	*	9.0
Grand Rapids	14	23	32	25	5	1	*	*	*	9.8
MINN. Duluth	6	15	33	31	11	4	1	*	*	12.6
Minneapolis	8	21	34	28	9	2	*	*	*	11.2
MISS. Jackson	33	25	26	14	2	*	*	*		7.1
MO. Kansas City	9	29	35	23	5	1	*			9.8
St. Louis	10	29	36	21	3	1	*	*	*	9.3
Springfield	4	13	34	32	13	3	1	*	*	12.9
MONT. Great Falls	7	19	24	24	15	9	3	1	*	13.9
NEBR. Omaha	12	17	29	28	11	3	*	*		11.6
NEV. Las Vegas	18	26	25	20	8	3	1	*	*	9.7
Reno	52	20	13	10	4	1	*	*	*	5.9
N. J. Newark	11	25	34	24	5	1	*	*	*	9.8
N. MEX. Albuquerque	17	36	26	13	5	2	*	*	*	8.6
N. Y. Albany	23	24	27	21	4	1	*	*		8.6
Binghamton	11	23	35	25	5	1	*	*	*	10.0
Buffalo	5	17	34	27	13	3	1	*	*	12.4
New York (Kennedy)	6	17	35	28	10	3	*	*	*	12.0
New York (La Guardia)	6	15	30	31	12	4	1	*	*	12.9
Rochester	8	22	34	25	9	2	1	*	*	11.2
Syracuse	14	27	30	23	5	1	*	*	*	9.7
N. C. Charlotte	20	32	31	14	2	*	*	*		7.9
Greensboro	20	32	31	14	2	*	*	*	*	8.0
Raleigh	18	33	34	14	2	*	*	*	*	7.7
Winston-Salem	19	22	33	21	4	1	*	*		9.0
N. DAK. Bismarck	14	20	27	24	12	3	1	*	*	11.2
Fargo	4	13	28	31	15	7	2	*	*	14.4
OHIO, Akron-Canton	7	25	35	26	5	1	*	*	*	10.4
Cincinnati	11	27	36	22	4	1	*	*	*	9.6
Cleveland	7	18	35	29	9	2	*	*	*	11.6
Columbus	26	23	29	18	4	1	*	*	*	8.2
Dayton	8	25	36	23	6	2	*	*	*	10.3
Youngstown	7	26	36	24	6	1	*	*	*	10.3
OKLA. Oklahoma City	2	11	34	34	13	6	1	*	*	14.0
OKLA. (Cont.) Tulsa	9	24	34	26	7	1	*	*	*	10.6
OREG. Medford	47	31	14	6	2	*	*	*	*	4.6
Portland	28	27	25	16	4	1	*	*	*	7.7
Salem	25	32	28	13	2	*	*			7.1
PA. Harrisburg	28	31	25	13	3	1	*	*		7.3
Philadelphia	11	27	35	21	5	1	*	*	*	9.6
Pittsburgh	12	26	34	22	4	1	*	*		9.4
Scranton	11	33	35	18	2	*	*	*		8.8
R. I. Providence	11	20	32	28	7	2	*	*	*	10.7
S. C. Charleston	12	28	35	19	4	1	*	*		9.2
Columbia	25	35	26	12	2	*	*			7.0
S. DAK. Huron	10	18	29	29	10	3	1	*	*	11.9
Rapid City	15	22	28	21	10	4	1	*	*	11.0
TENN. Chattanooga	39	25	24	11	1	*	*			6.1
Knoxville	29	29	25	12	4	1	*	*	*	7.5
Memphis	14	26	34	20	5	1	*	*		9.4
Nashville	27	31	25	14	2	*	*	*	*	7.2
TEX. Amarillo	5	15	32	32	12	4	1	*	*	12.9
Austin	13	25	34	23	5	1	*	*		9.7
Brownsville	10	17	25	30	14	3	*	*	*	12.3
Corpus Christi	11	16	26	33	12	2	*	*		11.9
Dallas	9	21	32	28	9	1	*	*		11.0
El Paso	10	22	32	22	9	4	1	*	*	11.3
Ft. Worth	4	14	34	34	10	3	*	*	*	12.5
Galveston	4	13	39	33	10	2	1	*	*	12.5
Houston	6	18	36	28	10	2	*	*	*	11.8
Laredo	6	15	32	34	12	1	*	*		12.3
Lubbock	4	11	33	34	13	5	1	*	*	13.6
Midland	9	22	38	26	4	1	*	*		10.1
San Antonio	18	23	32	22	4	1	*	*		9.3
Waco	3	14	36	35	10	2	*	*		12.5
Wichita Falls	5	22	41	27	5	1	*	*	*	10.5
UTAH, Salt Lake City	12	33	36	14	4	1	*	*	*	8.7
VT. Burlington	24	24	28	22	2	*	*			8.3
VA. Norfolk	14	23	30	25	6	1	*	*	*	10.2
Richmond	14	37	36	11	1	*	*	*		7.8
Roanoke	31	22	23	17	5	2	*	*		8.3
WASH. Seattle-Tacoma AP	13	16	35	26	8	2	*	*	*	10.7
Spokane	17	38	27	14	3	1	*	*		8.1
W. VA. Charleston	29	37	25	8	1	*	*			6.2
WIS. Green Bay	8	22	32	26	10	2	*	*	*	11.2
Madison	15	22	30	23	7	2	*	*	*	10.1
Milwaukee	8	17	31	30	11	3	1	*	*	12.1
WYO. Casper	8	16	27	27	13	7	2	*	*	13.3
PACIFIC, Wake Island	1	6	27	48	17	2	*	*		14.6
P. R. San Juan	15	28	27	25	4	*	*	*	*	9.1

Source: Climatography of the United States Series 82; Decennial Census of the United States Climate -- Summary of Hourly Observations, 1951-60 (Table B)

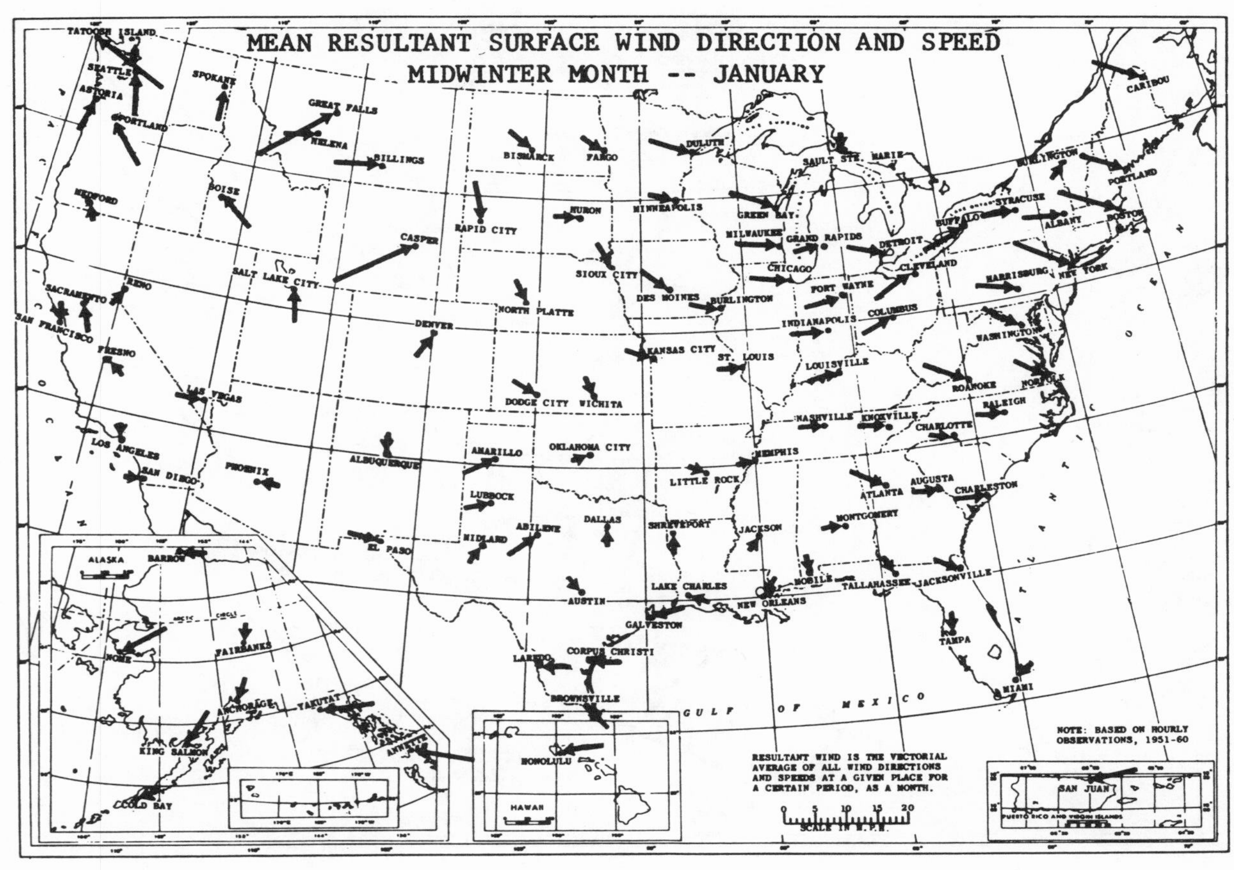
MEAN RESULTANT SURFACE WIND DIRECTION AND SPEED
MIDWINTER MONTH -- JANUARY
NOTE: BASED ON HOURLY OBSERVATIONS, 1951-60
RESULTANT WIND IS THE VECTORIAL AVERAGE OF ALL WIND DIRECTIONS AND SPEEDS AT A GIVEN PLACE FOR A CERTAIN PERIOD, AS A MONTH.
SCALE IN M.P.H.
GULF OF MEXICO

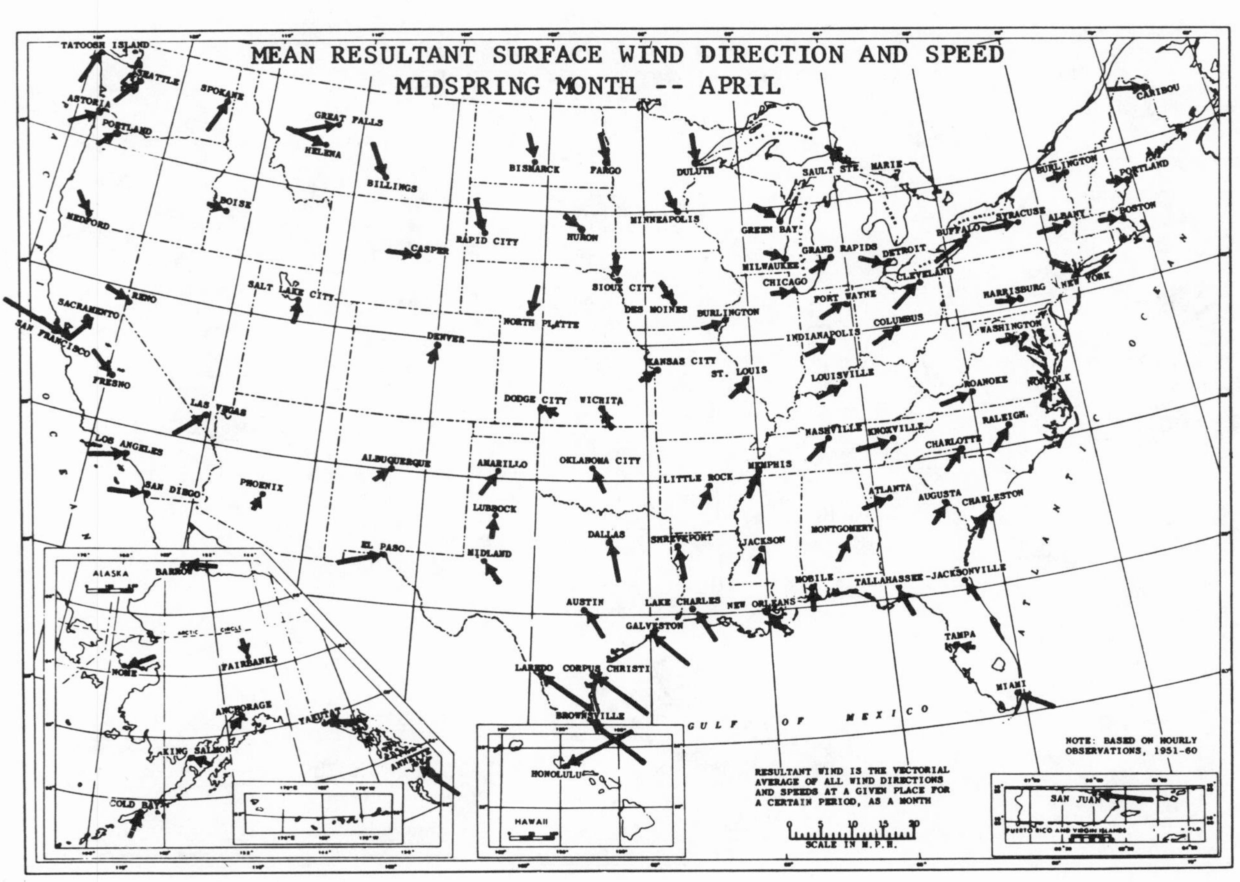
MEAN RESULTANT SURFACE WIND DIRECTION AND SPEED
MIDSPRING MONTH -- APRIL
NOTE: BASED ON HOURLY OBSERVATIONS, 1951-60
RESULTANT WIND IS THE VECTORIAL AVERAGE OF ALL WIND DIRECTIONS AND SPEEDS AT A GIVEN PLACE FOR A CERTAIN PERIOD, AS A MONTH
SCALE IN M.P.H.
GULF OF MEXICO

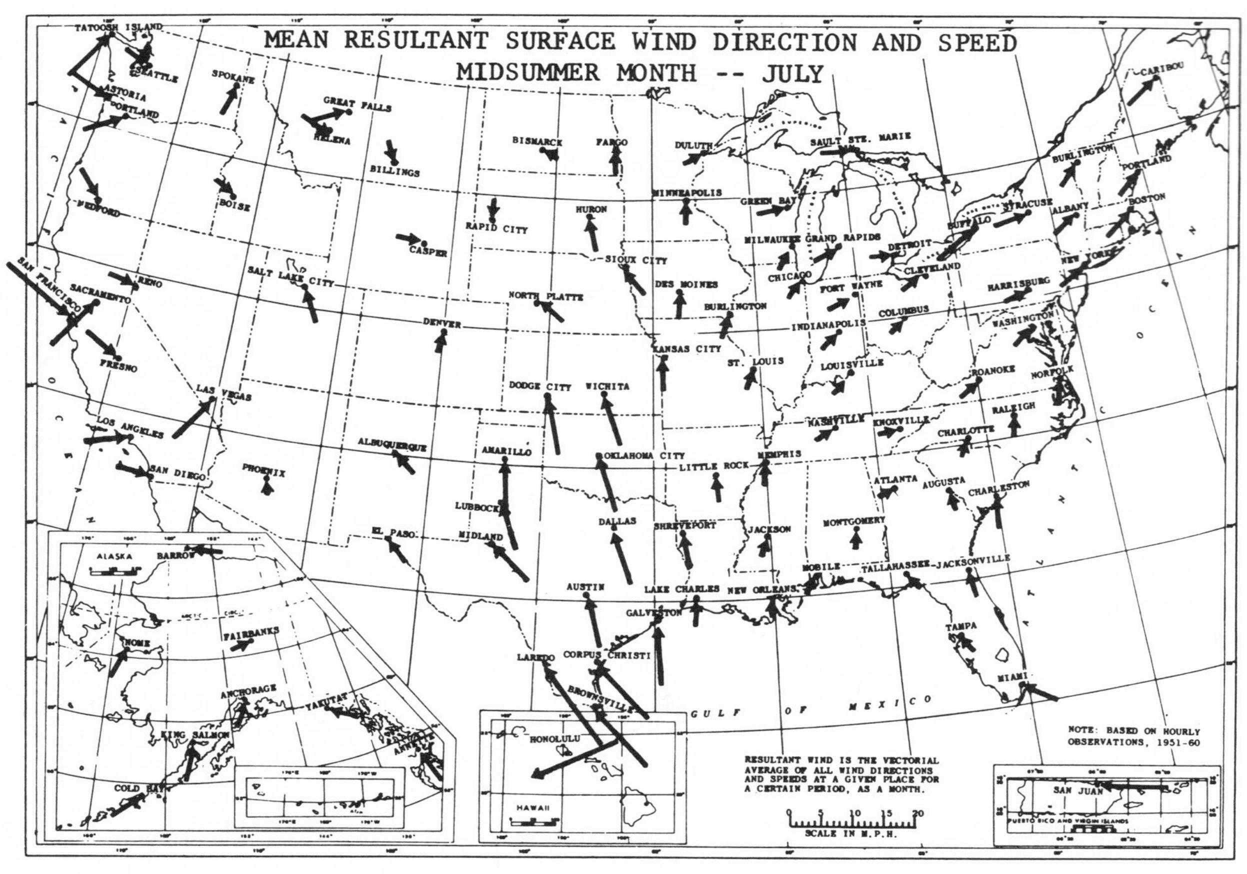
MEAN RESULTANT SURFACE WIND DIRECTION AND SPEED
MIDSUMMER MONTH -- JULY
TATOOSH ISLAND
SEATTLE
SPOKANE
ASTORIA
PORTLAND
GREAT FALLS
HELENA
BILLINGS
BISMARCK
FARGO
DULUTH
SAULT STE. MARIE
CARIBOU
BURLINGTON
PORTLAND
MEDFORD
BOISE
MINNEAPOLIS
GREEN BAY
SYRACUSE
ALBANY
BOSTON
HURON
RAPID CITY
CASPER
BUFFALO
MILWAUKEE
GRAND RAPIDS
DETROIT
NEW YORK
SAN FRANCISCO
SACRAMENTO
RENO
SALT LAKE CITY
SIOUX CITY
DES MOINES
CHICAGO
CLEVELAND
HARRISBURG
NORTH PLATTE
FORT WAYNE
BURLINGTON
COLUMBUS
DENVER
INDIANAPOLIS
WASHINGTON
FRESNO
KANSAS CITY
ST. LOUIS
LOUISVILLE
ROANOKE
NORFOLK
LAS VEGAS
DODGE CITY
WICHITA
LOS ANGELES
NASHVILLE
KNOXVILLE
RALEIGH
CHARLOTTE
ALBUQUERQUE
AMARILLO
OKLAHOMA CITY
LITTLE ROCK
MEMPHIS
SAN DIEGO
PHOENIX
ATLANTA
AUGUSTA
CHARLESTON
LUBBOCK
DALLAS
SHREVEPORT
JACKSON
MONTGOMERY
EL PASO
MIDLAND
MOBILE
TALLAHASSEE
JACKSONVILLE
AUSTIN
LAKE CHARLES
NEW ORLEANS
GALVESTON
TAMPA
LAREDO
CORPUS CHRISTI
MIAMI
BROWNSVILLE
GULF OF MEXICO
ALASKA
BARROW
NOME
FAIRBANKS
ANCHORAGE
YAKUTAT
KING SALMON
ANNETTE
COLD BAY
HONOLULU
HAWAII
NOTE: BASED ON HOURLY OBSERVATIONS, 1951-60
RESULTANT WIND IS THE VECTORIAL AVERAGE OF ALL WIND DIRECTIONS AND SPEEDS AT A GIVEN PLACE FOR A CERTAIN PERIOD, AS A MONTH.
0 5 10 15 20
SCALE IN M.P.H.
SAN JUAN
PUERTO RICO AND VIRGIN ISLANDS

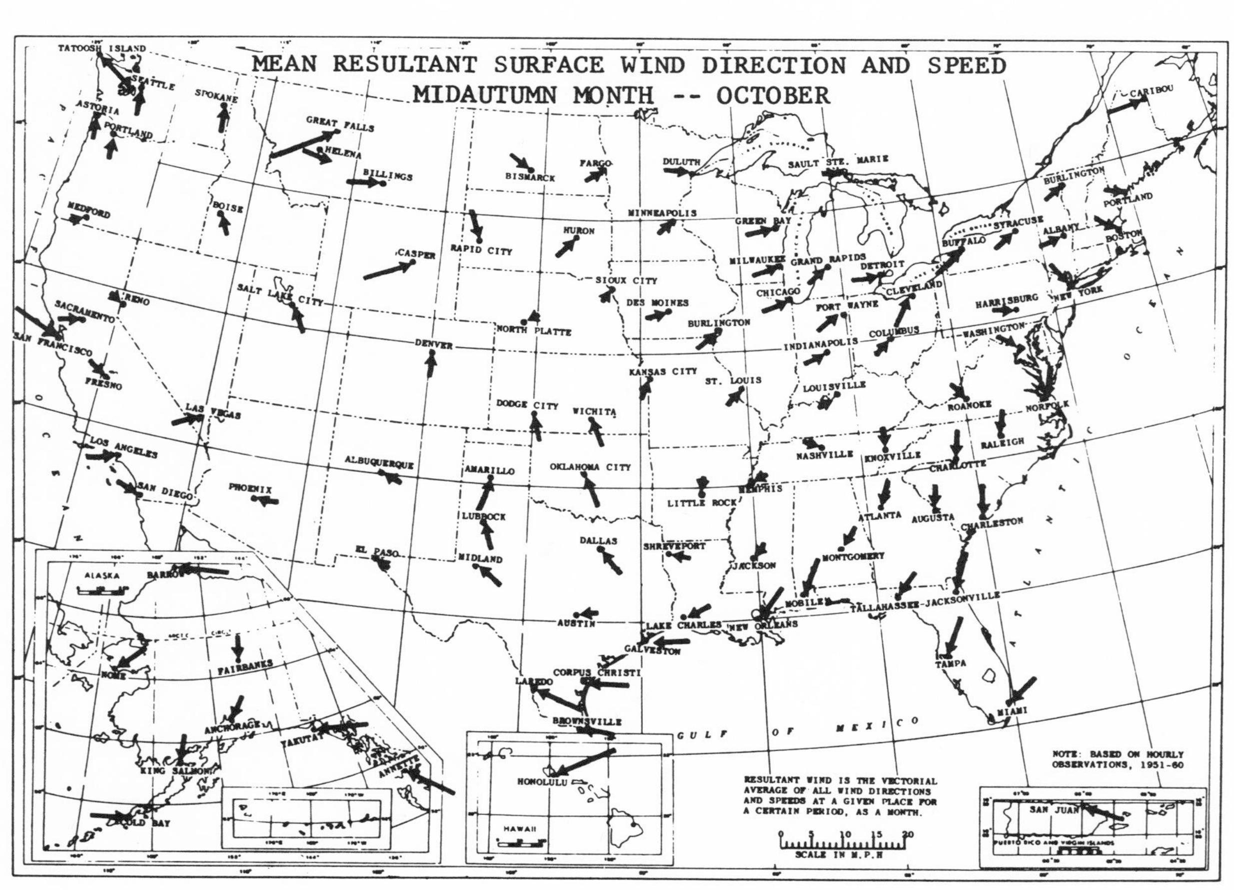
MEAN RESULTANT SURFACE WIND DIRECTION AND SPEED
MIDAUTUMN MONTH -- OCTOBER
TATOOSH ISLAND
SEATTLE
SPOKANE
ASTORIA
PORTLAND
GREAT FALLS
HELENA
BILLINGS
BISMARCK
FARGO
DULUTH
SAULT STE. MARIE
CARIBOU
BURLINGTON
PORTLAND
MEDFORD
BOISE
MINNEAPOLIS
GREEN BAY
SYRACUSE
ALBANY
BOSTON
HURON
RAPID CITY
CASPER
BUFFALO
MILWAUKEE
GRAND RAPIDS
DETROIT
SIOUX CITY
SACRAMENTO
RENO
SALT LAKE CITY
DES MOINES
CHICAGO
CLEVELAND
HARRISBURG
NEW YORK
SAN FRANCISCO
NORTH PLATTE
BURLINGTON
FORT WAYNE
INDIANAPOLIS
COLUMBUS
WASHINGTON
DENVER
FRESNO
KANSAS CITY
ST. LOUIS
LOUISVILLE
ROANOKE
NORFOLK
LAS VEGAS
DODGE CITY
WICHITA
LOS ANGELES
NASHVILLE
KNOXVILLE
RALEIGH
ALBUQUERQUE
AMARILLO
OKLAHOMA CITY
CHARLOTTE
SAN DIEGO
PHOENIX
LITTLE ROCK
MEMPHIS
LUBBOCK
ATLANTA
AUGUSTA
CHARLESTON
DALLAS
SHREVEPORT
EL PASO
JACKSON
MONTGOMERY
MIDLAND
MOBILE
TALLAHASSEE-JACKSONVILLE
AUSTIN
LAKE CHARLES
NEW ORLEANS
GALVESTON
TAMPA
LAREDO
CORPUS CHRISTI
MIAMI
BROWNSVILLE
GULF OF MEXICO
ALASKA
BARROW
NOME
FAIRBANKS
ANCHORAGE
YAKUTAT
KING SALMON
ANNETTE
COLD BAY
HONOLULU
HAWAII
NOTE: BASED ON HOURLY OBSERVATIONS, 1951-60
RESULTANT WIND IS THE VECTORIAL AVERAGE OF ALL WIND DIRECTIONS AND SPEEDS AT A GIVEN PLACE FOR A CERTAIN PERIOD, AS A MONTH.
0 5 10 15 20
SCALE IN M.P.H.
SAN JUAN
PUERTO RICO AND VIRGIN ISLANDS

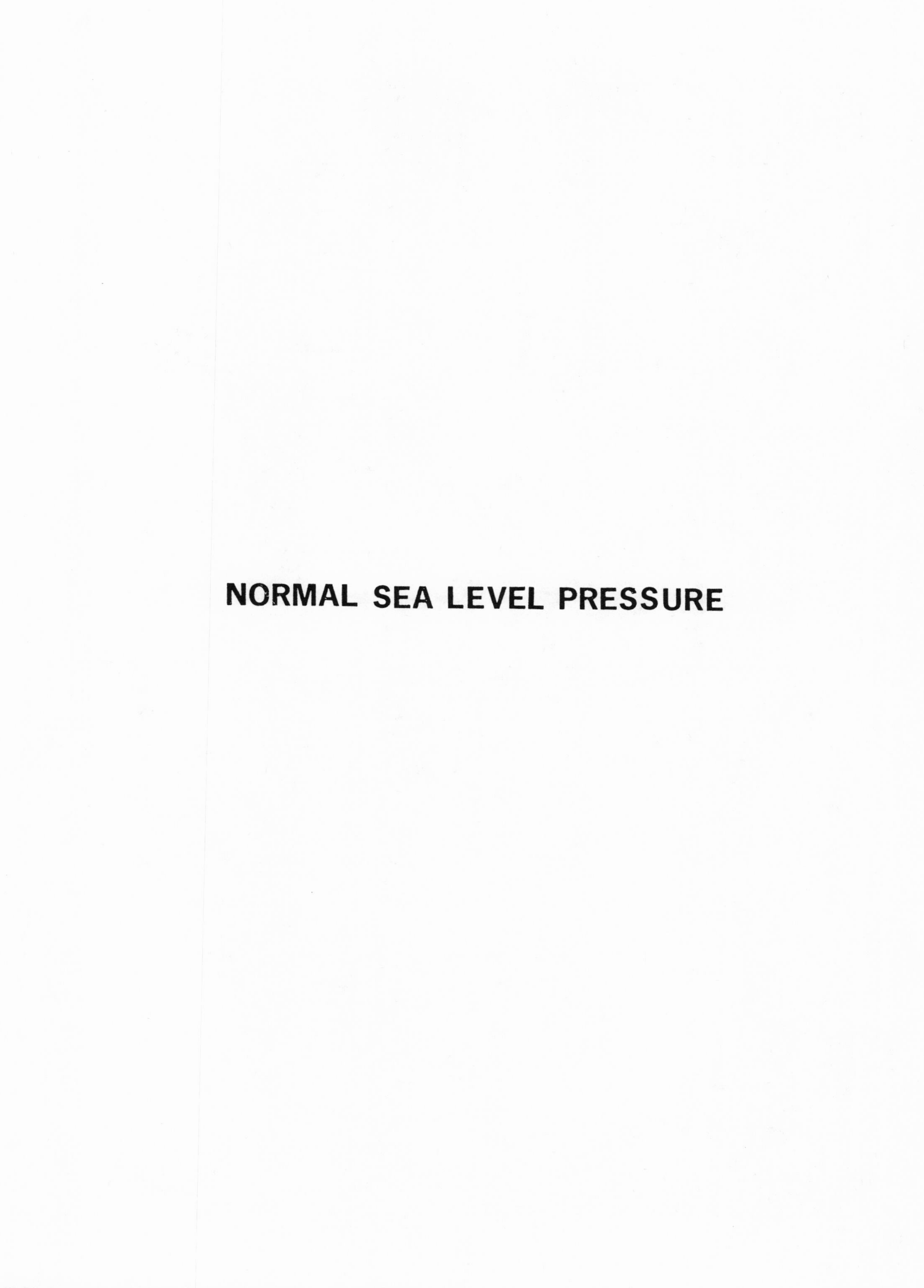

NORMAL SEA LEVEL PRESSURE

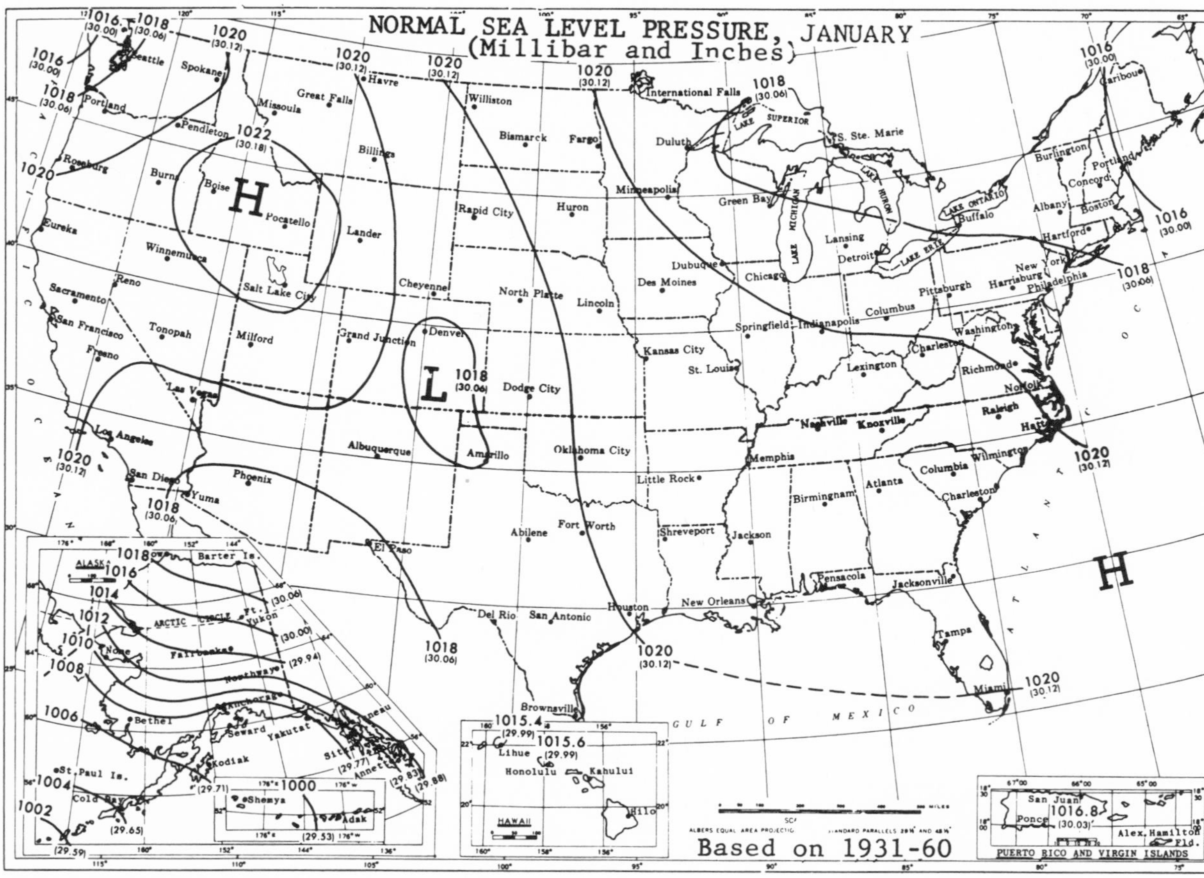

NORMAL SEA LEVEL PRESSURE, JANUARY
(Millibar and Inches)

Based on 1931-60

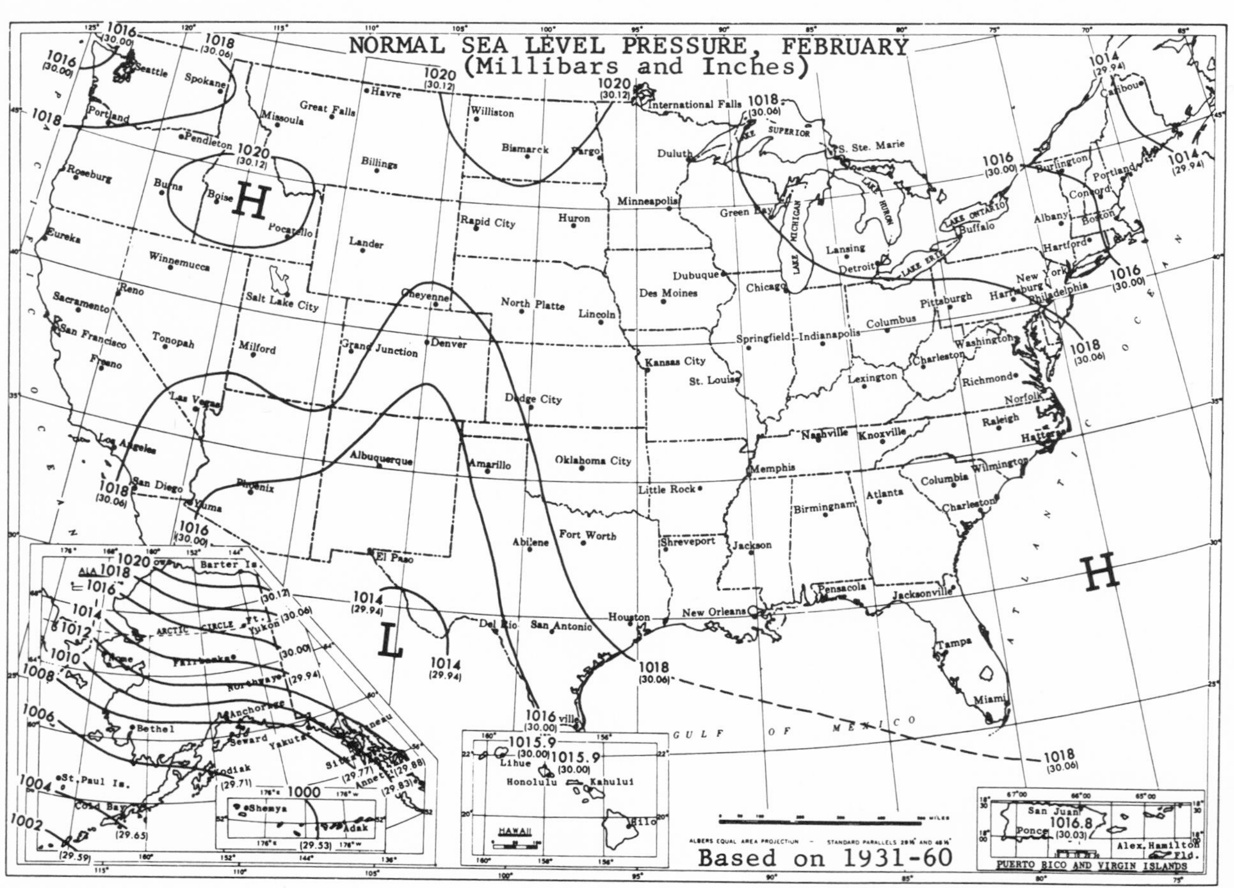

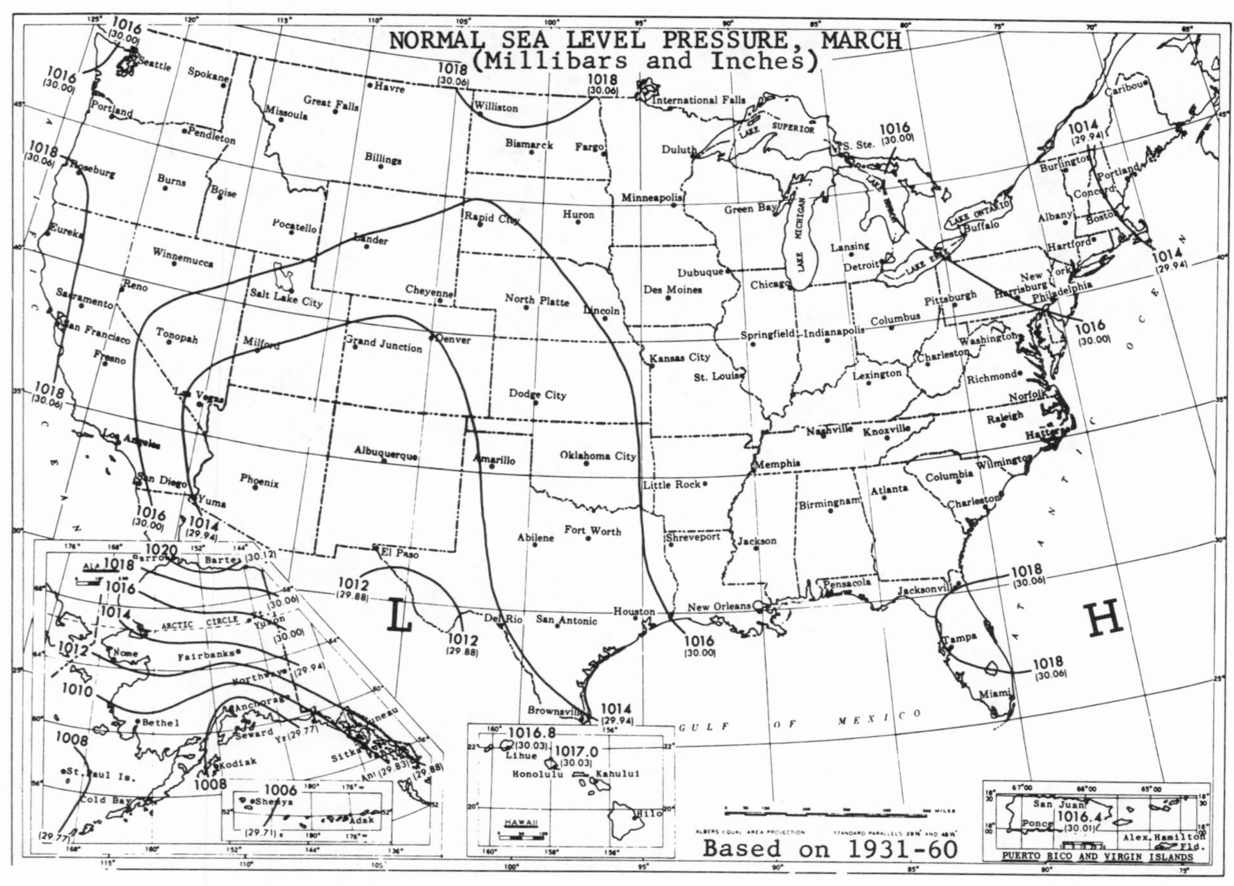
NORMAL SEA LEVEL PRESSURE, MARCH
(Millibars and Inches)
Based on 1931-60
GULF OF MEXICO
HAWAII
PUERTO RICO AND VIRGIN ISLANDS

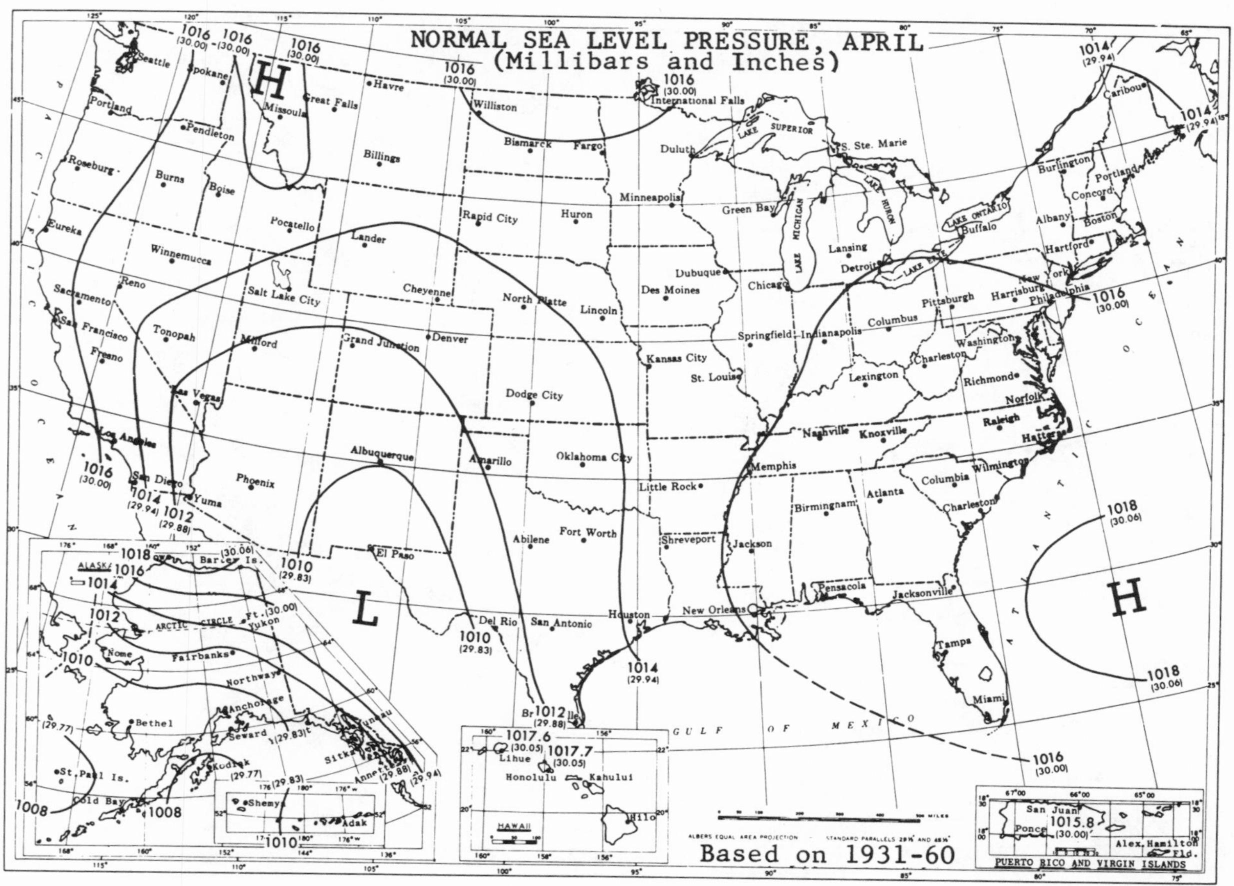
NORMAL SEA LEVEL PRESSURE, APRIL
(Millibars and Inches)
Based on 1931-60
GULF OF MEXICO
HAWAII
PUERTO RICO AND VIRGIN ISLANDS

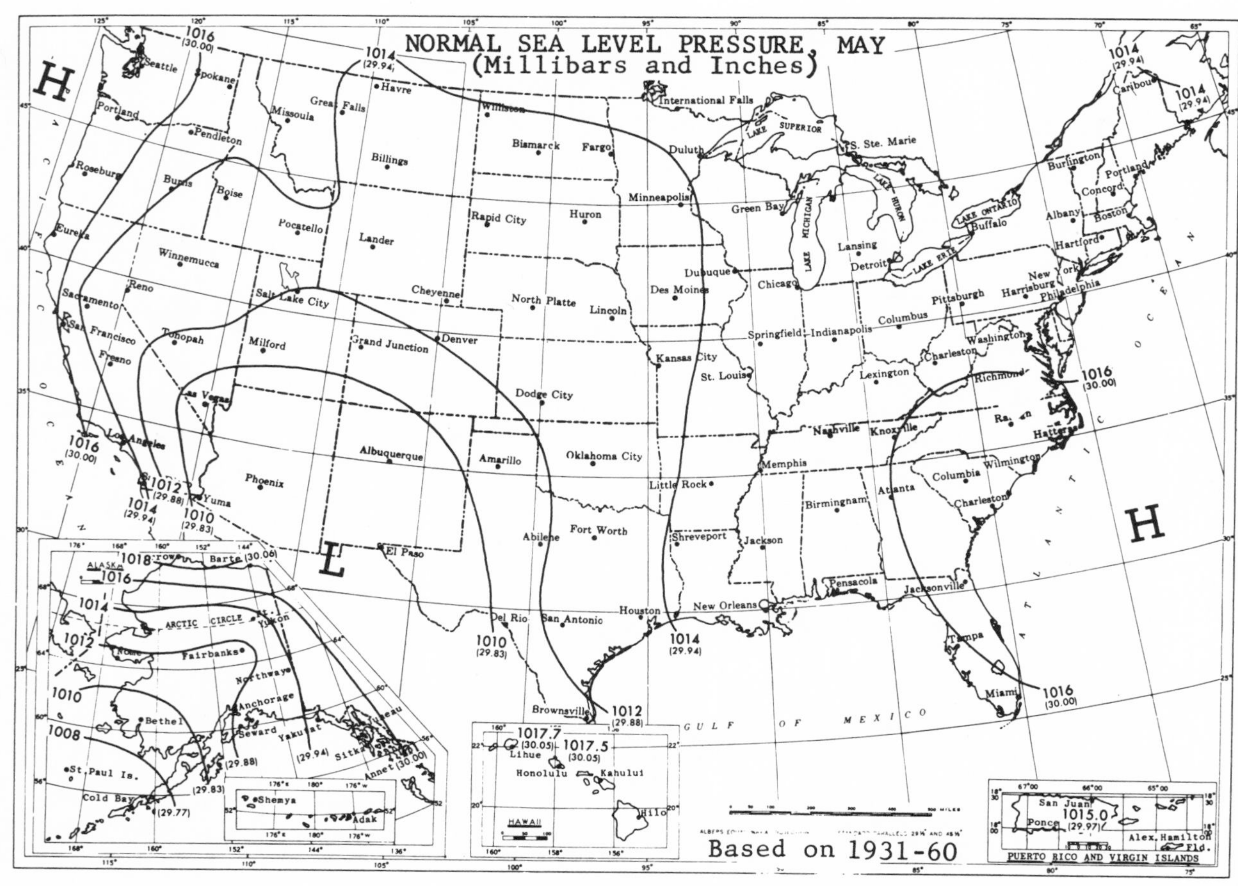
NORMAL SEA LEVEL PRESSURE, MAY
(Millibars and Inches)
H
L
H
1016 (30.00)
1014 (29.94)
1012 (29.88)
1010 (29.83)
1016 (30.00)
1014 (29.94)
1016 (30.00)
Seattle
Spokane
Portland
Pendleton
Roseburg
Burns
Boise
Eureka
Winnemucca
Reno
Sacramento
San Francisco
Fresno
Tonopah
Milford
Las Vegas
Los Angeles
Yuma
Phoenix
Havre
Great Falls
Missoula
Billings
Pocatello
Lander
Salt Lake City
Cheyenne
Grand Junction
Denver
Albuquerque
El Paso
Williston
Bismarck
Fargo
Rapid City
Huron
North Platte
Lincoln
Dodge City
Amarillo
Oklahoma City
Abilene
Fort Worth
Del Rio
San Antonio
Houston
Brownsville
International Falls
Duluth
Minneapolis
Des Moines
Dubuque
Kansas City
St. Louis
Springfield
Chicago
Green Bay
Lansing
Detroit
Indianapolis
Columbus
Lexington
Nashville
Knoxville
Memphis
Little Rock
Shreveport
Jackson
Birmingham
Atlanta
New Orleans
Pensacola
Jacksonville
Tampa
Miami
S. Ste. Marie
Buffalo
Pittsburgh
Harrisburg
Philadelphia
New York
Hartford
Boston
Albany
Concord
Portland
Burlington
Caribou
Washington
Charleston
Richmond
Raleigh
Hatteras
Wilmington
Columbia
LAKE SUPERIOR
LAKE MICHIGAN
LAKE HURON
LAKE ONTARIO
LAKE ERIE
GULF OF MEXICO
ATLANTIC OCEAN
PACIFIC OCEAN
ALASKA
Barrow
Barter
ARCTIC CIRCLE
Nome
Fairbanks
Northway
Anchorage
Bethel
Seward
Yakutat
Sitka
Annette
St. Paul Is.
Cold Bay
Shemya
Adak
1018
1016
1014
1012
1010
1008
(30.06)
(29.94)
(29.88)
(29.83)
(29.77)
(30.00)
HAWAII
Lihue
Honolulu
Kahului
Hilo
1017.7 (30.05)
1017.5 (30.05)
PUERTO RICO AND VIRGIN ISLANDS
San Juan
Ponce
1015.0 (29.97)
Alex. Hamilton Fld.
Based on 1931-60

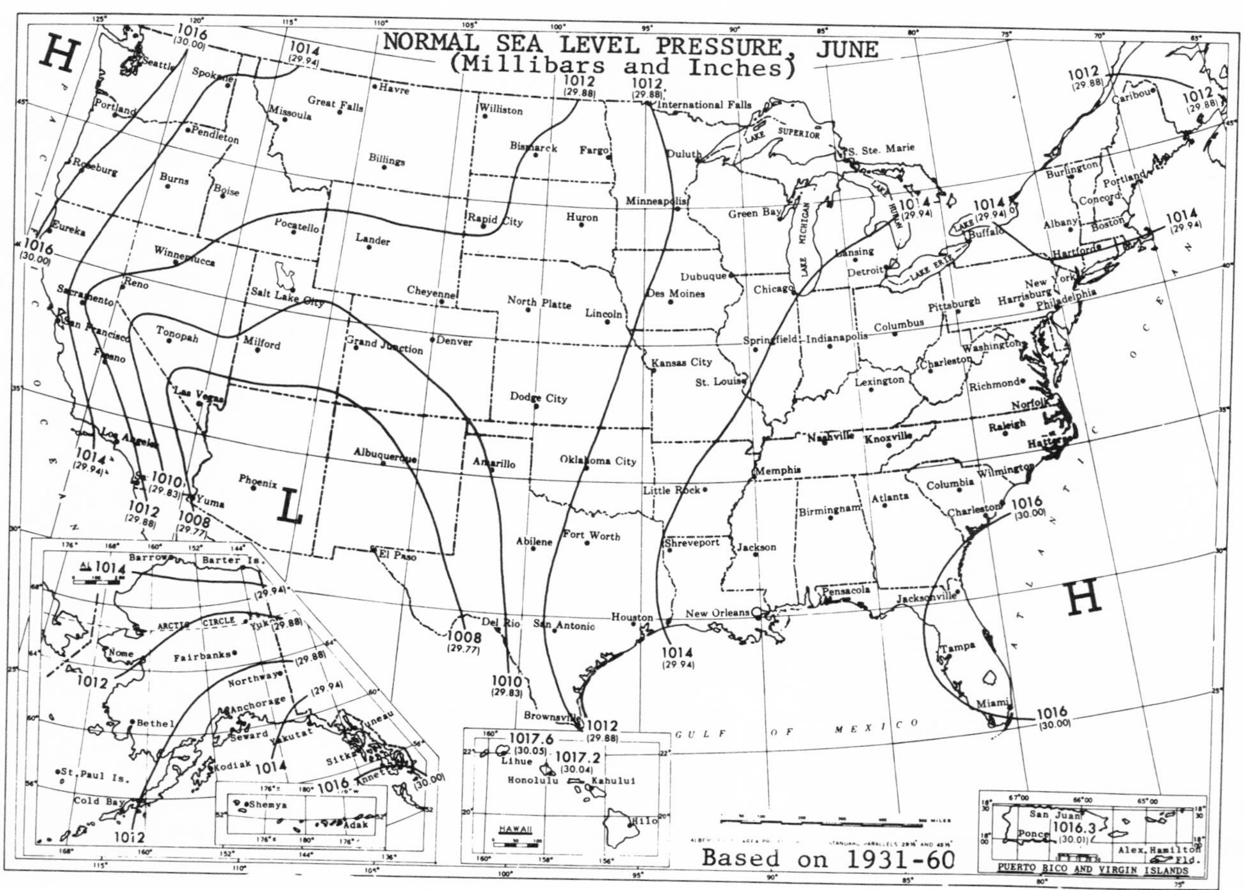
NORMAL SEA LEVEL PRESSURE, JUNE
(Millibars and Inches)
H
L
H
1016 (30.00)
1014 (29.94)
1012 (29.88)
1012 (29.88)
1016 (30.00)
1014 (29.94)
1010 (29.83)
1012 (29.88)
1008 (29.77)
1008 (29.77)
1010 (29.83)
1012 (29.88)
1014 (29.94)
1016 (30.00)
1012 (29.88)
1012 (29.88)
1014 (29.94)
1014 (29.94)
1014 (29.94)
1016 (30.00)
Seattle
Spokane
Portland
Pendleton
Roseburg
Burns
Boise
Eureka
Winnemucca
Reno
Sacramento
San Francisco
Fresno
Tonopah
Milford
Las Vegas
Los Angeles
Yuma
Phoenix
Havre
Great Falls
Missoula
Billings
Pocatello
Lander
Salt Lake City
Cheyenne
Grand Junction
Denver
Albuquerque
El Paso
Williston
Bismarck
Fargo
Rapid City
Huron
North Platte
Lincoln
Dodge City
Amarillo
Oklahoma City
Abilene
Fort Worth
Del Rio
San Antonio
Houston
Brownsville
International Falls
Duluth
Minneapolis
Des Moines
Dubuque
Kansas City
St. Louis
Springfield
Chicago
Green Bay
Lansing
Detroit
Indianapolis
Columbus
Lexington
Nashville
Knoxville
Memphis
Little Rock
Shreveport
Jackson
Birmingham
Atlanta
New Orleans
Pensacola
Jacksonville
Tampa
Miami
S. Ste. Marie
Buffalo
Pittsburgh
Harrisburg
Philadelphia
New York
Hartford
Boston
Albany
Concord
Portland
Burlington
Caribou
Washington
Charleston
Richmond
Norfolk
Raleigh
Hatteras
Wilmington
Columbia
LAKE SUPERIOR
LAKE MICHIGAN
LAKE HURON
LAKE ERIE
GULF OF MEXICO
Barrow
Barter Is.
ARCTIC CIRCLE
Yukon
Nome
Fairbanks
Northway
Anchorage
Bethel
Seward
Yakutat
Sitka
Kodiak
Annette
Juneau
St. Paul Is.
Cold Bay
Shemya
Adak
1014
1012
1014
1016
1012
(29.94)
(29.88)
(29.88)
(29.94)
(30.00)
HAWAII
Lihue
Honolulu
Kahului
Hilo
1017.6 (30.05)
1017.2 (30.04)
PUERTO RICO AND VIRGIN ISLANDS
San Juan
Ponce
1016.3 (30.01)
Alex. Hamilton Fld.
Based on 1931-60

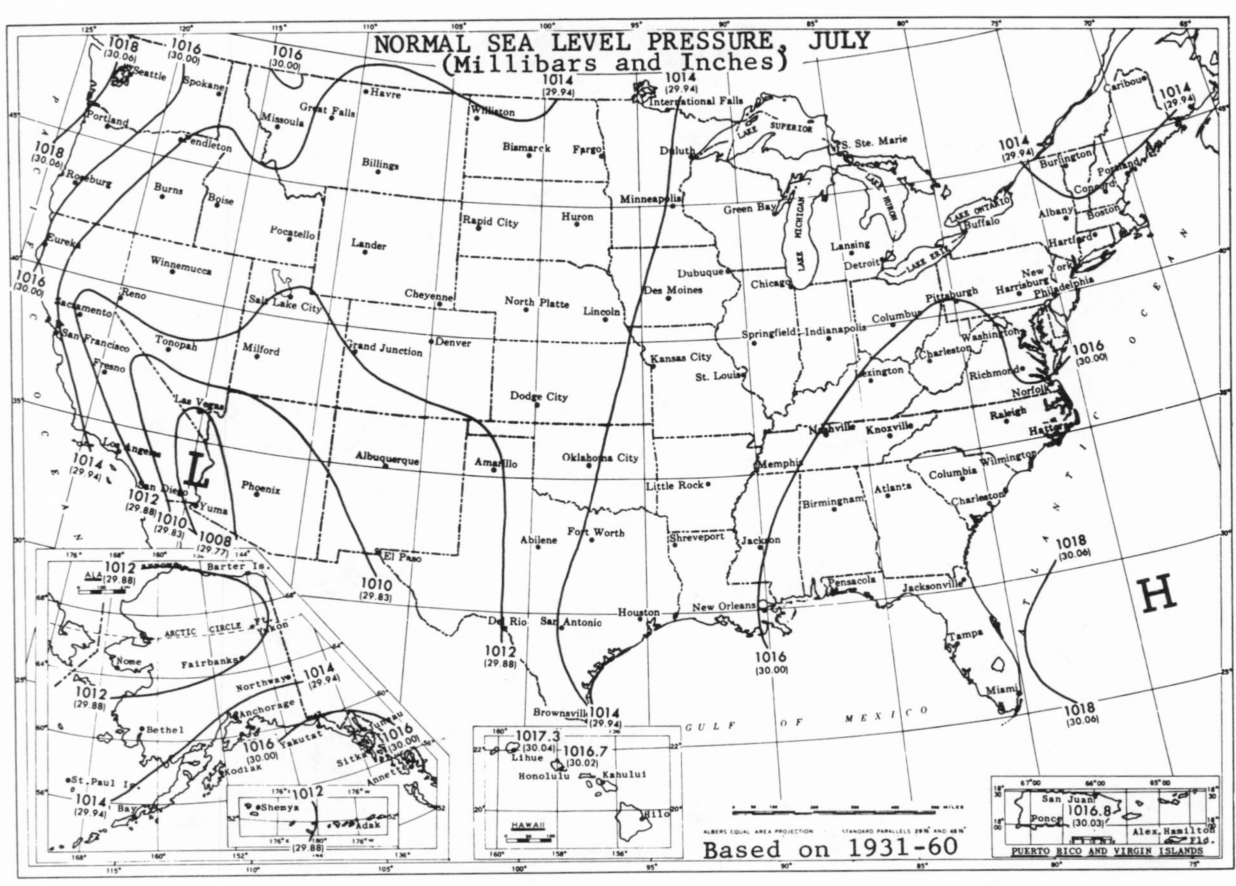
NORMAL SEA LEVEL PRESSURE, JULY
(Millibars and Inches)
1018 (30.06)
1016 (30.00)
1014 (29.94)
1012 (29.88)
1010 (29.83)
1008 (29.77)
L
H
Seattle
Spokane
Portland
Pendleton
Missoula
Great Falls
Havre
Williston
International Falls
Bismarck
Fargo
Duluth
Billings
Roseburg
Burns
Boise
Pocatello
Lander
Rapid City
Huron
Minneapolis
Green Bay
S. Ste. Marie
Eureka
Winnemucca
Reno
Sacramento
San Francisco
Fresno
Tonopah
Salt Lake City
Milford
Cheyenne
North Platte
Lincoln
Des Moines
Dubuque
Chicago
Lansing
Detroit
Buffalo
Albany
Boston
Burlington
Portland
Concord
Caribou
Hartford
New York
Philadelphia
Harrisburg
Pittsburgh
Columbus
Indianapolis
Springfield
Grand Junction
Denver
Kansas City
St. Louis
Lexington
Charleston
Washington
Richmond
Norfolk
Las Vegas
Dodge City
Raleigh
Hatteras
Nashville
Knoxville
Los Angeles
San Diego
Yuma
Phoenix
Albuquerque
Amarillo
Oklahoma City
Little Rock
Memphis
Wilmington
Columbia
Atlanta
Birmingham
Charleston
Abilene
Fort Worth
Shreveport
Jackson
El Paso
Pensacola
Jacksonville
Del Rio
San Antonio
Houston
New Orleans
Tampa
Miami
Brownsville
GULF OF MEXICO
PACIFIC OCEAN
ATLANTIC OCEAN
LAKE SUPERIOR
LAKE MICHIGAN
LAKE HURON
LAKE ONTARIO
LAKE ERIE
ALASKA
Barter Is.
ARCTIC CIRCLE
Ft. Yukon
Nome
Fairbanks
Northway
Anchorage
Bethel
Yakutat
Juneau
Sitka
Annette
Kodiak
St. Paul Is.
Cold Bay
Shemya
Adak
HAWAII
1017.3 (30.04)
1016.7 (30.02)
Lihue
Honolulu
Kahului
Hilo
ALBERS EQUAL AREA PROJECTION
Based on 1931-60
San Juan
Ponce
1016.8 (30.03)
Alex. Hamilton Fld.
PUERTO RICO AND VIRGIN ISLANDS

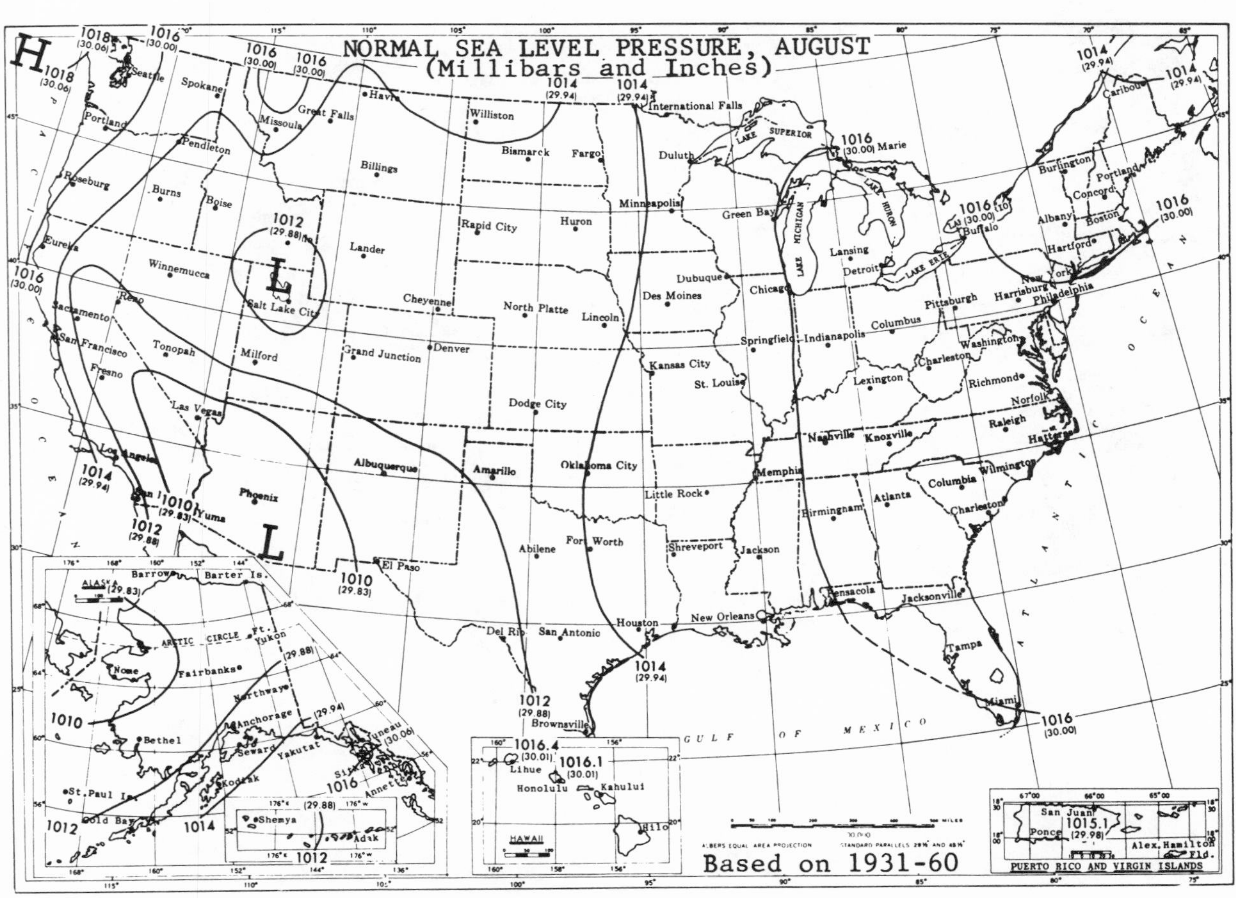
NORMAL SEA LEVEL PRESSURE, AUGUST
(Millibars and Inches)
1018 (30.06)
1016 (30.00)
1014 (29.94)
1012 (29.88)
1010 (29.83)
L
H
Seattle
Spokane
Portland
Pendleton
Missoula
Great Falls
Havre
Williston
International Falls
Bismarck
Fargo
Duluth
Billings
Roseburg
Burns
Boise
Lander
Rapid City
Huron
Minneapolis
Green Bay
Eureka
Winnemucca
Reno
Sacramento
San Francisco
Fresno
Tonopah
Salt Lake City
Milford
Cheyenne
North Platte
Lincoln
Des Moines
Dubuque
Chicago
Lansing
Detroit
Buffalo
Albany
Boston
Burlington
Portland
Concord
Caribou
Hartford
New York
Philadelphia
Harrisburg
Pittsburgh
Columbus
Indianapolis
Springfield
Grand Junction
Denver
Kansas City
St. Louis
Lexington
Charleston
Washington
Richmond
Norfolk
Las Vegas
Dodge City
Raleigh
Hatteras
Nashville
Knoxville
Los Angeles
Yuma
Phoenix
Albuquerque
Amarillo
Oklahoma City
Little Rock
Memphis
Wilmington
Columbia
Atlanta
Birmingham
Charleston
Abilene
Fort Worth
Shreveport
Jackson
El Paso
Pensacola
Jacksonville
Del Rio
San Antonio
Houston
New Orleans
Tampa
Miami
Brownsville
GULF OF MEXICO
PACIFIC OCEAN
ATLANTIC OCEAN
LAKE SUPERIOR
LAKE MICHIGAN
LAKE HURON
LAKE ERIE
ALASKA
Barrow
Barter Is.
ARCTIC CIRCLE
Ft. Yukon
Nome
Fairbanks
Northway
Anchorage
Seward
Bethel
Yakutat
Juneau
Sitka
Annette
Kodiak
St. Paul Is.
Cold Bay
Shemya
Adak
HAWAII
1016.4 (30.01)
1016.1 (30.01)
Lihue
Honolulu
Kahului
Hilo
ALBERS EQUAL AREA PROJECTION
Based on 1931-60
San Juan
Ponce
1015.1 (29.98)
Alex. Hamilton Fld.
PUERTO RICO AND VIRGIN ISLANDS

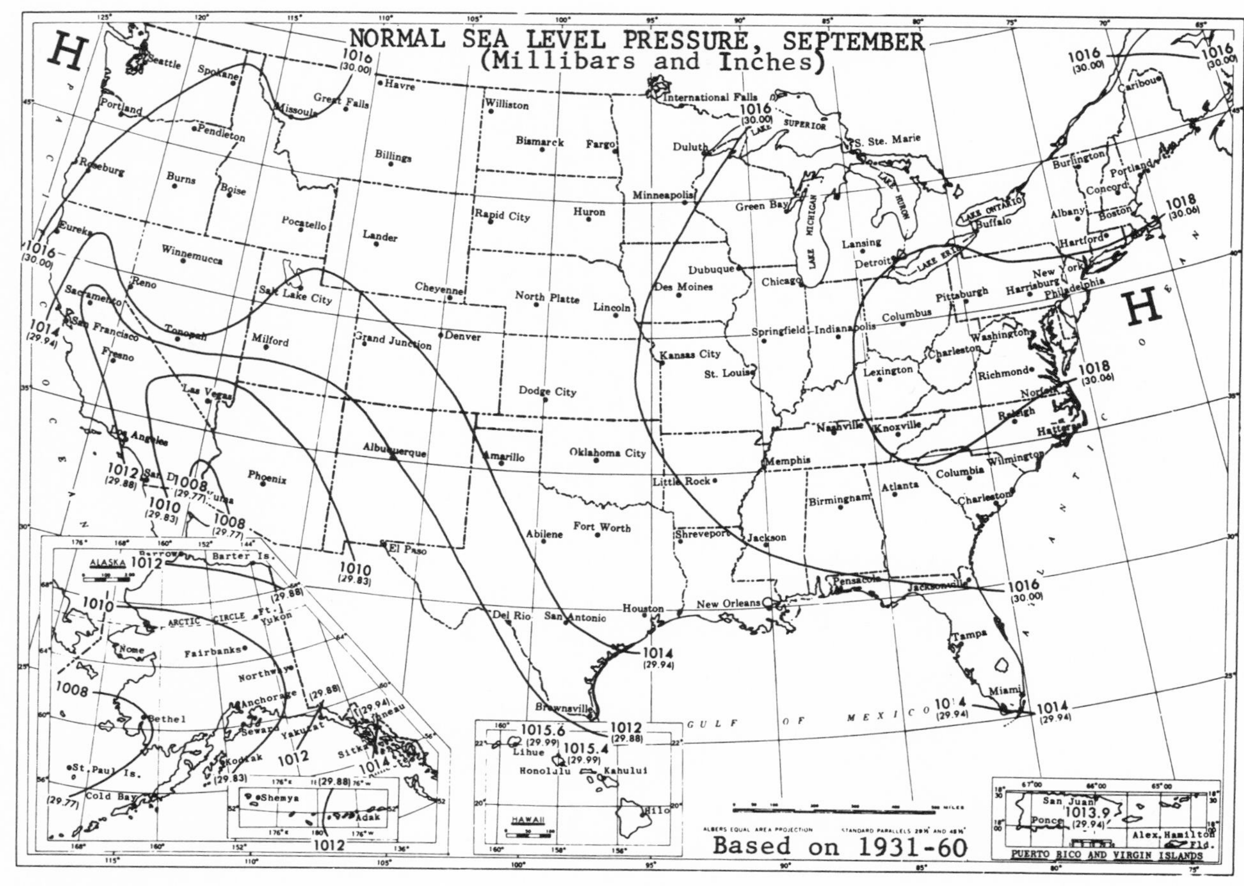
NORMAL SEA LEVEL PRESSURE, SEPTEMBER
(Millibars and Inches)
Based on 1931-60

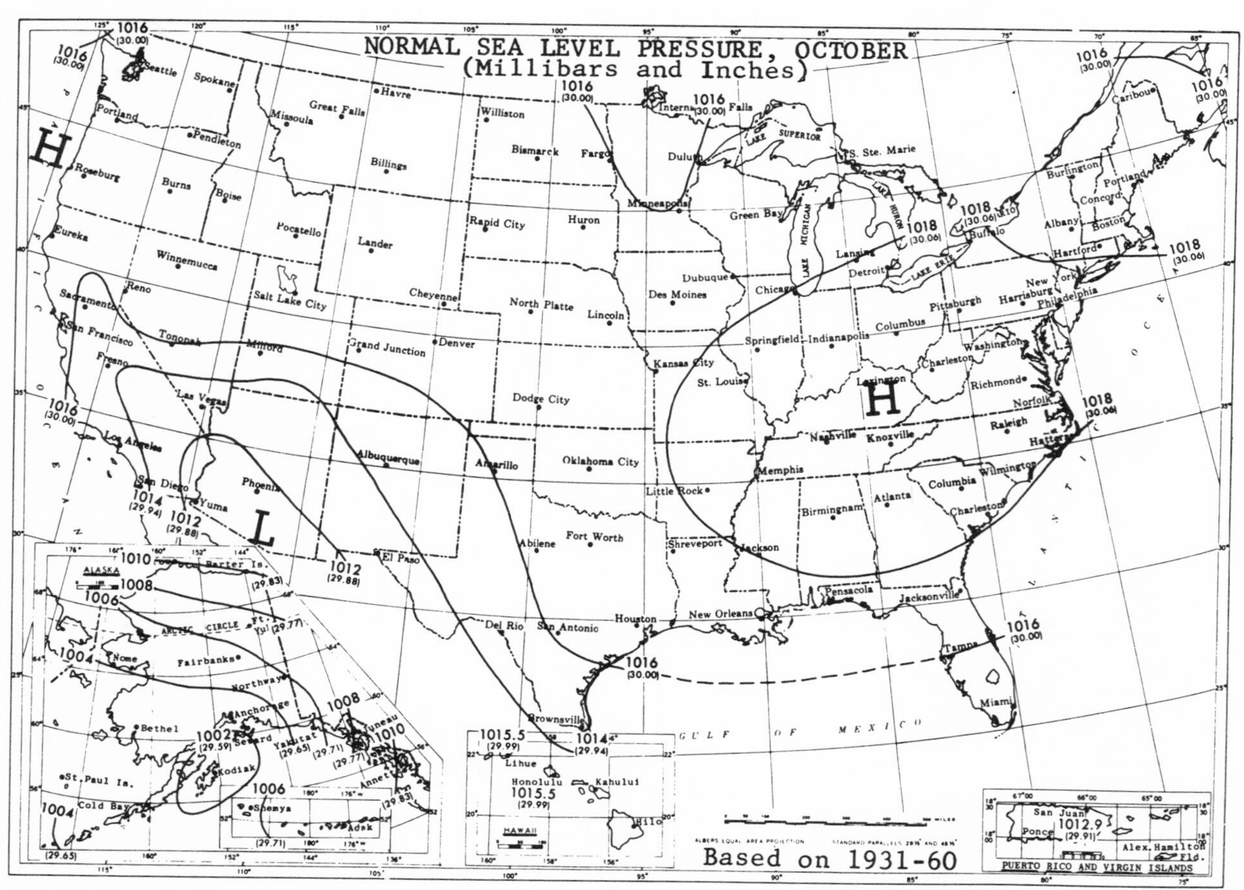
NORMAL SEA LEVEL PRESSURE, OCTOBER
(Millibars and Inches)
Based on 1931-60

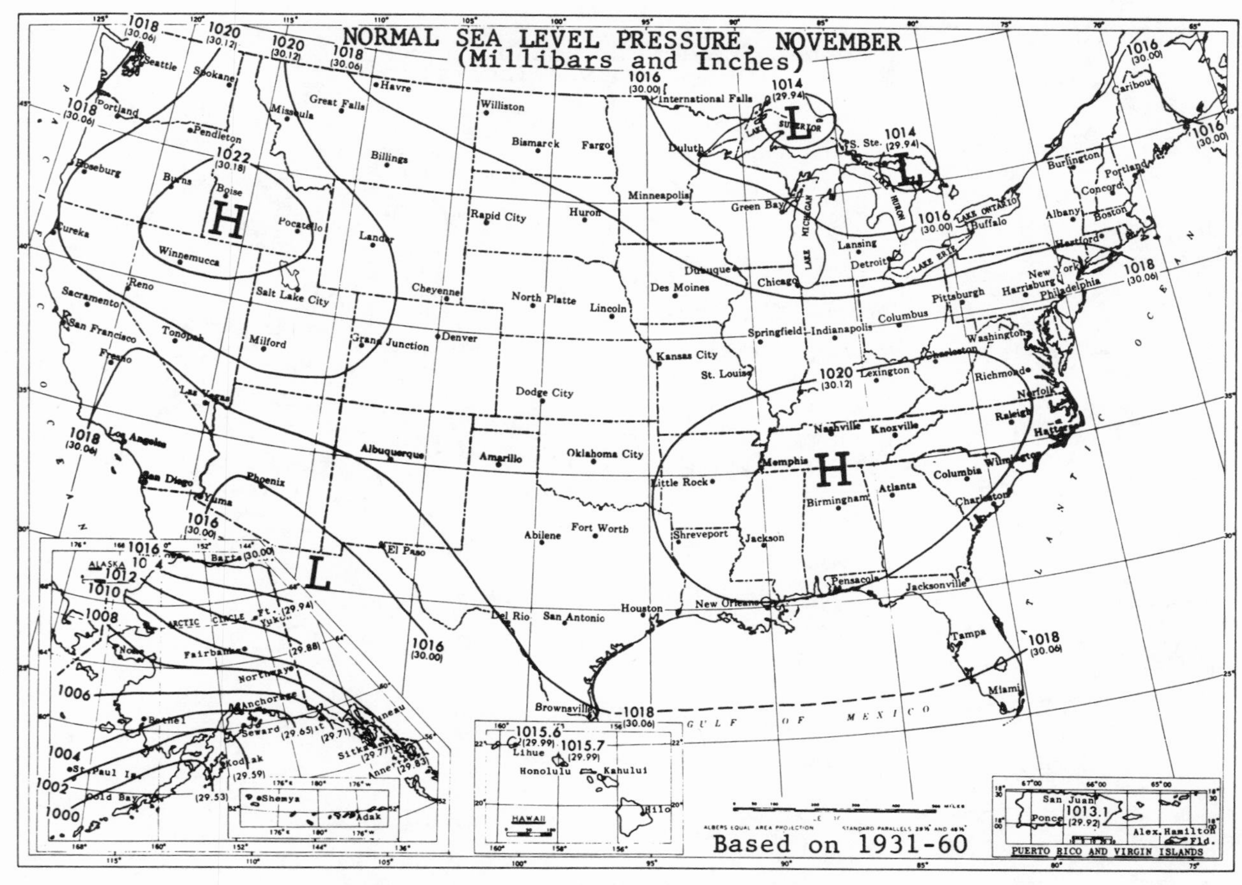
NORMAL SEA LEVEL PRESSURE, NOVEMBER
(Millibars and Inches)
Based on 1931-60
GULF OF MEXICO
HAWAII
PUERTO RICO AND VIRGIN ISLANDS
ALASKA

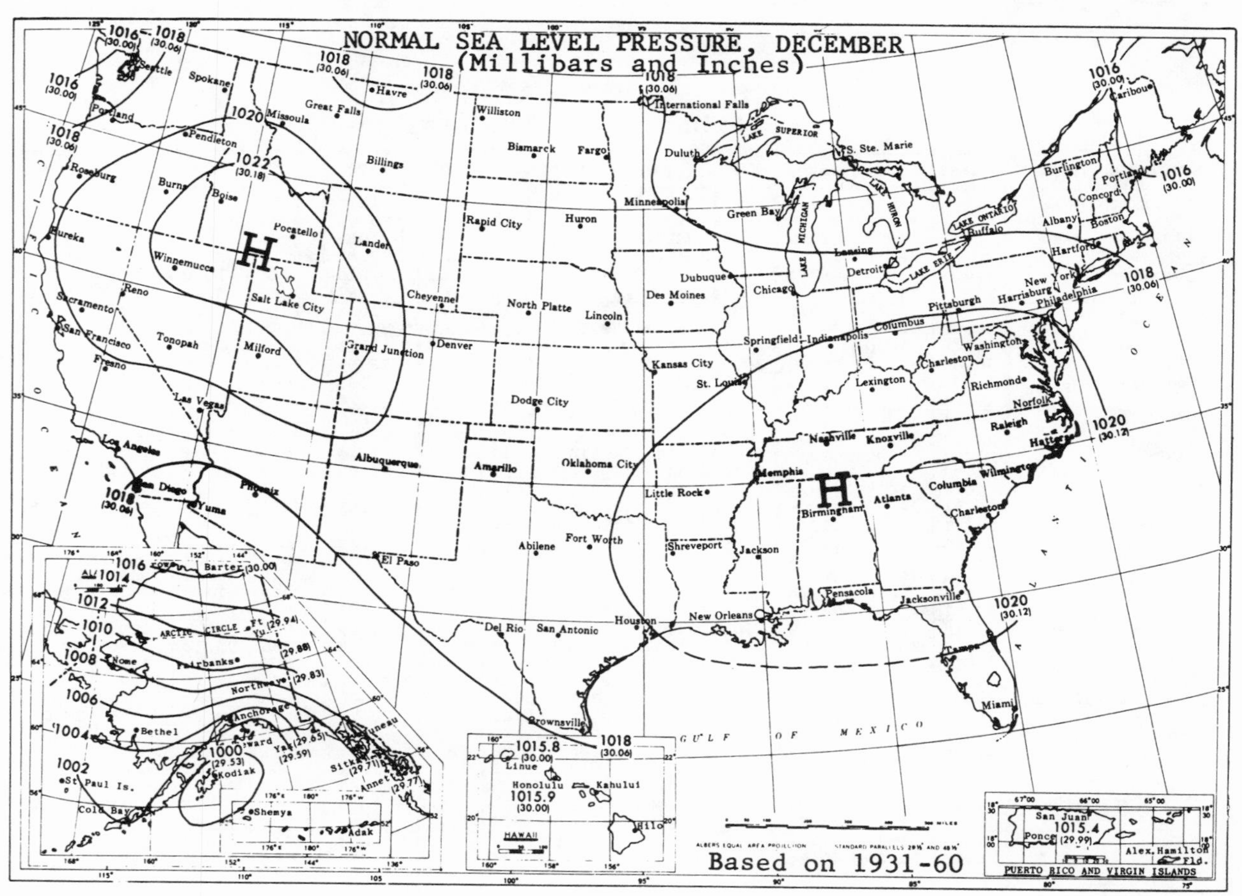
NORMAL SEA LEVEL PRESSURE, DECEMBER
(Millibars and Inches)
Based on 1931-60
GULF OF MEXICO
HAWAII
PUERTO RICO AND VIRGIN ISLANDS

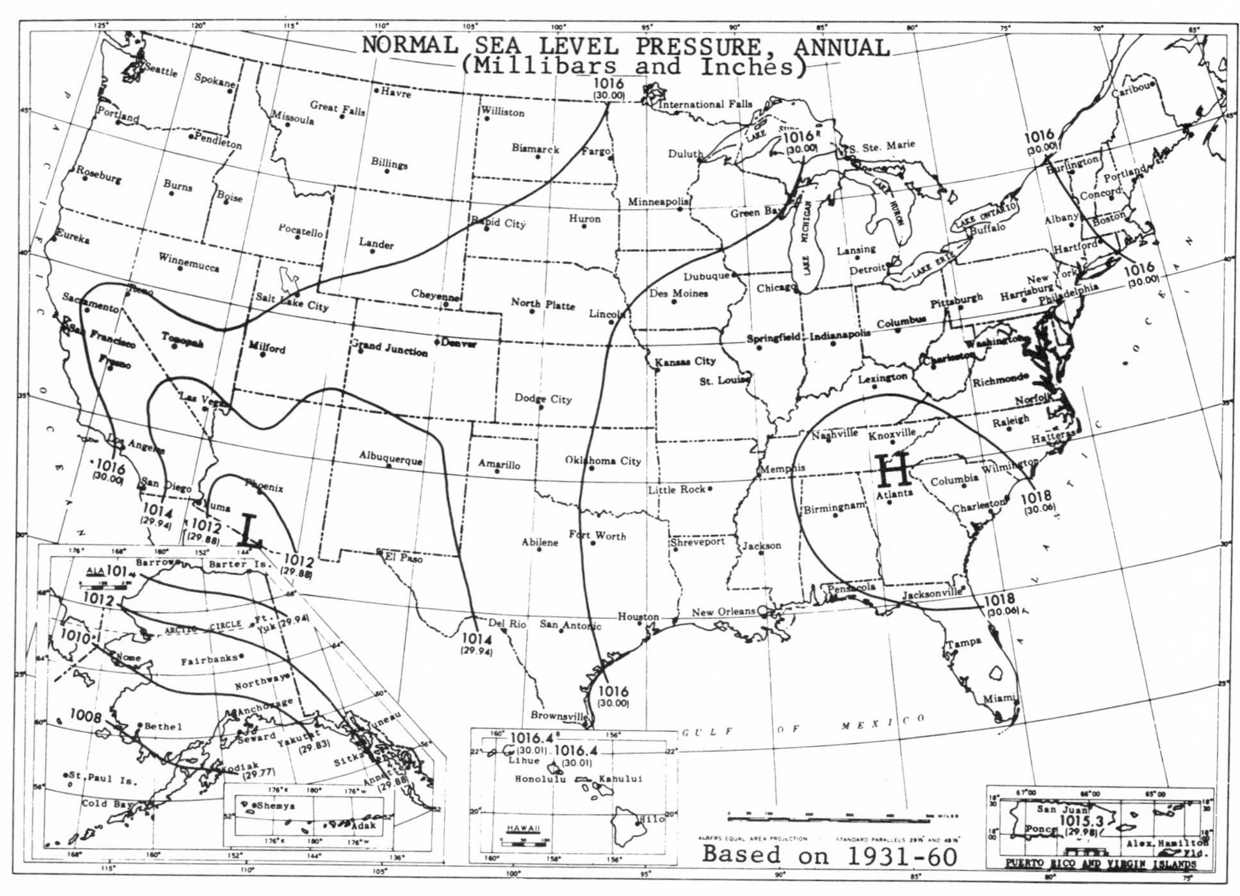

MEAN SEA LEVEL PRESSURE
(Millibars and Inches)

STATE AND STATION	JAN.	FEB.	MAR.	APR.	MAY	JUNE	JULY	AUG.	SEPT.	OCT.	NOV.	DEC.	ANNUAL
ALA. Birmingham	1021	1019	1018	1017	1016	1016	1017	1017	1017	1019	1021	1021	1018
"	30.15	30.19	30.06	30.03	30.00	30.00	30.03	30.03	30.03	30.09	30.15	30.15	30.06
ALASKA Anchorage	1007	1008	1008	1009	1012	1014	1014	1012	1009	1002	1004	1002	1009
"	29.74	29.77	29.77	29.80	29.88	29.94	29.94	29.88	29.80	29.59	29.65	29.59	29.80
Annette	1010	1012	1011	1014	1016	1016	1018	1017	1015	1011	1010	1008	1013
"	29.83	29.88	29.86	29.94	30.00	30.00	30.06	30.03	29.97	29.86	29.83	29.77	29.91
Barrow	1019	1021	1021	1018	1018	1015	1012	1011	1013	1010	1016	1016	1016
"	30.09	30.15	30.15	30.06	30.06	29.97	29.88	29.86	29.91	29.83	30.00	30.00	30.00
Barter Island	1019	1020	1019	1018	1018	1015	1012	1011	1013	1010	1016	1016	1016
"	30.09	30.12	30.09	30.06	30.06	29.97	29.88	29.86	29.91	29.83	30.00	30.00	30.00
Bethel	1007	1008	1010	1009	1009	1012	1013	1011	1008	1003	1005	1005	1008
"	29.74	29.77	29.83	29.80	29.80	29.88	29.91	29.86	29.77	29.62	29.68	29.68	29.77
Cold Bay	1004	1005	1010	1008	1007	1012	1016	1013	1008	1003	999	1002	1007
"	29.65	29.68	29.83	29.77	29.74	29.88	30.00	29.91	29.77	29.62	29.50	29.59	29.74
Cordova	1005	1006	1007	1009	1013	1014	1015	1014	1010	1003	1003	1001	1008
"	29.68	29.71	29.74	29.80	29.91	29.94	29.97	29.94	29.83	29.62	29.62	29.56	29.77
Fairbanks	1014	1014	1013	1011	1011	1011	1012	1012	1010	1006	1011	1010	1011
"	29.94	29.94	29.91	29.86	29.86	29.86	29.88	29.88	29.83	29.71	29.86	29.83	29.86
Juneau	1011	1012	1011	1012	1015	1015	1017	1016	1014	1009	1009	1008	1012
"	29.86	29.88	29.86	29.88	29.97	29.97	30.03	30.00	29.94	29.80	29.80	29.77	29.88
King Salmon	1007	1006	1009	1008	1009	1012	1014	1012	1009	1002	1002	1003	1008
"	29.74	29.71	29.80	29.77	29.80	29.88	29.94	29.88	29.80	29.59	29.59	29.62	29.77
Kotzebue	1015	1015	1014	1013	1013	1012	1011	1009	1010	1005	1010	1011	1012
"	29.97	29.97	29.94	29.91	29.91	29.88	29.86	29.80	29.83	29.68	29.83	29.86	29.88
McGrath	1013	1013	1012	1010	1011	1012	1013	1011	1009	1004	1008	1009	1010
"	29.91	29.91	29.88	29.83	29.86	29.88	29.91	29.86	29.80	29.65	29.77	29.80	29.83
Nome	1011	1011	1012	1011	1011	1012	1012	1009	1008	1004	1007	1008	1010
"	29.86	29.86	29.88	29.86	29.86	29.88	29.88	29.80	29.77	29.65	29.74	29.77	29.83
St. Paul Island	1005	1004	1008	1007	1007	1011	1013	1011	1008	1003	1003	1002	1007
"	29.68	29.65	29.77	29.74	29.74	29.86	29.91	29.86	29.77	29.62	29.62	29.59	29.74
Shemya Island	999	999	1006	1011	1007	1012	1011	1011	1013	1006	1000	999	1006
"	29.50	29.50	29.71	29.86	29.74	29.88	29.86	29.86	29.91	29.71	29.53	29.50	29.71
Yakutat	1007	1009	1008	1011	1014	1015	1017	1015	1012	1005	1005	1004	1010
"	29.74	29.80	29.77	29.86	29.94	29.97	30.03	29.97	29.88	29.68	29.68	29.65	29.83
ARIZ. Phoenix	1018	1016	1013	1011	1008	1007	1008	1009	1009	1012	1016	1018	1012
"	30.06	30.00	29.91	29.86	29.77	29.74	29.77	29.80	29.80	29.88	30.00	30.06	29.88
Yuma	1018	1017	1014	1011	1009	1007	1008	1009	1008	1011	1016	1019	1012
"	30.06	30.03	29.94	29.86	29.80	29.74	29.77	29.80	29.77	29.86	30.00	30.09	29.88

MEAN SEA LEVEL PRESSURE
(Millibars and Inches)

STATE AND STATION	JAN.	FEB.	MAR.	APR.	MAY	JUNE	JULY	AUG.	SEPT.	OCT.	NOV.	DEC.	ANNUAL
ARK. Little Rock	1021	1020	1017	1015	1015	1014	1016	1015	1016	1018	1020	1021	1017
" "	30.15	30.12	30.03	29.97	29.97	29.94	30.00	29.97	30.00	30.06	30.12	30.15	30.03
CALIF. Eureka	1020	1019	1019	1019	1019	1018	1018	1017	1016	1018	1020	1020	1019
"	30.12	30.09	30.09	30.09	30.09	30.06	30.06	30.03	30.00	30.06	30.12	30.12	30.09
Los Angeles	1019	1018	1016	1015	1014	1013	1013	1013	1012	1015	1017	1018	1015
" "	30.09	30.06	30.00	29.97	29.94	29.91	29.91	29.91	29.88	29.97	30.03	30.06	29.97
Mt. Shasta	1020	1020	1018	1016	1016	1015	1015	1015	1016	1018	1021	1022	1018
" "	30.12	30.12	30.06	30.00	30.00	29.97	29.97	29.97	30.00	30.06	30.15	30.18	30.06
Oakland	1020	1019	1018	1017	1016	1014	1014	1014	1014	1016	1019	1020	1017
"	30.12	30.09	30.06	30.03	30.00	29.94	29.94	29.94	29.94	30.00	30.09	30.12	30.03
Sacramento	1020	1019	1018	1016	1014	1012	1012	1012	1012	1015	1019	1020	1016
"	30.12	30.09	30.06	30.00	29.94	29.88	29.88	29.88	29.88	29.97	30.09	30.12	30.00
San Diego	1018	1018	1017	1015	1014	1013	1013	1013	1012	1014	1017	1018	1015
" "	30.06	30.06	30.03	29.97	29.94	29.91	29.91	29.91	29.88	29.94	30.03	30.06	29.97
San Francisco	1020	1019	1018	1017	1016	1014	1014	1014	1014	1016	1019	1020	1017
" "	30.12	30.09	30.06	30.03	30.00	29.94	29.94	29.94	29.94	30.00	30.09	30.12	30.03
Santa Maria	1020	1019	1018	1017	1016	1014	1015	1015	1014	1016	1018	1020	1017
" "	30.12	30.09	30.06	30.03	30.00	29.94	29.97	29.97	29.94	30.00	30.06	30.12	30.03
COLO. Denver	1018	1016	1014	1013	1012	1011	1014	1014	1014	1016	1018	1018	1015
"	30.06	30.00	29.94	29.91	29.88	29.86	29.94	29.94	29.94	30.00	30.06	30.06	29.97
Grand Junction	1021	1018	1014	1012	1011	1010	1012	1013	1013	1017	1021	1022	1015
" "	30.15	30.06	29.94	29.88	29.86	29.83	29.88	29.91	29.91	30.03	30.15	30.18	29.97
CONN. New Haven	1016	1016	1015	1015	1015	1014	1015	1016	1018	1018	1017	1017	1016
" "	30.00	30.00	29.97	29.97	29.97	29.94	29.97	30.00	30.06	30.06	30.03	30.03	30.00
D. C. Washington	1020	1019	1017	1016	1016	1015	1016	1016	1019	1019	1020	1020	1018
"	30.12	30.09	30.03	30.00	30.00	29.97	30.00	30.00	30.09	30.09	30.12	30.12	30.06
FLA. Jacksonville	1021	1020	1018	1018	1017	1016	1018	1017	1016	1017	1020	1021	1018
"	30.15	30.12	30.06	30.06	30.03	30.00	30.06	30.03	30.00	30.03	30.12	30.15	30.06
Key West	1019	1018	1017	1016	1015	1015	1017	1015	1013	1014	1017	1019	1016
" "	30.09	30.06	30.03	30.00	29.97	29.97	30.03	29.97	29.91	29.94	30.03	30.09	30.00
Miami	1020	1019	1018	1017	1016	1016	1017	1016	1014	1015	1017	1019	1017
"	30.12	30.09	30.06	30.03	30.00	30.00	30.03	30.00	29.94	29.97	30.03	30.09	30.03
Pensacola	1021	1020	1018	1017	1016	1016	1016	1016	1016	1018	1020	1020	1018
"	30.15	30.12	30.06	30.03	30.00	30.00	30.00	30.00	30.00	30.06	30.12	30.12	30.06
GA. Atlanta	1021	1019	1017	1017	1016	1016	1017	1016	1017	1019	1021	1021	1018
"	30.15	30.09	30.03	30.03	30.00	30.00	30.03	30.00	30.03	30.09	30.15	30.15	30.06
Macon	1020	1018	1017	1016	1016	1015	1016	1015	1016	1017	1020	1021	1017
"	30.12	30.06	30.03	30.00	30.00	29.97	30.00	29.97	30.00	30.03	30.12	30.15	30.03
HAWAII Hilo	1016	1016	1017	1018	1018	1018	1017	1016	1015	1016	1016	1016	1017
"	30.00	30.00	30.03	30.06	30.06	30.06	30.03	30.00	29.97	30.00	30.00	30.00	30.03
Honolulu	1016	1016	1017	1018	1018	1017	1017	1016	1015	1016	1016	1016	1016
"	30.00	30.00	30.03	30.06	30.06	30.03	30.03	30.00	29.97	30.00	30.00	30.00	30.00
Lihue	1015	1016	1017	1018	1018	1018	1017	1016	1016	1016	1016	1016	1016
"	29.97	30.00	30.03	30.06	30.06	30.06	30.03	30.00	30.00	30.00	30.00	30.00	30.00
IDAHO Boise	1023	1020	1017	1015	1013	1012	1012	1012	1015	1018	1022	1023	1017
"	30.21	30.12	30.03	29.97	29.91	29.88	29.88	29.88	29.97	30.06	30.18	30.21	30.03
ILL. Cairo	1021	1020	1017	1016	1015	1014	1016	1016	1017	1019	1020	1021	1017
"	30.15	30.12	30.03	30.00	29.97	29.94	30.00	30.00	30.03	30.09	30.12	30.15	30.03
Chicago	1019	1019	1016	1016	1015	1014	1015	1016	1017	1018	1018	1019	1017
"	30.09	30.09	30.00	30.00	29.97	29.94	29.97	30.00	30.03	30.06	30.06	30.09	30.03
IOWA Des Moines	1020	1020	1017	1015	1014	1013	1015	1015	1017	1018	1018	1020	1017
" "	30.12	30.12	30.03	29.97	29.94	29.91	29.97	29.97	30.03	30.06	30.06	30.12	30.03
KANS. Concordia	1020	1018	1016	1014	1013	1011	1014	1013	1015	1017	1019	1019	1016
"	30.12	30.06	30.00	29.94	29.91	29.86	29.94	29.91	29.97	30.03	30.09	30.09	30.00
Dodge City	1020	1018	1015	1014	1012	1011	1013	1013	1015	1017	1019	1019	1016
" "	30.12	30.06	29.97	29.94	29.88	29.86	29.91	29.91	29.97	30.03	30.09	30.09	30.00
Topeka	1021	1020	1016	1015	1014	1013	1015	1015	1016	1018	1018	1020	1017
"	30.15	30.12	30.00	29.97	29.94	29.91	29.97	29.97	30.00	30.06	30.06	30.12	30.03
Wichita	1020	1019	1015	1014	1013	1012	1014	1014	1015	1017	1019	1019	1016
"	30.12	30.09	29.97	29.94	29.91	29.88	29.94	29.94	29.97	30.03	30.09	30.09	30.00
LA. New Orleans	1021	1019	1017	1016	1015	1015	1016	1016	1015	1017	1020	1021	1017
" "	30.15	30.09	30.03	30.00	29.97	29.97	30.00	30.00	29.97	30.03	30.12	30.15	30.03
Shreveport	1021	1019	1016	1015	1014	1014	1015	1015	1016	1018	1020	1020	1017
"	30.15	30.09	30.00	29.97	29.94	29.94	29.97	29.97	30.00	30.06	30.12	30.12	30.03
MAINE Caribou	1015	1014	1013	1014	1014	1013	1014	1015	1017	1017	1016	1015	1015
"	29.97	29.94	29.91	29.94	29.94	29.91	29.94	29.97	30.03	30.03	30.00	29.97	29.97
MASS. Boston	1017	1015	1014	1015	1015	1014	1015	1016	1018	1018	1017	1017	1016
"	30.03	29.97	29.94	29.97	29.97	29.94	29.97	30.00	30.06	30.06	30.03	30.03	30.00
Blue Hill	1017	1015	1014	1015	1015	1014	1015	1016	1018	1018	1017	1017	1016
" "	30.03	29.97	29.94	29.97	29.97	29.94	29.97	30.00	30.06	30.06	30.03	30.03	30.00
Nantucket	1016	1015	1014	1015	1015	1014	1015	1016	1018	1018	1017	1017	1016
"	30.00	29.97	29.94	29.97	29.97	29.94	29.97	30.00	30.06	30.06	30.03	30.03	30.00
MICH. Alpena	1018	1017	1016	1014	1015	1014	1016	1016	1017	1017	1015	1016	1016
"	30.06	30.03	30.00	29.94	29.97	29.94	30.00	30.00	30.03	30.03	29.97	30.00	30.00
Marquette	1018	1017	1016	1014	1014	1013	1015	1015	1015	1016	1014	1016	1015
"	30.06	30.03	30.00	29.94	29.94	29.91	29.97	29.97	29.97	30.00	29.94	30.00	29.97
Sault Ste. Marie	1017	1017	1016	1016	1015	1014	1015	1016	1017	1017	1015	1017	1016
" " "	30.03	30.03	30.00	30.00	29.97	29.94	29.97	30.00	30.03	30.03	29.97	30.03	30.00
MINN. Duluth	1018	1018	1017	1016	1015	1013	1015	1015	1016	1016	1016	1018	1016
"	30.06	30.06	30.03	30.00	29.97	29.91	29.97	29.97	30.00	30.00	30.00	30.06	30.00
Internat'l Falls	1019	1019	1017	1016	1015	1013	1014	1014	1015	1016	1016	1018	1016
" "	30.09	30.09	30.03	30.00	29.97	29.91	29.94	29.94	29.97	30.00	30.00	30.06	30.00
Minpls-St. Paul	1020	1019	1018	1015	1014	1013	1015	1015	1016	1017	1016	1018	1016
" "	30.12	30.09	30.06	29.97	29.94	29.91	29.97	29.97	30.00	30.03	30.00	30.06	30.00
St. Cloud	1019	1019	1017	1015	1013	1012	1014	1014	1015	1016	1017	1018	1016
"	30.09	30.09	30.03	29.97	29.91	29.88	29.94	29.94	29.97	30.00	30.03	30.06	30.00
MISS. Vicksburg	1021	1019	1017	1017	1015	1015	1016	1016	1016	1018	1020	1020	1018
"	30.15	30.09	30.03	30.03	29.97	29.97	30.00	30.00	30.00	30.06	30.12	30.12	30.06
MO. Columbia	1020	1019	1016	1015	1014	1013	1015	1015	1017	1018	1019	1020	1017
"	30.12	30.09	30.00	29.97	29.94	29.91	29.97	29.97	30.03	30.06	30.09	30.12	30.03
St. Louis	1020	1019	1016	1015	1015	1014	1015	1016	1017	1018	1020	1020	1017
"	30.12	30.09	30.00	29.97	29.97	29.94	29.97	30.00	30.03	30.06	30.12	30.12	30.03

MEAN SEA LEVEL PRESSURE - Continued
(Millibars and Inches)

STATE AND STATION	JAN.	FEB.	MAR.	APR.	MAY	JUNE	JULY	AUG.	SEPT.	OCT.	NOV.	DEC.	ANNUAL
MONT. Great Falls	1020	1019	1017	1015	1014	1013	1014	1013	1016	1017	1019	1019	1016
" "	30.12	30.09	30.03	29.97	29.94	29.91	29.94	29.91	30.00	30.03	30.09	30.09	30.00
Havre	1020	1020	1017	1015	1014	1013	1013	1014	1015	1016	1018	1018	1016
"	30.12	30.12	30.03	29.97	29.94	29.91	29.91	29.94	29.97	30.00	30.06	30.06	30.00
Helena	1021	1020	1017	1016	1014	1014	1015	1014	1016	1019	1021	1021	1017
"	30.15	30.12	30.03	30.00	29.94	29.94	29.97	29.94	30.00	30.09	30.15	30.15	30.03
NEBR. North Platte	1020	1019	1015	1014	1013	1011	1013	1013	1015	1017	1019	1019	1016
" "	30.12	30.09	29.97	29.94	29.91	29.86	29.91	29.91	29.97	30.03	30.09	30.09	30.00
Omaha	1021	1019	1017	1014	1014	1012	1015	1014	1016	1018	1018	1019	1017
"	30.15	30.09	30.03	29.94	29.94	29.88	29.97	29.94	30.00	30.06	30.06	30.09	30.03
NEV. Ely	1021	1019	1016	1013	1012	1012	1013	1013	1014	1017	1021	1022	1016
"	30.15	30.09	30.00	29.91	29.88	29.88	29.91	29.91	29.94	30.03	30.15	30.18	30.00
Las Vegas	1019	1017	1013	1010	1008	1006	1008	1008	1009	1013	1018	1020	1012
"	30.09	30.03	29.91	29.83	29.77	29.71	29.77	29.77	29.80	29.91	30.06	30.12	29.88
Winnemucca	1022	1019	1017	1015	1013	1012	1012	1013	1015	1018	1022	1022	1017
"	30.18	30.09	30.03	29.97	29.91	29.88	29.88	29.91	29.97	30.06	30.18	30.18	30.03
N. Mex. Albuquerque	1019	1015	1012	1010	1008	1007	1011	1011	1012	1015	1018	1020	1013
" "	30.09	29.97	29.88	29.83	29.77	29.74	29.86	29.86	29.88	29.97	30.06	30.12	29.91
N. Y. Albany	1017	1016	1015	1014	1014	1014	1015	1015	1018	1018	1017	1018	1016
"	30.03	30.00	29.97	29.94	29.94	29.94	29.97	29.97	30.06	30.06	30.03	30.06	30.00
Buffalo	1018	1017	1016	1016	1015	1014	1015	1017	1018	1018	1017	1018	1017
"	30.06	30.03	30.00	30.00	29.97	29.94	29.97	30.03	30.06	30.06	30.03	30.06	30.03
New York	1018	1017	1015	1016	1015	1014	1015	1016	1018	1018	1018	1019	1017
"	30.06	30.03	29.97	30.00	29.97	29.94	29.97	30.00	30.06	30.06	30.06	30.09	30.03
N.C. Asheville	1021	1019	1017	1017	1016	1016	1017	1017	1019	1020	1021	1021	1018
"	30.15	30.09	30.03	30.03	30.00	30.00	30.03	30.03	30.09	30.12	30.15	30.15	30.06
Cape Hatteras	1020	1018	1017	1017	1016	1015	1017	1016	1017	1018	1019	1020	1018
"	30.12	30.06	30.03	30.03	30.00	29.97	30.03	30.00	30.03	30.06	30.09	30.12	30.06
N. DAK. Bismarck	1020	1020	1018	1016	1014	1012	1013	1013	1015	1016	1018	1019	1016
"	30.12	30.12	30.06	30.00	29.94	29.88	29.91	29.91	29.97	30.00	30.06	30.09	30.00
OHIO Cincinnati	1020	1019	1017	1015	1015	1015	1016	1016	1018	1019	1019	1020	1017
"	30.12	30.09	30.03	29.97	29.97	29.97	30.00	30.00	30.06	30.09	30.09	30.12	30.03
Columbus	1020	1019	1017	1016	1015	1015	1016	1017	1018	1019	1019	1020	1018
"	30.12	30.09	30.03	30.00	29.97	29.97	30.00	30.03	30.06	30.09	30.09	30.12	30.06
Dayton	1020	1019	1017	1016	1015	1015	1016	1017	1018	1019	1019	1020	1018
"	30.12	30.09	30.03	30.00	29.97	29.97	30.00	30.03	30.06	30.09	30.09	30.12	30.06
OKLA. Oklahoma City	1020	1019	1016	1013	1013	1012	1014	1014	1015	1017	1019	1020	1016
" "	30.12	30.09	30.00	29.91	29.91	29.88	29.94	29.94	29.97	30.03	30.09	30.12	30.00
ORE. Medford	1021	1019	1018	1017	1016	1015	1015	1014	1015	1018	1021	1021	1018
"	30.15	30.09	30.06	30.03	30.00	29.97	29.97	29.94	29.97	30.06	30.15	30.15	30.06
Portland	1019	1018	1018	1018	1017	1017	1017	1017	1016	1017	1019	1018	1018
"	30.09	30.06	30.06	30.06	30.03	30.03	30.03	30.03	30.00	30.03	30.09	30.06	30.06
PA. Philadelphia	1017	1017	1015	1015	1015	1014	1016	1016	1018	1018	1018	1019	1017
"	30.03	30.03	29.97	29.97	29.97	29.94	30.00	30.00	30.06	30.06	30.06	30.09	30.03
Pittsburgh	1019	1018	1016	1016	1016	1015	1016	1017	1019	1019	1019	1020	1018
"	30.09	30.06	30.00	30.00	30.00	29.97	30.00	30.03	30.09	30.09	30.09	30.12	30.06
R. I. Block Is.	1017	1016	1014	1015	1015	1014	1015	1016	1018	1018	1017	1017	1016
" "	30.03	30.00	29.94	29.97	29.97	29.94	29.97	30.00	30.06	30.06	30.03	30.03	30.00
S. C. Charleston	1021	1019	1018	1017	1016	1016	1017	1017	1017	1018	1020	1021	1018
"	30.15	30.09	30.06	30.03	30.00	30.00	30.03	30.03	30.03	30.06	30.12	30.15	30.06
Columbia	1021	1019	1017	1017	1016	1015	1017	1016	1018	1019	1020	1021	1018
"	30.15	30.09	30.03	30.03	30.00	29.97	30.03	30.00	30.06	30.09	30.12	30.15	30.06
S. Dak. Huron	1020	1020	1017	1015	1013	1012	1013	1013	1015	1016	1018	1019	1016
"	30.12	30.12	30.03	29.97	29.91	29.88	29.91	29.91	29.97	30.06	30.06	30.09	30.00
Rapid City	1019	1019	1015	1014	1014	1012	1014	1014	1015	1017	1019	1019	1016
" "	30.09	30.09	29.97	29.94	29.94	29.88	29.94	29.94	29.97	30.03	30.09	30.09	30.00
TENN. Bristol	1021	1020	1017	1016	1015	1015	1016	1016	1016	1018	1020	1021	1017
"	30.15	30.12	30.03	30.00	29.97	29.97	30.00	30.00	30.00	30.06	30.12	30.15	30.03
Nashville	1021	1020	1017	1016	1016	1015	1016	1016	1018	1019	1021	1021	1018
"	30.15	30.12	30.03	30.00	30.00	29.97	30.00	30.00	30.06	30.09	30.15	30.15	30.06
TEX. Abilene	1019	1018	1014	1013	1011	1011	1013	1013	1014	1016	1019	1019	1015
"	30.09	30.06	29.94	29.91	29.86	29.86	29.91	29.91	29.94	30.00	30.09	30.09	29.97
Brownsville	1018	1017	1014	1013	1012	1012	1014	1013	1013	1015	1018	1018	1015
"	30.06	30.03	29.94	29.91	29.88	29.88	29.94	29.91	29.91	29.97	30.06	30.06	29.97
El Paso	1018	1015	1012	1010	1009	1008	1011	1011	1011	1014	1018	1018	1013
" "	30.06	29.97	29.88	29.83	29.80	29.77	29.86	29.86	29.86	29.94	30.06	30.06	29.91
Galveston	1020	1018	1016	1015	1014	1015	1016	1015	1015	1016	1020	1020	1017
"	30.12	30.06	30.00	29.97	29.94	29.97	30.00	29.97	29.97	30.00	30.12	30.12	30.03
Midland	1018	1016	1013	1012	1011	1010	1013	1013	1014	1016	1019	1019	1014
"	30.06	30.00	29.91	29.88	29.86	29.83	29.91	29.91	29.94	30.00	30.09	30.09	29.94
San Antonio	1020	1018	1015	1013	1012	1012	1014	1014	1014	1016	1019	1019	1016
" "	30.12	30.06	29.97	29.91	29.88	29.88	29.94	29.94	29.94	30.00	30.09	30.09	30.00
UTAH Salt Lake City	1022	1020	1016	1013	1012	1010	1012	1012	1013	1017	1022	1022	1016
" " "	30.18	30.12	30.00	29.91	29.88	29.83	29.88	29.88	29.91	30.03	30.18	30.18	30.00
VT. Burlington	1018	1016	1015	1015	1014	1013	1014	1015	1017	1017	1017	1017	1016
"	30.06	30.00	29.97	29.97	29.94	29.91	29.94	29.97	30.03	30.03	30.03	30.03	30.00
VA. Lynchburg	1020	1019	1017	1016	1016	1016	1016	1017	1019	1020	1020	1021	1018
"	30.12	30.09	30.03	30.00	30.00	30.00	30.00	30.03	30.09	30.12	30.12	30.15	30.06
WASH. Olympia	1018	1017	1017	1018	1017	1017	1018	1017	1017	1017	1019	1017	1017
"	30.06	30.03	30.03	30.06	30.03	30.03	30.06	30.03	30.03	30.03	30.09	30.03	30.03
Spokane	1020	1018	1016	1016	1014	1014	1014	1014	1016	1018	1021	1020	1017
"	30.12	30.06	30.00	30.00	29.94	29.94	29.94	29.94	30.00	30.06	30.15	30.12	30.03
Tatoosh Is.	1015	1016	1016	1017	1018	1018	1019	1018	1017	1016	1017	1015	1017
" "	29.97	30.00	30.00	30.03	30.06	30.06	30.09	30.06	30.03	30.00	30.03	29.97	30.03
Walla Walla	1020	1019	1017	1016	1015	1014	1014	1014	1015	1018	1021	1020	1017
" "	30.12	30.09	30.03	30.00	29.97	29.94	29.94	29.94	29.97	30.06	30.15	30.12	30.03
WIS. Madison	1019	1019	1016	1015	1015	1014	1015	1016	1017	1018	1017	1019	1017
"	30.09	30.09	30.00	29.97	29.97	29.94	29.97	30.00	30.03	30.06	30.03	30.09	30.03
WYO. Sheridan	1020	1019	1017	1015	1014	1012	1014	1014	1015	1017	1019	1019	1016
"	30.12	30.09	30.03	29.97	29.94	29.88	29.94	29.94	29.97	30.03	30.09	30.09	30.00
P. R. San Juan	1017	1017	1016	1016	1015	1016	1017	1015	1014	1013	1013	1015	1015
" "	30.03	30.03	30.00	30.00	29.97	30.00	30.03	29.97	29.94	29.91	29.91	29.97	29.97

BASED ON 30-YEAR PERIOD, 1931-60